Smart Solutions to Climate Change

Comparing Costs and Benefits

Edited by

BJØRN LOMBORG

CAMBRIDGE
UNIVERSITY PRESS

CAMBRIDGE UNIVERSITY PRESS
Cambridge, New York, Melbourne, Madrid, Cape Town, Singapore,
São Paulo, Delhi, Dubai, Tokyo, Mexico City

Cambridge University Press
The Edinburgh Building, Cambridge CB2 8RU, UK

Published in the United States of America by Cambridge University Press, New York

www.cambridge.org
Information on this title: www.cambridge.org/9780521138567

First published 2010

Printed in the United Kingdom at the University Press, Cambridge

A catalogue record for this publication is available from the British Library

ISBN 978-0-521-76342-4 Hardback
ISBN 978-0-521-13856-7 Paperback

Smart Solutions to Climate Change

The failure of the Copenhagen climate conference in December 2009 revealed major flaws in the way the world's policy makers have attempted to prevent dangerous levels of increases in global temperatures. The expert authors in this specially commissioned collection focus on the likely costs and benefits of a very wide range of policy options, including geoengineering; mitigation of CO_2, CH_4, and "black carbon"; expanding forest Carbon Sequestration; R&D of low-carbon energy; and encouraging green technology transfer. For each policy, the authors outline all of the costs, benefits, and likely outcomes, in fully referenced, clearly presented chapters accompanied by shorter, critical alternative Perspective papers. To further stimulate debate, an Expert Panel of economists, including three Nobel laureates, evaluates and ranks the attractiveness of the policies.

This authoritative and thought-provoking book will challenge readers to form their own conclusions about the best ways to respond to global warming.

BJØRN LOMBORG is Director of the Copenhagen Consensus Center and Adjunct Professor in the Department of Management, Politics and Philosophy at Copenhagen Business School. He is the author of the controversial bestseller, *The Skeptical Environmentalist* (Cambridge, 2001), and was named as one of the top 100 intellectuals by *Foreign Policy* and *Prospect* magazines in 2008.

Contents

Figures

Tables

Contributors

Experts

Jagdish N. Bhagwati is University Professor at Columbia University and Senior Fellow in International Economics at the Council on Foreign Relations. He has been Economic Policy Adviser to Arthur Dunkel, Director General of GATT (1991–3), Special Adviser to the United Nations on Globalization, and External Adviser to the WTO. He has served on the Expert Group appointed by the Director General of the WTO on the Future of the WTO and the Advisory Committee to Secretary General Kofi Annan on the NEPAD process in Africa, and was also a member of the Eminent Persons Group under the chairmanship of President Fernando Henrique Cardoso on the future of UNCTAD. Five volumes of his scientific writings and two of his public policy essays have been published by MIT Press. The recipient of six Festschrifts in his honor, he has also received several prizes and honorary degrees, including awards from the governments of India (Padma Vibhushan) and Japan (Order of the Rising Sun, Gold and Silver Star). Professor Bhagwati's latest book is *In Defense of Globalization* (2004).

Finn E. Kydland is the Henley Professor of Economics at the University of California, Santa Barbara, and Director of The Laboratory for Aggregate Economics and Finance. Professor Kydland was awarded the 2004 Nobel Prize in Economics jointly with Edward C. Prescott of Arizona State University, "for their contribution to dynamic macroeconomics: the time consistency of economic policy and the driving forces behind business cycles." Their analysis of economic policy and the driving forces behind business cycles transformed economic research and greatly influenced the practice of economic policy in general, and monetary policy in particular. In two joint papers, in 1977 and 1982, Kydland and Prescott offered new approaches to the analysis of macroeconomic developments. Kydland joined the UC Santa Barbara faculty on July 1, 2004. He previously taught at Carnegie Mellon University, where he earned his PhD. He is also an Adjunct Professor at NHH, Norway, and consults as a Research Associate at the Federal Reserve Bank of Dallas

Thomas C. Schelling is Distinguished University Professor, Emeritus at the University of Maryland. For twenty years he was the Lucius N. Littauer Professor of Political Economy at the John F. Kennedy School of Government. Schelling was awarded the Nobel Memorial Prize in Economic Sciences jointly with Robert Aumann in 2005 for "having enhanced our understanding of conflict and cooperation through game-theory analysis." He has been elected to the National Academy of Sciences, the Institute of Medicine, and the American Academy of Arts and Sciences. In 1991, he was President of the American Economic Association, of which he is a distinguished fellow. He was the recipient of the Frank E. Seidman Distinguished Award in Political Economy and the National Academy of Sciences award for Behavioral Research Relevant to the Prevention of Nuclear War. He served in the Economic Cooperation Administration in Europe, and has held positions in the White House and Executive Office of the President, Yale University, the RAND Corporation and the Department of Economics and Center for International Affairs at Harvard University. He has published on military strategy and arms control, energy and environmental policy, climate change, nuclear proliferation, terrorism, organized crime, foreign aid and international trade, conflict and bargaining theory, racial segregation and integration, the military draft, health policy, tobacco and drugs policy, and ethical issues in public policy and in business.

Vernon L. Smith was awarded the Nobel Prize in Economic Sciences in 2002 for his groundbreaking work in experimental economics. Smith has joint appointments with the Argyros School of Business and Economics, and the School of Law at Chapman University, and he is part of a team that will create and run the new Economic Science Institute at Chapman. Smith has authored or co-authored more than 250 articles and books on capital theory, finance, natural resource economics, and experimental economics. He serves or has served on the board of editors of the *American Economic Review*, *The Cato Journal*, the *Journal of Economic Behavior and Organization*, the *Journal of Risk and Uncertainty*, *Science*, *Economic Theory*, *Economic Design*, *Games and Economic Behavior*, and the *Journal of Economic Methodology*. Smith is a distinguished fellow of the American Economic Association, an Andersen Consulting Professor of the Year, and the 1995 Adam Smith Award recipient conferred by the Association for Private Enterprise Education. He was elected a member of the National Academy of Sciences in 1995, and received CalTech's Distinguished Alumni Award in 1996. He has served as a consultant on the privatization of electric power in Australia and New Zealand and participated in numerous private and public discussions of energy deregulation in the USA. In 1997 he served as a Blue Ribbon Panel Member, National Electric Reliability Council. Smith completed his undergraduate degree in electrical engineering at CalTech, his master's degree in economics at the University of Kansas, and his PhD in economics at Harvard University.

Nancy L. Stokey is the Frederick Henry Prince Distinguished Service Professor of Economics at the University of Chicago. She earned her BA in economics from the University of Pennsylvania in 1972 and her PhD from Harvard University in 1978. Stokey has published significant research in the areas of economic growth and development, as well as papers on economic history ("A Quantitative Model of the British Industrial Revolution: 1780–1850," 2001) and economic theory ("Dynamic Pro-gramming with Homogeneous Functions," 1998, co-authored with Fernando Alvarez). She is the co-developer, with Paul Milgrom, of the no-trade theorem, a counter-intuitive proposition in financial economics, and offered the first rigorous proof of the Coase conjecture on durable goods pricing by a monopolist. She co-authored, with Nobel Prize laureates Robert Lucas, Jr. and Edward Prescott, a book on recursive methods in economic dynamics that is widely used by research economists and graduate students. Stokey is a member of the National Academy of Sciences and of the American Academy of Arts and Sciences, and she was a vice-president of the American Economic Association between 1996 and 1997. She has served as co-editor of *Econometrica* and the *Journal of Political Economy*.

Chapter Authors

Robert E. Baron is a senior consultant at Charles River Associates International. He has thirty years' experience in the energy industry, related to natural gas, petroleum, and electric utilities. Previously he was vice-president at ICF Consulting, a principal at DRI/McGraw-Hill, a manager at British Petroleum, and a research scientist at the MIT Energy Laboratory.

J. Eric Bickel is an assistant professor in the Graduate Program in Operations Research and Industrial Engineering at The University of Texas at Austin and a fellow in the Center for International Energy and Environmental Policy.

Francesco Bosello is a researcher at the Department of Economics, Business and Statistics, University of Milan, Italy, and senior researcher at the Fondazione Eni Enrico Mattei in Milan. He graduated in economics at the University Ca'Foscari of Venice, then achieved the title of Master of Science in Economics from University College, University of London (UK) and a PhD in Economics from the University of Venice. Previously he has been Visiting Scientist at the International Center for Theoretical Physics Abdus Salam in Trieste and Research Assistant at the University of Venice.

Carlo Carraro is Professor of Environmental Economics at the University of Venice. He holds a PhD from Princeton University and has been Vice-Provost for Research Management and Policy of the University of Venice (2000–6) and Chairman of the Department of Economics (2006–8). Carraro is Director of the Sustainable Development Programme of the Fondazione Eni Enrico Mattei and Director of the Climate Impacts and Policy Division of the Euro Mediterranean Centre for Climate Change (CMCC).

Enrica De Cian is a junior researcher at the Fondazione Eni Enrico Mattei in Venice. In 2008 she graduated in Economics and Organization at the School of Advanced Studies in Venice. In 2006–7, she was Visiting Student at the MIT Joint Program on the Science and Policy of Global Change, at MIT. Her main research interests are technological change and the environment; mitigation and adaptation policies; and climate coalitions.

Isabel Galiana is a PhD candidate in the Department of Economics at McGill University, Montreal, specializing in modeling of climate change policy and international economics.

Christopher Green is a professor of economics in the Department of Economics at McGill University. His main teaching and research fields are industrial organization; public policies toward business; and environmental economics, in particular the economics of climate change.

Claudia Kemfert is Head of the Department of Energy, Transportation, Environment at the German Institute of Economic Research (DIW, Berlin), and Professor of Energy Economics and Sustainability at the Hertie School of Governance (HSoG). From 2004 to 2009, she was Professor of Environmental Economics at Humboldt-University Berlin. She is an expert in the areas of energy and climate economics.

Lee Lane is a resident fellow at AEI and co-director of AEI's geoengineering (GE) project. Lane has been a consultant to CRA International, and Executive Director of the Climate Policy Center.

W. David Montgomery is Vice-President at Charles River Associates International. He is an internationally recognized expert on economic issues associated with climate change policy, and his work on these topics has been published frequently in peer-reviewed journals. He was a principal lead author of the Second Assessment Report of the Intergovernmental Panel on Climate Change (IPCC, 1995), Working Group III, and he has authored a number of peer-reviewed publications on climate policy over the past twenty years.

Wolf-Peter Schill is Research Associate at the Department of Energy, Transportation, Environment at the German Institute of Economic Research (DIW, Berlin). His fields of interest are economics of climate change, renewable energy, and electricity markets.

Brent Sohngen is Professor at The Ohio State University's Department of Agricultural, Environmental, and Development Economics, and University Fellow at Resources for the Future. Professor Sohngen develops economic models of land-use and land-cover change for climate policy analysis, and he studies the economics of non-point-source pollution control.

Richard S.J. Tol is a research professor at the Economic and Social Research Institute (Dublin) and the Professor of the Economics of Climate Change at the Free University of Amsterdam. An economist and statistician, his work focuses on impacts of climate change, international climate policy, tourism, and land and water use. He is ranked among the 200 best economists in the world. His recent publications include *Environmental Crisis: Science and Policy* (2007) and *Economic Analysis of Land Use in Global Climate Change* (2008).

Sugandha D. Tuladhar is an associate principal at Charles River Associates International. He specializes in computable general equilibrium (CGE) model development and its application. As a general equilibrium modeler with extensive econometric and programming skills, Tuladhar's work focuses on the global impact of environmental and climate policy changes.

Zili Yang is Professor of Economics at the State University of New York at Binghamton. His focus is on resource and environmental economics, energy economics, and public economics.

Perspective Paper Authors

David Anthoff is an environmental economist working on climate change. He is a postdoctoral associate of the Economic and Social Research Institute, Dublin, and a freelance consultant on climate change issues. He holds a PhD in Economics from Hamburg University and the International Max Planck Research School on Earth System Modelling, a MSc in Environmental Change and Management from the University of Oxford and a MPhil in Philosophy from LMU, Munich. In the autumn of 2008, he was a visiting research fellow at the Smith School of Enterprise and the Environment, University of Oxford.

Valentina Bosetti holds a PhD in Computational Mathematics and Operation Research from the Università Statale of Milan and a Masters Degree in Environmental and Resources Economics from University College, University of London. At FEEM since 2003, she works as a modeler for the Sustainable Development Program, leading the Climate Change topic and coordinating a research group on numerical analysis of carbon mitigation options and policies. She has also collaborated with a number of other institutes such as the Euro-Mediterranean Center on Climate Change, the NOAA, and Italian Universities. Her main research interest is socio-economic modeling of climate change, with particular emphasis on innovation, uncertainty, and irreversibility.

Samuel Fankhauser is a principal fellow at the Grantham Research Institute on Climate Change and the Environment at the London School of Economics He also serves as Chief Economist for Globe International, the international legislators' forum, and is a member of the Committee on Climate Change, an independent public body that advises the UK government on its carbon targets. Dr. Fankhauser served on the 1995, 2001, and 2007

assessments of the Intergovernmental Panel on Climate Change (IPCC).

Sabine Fuss is a research scholar at the Forestry Program of the International Institute of Systems Analysis (IIASA) in Laxenburg, Austria.

Andrew P. Grieshop is a postdoctoral research fellow at the Institute for Resources, Environment, and Sustainability at the University of British Columbia, where he works at the intersection of energy consumption, atmospheric sciences, and public policy, focusing on issues in developing countries. His doctoral research, completed as part of the Center for Atmospheric Particle Studies at Carnegie Mellon University, examined the properties and atmospheric evolution of carbonaceous aerosols.

Fredrik Hedenus is a researcher at the Department of Physical Resource Theory at Chalmers University of Technology. His research interests include greenhouse gas (GHG) emissions from food, energy system modelling, energy security, and technological change.

Daniel J.A. Johansson is a postdoctoral student at the Division of Physical Resource Theory, Department of Energy and Environment at Chalmers University of Technology, Gothenburg. Johansson's main research interests are related to energy and climate change economics and policy.

Frank Jotzo is Research Fellow at the College of Asia and the Pacific, Australian National University, and Deputy Director of the ANU Climate Change Institute. He specializes in the economics and policy of climate change. He has worked and published on these topics and other aspects of international and development economics since 1998, and has advised several governments on climate policy.

Milind Kandlikar is an associate professor at the Liu Institute for Global Issues, University of British Columbia, Vancouver, Canada. His work focuses on the intersection of technology innovation, human development, and the global environment. He has published extensively on the science and policy of climate change.

Onno Kuik is a researcher at the Institute for Environmental Studies, Vrije Universiteit, Amsterdam. He has worked for the past twenty years on the economic analysis of environmental policy. He received the *Energy Journal*'s 2003 Best Paper Award for the paper "Trade Liberalization and Carbon Leakage," co-authored by Reyer Gerlagh.

Gregory Nemet is an assistant professor at the University of Wisconsin in the La Follette School of Public Affairs and the Nelson Institute for Environmental Studies. He is also a member of the University's Energy Sources and Policy Cluster and a senior fellow at the University's Center for World Affairs and the Global Economy. His research and teaching focus on improving understanding of the environmental, social, economic, and technical dynamics of the global energy system. A central focus of his research involves empirical analysis of the processes of innovation and technological change.

Roger A. Pielke, Jr. is a professor in the Environmental Studies Program at University of Colorado and a Fellow of the Cooperative Institute for Research in Environmental Sciences. In 2006, he received the Eduard Brückner Prize for outstanding achievement in interdisciplinary climate research.

David Popp is Associate Professor of Public Administration in the Center for Policy Research of the Maxwell School at Syracuse University. He is also a research associate of the National Bureau for Economic Research (NBER). He is an economist with research interests in environmental policy and the economics of technological change. Much of his research focuses on the links between environmental policy and innovation. He is particularly interested in how environmental and energy policies shape the development of new technologies that may be relevant for combating climate change.

Conor C.O. Reynolds is a doctoral candidate at the Institute for Resources, Environment and Sustainability at the University of British Columbia. He conducts research on the climate and air-quality impacts of transportation. Reynolds' doctoral dissertation assesses strategies to reduce transportation emissions in rapidly industrializing countries, such as the use of alternative fuels.

Roberto Roson has been a visiting fellow at the Free University of Amsterdam, at the University of Warwick, and at S. Francisco Xavier University (Sucre, Bolivia). He is currently Associate Professor at Ca'Foscari University, Venice, where he teaches industrial organization, international and antitrust economics. He is currently collaborating with the Euro-Mediterranean Centre for Climate Change, IEFE–Bocconi University and the World Bank.

Anne E. Smith is Vice-President and Practice Leader of Climate & Sustainability at Charles River Associates International. She is an expert in environmental policy assessment and corporate compliance strategy planning and has made major analysis contributions to most major air quality policy issues. She has a PhD in economics from Stanford University.

Acknowledgments

This book presents research and findings from the Copenhagen Consensus on Climate project which was made possible by funding from the Danish government. I am grateful for the commitment of current and past staff members at the Copenhagen Consensus Center: Cecilie Olsen, Gabriela Garza de Linde, Henrik Meyer, Katja Johansen, Kristine Ohrt, Maria Jakobsen, Tommy Petersen, and Zenia Stampe. I would particularly like to thank the members of the project team for their exceptional dedication and enthusiasm: Anders Møller, Anita Overholt Nielsen, David Young, Kasper Thede Anderskov, Maja Makwarth, Sasha Larsen Beckmann, Ulrik Larsen, and Zsuzsa Horvath. And I would especially like to express my sincere appreciation to the authors of the outstanding research in this volume, and to the remarkable members of the Copenhagen Consensus Expert Panel.

Abbreviations and acronyms

ABC	Atmospheric Brown Clouds	EMF	Energy Modeling Forum
ABI	Association of British Insurers	ENSO	El Niño-southern oscillation
AC	air capture	EPA	Environmental Protection Agency (USA)
AC	average cost		
ACEI	Autonomous Carbon Efficiency Improvement	EPO	European Patent Office
		EST	environmentally sound technologies
AEEI	Autonomous Energy Efficiency Improvement	ETS	emission trading scheme
		ETS	Emission Trading System
AETG	advanced energy technology gap	FDI	foreign direct investment
APP	Asia Pacific Partnership	FEEM	Fondazione Eni Enrico Mattei
B/C	benefit-cost	FSU	former Soviet Union
BAU	business-as-usual	FUND	Climate Framework for Uncertainty, Negotiation, and Distribution
BCA	benefit-cost analysis		
BCR	benefit-cost ratio	G4M	Global Forest Model
CBA	cost-benefit analysis	GAINS	Greenhouse Gas and Air Pollution Interactions and Synergies
CCN	cloud condensation nucleae		
CCS	carbon capture and storage	GCC	global climate change
CCSP	Climate Change Science Program	GCM	general circulation model
CDM	clean development mechanism	GCP	Global Carbon Project
CE	climate engineering	GDP	gross domestic product
CES	constant elasticity of substitution	GE	geoengineering
CFC	chlorofluorocarbon	Gg	giga-gram
CGE	computable general equilibrium	GHG	greenhouse gas
CIO	carbon intensity of output	GLOBIOM	Global Biomass Optimization Model
CIR^{3}D	carbon intensity-reducing return to R&D investment		
		GtC	gigatons of carbon
CIS	Commonwealth of Independent States	GTP	Global Temperature Potential
		GWe	giga-watt electric
CNG	compressed natural gas	GWP	global warming potential
COP	Conference of Parties	GWP	gross world product
CRP	Conservation Reserve Program (USA)	HDDV	heavy-duty diesel vehicles
		HFC	hydrofluorocarbon
CV	current value	IA	impact assessment
DALY	disability adjusted life year	IA	integrated assessment
DICE	Dynamic Integrated model of Climate and the Economy	IAM	integrated assessment model
		ICES	Intertemporal Computable Equilibrium System
EDF	Environmental Defense Fund		
EI	energy intensity of output	IEA	international environmental agreement
EJ	exajoules		

IEA WEO	International Energy Agency World Energy Outlook	OHI	other high-income countries
IFR	integral fast reactor	OLS	ordinary least squares
IGCC–CCS	integrated coal gasification combined cycle power plant	PDF	probability density function
		PFC	perfluorocarbons
IIASA	International Institute for Applied Systems Analysis	PM	particulate matter
		PPP	purchasing power parity
IPCC	Intergovernmental Panel on Climate Change	ppb	parts per billion
		ppm	parts per million
IRR	internal interest rate	ppmv	parts per million by volume
ISM	Indian summer monsoon	PV	present value
ITC	induced technological change	QALY	quality of life
LBD	learning-by-doing	R&D	research and development
LBR	learning-by-researching	RCIO	rate of decline in the carbon intensity of output
LDC	least developed country		
LFG	landfill gas	REDD	Reducing Emissions from Deforestation and Forest Degradation (avoided deforestation)
LLF	long-lived fluorinated gases		
LNG	liquefied natural gas		
LPG	liquefied petroleum gas	RICE	Regional Dynamic Integrated model of Climate and the Economy
MAC	marginal abatement cost		
MC	marginal costs	SCC	social costs of carbon
MEF	Major Economies Forum	SF_6	sulfur hexafluoride
MER	market exchange rates	SLF	short-lived fluorinated gases
MFI	micro-finance institution	SPM	Summary for Policy Makers (IPCC)
MMV	measurement, monitoring, and verification	SRES	Special Report on Emission Scenarios
MT	million ton	SRM	solar radiation management
MTC	metric ton of carbon	SSA	sub-Saharan Africa
NAPA	National Adaptation Programmes of Action	SYR	Synthesis Report (IPCC)
		TFP	total factor productivity
NAS	National Academy of Sciences	Tg	teragram
NC	No Controls	THC	Thermohaline Circulation
NCAR	The National Center for Atmospheric Research	TOA	top of the atmosphere
		TT	technology transfer
NGO	non-governmental organization	TW	terawatt
NIS	newly independent states	UNFCCC	United Nations Framework Convention on Climate Change
NPV	net present value		
NRTEE	National Roundtable on the Environment and the Economy (Canada)	UV	ultraviolet
		VOC	volatile organic compounds
		VOI	value of information
O&M	operation and maintenance	VSL	value of a statistical life
OC	Optimal Controls	WHO	World Health Organization
OC	organic carbon	WM	Waxman–Markey (bill)
ODA	overseas development aid	WTAC	willingness to accept compensation
OECD	Organization for Economic Cooperation and Development	WTP	willingness-to-pay
		YLD	years lost due to disability
		YLL	years of life lost

Introduction

BJØRN LOMBORG

The risks of unchecked global warming are now widely acknowledged: a rise in sea levels threatening the existence of some low-lying coastal communities; pressure on freshwater resources, making food production more difficult in some countries and possibly becoming a source of societal conflict; changing weather patterns providing favorable conditions for the spread of malaria. To make matters worse, the effects will be felt most in those parts of the world which are home to the poorest people who are least able to protect themselves and who bear the least responsibility for the build-up of greenhouse gases (GHGs). Concern has been great, but humanity has so far done very little that will actually prevent these outcomes. Carbon emissions have kept increasing, despite repeated promises of cuts.

As I wrote in *The Skeptical Environmentalist* (Lomborg 2001), man-made global warming exists. There is still meaningful and important work going on looking at the range of outcomes that we should expect but it is vital to emphasize the consensus on the most important scientific questions. We have long moved on from any mainstream disagreements about the science of climate change. The crucial, relevant conversation of today is about what to do about climate change – the economics of our response.

Finding a better response to global warming has become all the more important as the current political approach – seen at summits in Rio de Janeiro, Kyoto, and Copenhagen, has seemingly run aground. The failure of the Copenhagen Climate Summit in December 2009 was a great disappointment for the millions who had hoped for strong and meaningful action on global warming.

After Copenhagen, political leaders looked for sources of blame. China bore the brunt of western anger, while many declared that the UN negotiation process needed to be reformed.

It is more constructive to consider the range of policy responses that we have, and to identify what we can do in different areas. Economic research serves to underscore some of the hurdles before us – but it also highlights very promising avenues for exploration. It would be morally unconscionable to spend enormous sums of money making a minor difference to long-term global warming and human well-being if we could achieve a lot more impact on the climate – and leave future generations better off – with a smaller investment on smarter solutions.

The research presented in this volume was drafted by expert economists for the Copenhagen Consensus on Climate project, which utilizes a process that was first designed to prioritize global opportunities. The approach is simple, and is founded on the belief that basic principles of economics can be used to help any nation or organization to spend its money to achieve the most "good" possible.

In 2004 and 2008, the Center gathered research on ten key global challenges – from malnutrition to terrorism – and commissioned a panel of expert economists to rank the investments. The research from the Copenhagen Consensus 2004 and the Copenhagen Consensus 2008 is available in *Global Crises, Global Solutions* (Cambridge University Press, 1st edn., 2005, 2nd edn., 2009).

These projects attracted attention from all around the world. Denmark's government spent millions more on HIV/Aids projects, which topped the economists' "to do" list in 2004. Micronutrient programs in Africa and elsewhere received significant attention and greater resources after they topped the list in 2008.

The Copenhagen Consensus prioritization process has also been carried out with UN ambassadors from twenty-three nations including China, India and the USA, and for Caribbean and Latin

American problems. The research for the latter is available in *Latin American Development Priorities* (Cambridge University Press, 2009).

These projects showed that an informed ranking of solutions to the world's big problems is possible, and that cost-benefit analyses (CBAs) – much maligned by some – can lead to a clear and compassionate focus on the most effective ways to respond to the real problems of the world's most afflicted people.

Climate change is undoubtedly one of the chief concerns facing the world today. It has attracted top-level political concern and repeated efforts to form a global consensus on carbon cuts. But many questions have remained unaddressed and unanswered. Should politicians continue with plans to make carbon-cutting promises that, on past experience, are unlikely to be fulfilled? What could be achieved by planting more trees, cutting methane (CH_4), or reducing black soot emissions? Is it sensible to focus on a technological solution to warming? Or should we focus to adapt to a warmer world?

The research presented in this volume addresses these questions, along with how much each approach would cost and how much it would help in tackling climate change. Most importantly, the research presented together answers a fundamental question that we often overlook: not *if* we should do something about global warming, but rather *how best* to go about it. The starting point for every chapter is that global warming is a challenge that humanity must confront.

Just as with any other problem we face, there are many possible remedies, and some of them are much better than others. Not just cheaper, but more effective, more efficient, and – crucially – more likely to actually happen.

This book presents some of the recommended responses to global warming by experts in each field. There is a range of fresh thinking and new approaches. In these pages, for example, you will find one of the first – and certainly the most comprehensive – CBA of climate engineering (CE) options.

For each topic in this book – whether it is CO_2 mitigation, adaptation, or technology transfer (TT) – you will find at least two responses. This is because we commissioned a secondary group

of qualified economists to provide a critique on the assumptions made in each chapter, for every topic. The "Alternative Perspectives" papers provide another way of looking at the costs, benefits, and risks of a particular response to climate change, and highlight the areas where expert opinion diverges. Some of them also provide an alternative solution, complete with estimates on costs and benefits.

For the topic of Climate Engineering, J. Eric Bickel and Lee Lane (chapter 1) offer an assessment of the potential benefits and costs of such engineering, examining two families of technologies – solar radiation management (SRM) and air capture (AC). Among other findings, they conclude that large potential net benefits of SRM mean that there is strong evidence for researching this technology further in the short term.

Two authors offer different perspectives on CE. Roger A. Pielke, Jr. (Perspective paper 1.1) argues that Bickel and Lane's analysis of SRM is not grounded in a realistic set of assumptions about how the global earth system actually works. He agrees that there is justification for continued research into technologies of SRM, but finds that this judgment does not follow from a CBA. Pielke also summarizes an analysis of the potential role for AC technologies to play in the de-carbonization of the global economy, and argues that since the costs of AC are directly comparable with major global assessments of the costs of conventional mitigation policies, AC also deserves to receive further study.

Anne E. Smith (Perspective paper 1.2) overlays Bickel and Lane's work with a consideration of the potential unintended consequences from CE, and extends it by calculating the value of information (VOI) from research and development (R&D). She then goes further and takes a critical look at the theoretical assumptions underpinning the standard formula for VOI.

On carbon emission reductions, Richard S.J. Tol (chapter 2) examines the costs and benefits of cutting carbon under different scenarios, and finds that while a well-designed, gradual policy of carbon cuts could substantially reduce emissions at low cost to society, ill-designed policies, or policies that seek to do too much too soon, can be orders of magnitude more expensive. He notes that while the academic literature has focused on the former, "policy makers have opted for the latter." Tol specifically considers five policies for carbon dioxide emission reduction. His findings include the point that very stringent targets, such as the EU's target of keeping temperature rises under 2°C, may be very costly or even infeasible, while suboptimal policy design would substantially add to the costs of emission abatement.

Onno Kuik (Perspective paper 2.1) is in agreement with most of what is written by Tol on the state of the art of economic research into the impacts of climate change and climate change policies, but highlights a complementary approach based on a direct elicitation of preferences for climate change.

Roberto Roson (Perspective paper 2.2) notes that Tol's chapter is largely based on the Climate Framework for Uncertainty, Negotiation, and Distribution (FUND) model and the results of a set of simulation exercises where a number of policy options are explored and assessed. Roson points out a series of limitations of this model. However, he concludes that when considering the simulation scenarios, "we could have got about the same findings with a different model."

Brent Sohngen (chapter 3) looks at forestry carbon sequestration and indicates that if society follows an "optimal" carbon abatement policy, as defined in Nordhaus (2009), forestry could accomplish roughly 30% of total abatement over the century, while if society places strict limits on emissions in order to meet a 2°C temperature increase limitation, then the component that forestry provides lowers overall abatement costs by as much as 50%. Sabine Fuss (Perspective paper 3.1) is in broad agreement with Sohngen's analysis of costs and benefits. She concludes, like Sohngen, that forest carbon will be needed as part of a strategy to mitigate climate change.

Robert E. Baron, W. David Montgomery, and Sugandha D. Tuladhar point out in chapter 4 that a significant share of current net warming is attributable to black carbon. Black carbon is essentially the soot produced through diesel emissions, and – in developing countries – people burning organic matter to cook food and stay warm. It can be eliminated with cleaner fuels and new cooking technologies. Sooty pollution from indoor fires claims

several million lives each year so reducing black carbon would also be a life-saver. Black carbon can be controlled in developing countries through the implementation of cleaner fuels, new cooking technologies, and changing crop management practices. The authors present potential ways to implement these policies, and provide cost-benefit (C/B) estimates that indicate that spending around $359 million could slash around 19% of black carbon emissions. Milind Kandlikar, Conor C.O. Reynolds, and Andrew P. Grieshop (Perspective paper 4.1) argue that it is important to recognize that black carbon reductions are not a substitute for reductions in emissions of carbon dioxide (CO_2), but that the two approaches must be applied together.

Claudia Kemfert and Wolf-Peter Schill (chapter 5) look at ways to mitigate CH_4, a major anthropogenic greenhouse gas (GHG), second only to CO_2 in its impact on climate change. They recommend an economically efficient global CH_4 mitigation portfolio for 2020 that includes the sectors of livestock and manure, rice management, solid waste, coal mine CH_4, and natural gas.

David Anthoff (Perspective paper 5.1) argues that joint methane (CH_4) and CO_2 emission mitigation is an optimal policy mix and leads to the highest net benefits, suggesting that an "either-or" approach between CO_2 or CH_4 emission mitigation forgoes at least some joint benefits. Daniel J.A. Johansson and Fredrik Hedenus (Perspective paper 5.2) note that the technical measures available to reduce emissions from livestock, the most important single sector emitting CH_4, are small. The combination of being a non-point emission source and having few technical abatement measures implies that output-based policies may be appropriate for reducing these emissions.

Francesco Bosello, Carlo Carraro, and Enrica De Cian (chapter 6) carry out an integrated analysis of both optimal carbon mitigation and adaptation at the global and regional level, and show that, compared to mitigation which reduces mainly future damages, adaptation is more rapidly effective for contrasting future and present damages. In particular, in a high-damage world (but without climate catastrophes), adaptation becomes the preferred strategy and this is reflected in an increasing BCR. They note that most adaptation expenditures need to be carried out in developing countries, but that the size of the required resources is likely to be well beyond their absorptive capacity. Therefore, international cooperation is necessary to successfully transfer resources and adaptation technology to developing countries.

Samuel Fankhauser (Perspective paper 6.1) argues that adaptation is now unavoidable, because there are no realistic mitigation policies that restrict warming to a level that does not require substantial adaptation. He notes that it is made more difficult by uncertainty about the exact nature of the expected change, which puts a premium on adaptations that yield early benefits or increase the flexibility of systems to react to unexpected change. Frank Jotzo (chapter 6.2) notes that economic analysis of adaptation is subject to the same complications and limitations that beset quantitative economic analysis of climate change mitigation. There is a long road ahead in improving the tools for economic modeling of adaptation, and the mitigation–adaptation nexus, and in the meantime the crucial question for policy makers is whether and where specific adaptation actions are beneficial, what new policies are needed to support adaptive action, and what existing policies need to be changed or scrapped.

Isabel Galiana and Christopher Green (chapter 7) examine a technology-led approach to climate policy. They write that the rationales for this approach include the huge energy technology challenge to stabilizing climate; the lack of readiness or scalability of current carbon emission-free energy technologies; the energy-intensive nature of growth in populous developing countries, especially in Asia; the economic and political limitations of a carbon pricing-led policy; and the large economic cost of "brute force" mitigation policies.

Valentina Bosetti (Perspective paper 7.1) finds that combining R&D and climate policies might lead to efficiency gains and help contain climate policy costs. Bosetti also specifically focuses on analyzing the costs and benefits of research and development in CO_2 capture and storage (CCS). This allows the continued use of fossil fuels while reducing the CO_2 emissions produced and may therefore be hugely helpful, especially in countries like China and India, that heavily rely on coal for the generation of electricity. Although uncertainties are present when dealing with R&D investments, Bosetti finds that a program aiming at decreasing capturing costs or increasing the CO_2 capture rate is shown to pass the cost-benefit (C/B) test, if a climate policy is in place.

Gregory Nemet (Perspective paper 7.2) agrees with Galiana and Green regarding the magnitude of the technological revolution required to address climate change, and the inability of on-the-shelf technologies to adequately fulfill the required technological change. Among other points, Nemet notes that a carbon price signal is insufficient to induce the technology development investments required to limit global temperature increase, and that the technology-led policy will shift the bulk of technological decision making from the private sector to the public sector.

Zili Yang (chapter 8) looks at technology transfer (TT): the process of sharing skills, knowledge, and technological breakthroughs among governments and other institutions to ensure that scientific and technological developments are accessible to a wider range of users. He finds that such transfers are an effective and necessary component when dealing with climate change, because international cooperation on both GHG mitigation and adaptation must involve transfer of technologies or dissemination of knowledge. David Popp (Perspective paper 8.1) critiques Yang's estimate of the potential of TT as a climate policy option, noting that Yang focuses on the direct gains from developed country financing of abatement in developing countries. Popp points out that there is an important secondary gain from TT – the potential for knowledge spillovers. He assesses the potential role that spillovers might play, and offers an assessment of the overall potential of international TT as a policy solution.

I believe that all of this research, in itself, provides a valuable contribution to and overview of today's discussions on global warming policy. But it is vital that we test and debate the experts' recommendations, and identify the most attractive possibilities for policy makers to further explore. That is why the Copenhagen Consensus process goes beyond just gathering new research.

As in the Copenhagen Consensus 2004 and 2008 projects, an Expert Panel of economists – including three Nobel laureates – examined all of the research presented here. The five-strong Expert Panel for the Copenhagen Consensus on Climate engaged with all of the chapter and Perspective paper authors and came to their own conclusions about the merits of each suggested solution.

The Expert Panel discussed and debated all of the possibilities raised in the research, in sessions designed to promote free debate. They weighed up each solution that you will find in this book, and compared it to the other options. The Expert Panel was tasked with answering the question:

If the global community wants to spend up to, say, $250 billion per year over the next 10 years to diminish the adverse effects of climate changes, and to do most good for the world, which solutions would yield the greatest net benefits?

Later in the book, you will find their answers to that question, along with their individual explanations of how it was reached. They focused largely on estimates of costs and benefits, which is a transparent and practical way to show whether or not spending is worthwhile.

I invite you to read the research and the Expert Panel's findings, and form your own view on the best – and worst – ways we can respond to global warming. It is certainly time that we focused more on the solutions to this challenge.

Bibliography

Lomborg, B., 2001: *The Skeptical Environmentalist: Measuring the Real State of the World*, Cambridge University Press, Cambridge

Nordhaus, W.D., 2009: *A Question of Balance: Weighing the Options on Global Warming Policies*, Yale University Press, New Haven, CT

PART I

The Solutions

Climate Engineering

J. ERIC BICKEL AND LEE LANE*

Considering Climate Engineering as a Response to Climate Change

Climate Change and Benefit-Cost Analysis

The task of this chapter is to answer a question that has been posed as part of the Copenhagen Consensus exploration of climate policy. That question is:

> If the global community wants to spend up to, say, $250 billion per year over the next 10 years to diminish the adverse effects of climate changes, and to do most good for the world, which solutions would yield the greatest net benefits?

To address this question, we agreed to summarize the existing literature regarding the costs and benefits of geoengineering (GE), supplement these estimates where needed and feasible, and to provide benefit-cost ratios (BCRs) for at least two GE alternatives. Based on this analysis, the current chapter argues that some portion (0.3%) of the hypothetical $250 billion a year should be devoted to the task of researching and developing two GE areas: solar radiation management (SRM) and air capture (AC). As the reader will see, we argue that more emphasis should be placed on SRM but that AC merits some research support.

The reader should not interpret our focus on climate engineering (CE) as implying that other responses to climate change are unneeded. The proper mix and relative priority of various responses to climate change is in the purview of the expert panel, to which our chapter is one input. One might also note that, with but one exception, every scenario considered in this chapter is accompanied by greenhouse gas (GHG) control measures.

The US Environmental Protection Agency (EPA) describes GE as "the intentional modification of Earth's environment to promote habitability" (EPA 2009). Many experts prefer the term "climate engineering" (CE) as more accurately describing the most widely discussed current concepts of modifying climate to curtail the harmful effects of global warming, and we will adopt this term.

Following the Copenhagen Consensus project framework, this chapter applies benefit-cost analysis (BCA) to gain insight into the net economic benefits that society might achieve by deploying CE. A finding that net benefits may be large, but are uncertain, suggests that society should devote some current resources to researching and developing this capacity. Some people object to BCA, and to CE, on what they regard as ethical grounds. Ethical conjectures are notoriously resistant to empirical falsification, and this chapter will not attempt to join this debate. Instead, we adopt the viewpoint that climate-change policies, including the possible use of CE, should be designed to maximize the welfare of human beings over time. "Welfare," in this context, includes the consumption of both market and non-market goods, such as environmental services (Nordhaus 2008).

Other objections to BCA rest on more purely pragmatic grounds. BCA is often difficult to apply because either costs or benefits may be difficult, or maybe even impossible, to quantify with confidence. Analysts may be tempted to overlook or to assume away some of these hard-to-quantify factors in hopes of keeping the analysis tractable. To choose an example that this chapter will address, BCA often ignores transaction costs, and a whole

* The authors gratefully acknowledge Ken Caldeira and Lowell Wood for helpful discussions regarding the structure of a R&D program. The authors also thank Christian Bjørnskov, Roger Pielke, Jr., Anne E. Smith, and Vernon Smith for careful and challenging reviews. Finally, the authors acknowledge the valuable assistance of Dan Fichtler in preparing this chapter.

school of economics has grown up around the task of correcting the mistakes to which this simplification can sometimes lead (North 1990). Transaction costs are, indeed, hard to quantify. The existing climate policy literature has made no attempt in this direction, and this chapter will offer only a qualitative discussion of some salient points about the main issues. It suggests, however, that the transaction costs associated with SRM may be smaller than those that apply to some other climate strategies.

Likewise, we do not attempt to perform a probabilistic BCA, though one is clearly needed. We take this approach for two reasons. First, an important aspect of the Copenhagen Consensus project framework is ensuring a consistency among the chapters, which is harder to maintain in a probabilistic setting. Second, the state of knowledge about both the benefits of CE and its costs is primitive. Even base-case estimates for many important benefit and cost parameters are unknown. Thus, where the existing literature contains quantitative estimates, this chapter will select what we regard as the best available, with the caution that today's estimates are very much subject to change. Where possibly important factors have not been quantified, this analysis will point to their nature and discuss their potential significance.

In sum, we adopt what we hope readers will regard as a pragmatic approach to BCA. As one economist has observed, "everyone who urges a change in policy (or resists one) is at least implicitly comparing costs with benefits" (Cooper 2000). Making the basis of this comparison more explicit seems, on principle, likely to facilitate a more reasoned discourse.

The Budget Constraint and the Assumption of "Sensible" Policies

At this point, the Copenhagen Consensus budget constraint does not play much of a role in the issues raised by CE. Currently, CE is a concept deserving, we believe, R&D. It is not ready for deployment. How much money should go into the concept's exploration depends in part on the results of the initial research. However, the rudimentary state of knowledge about the concept suggests that an

investment of perhaps 0.3% ($750 million per year) of the global total proposed by the Copenhagen Consensus guidelines might be an appropriate average yearly expenditure for the first decade. As R&D progresses, and assuming that results were favorable, spending would increase from tens of millions of dollars in the early years to the low billions of dollars. Extended large-scale field tests might be needed for perhaps an additional five years. Thus, spending in the first decade would not approach the budget constraint, although deployment could involve costs in the tens to hundreds of billions.

The chapter focuses on a BCA of deploying CE beginning in 2025. That choice rests on the proposition that the very large net benefits found in this analysis of CE make a convincing case for incurring upfront costs to research, to develop, and to demonstrate the concept. The chapter, in this regard, does assume that future policies will be "sensible," in that it assumes that R&D of a concept promising large net benefits would lead, at some point, to an effort to realize those benefits in practice.

However, the analysis also considers some policies that are not sensible – or perhaps one should say that it looks at some policies that do not appear to be optimal within the framework of a somewhat blinkered BCA. The paper considers how these options might affect the performance of CE, and looks briefly at how CE might affect the results of a few badly structured GHG control regimes. Some consideration of non-optimal policies can offer useful insights about how CE might function in the real world in which policy choices are rarely optimal (North 1990).

Description of Human-Induced Climate Change

Greenhouse gases (GHGs) in the Earth's atmosphere cause the planet's surface to be about $30°C$ warmer than would otherwise be the case (Stocker 2003). These gases allow the passage of short-wave radiation (sunlight), but absorb long-wave radiation (heat) and radiate a fraction of it back to the Earth's surface (Trenberth et al. 2009). This fact has been well established for a very long time.

It is equally clear that human activities can add to the GHG stocks in the Earth's atmosphere. The burning of fossil fuels, deforestation, and agriculture and animal husbandry are all practices that have this effect (IPCC 2007). All else being equal, although all else may not be equal, higher GHG concentrations will raise global mean temperatures (IPCC 2007).

The policy implications of this relationship, though, remain far from clear. Hard-to-predict demographic and economic trends will influence future emissions. Technological change is also a powerful driver of emissions, and its future direction and pace are still more unclear than are those of population and output. How well or poorly will societies adapt to climate change? The answer remains in doubt, but it will greatly affect the size of the costs and benefits that societies will experience.

The state of climate science compounds the uncertainties (IPCC 2007). How an increment of GHG will impact future temperature remains the subject of lively dispute. Man-made GHG emissions may interact in poorly understood ways with clouds, eco-systems, ocean currents, chemical cycles, and myriad other factors. These interactions may produce non-linear effects. Some feedback loops may amplify the warming impetus of larger GHG stocks. Some may dampen it. Science understands some of the interactions well, but many remain murky.

Even more doubts shadow predictions of what to expect from whatever warming does occur. Some experts believe that the climate system includes "tipping points" at which temperature, or other factors, may generate rapid and potentially very destructive changes. Where these tipping points may lie, how many (or few) of them there may be, whether they are near or far in time from the present, what happens if they are crossed – all these questions are unanswered.

The trajectory of GHG emissions also depends on future policy choices by many nation-states, and how their policies evolve. On this score, the historical record is clear:

The year 2008 marks the 20th anniversary of the first meeting of the IPCC, the international body established by the UN to solve the problem of warming. The "progress" to date has been almost purely rhetorical. Currently, according to the US Energy Information Agency, global emissions

of CO_2 [carbon dioxide], the most important greenhouse gas, were over a third higher than they had been in 1988. The IPCC reports that the rise in atmospheric concentrations has accelerated through the last several decades. (Lane and Montgomery 2008)

In fact, global CO_2 emissions grew four times more quickly between 2000 and 2007 than they did between 1990 and 1999 (Global Carbon Project 2008).

Thus, twenty years of protracted diplomatic talk and laborious scientific study have so far failed to move the needle on emission rates. During this period, GHG output has fallen in some countries, but, where such declines have occurred, "underlying changes in economic structure may have played a bigger role than climate policy" (Lane and Montgomery 2008). For example, most Kyoto Protocol signatories are failing to reduce emissions, much less meet their targets (UNFCCC 2009). The reductions that were achieved were heavily concentrated in Central and Eastern Europe (CEE), whose economies contracted and were restructured after the fall of the Soviet Union (UNFCCC 2008). The overall trend remains clear, and the prospects daunting.

Three Aspects of GHG Emissions that Cause Concern

GHG emissions may actually cause three quite distinct kinds of problems. They differ in the likelihood of their occurrence, their probable timing, and the incidence of their costs and benefits.

Gradual climate change

Gradual warming is likely to unfold over long periods of time, but its pace may vary from decade to decade. The process is likely to bring both benefits and costs. Benefits will include some higher crop yields from longer growing seasons and CO_2 fertilization. Mortality from cold will be likely to fall, as will heating costs. At the same time, gradual warming will impose costs. Some crop yields will fall, sea levels will rise, some storms may grow in intensity, more intense heatwaves will occasion health problems and raise cooling costs, and in some cases the range of tropical diseases may widen. While societies will adapt, as they have to prior climate changes, adaptation will often not be free. Many poorer societies currently lack the human and physical capital required to make the needed changes. Some valuable unmanaged eco-systems may also fall short on adaptive capacity.

Geographically, the incidence of costs and benefits will vary. Benefits are likely to be concentrated in higher latitudes, whereas most costs are likely to appear in climates that were warmer to begin with. Over time, though, even in regions with initially cooler temperatures, costs will climb relative to benefits. Nonetheless, in the midst of these changes – some positive, some negative – much of the industrial sector is likely to be unaffected. The pace of economic growth is generally expected to outrun that of gradual climate change (Schelling 2002). Thus, if climate changes gradually, the harm that it could occasion would take place in the context of a growing global economy.

Rapid climate change

Rapid high-impact climate change might occur relatively swiftly and could produce very large social costs. The timing and probability of such change are speculative. However, the risk cannot be ruled out. One current worry is that man-made warming could trigger large-scale methane (CH_4) release from the Arctic and sub-Arctic tundra (Corell *et al.* 2008). CH_4, itself, is a powerful GHG. Hence, man-made warming might unleash a self-reinforcing process. This warming might, in turn, accelerate the melting of the Greenland and Antarctic ice sheets. The latter would hasten the rise in sea levels, possibly doing serious economic harm to coastal cities. Other speculation has focused on major shifts in the pattern of ocean currents. Such a shift might reorganize the distribution of temperatures and precipitation.

Compared to gradual climate change, rapid change scenarios promise little upside (Barrett 2007a). The mere fact that a change happens rapidly is likely to raise the costs of adapting to it, and rapid change is often assumed to be quite destructive, even though its probability is low and highly uncertain (Weitzman 2007).

Ocean acidification

Finally, the ocean becomes more acidic as it absorbs CO_2 from the atmosphere (Royal Society 2005). Some studies suggest that, over time, this process could disrupt marine eco-systems and perhaps cause economic harm (Royal Society 2005). This risk, whatever its severity, is not strictly speaking climate change, but it is another aspect of CO_2 discharges.

Acidification and warming are likely to interact. Acidification is believed to weaken the ability of coral reefs to recover from bouts of bleaching caused by warm ocean temperatures (Kleypas *et al.* 2006). Corals are productive and economically valuable, and acidification might also harm other species near the base of the ocean food chain. The severity of the problem is poorly understood at the moment, but is causing concern among some scientists.

The uncertain state of knowledge about acidification greatly complicates the task of formulating an efficient policy response to it. At least some analysis suggests that even the most severe GHG controls might fail to halt the destruction of most coral reefs. The CO_2 already in the atmosphere could cause enough acidification to destroy all (or most) of the existing reefs (Cao and Caldeira 2008). Conversely, novel GE technologies beyond the scope of this chapter might be able to reverse acidification, at least in some areas (Rau *et al.* 2007). At this point, then, acidification appears to be a potentially important matter, but its relevance to CE remains doubtful. SRM does not address ocean acidification, and, accordingly, our BCA gives SRM *no* credit for doing anything about it.

Time Scales

CO_2, once in the atmosphere, will remain there for a century or more. Attempts to abate GHG emissions are also subject to lengthy time lags. Major technological changes often take a long time to mature, and new technology is often slow to diffuse globally (Edgerton 2007). Electrification of the global economy has been in train for over one hundred years, and is still incomplete. The spread of electrification was spurred forward by the large net benefits that accrued to those investing in it.

GHG controls will demand still more far-reaching changes in technology. Developing many of these innovations, according to Secretary of Energy Steven Chu, must await the appearance of multiple major breakthroughs in basic science (Broder and Wald 2009). Effective GHG controls, moreover, will require almost world-wide efforts (Jacoby *et al.* 2008). The need for such wide-ranging change is likely to extend the amount of time that the process will require. That many low-GHG technologies cost more than those they seek to replace will further delay their spread. By inference, new laws and regulations will have to be adopted before such technologies can gain acceptance. A great deal of time may, therefore, separate the onset of serious efforts to limit emissions and the actual stabilization of climate.

This potential lag creates tension between the risk of rapid climate change and the slow speed with which GHG controls can take effect. Steep GHG cuts are substantially more costly than gradual ones (Richels *et al.* 2004). Yet, should it appear that a tipping point was imminent, controls might do little to stabilize the situation.

Climate Engineering

With these challenges as a backdrop it is easy to grasp why proposals to seriously study CE are gaining adherents. Both the National Academy of Sciences in the USA and the Royal Society in Britain are exploring the concept. The American Meteorological Society is also evaluating it. Such prominent scientists as Edward Teller, Paul Crutzen, Ralph Cicerone, Alan Robock, and Tom Wigley have highlighted the need for study, and John Holdren, President Obama's new science advisor, has said, "It's [CE] got to be looked at" (Borenstein 2009). Economists like Scott Barrett, William Nordhaus, Thomas Schelling, and Lawrence Summers have also suggested further exploration (Barrett 2007b; Summers 2007; Lane and Montgomery 2008).

Solar Radiation Management

Types of SRM options

SRM aims at offsetting the warming caused by the build-up of man-made GHGs in the atmosphere by reducing the amount of solar energy absorbed

by the Earth. As discussed above, GHGs in the atmosphere absorb long-wave radiation (thermal infrared or heat) and then radiate it all directions – including a fraction back to the Earth's surface. This creates an energy imbalance and rising temperatures. SRM does not attack the underlying cause of the warming, higher GHG concentrations. Rather, it seeks to reflect back into space a small part of the Sun's incoming short-wave radiation. In this way, temperatures are lowered even though GHG levels are elevated. At least some of the risks of global warming can, thereby, be counteracted (Lenton and Vaughan 2009).

Reflecting into space only 1–2% of the sunlight that strikes the Earth would cool the planet by an amount roughly equal to the warming that is likely from doubling the preindustrial levels of GHGs (Lenton and Vaughan 2009). Scattering this amount of sunlight appears to be possible. Past volcanic eruptions, for example, have shown that injecting relatively small volumes of matter into the upper atmosphere can cause discernable cooling. The 1991 eruption of Mount Pinatubo reduced global mean temperature by about 0.5°C (Lane *et al.* 2007).

Several concepts have been proposed for accomplishing SRM:

Shortwave geoengineering proposals . . . start with reflecting away (or shading out, as seen from Earth) a fraction of incoming solar radiation by placing objects in a solar orbit, e.g. at the inner Lagrange point (L1) (Angel 2006). Alternatively, sunshades could be placed in an Earth orbit (NAS 1992; Pearson *et al.* 2006). Once solar radiation enters the atmosphere, its reflection back to space could be enhanced by adding sulphate aerosol (Crutzen 2006) or manufactured particles (Teller *et al.* 1997, 2002) to the stratosphere. Adding such aerosols to the troposphere (NAS 1992) has been ruled out due to negative impacts on human health, the greater loading required than the equivalent intervention in the stratosphere, and the need for multiple injection locations (Crutzen 2006; MacCracken 2006). However, increasing the reflectivity of low level marine stratiform clouds by mechanical (Latham 1990) or biological (Wingenter *et al.* 2007) generation of cloud condensation nuclei (CCN) is being considered. Finally, the reflectivity of the Earth's surface could be increased, with recent proposals focused on the land surface, including albedo modification of deserts (Gaskill 2004), grasslands (Hamwey 2007), croplands (Ridgwell *et al.* 2009), human settlements (Hamwey 2007), and urban areas (Akbari *et al.* 2009). (Lenton and Vaughan 2009)

The various SRM options differ importantly in the scale of their promise and in the range of their possible use. For example, none of the concepts for modifying the albedo of the Earth's surface represents a global-level solution. Then too, objects on the Earth's surface get dirty, raising maintenance costs. There is also a risk that many of these options might disrupt surface eco-systems. Surface-level approaches may still be locally useful as a counter to the urban heat island effect. Hence, they may become niche technologies, and on that basis may warrant further study. They cannot, however, offer large net benefits on a global scale. This chapter will, therefore, address the sunshade, stratospheric aerosols and marine cloud whitening at greater length, because these concepts might be able to offset warming on a global scale.

SRM and institutions

Compared to GHG control options, SRM involves no infringement of economic freedom. An observation recently applied to AC applies at least equally to SRM: "Technological fixes do not offer a path to moral absolution, but to technical resolution. Indeed, one of the key elements of a successful technological fix is that it helps to solve the problem while allowing people to maintain the diversity of values and interests that impede other paths to effective action (Sarewicz and Nelson 2008). The institutional pros and cons of SRM are discussed at greater length in the section below describing its political transaction costs.

Air Capture

AC, the second family of CE concepts, would work on a different principle. It focuses on removing CO_2 from the atmosphere and securing it in land or sea-based sinks. Thus, AC, unlike SRM, ignores short-wave radiation. Instead, it attacks frontally

the impact of GHG concentrations on long-wave radiation:

> Air capture may be viewed as a hybrid of two related mitigation technologies. Like carbon sequestration in eco-systems, air capture removes CO_2 from the atmosphere, but it is based on large-scale industrial processes rather than on changes in land use, and it offers the possibility of near-permanent sequestration of carbon. (Keith and Ha-Duong 2003)

Like carbon capture and storage (CCS), AC involves long-term storage of CO_2, but AC removes the CO_2 directly from the atmosphere rather than from the exhaust streams of power plants and other stationary sources. AC may eventually be a useful option in coping with mobile GHG emission sources. As one expert describes the concept:

> For distributed, mobile sources like cars, on-board capture at affordable cost would not be feasible. Yet, in order to stabilize atmospheric levels of CO_2, these emissions, too, will need to be curtailed... extraction of CO_2 from air could provide a viable and cost-effective alternative to changing the transportation infrastructure to non-carbonaceous fuels. Ambient CO_2 in the air could be removed from natural airflow passing over absorber surfaces. The CO_2 captured would compensate for CO_2 emission from power generation two orders of magnitude larger than the power... Air extraction is an appealing concept, because it separates the source from disposal. One could collect CO_2 after the fact and from any source. Air extraction could reduce atmospheric CO_2 levels without making the existing energy or transportation infrastructure obsolete. There would be no need for a network of pipelines shipping CO_2 from its source to its disposal site. The atmosphere would act as a temporary storage and transport system. (Lackner *et al.* 2001)

A recent survey described a number of possible AC technologies (Pielke 2009). It noted that "The most straightforward means of AC is simply through photosynthesis." Thus, biomass could fuel power plants operating with carbon capture and storage systems. Similar concepts involve fertilizing the ocean using nitrogen, iron, or phosphorous as a route to increasing carbon storage in deep ocean sinks. Inadvertent phosphorous fertilization is already occurring (Lenton and Vaughan 2009).

Other approaches propose to use chemical reactions capable of capturing carbon from the air on an industrial scale (Keith *et al.* 2006). One such approach envisions capturing CO_2 in sodium hydroxide in cooling-tower-like structures. The chemicals required for this process are "inexpensive, abundant, and relatively benign" (Keith *et al.* 2006). In the view of some experts, a well-funded R&D program might make such a technology available on a large scale. Unfortunately the process requires relatively large energy inputs, which may also affect its monetary costs (Keith *et al.* 2006).

The institutional case for air capture

As compared to GHG controls, AC offers major institutional advantages. It circumvents many of the problems that are plaguing GHG controls. For example, with GHG controls, new technologies will compete with each other and with existing technologies. As a result, "we can expect ongoing technical and political debates about efficacy of specific technologies, as seen for biofuels today" (Sarewitz and Nelson 2008). There is, however, no unambiguous metric for evaluating the myriad rival GHG control technologies. In the absence of such a metric, debate tends to be protracted, and the task of GHG control is, therefore, largely impervious to R&D-generated solutions (Sarewitz and Nelson 2008).

The performance of AC, in contrast, is relatively straightforward: how much CO_2 does an AC technology capture, and at what cost? Furthermore, despite the cost challenges that it presents, AC is building on a base of existing scientific knowledge. CO_2 capture is clearly possible and several well-understood chemical processes exist for doing it. Finally, and perhaps most importantly, AC, like SRM, might be structured to have relatively minimal impacts on economic freedom (Sarewitz and Nelson 2008). (AC would not necessarily have this feature if it were financed through the offset provisions of a cap-and-trade scheme.) These potential institutional virtues argue strongly in favor of R&D effort aimed at lowering the costs of AC.

Air Capture and Solar Radiation Management

Although AC is often classified along with SRM as CE, the two concepts differ in important ways. First, as discussed above, they in disparate ways address warming: SRM directly reduces short-wave radiation, and AC directly reduces long-wave radiation through the removal of CO_2 from the atmosphere. Second, as we demonstrate below, some SRM technologies can affect temperature more quickly than AC. This feature may be particularly important if one is concerned about rapid warming and abrupt change. The time-lag involved with AC lowers its benefit. A recent paper compared AC to SRM on this dimension. It concluded:

> Thus, it would appear that only rapid, repeated, large-scale deployment of potent shortwave geoengineering options (e.g. stratospheric aerosols) could conceivably cool the climate to near its preindustrial state on the 2050 timescale. However, some carbon cycle geoengineering options could make a useful contribution of similar magnitude to identified mitigation "wedges" (Pacala and Socolow 2004). In the most optimistic scenarios, air capture and storage by BECS [Bio-energy with carbon storage], combined with afforestation and bio-char production appears to have the potential to remove $\sim$100 ppm of CO_2 from the atmosphere giving $\sim$$-1.3\,W\,m^{-2}$. Combined iron, nitrogen and phosphorus fertilisation of the ocean can only achieve a maximum $\sim$20 ppm CO_2 drawdown and $-0.24\,Wm^{-2}$ on the 2050 timescale. (Lenton and Vaughan 2009)

Thus, AC may have a useful role to play in climate policy. In fact, AC may offer major advantages relative to GHG controls. As noted, it seems well suited to the task of controlling mobile source emissions. Moreover, many institutional factors are likely to distort the application of GHG controls and to inhibit their spread (Lane and Montgomery 2008); AC might sidestep some of these factors.

Finally, SRM and AC appear to differ in terms of cost. As will be discussed later, some of the SRM concepts appear to have very low deployment costs. The costs of AC, on the other hand, may be on the order of $500 per metric ton of carbon (MTC), and are not competitive with near-term mitigation technologies such as CCS (Keith *et al.* 2006). The high costs of AC and the long time scales it would require to become effective are serious drawbacks relative to SRM. On the other hand, AC, by seeking to remove CO_2 from the atmosphere, reduces some of the risks that remain with SRM.

In effect, AC raises issues that differ fundamentally from those that surround SRM. The remainder of this chapter will focus primarily, although not exclusively, on the benefits and costs of SRM. We understand that the Perspective paper by Roger Pielke, Jr. (chapter 1.1), will more carefully address the costs and benefits of AC.

Definition and Description of SRM Solutions

Several technological concepts have been proposed as possible means of effecting SRM. Whatever technology might be used, there are also choices about the mode of deployment. The way these issues are resolved will affect both costs and benefits.

Three SRM Concepts Merit Evaluation

At least two of the available options appear to be promising candidates for affecting global climate: marine cloud whitening and stratospheric aerosols. A discussion of the space-based sunshade is included because the concept has been widely discussed. In this section, we define these technologies. We explore their benefits and costs in the subsequent two sections.

Marine cloud whitening

One current proposal envisions producing an extremely fine mist of sea water droplets. These droplets would be lofted upwards and would form a moist sea salt aerosol. The particles within the aerosol would be less than 1 micron in diameter. These particles would provide sites for cloud droplets to form once they rise to the marine cloud layer. The up-lofted droplets would add to the effects of natural sea salt and other small particles, which are called, collectively, cloud condensation

nuclei (Latham *et al.* 2008). The basic concept was succinctly described by one of its developers:

> Wind-driven spray vessels will sail back and forth perpendicular to the local prevailing wind and release micron-sized drops of seawater into the turbulent boundary layer beneath marine stratocumulus clouds. The combination of wind and vessel movements will treat a large area of sky. When residues left after drop evaporation reach cloud level they will provide many new cloud condensation nuclei giving more but smaller drops and so will increase the cloud albedo to reflect solar energy back out to space. (Salter *et al.* 2008)

Just as volcanoes have provided the natural experiment suggesting the efficacy of stratospheric aerosol, the long white clouds that form in the trails of exhausts from ship engines illustrate this concept. Sulfates in the ships' fuel provide extra condensation nuclei for clouds. Satellite images provide clear evidence that these emissions brighten the clouds along the ships' wakes:

> Since, in the scheme we propose, the aim is to increase the solar reflectivity of such low-level maritime clouds and since a fine salt aerosol provides an admirable replacement for the sulphates whose effectiveness is evident . . . , it seemed appropriate for the sprays to be dispersed from seagoing vessels (rather than, say, low-flying aircraft) and for the source of the sprays to be drawn from the ocean itself. (Salter *et al.* 2008)

The plan's developers conceive of an innovative system:

> Energy is needed to make the spray. The proposed scheme will draw all the energy from the wind . . . The [ships'] motion through the water will drive underwater "propellers" acting in reverse as turbines to generate electrical energy needed for spray production. Each unmanned spray vessel will have a global positioning system, a list of required positions and satellite communications to allow the list to be modified from time to time, allowing them to follow suitable cloud fields, migrate with the seasons and return to port for maintenance. (Salter *et al.* 2008)

Thus, the plan rests on an integrated system of technologies. One key to the system is the wind-driven rotor system developed in the early twentieth century by Anton Flettner. This system allows the ships to be powered by wind but to avoid the high handling and maintenance costs of sails. It also promises superior handling:

> The rotors allow a sailing vessel to turn about its own axis, apply "brakes" and go directly into reverse. They even allow self-reefing at a chosen wind speed. Flettner's rotor system weighed only one-quarter of the conventional sailing rig which it replaced. The rotor ships could sail 20° closer to the wind than unconverted sister ships. The heeling moment on the rotor flattened out in high wind speeds and was less than the previous bare rigging. With a wind on her quarter, the ship would heel into the wind. The only disadvantage of these vessels is that they have to tack to move downwind. Energy has to be provided for electric motors to spin the rotors, but this was typically 5–10 per cent of the engine power for a conventional ship of the same thrust. (Salter *et al.* 2008)

Clearly this power system offers significant advantages for the tasks implied by marine cloud whitening.

Preliminary calculations suggest that the marine clouds of the type considered by this approach contribute to cooling, and that augmenting this effect could, in theory, produce enough cooling to offset a doubling of atmospheric GHG concentrations. The logistical problems do not appear to be unmanageable. Analyses using the general circulation model of the Hadley Centre of the UK Meteorological Office offer quantitative support for the scheme's feasibility. Thus a recent study observed, of results produced using this model:

> These indicate that warming due to a doubling of the CO_2 content of the atmosphere could be roughly compensated for – when taking account of the negative forcing due to the production of anthropogenic aerosol to date – by a doubling of the droplet number concentration N_d in three extensive regions of maritime stratocumulus clouds (off the West coasts of Africa and North and South America), which together cover about 3% of the global surface. If the anthropogenic aerosol factor is discounted, N_d would need to be roughly quadrupled. If only clouds covering this specially selected 3% of the Earth's surface were modified, instead of all marine stratocumulus clouds, the critical value of

top-of-cloud albedo-change required to compensate for a doubling of CO_2 concentration would rise from 0.02 . . . to about 0.16. The associated values of enhanced [albedo] are within natural bounds. (Bower *et al.* 2006)

An important aspect of this result is the relatively low percentage of the total marine cloud cover that would have to be enhanced in order to produce the desired cooling.

The concept's developers have devised a deployment strategy, which they describe in the following terms:

> Suitable sites for spraying need plenty of incoming sunshine to give something to reflect. They must have a high fraction of low-level marine stratocumulus cloud. They should have few high clouds because these will reduce incoming energy and send the reflected energy down again. There should be reliable but not extreme winds to give spray vessels sufficient thrust. There should be a low density of shipping and icebergs. It helps to have a low initial density of cloud condensation nuclei because it is the fractional change that counts. This suggests sea areas distant from dirty or dusty land upwind. Owing to a possible anxiety over the effect of extra cloud condensation nuclei on rainfall, areas upwind of land with a drought problem should be avoided. (Salter *et al.* 2008)

A British effort is developing hardware with which to test the feasibility of this concept. This effort seeks to resolve a number of technological and scientific problems. Among the technical issues, two stand out:

> Two crucial technological questions so far unanswered are: (a) how do we produce the seawater aerosol of the required sizes and number concentrations? (b) How do we disseminate these particles to ensure that sufficient numbers of them enter the clouds to be adulterated? (Bower *et al.* 2006)

Stratospheric aerosols

Inserting aerosols into the stratosphere is another approach. Indeed this concept of SRM has probably been more widely discussed than any of the others. The reason for its prominence is not difficult to discern. The volcanic record offers a close and suggestive analogy to this approach. Examples include the eruptions of Tambora, Krakatau, El Chichón, and Pinatubo. Such eruptions loft particles into the atmosphere. The particles enhance Earth's brightness (i.e. raise planetary albedo). They scatter back into space some of the sunlight that would otherwise have been absorbed by, and warmed, the surface. As more sunlight is scattered back into space, the planet cools. The cooling from the large Pinatubo eruption that occurred in 1991 was especially well documented (Robock and Mao 1995).

The obvious question is whether it would be possible to emulate the cooling that tropical eruptions have so often produced. The goal would be to inject sub-micron-sized particles into the stratosphere. The particles would scatter sunlight back to space. Compared to volcanic ash, the particles would be much smaller in size. Particle size is important:

All matter scatters electromagnetic radiation. Small particles appear to be the most effective form for climate engineering. The goal is to maximize matter-radiation interaction favoring forms of the greatest electromagnetic cross-section for sunlight. Thus, the particles of greatest interest would be those with dimensions of the order of the wavelength of the optical radiation to be scattered, as such particles tend to scatter radiation with the highest specific efficiency and minimal mass usage. (Caldeira and Wood 2008)

Smaller particle sizes also offer the advantage of reflecting sunlight while not impeding the passage of long-wave radiation (Lenton and Vaughan 2009).

Smaller particles would also remain in the air masses into which they were injected for longer times than does most of the matter from volcanoes. Again, the goal is to decrease the mass that must be lofted in order to achieve cooling. Eventually, though, even the smaller particles would descend from the stratosphere into the lower atmosphere. Once there, they would precipitate out. The total mass of such particles would amount to the equivalent of a few percent of today's sulfur emissions from power plants (Lane et al. 2007).

Injecting the particles near the equator and at higher altitudes lengthens their life in the atmosphere. A longer atmospheric life reduces the total mass that must be put into the stratosphere in order to achieve a given change in global mean temperature. If adverse effects appeared following the intro-

duction of such a scheme, most of these effects would be expected to dissipate once the particles were removed from the stratosphere.

Sulfur dioxide (SO_2), as a precursor of sulfate aerosols, is a widely discussed option for the material to be injected. Other candidates include hydrogen sulfide (H_2S) and soot (Crutzen 2006). A fairly broad range of materials might be used as stratospheric scatterers:

Among dielectrics, many alternatives have been proposed (e.g. NAS 1992) and all appear to be fundamentally workable. Liquid SO_2 (or perhaps SO_3) appears to be optimized for mass efficiency, transport convenience and relative non-interference with all known processes of substantial biospheric significance, although fluidized forms of MgO, Al_2O_3 or SiO_2 (e.g., as hydroxides-in-water) seem competitive in most pertinent respects. (Caldeira and Wood 2008)

It might also be possible to develop engineered particles. Such particles might improve on the reflective properties and residence times now envisioned with dielectrics (Teller et al. 2003). Engineered particles, in comparison with sulfates or similar materials, would raise material costs per unit of weight, but the total mass needed to deflect the desired quantity of sunlight would fall. The feasibility of these concepts may hinge on the feasibility of fabricating materials able to maintain the desired optical properties in the atmosphere's chemically active environment.

As a matter of logistics, the challenge seems large, but manageable. The volumes of material needed annually do not appear to be prohibitively large. One estimate is that, with appropriately sized particles, material with a combined volume of about 800,000 m^3 would be sufficient. This volume roughly corresponds to that of a cube of material of only about 90 m on a side (Lane et al. 2007). The use of engineered particles could, in comparison with the use of sulfate aerosols, potentially reduce the mass of the particles by orders of magnitude (Teller et al. 2003).

Several proposed delivery mechanisms may be feasible (NAS 1992). The choice of the delivery system may depend on the intended purpose of the SRM program. In one concept, SRM could be deployed primarily to cool the Arctic. Such a deployment might be in response to a threat of CH_4 release or it could serve as a large-scale experiment moving toward a larger-scale deployment. With an Arctic deployment, large cargo planes or aerial tankers would be an adequate delivery system (Caldeira and Wood, personal communication, 2009). A global system would require particles to be injected at higher altitudes. Fighter aircraft, or planes resembling them, seem like plausible candidates. Another option envisions combining fighter aircraft and aerial tankers, and some thought has been given to balloons (Robock *et al*. 2009).

A space-based sunshade

A third approach that has also been widely discussed is the concept of an orbiting sunshade in space:

> The inner Lagrange point L1 is in an orbit with the same one-year period as the Earth, in-line with the sun at a distance where the penumbra shadow covers, and thus cools, the entire planet. A presentation on this concept proposed several approaches for overcoming the various engineering and economic challenges a sunshade presented although those challenges remain daunting. (Lane *et al*. 2007)

A new version of this concept has been proposed. It envisions a different system for implementing the actual scattering of the incoming sunlight:

> Previous L1 concepts have envisaged very large space structures. The alternative described here has many free-flyers located randomly within a cloud elongated along the L1 axis. The cloud cross-section would be comparable to the size of the Earth and its length much greater, approximately 100,000 km. This arrangement has many advantages. It would use small flyers in very large numbers, eliminating completely the need for on-orbit assembly or an unfolding mechanism. The requirements for station-keeping are reduced by removing the need for the flyers to be regularly arrayed or to transmit any signals. (Angel 2006)

The concept is immensely complex and intricate. It would include large-scale development and ground operations, as well as the flyer production and transportation. The plan entails infrastructure investments several orders of magnitude greater that the two previously discussed SRM technologies. As such, its fixed costs would be far higher than theirs and would probably be a much larger percentage of the total costs.

Such an approach does offer some advantages. Once in place, the sunshade could have a lifetime of many decades. In part because of damage from cosmic rays, current spacecraft such as communications satellites last for roughly twenty years when placed in high orbit. The flyers envisioned for the sunshade, however, are simpler than satellites and can be better protected against radiation damage. These features should allow them to achieve lifetimes greater than or equal to fifty years. Proponents believe that the sunshade could be stabilized by modulating solar pressure. This would avoid the need for expendable propellants and the need to lift their weight (Angel 2006).

Although this concept clearly entails a very large fixed cost, the program managers could halt cooling at any time. To do so, they would merely need to reorient the shield. Thus, in a physical sense, the plan is reversible. The impacts on the Earth may be less uncertain since the shield would only alter the flux of solar radiation (Govindasamy and Caldeira 2000). The sunshade would not change the composition of the atmosphere and ocean beyond their loading with GHGs (Lane *et al*. 2007). These factors lower the risk of an *ex post* decision to halt cooling. However, such a decision still cannot be ruled out given uncertainties about unwanted effects of cooling on the climate itself. Should such a policy reversal occur, the huge fixed costs of the sunshade would likely have to be almost entirely written off. Other disadvantages of the approach include the enormous area and mass required. These features make the concept technically very challenging to construct, and the concept raises daunting issues related to materials, launch costs, propulsion, and station-keeping.

The first two of these approaches, then, stratospheric aerosols and marine cloud whitening, appear to be the most promising. Research could, of course, uncover fatal flaws in either of them. Alternatively, a research effort might also bring forth other concepts. Until resources are committed to exploring this, it is hard to know how wide or narrow the range of feasible options actually is.

Possible Deployment Strategies

In addition to the question of which technology (or combination) might be best, having an SRM option would pose strategic choices and the transition from R&D to deployment may be a complex process.

The R&D stage[1]

With SRM, absent a big climate crisis, the risks of taking action will loom very large in the policy process. Politicians' fear of being linked to a disaster will ensure that extensive testing and evaluation will take place before deployment will be possible. Superficially, it might seem that, with enough money, rigorous testing and evaluation might be reconciled with a tight deployment schedule. In fact, some delay is simply built into the process.

One or more large field tests will almost certainly be required before full deployment will be either possible or desirable. These field tests will have to be conducted over at least a few years. The effects of a prolonged intervention may differ from those of a brief one. This argument has already surfaced with regard to the applicability of the Mount Pinatubo analogy. If anything out of the ordinary does happen during the field tests, and the odds are that something will happen someplace, it will be necessary to ascertain if it was linked to the experiment (Caldeira and Wood, personal communication, 2009). The inference seems clear: field tests are likely to consume a number of years – perhaps five to seven years would be a reasonable minimum.

Deployment options

As suggested by the discussion of R&D, the final stages of R&D may blend without very clear demar-

cation into the initial phases of deployment. Arctic cooling is one widely discussed option for a possible regional deployment of SRM (Caldeira and Wood 2008). Many climate concerns center on this region. Climate change there has appeared to be especially pronounced. Further, the Arctic seems to be potentially vulnerable. "Arctic sea-ice is disappearing at rates greater than previously observed or predicted (Kerr 2007) and the southern part of the Greenland ice sheet may be at risk of collapse (Christoffersen and Hambrey 2006)" (Caldeira and Wood 2008). Robock *et al.* (2008) have performed simulations that suggest that aerosol injections could maintain or increase Arctic sea ice. However, these simulations also suggest that cooling and possible side effects would not be confined to the Arctic.

In itself, disappearance of Arctic sea ice offers substantial benefits. It would shorten global trade routes and boost world trade by lowering transport costs. It would also open new opportunities for resource extraction. However, some scientists have speculated that current trends, were they to involve melting of the grounded glaciers of Greenland, might lead to a relatively rapid rise in global sea levels, the release of large quantities of CH_4 from arctic tundra, or even disruption of ocean currents (Stocker 2003; Gulledge 2008). These concerns have helped to fuel thinking about a regional cooling plan. Were such an effort to be undertaken, it would probably proceed in phases. Each phase could constitute an experiment producing information about the costs, benefits, and risks of further expansion. Taken as a whole, the effort might serve as a starting point for weighing the further expansion to lower latitudes.

SRM would also pose other risk management choices. Some have proposed, for example, that it should be deployed preemptively in conjunction with GHG controls. The goal would be to improve prospects of forestalling harmful climate change (Wigley 2006). Others regard such proposals as politically unrealistic and propose that SRM technologies be developed and held in readiness (Barrett 2007b). In this view, SRM would be deployed only in the event of evidence that very threatening climate change was happening, or was imminent. This second approach, in effect, accepts

---

[1] Since the completion of this chapter, a report discussing R&D issues surrounding CE has been published by the Novim Group. Please see Jason J. Blackstock, David S. Battisti, Ken Caldeira, Douglas M. Eardley, Jonathan I. Katz, David W. Keith, Aristides A. N. Patrinos, Daniel P. Schrag, Robert H. Socolow, and Steven E. Koonin, 2009: "CE responses to climate emergencies," http://novim.org/.

some additional risk of harm from stumbling on unseen tripwires in exchange for avoiding the risks of deploying SRM. Both approaches accept that a climate strategy will involve some mix of SRM and GHG limits (and perhaps AC). They differ in the assessment of the relative risks of SRM, on the one hand, and of the risks of unseen climate thresholds, on the other.

Another option might be to defer SRM deployment until background conditions have changed. For instance, deploying stratospheric aerosols before about mid-century entails some risk, although it may be a modest one, of slowing the recovery of stratospheric ozone (O_3) levels. Delaying SRM deployment until around mid-century would eliminate this risk. It might also greatly ease the task of reaching international consensus on deployment to wait until even high-latitude countries had exhausted the expectation of net benefits from warming.

In this chapter we analyze the preemptive use of CE in conjunction with GHG control measures (or, in one case, without them) on a global basis. We do not attempt to analyze the option of holding a CE capacity in reserve, though this is an important question. Our main analysis assumes that CE is deployed in 2025 and continued at least through the end date of the analysis. We chose 2025 as a benchmark in the belief that it might allow time for adequate research and testing. In addition, owing to the above-mentioned concerns regarding O_3 depletion and national interest conflicts, we include a briefer analysis of a 2055 starting date.

The Kinds of Costs Implied by SRM

The costs of SRM fall into three broad categories. These include the *direct* costs, such as the expense of developing and deploying SRM technology. They also encompass the *indirect* costs, which might be thought of as the harm that might result from using these technologies. Finally, they include the *transaction* costs entailed by SRM. These costs might include the resources consumed in bargaining to secure agreement to use SRM, or the costs of conflict that its use might occasion. Transaction costs also include routine considerations

such as the costs of monitoring and measuring the system's performance or nations' contributions to it.

Direct costs

Deploying SRM would entail direct costs – the resources consumed in building and operating the planes, balloons, ships, or satellites needed to reflect the desired amount of sunlight. It would also require resources to develop these systems and to assess their impacts and side effects.

Deployment costs   As estimated and discussed in the fourth section of this chapter, the deployment costs of the three technologies vary greatly. Of them, the sunshade is by far the most costly. For stratospheric aerosols, using conventional technologies, the total present value (PV) of deployment costs is less than $1 trillion. Marine cloud whitening costs are estimated to be almost an order of magnitude lower than either a sunshade or an aerosol injection. The primary reason for this difference is that the SRM intervention takes place near the Earth's surface and therefore requires less energy to deploy. This large cost edge comes with some possible penalty in the unevenness of the geographic distribution of the cooling effects.

For selectively cooling the Arctic, the costs appear to be much lower for either of the two best approaches. The lower altitude of the tropopause over the Arctic decreases the costs of delivering aerosols to the stratosphere in this region, and the smaller area to be covered diminishes costs. Benefits, however, would be reduced as well.

Development costs   At this point, no fully worked out concept for implementing SRM exists (Robock *et al.* 2009). Thus, all SRM concepts entail at least some R&D investment. Likewise, a fully worked out R&D program has not been developed. Therefore, in this section, we offer a preliminary sketch of how such a program may progress, and at what level it might be funded.

Some scientists propose a phased approach. In this notion, likely begin with modeling and "paper" studies. These activities have modest budgetary impacts. Laboratory testing would begin as work progressed. Depending on the results of the earlier exploration, field trials might eventually follow. Later, the process could lead to a major experiment perhaps at one-tenth the scale of full global deployment. Arctic cooling is a possible option for an experiment. Regional cooling might begin small and gradually increase in scale (Caldeira and Wood, personal communication, 2009).

The broader literature on the economics of innovation suggests that the process will involve more than a simple one-way progression from research to development to demonstration. Rather, problems will arise that are likely to require a looping back into more basic research (Nelson and Winter 1982). This pattern may be another factor in suggesting that R&D is likely to be time-consuming.

Defensive research, exploration of possible harmful effects from SRM, is likely to be a more important cost item than actual hardware development. Hardware development, in fact, may imply only modest cost. For stratospheric aerosols, concepts based on current technology would, with a few years of development effort, be capable of injecting the desired gases into the stratosphere (Robock *et al.* 2009). More advanced delivery systems would doubtless require more research, but they, too, do not demand breakthroughs in basic science (Robock *et al.* 2009). In the case of marine cloud whitening, the expected R&D costs are clearly quite low. Indeed, they appear to be almost negligible (Salter *et al.* 2008).

In contrast, careful monitoring of changes of the climate system response would demand a major effort. Questions would include what albedo changes were produced when, where, and in what spectral bands. The prominence of defensive research costs is warranted given the possibility that unwanted side effects could far exceed deployment costs. The next section on indirect costs will discuss at greater length the potential importance of side effects. Consistent with this emphasis, satellite costs might prove a major expense at the stage of field experiments. It may be possible, however, to limit these costs by using drones rather than satellites as platforms for the monitoring systems. The former may also offer more flexible deployment.

Whatever type of platform is chosen, monitoring will clearly be a major element in the concept's

R&D cost structure. This fact, in turn, argues that much of the total budget could properly fall within the purview of a broader climate research agenda. GHG control policy, adaptation, and CE strategies would all benefit from greater knowledge about the causes, detailed effects, and timing of future climate change. Many of the research projects needed to explore CE are likely to be more widely useful. At the same time, it may not be entirely safe to assume that future climate science spending patterns will, in fact, respond to these needs.

An SRM research program budget might start at about $10 million per year and then ramp up to $100 million per year as initial field tests began. For comparison, today, the US government alone is spending about $10 billion a year on climate science and technology. Costs would scale up again as the project started to engage in large-scale field tests. The estimates of early expenditures are slightly more aggressive than those generated by the US government program for CE research. This program was proposed during the Bush Administration but was not actually funded. In 2001, the interagency panel that devised this plan proposed a gradually rising budget that entailed a total five-year cost of $98 million (US DOE 2002). The Arctic cooling experiment might be feasible for annual costs of around $1 billion (Caldeira and Wood, personal communication, 2009).

Resource costs might exceed budgetary figures. Given the apparently high rates of return earned by R&D expenditures, the opportunity costs associated with specialized resource inputs into R&D may well exceed these resources' market prices (Nordhaus 2002). By inference, R&D investments that appear to be cost-beneficial may not be if a high proportion of their inputs are drawn from other high-payoff R&D. In the past, this kind of resource redeployment supplied about half of the inputs used to increase specific kinds of federal R&D spending (Cohen *et al.* 1991). It is nonetheless quite clear that R&D costs could have only a trivial impact on the benefit-cost ratios (BCRs) that will be presented later in this chapter.

The space sunshade concept is, of course, an exception. The project's scale and the relative novelty of the technologies it calls for seem to portend very substantial R&D costs. The high R&D

costs likely to be associated with the sunshade do not, though, imply that the more down to earth concepts would require comparable resource commitments, and if R&D resource constraints remain tight one obvious response might be to severely limit research on this concept or to omit it altogether.

Indirect costs

SRM is likely to involve indirect costs as well as direct ones. GHG controls will certainly incur such indirect costs (Barrett 2007b). In this respect, then, SRM resembles GHG curbs. The latter, for instance, might increase other kinds of emissions, cause leakage, cause harm from use of some biofuels, and, through its high direct costs, curtail societies' capacity to adapt to climate change.

With regard to the scale of the indirect costs of SRM, the literature offers virtually no guidance, limiting what can be said in this study. One might note in passing that little more is available for the task of assessing GHG controls, where many analyses have simply ignored indirect costs. However, indirect costs may be a larger share of SRM's total costs than they are with GHG controls and AC, given the much higher direct costs of the latter two strategies and the fact that these technologies, in some sense, return the Earth to a prior state. As a result, as one recent assessment noted, fears about indirect costs are probably a much more important impediment to SRM's acceptance than are concerns about its direct costs (Robock *et al.* 2009). It is, then, important to at least describe the main kinds of indirect costs that SRM might occasion.

Negative side effects   Changing global temperatures without lowering the level of GHG concentrations is a source of much of the concern about SRM. One risk is the possible lessening of rainfall; a possible weakening of the Indian or African monsoons is a particular worry. In the wake of the Pinatubo eruption, there was some diminution in rainfall, and some model results suggest this might be a result of SRM strategies as well (Robock 2008).

If SRM were to have this effect, the lost output might amount to a significant increase in the total costs of its use. Between 2001 and 2005,

Indian agriculture and forestry produced output worth between $96 billion and $135 billion, annually (UN Statistics Division, 2008). The monsoon-dependent part of Indian agriculture ranges from slightly less than half to about two-thirds. A 10% loss of the monsoon-dependent production might add $4.5 billion to $9 billion to the total cost of SRM. This figure is not intended to actually quantify the potential costs from this hypothetical effect. It is merely meant to signal that, if the effect is real, it could be important.

Although this point clearly warrants serious study, the underlying climate science, itself, remains unsettled: "Studies with general circulation models (GCMs) investigating the response of the ISM [Indian Summer Monsoon] to increased concentrations of GHGs and sulphate aerosols (Meehl and Washington 1993; Lal *et al.* 1995; Hu *et al.* 2000; May 2002) were so far not able to provide a clear answer" (Zickfeld *et al.* 2005). On the other hand, Robock *et al.* (2008) perform simulations that suggest the aerosol injections, which result in a decrease in radiative forcing of 1 W m^2, in conjunction IPCC's A1B emission scenario, decrease global precipitation by about 1.7% and adversely affect the Indian and African monsoons. Rasch *et al.* (2008), in which Robock is a contributor, qualify this result by noting

> The NCAR results [performed by Rasch] . . . suggest a general intensification in the hydrologic cycle in a doubled CO_2 world with substantial increases in regional maxima (such as monsoon areas) and over the tropical Pacific, and decreases in the subtropics. Geoengineering . . . reduces the impact of the warming substantially. The Rutgers simulations [Robock *et al.* 2008] show a somewhat different spatial pattern, but, again, the perturbations are much smaller than those evident in an "ungeoengineered world" with CO_2 warming.

> Robock *et al.* (2008) have emphasized that the perturbations that remain in the monsoon regions after geoengineering are considerable and expressed concern that these perturbations would influence the lives of billions of people. This would certainly be true. However, it is important to keep in mind that: (i) the perturbations after geoengineering are smaller than those without geoengineering; (ii) the remaining perturbations are less

than or equal to 0. 5 mm d^{-1} in an area where seasonal precipitation rates reach 6–15 mm d^{-1}; (iii) the signals differ between the NCAR and Rutgers simulations in these regions; and (iv) monsoons are a *notoriously* difficult phenomenon to model [emphasis in original].

In principle, if the monsoon effect were to be confirmed, it could constitute a significant cost item, but none of the existing studies made any effort to actually place a dollar value on the expected harm. As in many other instances, making a knowledgeable decision about SRM would require significant advances in more general climate science.

It should be noted that not all SRM technologies are equally at risk on this score. The more localized nature of marine cloud whitening may represent a positive advantage. With this system: "Owing to a possible anxiety over the effect of extra cloud condensation nuclei on rainfall, areas upwind of land with a drought problem should be avoided" (Salter *et al.* 2008). In a sense the more localized nature of marine cloud whitening operations is an offset to the potential disadvantages of the patchy effects of this approach.

Moreover in considering possible impacts on precipitation and other negative side effects, it is important to be clear about the relevant comparison:

> The choice is not between a climate-engineered world and a world without climate change; rather, it is between the former and the world that would prevail without climate engineering.

Work by Caldeira and Wood (2008) indicates that a high-temperature high-GHG world involves far larger changes in precipitation patterns than does a low-temperature high-GHG world, echoing the findings of Rasch *et al.* (2008). In other words, controlling temperature might at least limit the damage from climate change even if it did not entirely prevent it (Caldeira and Wood 2008). This point is directly relevant to the example of the ISM. Many models predict that warming itself is likely to cause severe problems for the Indian agriculture sector. It might, then, be fair to conclude: "While a major effort should be put into the study of all possible side effects of keeping sea temperatures at present values (or other values of our choosing), many of

the side effects appear to be benign and less dangerous than those of large, unbridled temperature rises" (Salter *et al.* 2008).

In addition to possible changes in precipitation, SRM may entail risks of other unwanted consequences. Some of these worries seem potentially much more serious. It is also possible that further research might uncover some hitherto unknown effect of SRM that could be harmful enough to render a technology, or even the whole concept, infeasible. Ozone depletion and the potential loss of protection from ultraviolet radiation that it provides, has been suggested as a possible side effect of SRM. This risk may be greatest until chlorine concentrations return to their 1980s levels, because sulfate aerosols added to the stratosphere may retard the O$_3$ layer's recovery (Tilmes *et al.* 2008). Again, this is a matter that would demand further study. Some have suggested that this risk, while real, may not be pronounced:

> This particular risk, however, is likely to be small . . . With current elevated chlorine loadings, O$_3$ loss would be enhanced. This result would delay the recovery of stratospheric O$_3$ slightly but only until anthropogenic chlorine loadings returned to levels of the 1980s (which are expected to be reached by the late 2040s). (Wigley 2006)

Rasch *et al.* (2008) note that while O$_3$ depletion may in fact take place, the attenuation of ultraviolet-B radiation by the sulfate cloud may offset this effect in terms of its impact on human health. Again, the studies that have raised the issue have not sought to quantify the dollar value of the possible harm.

Later in this chapter, we consider delaying the deployment of SRM until 2055 to further lower the impact on O$_3$ levels. In this case, we demonstrate that delaying SRM until 2055 still produces large benefits and results in almost the same maximum temperature change. Marine cloud whitening may be immune from this particular concern, as it does involve the injection of particles into the stratosphere.

Other concerns, while perhaps meriting some further study, appear to be less important. For instance, some concern had arisen over acid precipitation if SO$_2$ were injected into the stratosphere. These fears, though, appear to be exaggerated. Thus a recent study concluded: "Analysis of our results and comparison to the results of Kuylenstierna *et al.* (2001) and Skeffington (2006) lead to the conclusion that the additional sulfate deposition that would result from GE will not be sufficient to negatively impact most eco-systems, even under the assumption that all deposited sulfate will be in the form of sulfuric acid" (Kravitz *et al.* 2009).

Others have suggested that stratospheric aerosol injections would whiten skies, interfere with terrestrial astronomy, and reduce the efficiency of solar power (Robock 2008b). Just how significant these effects would be is an open question, particularly for the low levels of SRM that we consider in this chapter ("low" being less than a complete offsetting of CO_2 emissions). Furthermore, it may be the case that society will have to choose between whiter skies, for example, and accepting the risk of a planetary emergency.

The lost benefits of warming The most clear-cut indirect cost of SRM, although not perhaps the most important, would spring from the loss of some of the benefits of warming. In most temperate countries, warming will bring at least some benefits as well as costs, and for some countries benefits may well exceed costs – at least for a number of decades (Nordhaus 2008). Cooling the climate relative to the business-as-usual trend (BAU), would sacrifice these benefits. By inference, the SRM that occurred while warming was still producing significant net benefits in some localities would vitiate these potential gains, and interests that incurred net losses as a result might well object. These costs would occur regardless of whether aerosols or cloud whitening produces the cooling. Indeed, were GHG controls able to cool the planet as rapidly as some might wish, they too would be likely to provoke resistance on this score. Cooling the Earth inherently brings losses as well as gains.

The political transaction costs of SRM

Policy making is subject to transaction costs. Implementing any policy entails costs of striking a bargain, assessing compliance with it, and enforcing its terms. Bargaining can be costly in its own right. Moreover, political structures and rules can sometimes block or distort the choice of the best response

to a problem. The resulting lost benefits can be thought of as part of the transaction costs of adopting a policy. Transaction cost levels, however, may vary among climate policy options. Some policies, for instance, may offer more tempting targets than others for pork barrel politics or for other forms of rent seeking. Ultimately, one would like to compare the transaction costs of SRM with those of other responses to climate change. *Ex ante*, though, such comparisons are necessarily speculative. No one can yet know how the process will distort the various options.

The level of conflict over SRM Conflicting interests tend to drive up the costs of reaching and maintaining an agreement. Clearly, nations differ as to their perceived interests in curtailing warming. Divergent national interests have helped to push the transaction costs of a global bargain on GHG controls above its expected benefits (Bial *et al.* 2000).

Nations are also likely to differ over SRM. Absent the presence of a global climate crisis, an early move to deploy SRM is likely to generate conflict (Victor *et al.* 2009). However, in contrast to GHG control, SRM does not require active efforts from all, or even from several, nations. It merely requires acquiescence from all major powers (Barrett 2007b). The latter test is easier than the former (Barrett 2009). Further, the net benefits of SRM, unlike those of GHG curbs, may be large enough to buy off the few states with both the power to deter the use of SRM and an interest in opposing it. Side payments of the type envisioned here may have ambiguous impacts on their recipients (Easterly 2006). Nonetheless, if SRM can indeed produce large global net benefits, the resources needed to buy acquiescence are likely to be available.

Within nation-states, SRM draws strong ethical objections in some quarters. Many environmental advocacy groups passionately oppose even researching it (Tetlock and Oppenheimer 2008). Their resistance might well take the form of litigation, which would add to the transaction costs of SRM. Alternatively, they might, at some point, be able to win commitments to GHG control measures with net costs as part of the political price for a decision to advance SRM. Hypothetically, as suggested by the Nordhaus analysis of the Gore and Stern proposals, the global costs of meeting such demands might run into trillions of dollars. In reality, important economies remain largely beyond the influence of environmental advocacy groups, a fact that is likely to limit the effects of the environmentalists' demands. One possible result may be to ensure that nations with relatively weak environmental lobbies will take the lead in researching and deploying SRM. That prospect may, in turn, curtail the green advocacy groups' influence on policy, even in countries where they enjoy higher levels of support.

Rent seeking   While governments provide public goods, they also often become vehicles for the pursuit of unearned income (rent seeking). Rent seeking usually ends up consuming resources and leaving society as a whole worse off (Olson 1982). In other cases, groups acting out of ideology may be able to impose policies that raise social costs (North 1990). Rent seeking and ideology can cause actual policies to diverge widely from those that would maximize economic well-being. Nothing precludes policies that do net harm, and their adoption is common (North 2005). Many specific GHG controls proposals show that such harmful choices are common within the realm of climate policy (Lane and Montgomery 2008).

SRM policy may not, though, be an especially apt vehicle for rent seeking. At the moment, it appears that SRM technologies might be deployed by relatively few planes, balloons, sailing ships, or other fairly low-cost systems. To be sure, all these concepts need development before they could be used, but none, at least compared to GHG controls, currently seems to require especially high costs to develop or even to deploy. SRM may, therefore, be so low cost that would-be suppliers have little interest in attempting to distort decisions about which systems to develop or deploy. Further, in the USA at least, legislators may feel that SRM offers rewards in local jobs and spending that are too small to justify the effort needed to steer development efforts to their home districts. The same may not be true of AC, which would encompass a massive industrial operation that could conceivably be located anywhere.

The converse may be that SRM is too efficient for its own political good. The US Congress tends to

fund R&D based largely on the "distributive benefits" that it offers, i.e. on the value of its development costs as a source of local spending and jobs (Cohen and Noll 1991). SRM may not be very rewarding from this point of view, and would-be input suppliers have not so far committed resources to urging government support for its development.

The costs of discontinuity: and the implied risks of avoiding it   Finally, governments have often found it difficult to bind themselves and their successors to future actions (North *et al.* 2009). At least in theory, this fact poses a difficulty for SRM (and for GHG controls, for that matter). In the event that global GHG controls remain patchy and ineffectual for an extended time, a country that substituted SRM for adaptation, over a long period, would face high costs were it to later halt SRM. To avoid such costs, a nation embarking on SRM would have to be able and willing to commit to conducting it for a very long period (Matthews and Caldeira 2007). This consideration seems likely to lead to a great deal of cautious and careful research before an initial deployment might be politically acceptable. Further, the high cost of rapid rebound warming would itself deter a state from frivolously abandoning a long-running system (Barrett 2007b). If a decision to stop an SRM program was made, a strong argument could be made for phasing it out gradually. In effect, the high costs of halting SRM may encourage policy continuity without any special institutional arrangements designed to guarantee it.

Continuity, though, carries a downside. Harmful side effects of SRM might appear only after decades. In that case, governments might have few alternatives to either accepting the side effects or incurring the costs of the discontinuity. Physically the SRM program could be rapidly turned off. Practically, this step might be expensive (Goes *et al.* 2009).

Nature of the Potential Benefits of SRM

Some benefits of SRM stem from its possible direct impact on climate change and on the costs of coping with it. Some forms of SRM might also produce other desirable effects. The latter, should they materialize, would also be relevant.

Direct Benefits of Climate Engineering

Developing a capacity to geoengineer climate would produce three types of potential benefits. First, it would allow societies to avoid some of the damages from climate change. In particular, SRM may offer real advantages as a means of averting potential harm from catastrophic climate change but, in some deployment modes, it might also lower the harm from continuous climate change. Second, it would allow lower mitigation costs to produce a given level of protection from the harm of climate change. Third, it would lower adaptation costs.

In the next section we estimate these benefits for SRM and AC through the use of the Dynamic Integrated model of Climate and the Economy (DICE) (Nordhaus 2008). Differing assumptions about possible GHG control regimes change the size of the benefits from the use of SRM. It also shifts the mix between the first and second type.

Indirect Benefits of Climate Engineering

With stratospheric aerosols it may be possible to lower UV radiation striking the surface of the Earth. This effect would lower skin cancer rates and increase agricultural productivity. The potential savings are estimated to be large, perhaps on the order of $1 trillion per year (Teller *et al.* 2003), although some scientists have challenged the proposition that scatterers can be designed and maintained in the atmosphere to operate with the required degree of precision (Lane *et al.* 2007).

Geoengineering Benefit Estimates

The primary challenge in estimating CE BCRs is that rigorous benefit and cost estimates, and the surrounding uncertainty, do not exist. This is a deficiency that we attempt to remedy to some extent when considering benefits. On the cost side, too, we provide new analysis, but rely on published estimates to a much larger extent.

To date, the primary studies of CE's benefits (Crutzen 2006; Wigley 2006; Caldeira and Wood 2008) measure benefits in terms of CE's ability to

alter climate parameters such as global mean temperature change and sea level rise, rather than in economic terms. Nordhaus (1994) is a notable, and early, exception. He found that costless CE, which completely offset global warming, had a net benefit of almost $9 trillion (2005 $), which was DICE's estimate of the present value (PV) of climate damages at that time. While Nordhaus' assessment is helpful, we require an updated estimate and seek to understand the benefit of more modest CE deployments – for example, those that ameliorate, but do not completely offset, the effects of climate change.

Likewise, our cost estimates are preliminary. In addition, most estimates were developed under the assumption of a deployment that would offset the warming associated with a doubling of CO_2 concentrations. Again, we seek cost estimates that consider lesser interventions.

Therefore, in this chapter, we have undertaken a new study of CE benefits and costs. To make our analysis as general as possible, we estimate the economic benefit of a generic CE technology that would be able either to reduce radiative forcing directly, such as SRM, or to remove CO_2 from the atmosphere permanently, such as AC. We use the DICE-2007 model (Nordhaus 2008) to estimate the benefits of CE. DICE is a well-established integrated-assessment climate change model, which allows our results to be placed within the context of existing economic analyses of climate change. Furthermore, and perhaps more importantly, the use of DICE allows us to estimate the impact of CE on key policy variables such as emissions control rates and carbon taxes. Of course, our use of DICE entails the acceptance of DICE's assumptions and limitations. The reader should not take this as an indication of our agreement with these assumptions, or the degree to which we believe that DICE faithfully models important aspects of natural and human systems. DICE is a model, a very useful model in our opinion but, as a model, it is necessarily an imperfect reflection of reality. Recent meta-analysis has confirmed that one of DICE's primary outputs, the social cost of carbon, is in the "mainstream" of peer-reviewed estimates (Tol 2008).

As the reader will see, we consider several CE deployment examples. This analysis helps to build intuition and insight. We do not attempt to analyze specialized strategies or determine the "optimal" use of CE. For our cost estimates, we start with published studies of particular CE technologies. We then scale these for our level of deployment.

The DICE Model, Discount Rates, and DALYs

DICE is an optimal-economic-growth model that relates economic growth to emissions of CO_2, CO_2 emissions to temperature, and temperature to climate damage. In so doing, DICE solves for the optimal emissions control rate, CO_2 emissions, temperature, abatement costs, and climate damages, among other variables, in each decade for the next 600 years (2005 to 2605). We, however, limit our analysis to 200 years (2005 to 2205). We selected a study period of 200 years because, as the reader will see, the climate system reaches equilibrium under constant forcing over this time scale. This is not the case with other natural choices such as 100 years. In addition, maximum temperature changes are also obtained within this time frame, and we wish to investigate the impact of SRM on this parameter. Because of time discounting, most of the net benefits occur over the next 100 years and therefore a 100-year study period would not materially change our results.

Since DICE endogenously determines CO_2 emissions, we do not consider particular emissions scenarios. Instead, we will consider the impacts of CE in three different emission controls environments: no controls (a lack of emissions controls), optimal controls, and limiting temperature change to 2°C. In addition, we will analyze the effect of SRM on a policy based on a very low discount rate like that assumed in *The Stern Review* (Stern 2007). Finally, we assess the impacts of combining optimal controls and a delayed CE. Such delay might occur owing to concerns regarding stratospheric O_3 depletion or some high-latitude countries' reluctance to halt warming in the next few decades.

Like emissions, DICE endogenously determines the real return on capital based on the pure rate of social time preference (ρ) and the marginal utility of consumption elasticity (α). These two parameters, related through a Ramsey growth model, are calibrated ($\rho = 1.5\%$, $\alpha = 2$) to match the empirical real return on capital, which was estimated to be 5.5% per annum (Nordhaus 2008). We use this endogenously determined return to calculate present values. This has the following benefits. First, it facilitates the comparison of our results to those of Nordhaus (2008). Second, using a different discount rate would be internally inconsistent with the DICE model (and the real economy). While our real discount rate varies, it averages about 5.5% for the first 50 years and about 4.5% over our 200-year study period. As a shorthand, we will refer to this as the "market discount rate" scenario. We will, in addition, analyze a low-discount rate scenario of approximately 2% real, based on *The Stern Review*.

Nordhaus and Boyer (2000) estimate that the health impacts of climate change amount to approximately 7% of the total global damages. Thus, we do not attempt to alter DICE's damage equation or its assumptions regarding the value of disability adjusted life years (DALYs). Doing so would require a significant refitting of the DICE model and is unlikely to alter our main conclusions.

Changes Made to DICE

In order to estimate the benefits of CE, we must make a few changes to DICE. These include changes to DICE's radiative forcing and carbon cycle equations.

We begin by modifying DICE's radiative forcing equation to allow for inclusion of an additional external forcing component, $SRM(t)$, which we take to be the negative forcing due to SRM. The radiative forcing (W m^{-2}) at the tropopause for period t (a decade in the DICE model) is

$$F(t) = \eta \log_2 \frac{M_{AT}(t)}{M_{AT}(1750)} + F_{EX}(t) - SRM(t).$$

(1)

$M_{AT}(t)$ is the atmospheric concentration of CO_2 in gigatons of carbon (GtC) at the beginning of period t and $M_{AT}(1750)$ is the preindustrial atmospheric concentration of CO_2, taken to be the concentration in the year 1750. DICE sets the 1750 CO_2

concentration at 596.4 GtC ($\sim$280 ppm)[2]. η is the radiative forcing for a doubling of CO_2 concentrations and is assumed to be 3.8 W m^{-2}. $F_{EX}(t)$ represents the forcing of non-CO_2 GHGs such as CH_4 and the negative forcing of aerosols. $SRM(t)$ is the change in the radiative forcing (W m^{-2}) in period t due to SRM. Thus, our modeling of SRM is consistent with DICE's treatment of aerosols. We do not require that the quantity of SRM be constant, but will focus on this case.

We next modify DICE's atmospheric CO_2 concentration equation. The mass of carbon contained in the atmosphere (GtC) at the beginning of period t is:

$$M_{AT}(t) = E(t-1) + \phi_{11}M_{AT}(t-1)$$
$$+ \phi_{21}M_{UP}(t-1) - AC(t-1). \quad (2)$$

$E(t-1)$ is the mass of carbon that enters the atmosphere due to land-use changes. $M_{UP}(t-1)$ is the mass of carbon contained in the biosphere and upper ocean at the beginning of period $t-1$. ϕ_{11} is the fraction of carbon that remains in the atmosphere between periods $t-1$ and t. ϕ_{21} is the fraction of carbon that flows from the biosphere and upper ocean to the atmosphere between periods $t-1$ and t. $AC(t-1)$ is the mass of carbon permanently removed from the atmosphere during period $t-1$ via AC, which we assume occurs concurrently with emissions. Again, this amount is not restricted to be constant, but we will focus on this case.

DICE uses a two-stratum model of the climate system. The first stratum is the atmosphere, land, and upper ocean. The second stratum is the deep ocean. DICE models the global mean temperature of the first stratum, T_{AT}, as a function of the radiative forcing at the tropopause, the temperature of the atmosphere in the previous period, and the temperature of the lower oceans, T_{LO}, in the previous period. In particular,

$$T_{AT}(t) = T_{AT}(t-1) + \xi_1 [F(t) - \xi_2 T_{AT}(t-1)$$
$$- \xi_3 [T_{AT}(t-1) - T_{LO}(t-1)]] . \quad (3)$$

ξ_2 is the climate feedback parameter, which is equal to the radiative forcing for a doubling of

CO_2 concentrations, η, divided by the temperature increase for a doubling of CO_2, which DICE assumes is 3°C. ξ_2 is therefore equal to 1.27 (3.8/3.0). In equilibrium, (3) implies that the impact of a change in radiative forcing is $\Delta T = \xi_2^{-1}\Delta F \approx$ (1.27)$^{-1}\Delta F \approx 0.79\Delta F$. Similarly, the negative radiative forcing at the troposphere required to offset a temperature increase of 3°C is $\Delta F \approx 1.27 \cdot 3 = 3.8$ W m^{-2}. Nordhaus (1994) has shown that DICE's simple climate model faithfully represents the aggregate results of larger GCMs on a decadal time-scale. It may not, however, be able to represent more rapid temperature changes. We do not alter DICE's temperature equation and therefore might underestimate the effect of strong negative or positive forcing. We also note that DICE's use of a two-stratum climate model does not account for vertical differences in radiative forcing, which may be important (NRC 2005). We do not, however, believe that these limitations limit the usefulness of DICE within the present context or undermine our argument that CE merits additional research.

DICE assumes that climate damages are a quadratic function of temperature. Damages are measured as the loss in global output. The damage in period t is

$$D(t) = \psi_1 T_{AT}(t) + \psi_2 T_{AT}(t)^2, \quad (4)$$

where ψ_1 and ψ_2 are chosen to fit the literature regarding climate impacts. The particular limitation of (4) is that damage is not a function of the rate of temperature change, which could be important in the case of SRM, and possibly AC. We do not alter DICE's damage function and therefore may underestimate the benefit of rapid cooling or the cost of rapid warming.

Solar Radiation Management

We analyze three generic SRM strategies that entail deploying either 1 W m^{-2}, 2 W m^{-2}, or 3 W m^{-2} of negative forcing beginning in 2025 and continuing through 2605. We refer to these as SRM 1, SRM 2, and SRM 3, respectively. As mentioned above, we consider three different emission controls

---

[2] Assuming 2.2 GtC per 1 ppm.

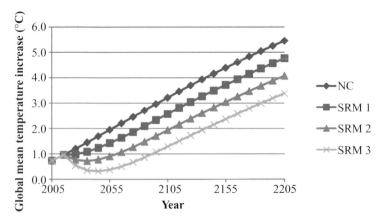

Figure 1.1 *NC temperature changes (°C) with the deployment of SRM (2005–2205)*

environments: no controls, optimal controls, and limiting temperature change to 2°C.

No Controls

We begin by analyzing the use of SRM under an assumption of no reductions in GHG emissions, referred to as No Controls (NC). Running the DICE model without any CE produces a 200-year PV of climate damages of $21.7 trillion (all dollar amounts are in 2005 US $), compared to a 600-year PV of $22.6 trillion reported by Nordhaus (2008).

The impact of deploying SRM on temperature is shown in figure 1.1. Increasing the quantity of SRM shifts temperature increases into the future. In fact, each W m^{-2} of SRM shifts the higher temperatures due to elevated GHG concentrations out about thirty years. Thus, if society sought to avoid the amount of harm that would be caused by reaching a given temperature level, deploying 3 W m^{-2} of SRM would buy almost 100 years of time in which to develop the less costly low-carbon energy sources needed to reach that goal.

Figure 1.2a displays the change in temperature relative to NC that is accomplished by each SRM strategy. SRM 1, for example, approaches −0.79°C, which is expected from the equilibrium relationship $\Delta T = \xi_2^{-1} \Delta F \approx 0.79 \Delta F$. Figure 1.2b displays the equilibrium negative radiative forcing that is equivalent to the temperature decreases in the top panel. We see that, for example, a constant 3 W m^{-2} of SRM will not produce negative forcing equivalent to an equilibrium forcing of −3 W m^{-2}. This result obtains because of lags in the climate system (e.g. see (3)) and the fact that the climate is being forced away from equilibrium through continued carbon loading under the NC scenario.

The damages imposed by climate change are lessened because of the reduction in temperature increases. The PV of climate damages is reduced from $21.7 trillion to $14.2 trillion through the deployment of 1 W m^{-2} of SRM. Thus, the benefit of 1 W m^{-2} of SRM is $7.5 trillion, which is a cost reduction of about 34%. To place SRM 1 in perspective, 1 W m^{-2} is about 0.3% of the incoming solar radiation of 341 W m^{-2} (Kiehl and Trenberth 1997; Trenberth *et al.* 2009). Table 1.1 summarizes the benefit of each SRM strategy. For example, 3 W m^{-2} of SRM reduces climate damages by almost $17 trillion or 78%. Clearly, SRM hold the potential to significantly reduce climate damages.

Table 1.1 **Benefit of SRM under NC with 2025 start ($ are trillion US $2005)**

SRM strategy	PV of climate damages ($)	Benefit of SRM ($)	Cost reduction (%)
SRM 0 (NC)	21.7	0	0
SRM 1	14.2	7.5	34
SRM 2	8.6	13.1	60
SRM 3	4.9	16.8	78

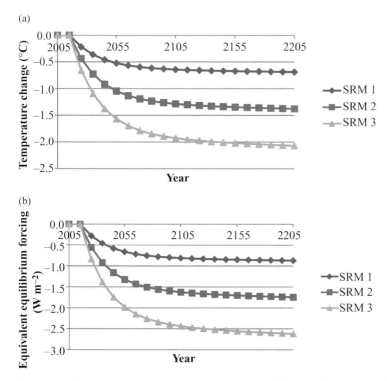

(a)

(b)

Figure 1.2 *Difference in temperature (a) relative to NC and (b) the equivalent equilibrium radiative forcing (2005–2205)*

We notice that the marginal benefit of SRM is decreasing. For example, SRM 2 does not have twice the benefit of SRM 1. This occurs because of the quadratic nature of the DICE damage function, (4); as temperature is reduced, the next incremental change has less of an impact on damages. Further, the benefits estimated here exceed the value for CE presented by Nordhaus (1994). This result stems from the fact that the PV of climate damages reported at that time were $9 trillion (2005 $), compared to $21.7 trillion (2005 $) reported in Nordhaus (2008).[3]

A possible concern with the use of SRM under a NC scenario, identified in the second section of this chapter, is that the atmosphere would continue to be loaded with CO_2 and that temperatures may increase rapidly if SRM were ended (Wigley 2006). As can be seen in the top panel of figure 1.2, in 2105 SRM 1, SRM 2, and SRM 3 would offset approximately 0.6°C, 1.3°C, and 1.9°C, respectively.

Optimal Controls

We next analyze the use of SRM under a scenario of optimal emissions as determined by DICE, which we refer to as Optimal Controls (OC). In this case, DICE determines the optimal level of abatement via emissions reductions. Therefore, in this case, we are investigating the combined use of CE and (optimal) abatement. The 200-year PV of climate damages under OC is $16.2 trillion. The PV of the abatement costs is $2.0, bringing the total cost of climate change, under optimal emissions, to $18.2 trillion, compared to a 600-year PV of $19.5 trillion reported by Nordhaus (2008). The impact of deploying SRM on temperature is shown in figure 1.3. Under SRM 1, temperatures peak in 2205. Under SRM 2 and SRM 3 temperatures increases peak at 2.7°C in 2215 and 2.4°C in 2235 (not shown), respectively. In this case, each W m^{-2}

[3] Please see Nordhaus (2008: 112) for a discussion of differences between DICE-1999 and DICE-2007.

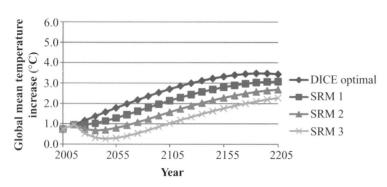

Figure 1.3 *OC temperature changes (°C) with the deployment of SRM (2005–2205)*

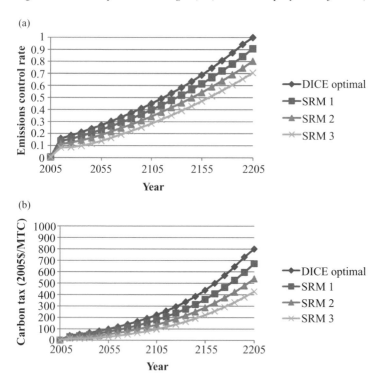

Figure 1.4 *(a) Optimal emissions controls and (b) carbon taxes with the use of SRM (2005–2205)*

of SRM shifts increased temperatures out about thirty-five years. We see that the combined use of CE and optimal abatement is nearly able to hold the temperature change to 2°C.

Figure 1.4 displays optimal emissions control rates (1.4a) and carbon taxes (1.4b) as a function of the SRM level.[4] We see that the use of SRM on

the levels we are considering here will not replace emissions reductions or carbon taxes. However, each W m^{-2} of SRM delays a given emissions reductions level or carbon tax by about twenty-five years. Thus, 3 W m^{-2} of SRM would forestall the level of emission reductions produced by DICE by about seventy-five years. We should also note that optimal emission control rates are affected in the years before SRM is deployed. For example, as shown in the top panel of figure 1.4,

---

[4] Divide by 3.66 to place the carbon taxes in terms of $ per MT of CO_2.

Table 1.2 Benefit of SRM under OC with 2025 start ($ are trillion US $2005)

SRM strategy	PV of climate damages ($)	PV of abatement costs ($)	PV of climate damages and abatement costs ($)	Benefit of SRM ($)	Cost reduction (%)
SRM 0	16.2	2.0	18.2	0	0
SRM 1	10.5	1.4	11.9	6.3	35
SRM 2	6.3	0.9	7.2	11.0	60
SRM 3	3.7	0.5	4.2	14.0	77

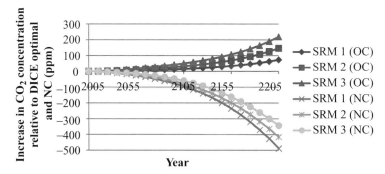

Figure 1.5 *Difference in SRM CO_2 concentrations compared to OC and NC (2005–2205)*

emissions control rates differ in 2015, even though SRM is not deployed until 2025. This occurs because DICE has perfect foresight and "knows" that SRM will be deployed in the future. This feature, common in modeling, may, of course, simulate actual social decision making rather poorly.

As was the case with NC, the reduction in temperature reduces the damage imposed by climate change. The PV of climate damages and abatement costs is reduced from $18.2 trillion to $11.9 trillion through the deployment of 1 W m^{-2} of SRM – a benefit of $6.3 trillion or about a 35% cost reduction. This benefit is about $1 trillion less than the case of NC because emission reductions are also avoiding damages in the optimal scenario. Table 1.2 summarizes the benefit of each SRM strategy. We see again that SRM can significantly reduce climate damages. For example, 3 W m^{-2} of SRM reduces climate damages and abatement costs by $14 trillion, or 77%. In percentage terms, the benefits are almost evenly split between reductions in climate damages and reductions in abatement costs. For example, SRM 3 reduces climate damages by 77% and abatement costs by 78%.

Carbon loading   A concern expressed regarding the use of SRM is the continued carbon loading of the atmosphere. To get a rough sense for this risk we analyze the CO_2 concentrations when SRM is employed compared to the situation when it was not employed. For example, the 2105 CO_2 concentration under SRM 3 with optimal abatement is 631 ppm, compared to 581 ppm in the OC case without SRM – a difference of 50 ppm or a ratio of 1.086. Thus, the "latent forcing" due solely to the increase in carbon loading is 0.36°C ($3.8 \cdot 0.79 \cdot \log_2(1.086)$). This should be compared to the SRM 3 temperature decrease in 2105, which is 1.7°C (see figure 1.3).

Thus, we see that under OC the primary risk in ending an SRM program stems from stopping the negative forcing of the SRM itself, rather than from its secondary effect on CO_2 emissions. Conversely, the 2205 CO_2 concentration under NC is 1189 ppm, which is 488 ppm greater than OC with SRM 1. These differences are shown relative to OC and NC for SRM 1, SRM 2, and SRM 3 in figure 1.5. In 2105 the increased loading due to the availability of SRM is between 15 and 50 ppm, relative to

Table 1.3 Benefit of SRM under OC with 2055 start ($ are trillion US $2005)

SRM strategy	PV of climate damages ($)	PV of abatement costs ($)	PV of climate damages and abatement costs ($)	Benefit of SRM ($)	Cost reduction (%)
SRM 0	16.2	2.0	18.2	0	0
SRM 1	12.7	1.6	14.3	3.9	22
SRM 2	10.0	1.1	11.1	7.1	39
SRM 3	8.0	0.7	8.7	9.5	52

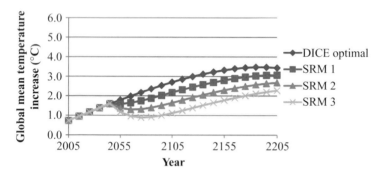

Figure 1.6 *OC temperature changes (°C) with 2055 deployment of SRM (2005–2205)*

optimal controls. This loading is not as pronounced as the case where one assumes that SRM completely displaces GHG reductions (e.g. Goes *et al.* 2009). Such an assumption regarding SRM deployment is a modeling choice, rather than an inherent feature of the technology. Furthermore, it is an assumption that does not correspond to the manner in which we and others (e.g. Wigley 2006) have suggested that SRM might be used.

Delayed start The reasons that society might wish to delay the use of SRM have been discussed earlier in this chapter. As noted there, both Crutzen (2006) and Wigley (2006) argue that the O_3 risk, while warranting further study, may not be significant. However, delay would dispel whatever fears might exist on these grounds, and it would also greatly dampen the risks that SRM might do net harm to high-latitude nations. Both of these effects would lower the transaction costs of deploying SRM. Society might then wish to wait until the middle of this century before beginning SRM. The benefit of starting a SRM program in 2055 instead of 2015 is $3.9, $7.1, and $9.5 trillion for SRM 1, SRM 2, and SRM 3, respectively. The impact on

temperature in this case is shown in figure 1.6. Comparing figure 1.6 and figure 1.3, we see that while delay lowers benefits, it has no discernible effect on maximum temperature changes.

Table 1.3 details the impact of a delayed start on the PV of climate damages and abatement costs. Comparing table 1.2 and table 1.3 we see that the delay primarily increases the PV of climate damages.

Temperature constraints

We now consider the case of using SRM to lessen the cost of meeting temperature constraints. Specifically, we assume that society chooses to constrain the increase in global mean temperature to no more than 2°C, noting that many governments have embraced this target. Constrained by this target, the optimal GHG control policy would result in a 200-year PV of climate damages of $11.9 trillion. The PV of the abatement costs would be $10.9 trillion. Thus, the total cost of climate change would be $22.8 trillion, compared to a 600-year PV of $24.4 trillion reported by Nordhaus (2008). We see that limiting temperature change to be no more than 2°C would cause a loss of $4.6 trillion when compared

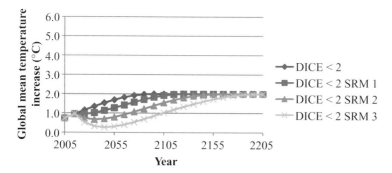

Figure 1.7 *Temperature changes (°C) with the deployment of SRM under a 2°C temperature constraint (2005–2205)*

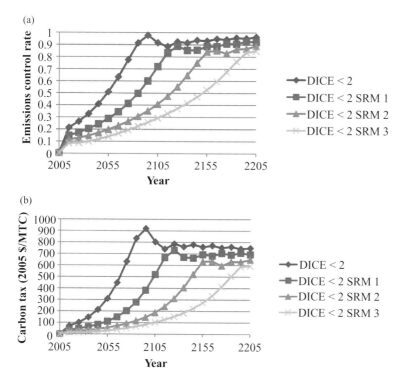

Figure 1.8 *(a) Optimal emissions rates and (b) carbon taxes under a 2°C constraint (2005–2205)*

to OC and would be $1.1 trillion *worse* than NC, i.e. it would be worse than doing nothing at all. Compared to OC, limiting the increase in temperature reduces damages by $4.3 trillion, but incurs $8.9 trillion more in abatement costs. Compared to NC, this strategy reduces climate damage by $9.8 trillion, but requires $10.9 trillion in abatement.

The impact of deploying SRM on temperature in this case is shown in figure 1.7. Without SRM, the

$2°C$ constraint is reached in 2095. With 1 W m^{-2}, 2 W m^{-2}, or 3 W m^{-2} or SRM the temperature constraint is reached in 2125 ($+$ 30 years), 2165 ($+$ 70 years), and 2205 ($+$ 110 years), respectively. Again, SRM could give society more time to make the technological change required to limit the harm from a global temperature increase.

Figure 1.8 displays optimal emissions control rates (1.8a) and carbon taxes (1.8b) as a function

Table 1.4 Benefit of SRM under 2°C constraint with 2025 start ($ are trillion US $2005)

SRM strategy	PV of climate damages ($)	PV of abatement costs ($)	PV of climate damages and abatement costs ($)	Benefit of SRM ($)	Cost reduction (%)
SRM 0	11.9	10.9	22.8	0	0
SRM 1	9.0	4.0	13.0	9.8	43
SRM 2	6.0	1.5	7.5	15.3	67
SRM 3	3.6	0.6	4.2	18.6	81

of SRM level.[5] SRM significantly alters the timing of required emissions controls and carbon taxes. For example, in order to meet a 2°C temperature constraint, an emissions reduction of 20% would have to be in force by 2015. Under SRM 3 this level of emissions reductions would not be required for another seventy years. Likewise, a carbon tax exceeding $100 per MTC (2005 $) would not be necessary until 2105.

The PV of climate damages and abatement costs is reduced from $22.8 trillion to $13.0 trillion through the deployment of 1 W m^{-2} of SRM – a benefit of $9.8 trillion, or about 43%. This benefit is over $3.5 trillion more than the OC case (benefit of $6.3 trillion); SRM is worth more under a temperature constraint because it lessens the need to perform costly abatement in the near term. Table 1.4 summarizes the benefit of each SRM strategy. The benefits of SRM in this case are quite significant, with the largest percentage gains coming from reduced abatement costs. For example, while SRM 1 reduces the PV of damages and abatement costs by 43%, it reduces the PV of abatement by 63% (from $10.9 to $4.0). It is also noteworthy that a 2°C temperature constraint with 1 W m^{-2} of SRM is more than $5 trillion better than DICE's OC ($18.2 trillion) and over $1 trillion *better* than NC with 1 W m^{-2} of SRM ($14.2 trillion). This later result is surprising given that, without SRM,

a 2°C constraint is worse than doing nothing. This result is obtained because SRM holds temperatures in check, avoiding climate damages, while society builds the capital and technology necessary to achieve emissions reductions at lower cost. The policy lesson, of course, is that SRM can lower the costs of pursuing non-optimal GHG control strategies, not that non-optimal strategies are harmless. SRM with a 2°C constraint is still worse than SRM with OC.

Low-discount rate (*The Stern Review*)

In order to match the assumptions made by *The Stern Review*, Nordhaus (2008) sets the time preference to 0% and the consumption elasticity to 1.0. These assumptions imply a real rate of return of about 2%. We adopt this case as our low-discount rate scenario. It is important to note, however, that these assumptions do not match empirical returns on capital and therefore the benefit of investments in the real economy.

The low discounting greatly amplifies future damages and *The Stern Review* assumptions result in an emissions reduction of 50% and a carbon tax of over $300 per MTC beginning in 2015. In order to compare these strong abatement measures, resulting from a low discount rate, to our other scenarios we, following Nordhaus, find the PV using our previous 5.5% real rate.[6] The 200-year PV of climate damages is $9.2 trillion, about $7 trillion less than OC. However, the PV of *The Stern Review*'s abatement costs is $22.1 trillion, about $20 trillion more than OC.

Employing 1 W m^{-2} of SRM reduces the PV of climate damages and abatement costs to $19 trillion ($6.3 trillion in climate damages and $12.7 trillion in abatement costs), which is a benefit

---

[5] The unstable behavior of these graphs occurs once the temperature constraint has been reached and DICE alternates between relaxing emissions constraints only to have it impose them again as temperature rises.

[6] This is, of course, not internally consistent with DICE. However, the present values calculated using 5.5% can be thought of as how much capital would be required to finance *The Stern Review*'s recommendations in the real economy.

Table 1.5 Benefit of SRM under low-discount-rate scenario with 2025 start ($ are trillion US $2005)

SRM strategy	PV of climate damages ($)	PV of abatement costs ($)	PV of climate damages and abatement costs ($)	Benefit of SRM ($)	Cost reduction (%)
SRM 0	9.2	22.1	31.2	0	0
SRM 1	6.3	12.7	19.0	12.2	39
SRM 2	4.0	7.1	11.1	20.1	64
SRM 3	2.6	3.8	6.4	24.8	79

of over $12 trillion. Surprisingly, the PV of *The Stern Review* policy with 1 W m^{-2} of SRM is very close to that of DICE Optimal Controls. Clearly, SRM holds the potential to mitigate damage to the environment induced by global warming and damage to the economy as the result of poor policy. Table 1.5 summarizes the benefit of each SRM scenario.

Air Capture

As mentioned in the introduction, AC technologies do not appear as promising as SRM from a technical or a cost perspective. For this reason, we have focused primarily on SRM. However, it is useful to contrast the potential net benefits of SRM and AC.

As a point of comparison, we begin by determining the level of AC that would have the same *economic benefit* as SRM 1. As described above, we modify DICE's carbon-cycle model to allow for the permanent removal of CO_2 from the atmosphere. After multiple DICE runs, we find that capturing and sequestering 5.5 GtC of CO_2 per year has approximately the same benefit as one W m^{-2} of SRM. We will refer to this AC scenario as AC 5.5. Specifically, the benefit of AC 5.5 is $5.5 trillion, compared to $6.3 trillion for SRM 1. To place this number in perspective, current global CO_2 emissions are around 8.5 GtC per year. Thus, AC 5.5 is equivalent to removing and sequestering almost 65% of current emissions. In other words, when considering only the impact on temperature, capturing and sequestering almost 65% of global annual CO_2 emissions has about the same economic benefit of reducing the solar flux by 0.3%.

As discussed earlier, Keith *et al.* (2006) have estimated that the cost of AC using current technology is $500 MTC^{-1} and might be driven below $200 MTC^{-1} over the next century. They caution however that these estimates could be off by a "factor of three." Pielke (2009) notes that at $500 MTC^{-1} AC costs about $1 *trillion per ppm*. Pielke (2009) goes on to cite Klaus Lackner as estimating the current cost to be $360 MTC^{-1} and that eventually it might fall to $100 MTC^{-1}. At $500 MTC^{-1} AC 5.5 would cost $2.75 trillion *per year*. The 200-year PV of this cost is almost $30 trillion, yielding a B/C ratio (BCR) of 0.20. At a cost of $100 MTC^{-1} the present cost of AC 5.5 is $5.6 trillion, approximately equal to its benefit. This example reveals the tremendous cost challenge faced by AC technologies. As we show in the next section, SRM might be able to achieve this same benefit for less than $0.5 trillion.

Given the benefit-cost (B/C) framework of this Copenhagen Consensus study, we will not consider even higher levels of AC. However, as another point of reference, consider the scenario suggested by Pielke (2009) of the capturing and sequestering all US auto emissions, which total 0.48 GtC annually. We round this and consider an AC 0.5 strategy. As a reminder, the optimal controls 200-year PV of climate damages and abatement costs is $18.2 trillion. Under AC 0.5 with optimal controls, the PV of climate damages and abatement costs is reduced to $17.7 trillion – a benefit of $0.5 trillion or savings of about 3%. Pielke (2009) estimates that it would cost $0.240 trillion per year (at $500 MTC^{-1}) to capture and store 0.48 GtC annually. Thus, costs would exceed the complete 200-year benefit of AC 0.50 in only two years.

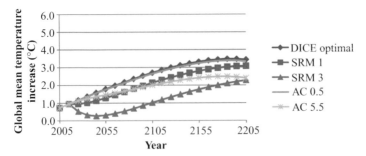

Figure 1.9 *OC temperature changes (°C) with the deployment of AC (2005–2205)*

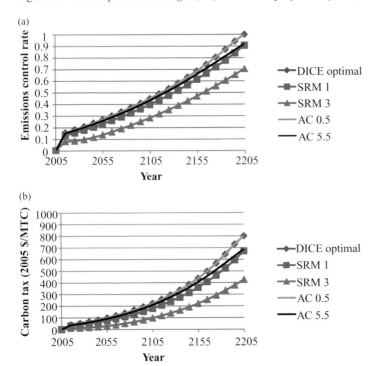

Figure 1.10 *(a) Optimal emissions controls and (b) carbon taxes with the use of AC (2005–2205)*

These results should not be surprising given current cost estimates and AC's benefit profile. The impact of deploying AC on temperature is shown in figure 1.9. As was the case with SRM, AC does delay temperature increases. However, the patterns of performance are quite different. We have added SRM 1 and SRM 3 to figure 1.9 as a reference. AC's impact is delayed because of lags in the climate system. We see that SRM 1 outperforms AC 5.5 until 2085 (+ 70 years) and AC 0.5 through at least 2205 (+ 200 years). SRM 3 outperforms all three AC scenarios over the next 200 years. In addition, the levels of AC we consider here are unable

to hold temperatures changes below 2°C and AC 0.5 offers almost no temperature benefit relative to OC, hence its near zero economic benefit.

Figure 1.10 displays optimal emissions control rates (1.10a) and carbon taxes (1.10b) as a function of AC level. Importantly, we see that the use of AC has almost no effect on the optimal emissions control rate or the carbon tax – especially in the short term, because of AC's delayed effect. None of the AC scenarios considered can match the impact that even SRM 1 has on these policy variables.

This provides another lens through which to view AC's cost challenges. As discussed above, AC costs

Table 1.6 Benefit of AC under OC ($ are trillion US $2005)

SRM strategy	PV of climate damages ($)	PV of abatement costs ($)	PV of climate damages and abatement costs ($)	Benefit of AC ($)	Cost reduction (%)
AC 0	16.2	2.0	18.2	0	0
AC 0.5	15.7	2.0	17.7	0.5	3
AC 5.5	10.9	1.8	12.7	5.5	25

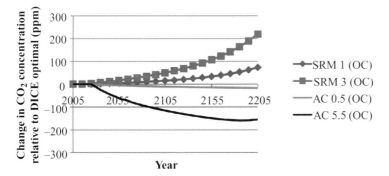

Figure 1.11 *Difference in AC and SRM CO_2 concentrations compared to OC (2005–2205)*

are on the order of $500 MTC^{-1} and might fall to 100 MTC^{-1} over a century. These costs exceed the optimal carbon tax (or the social cost of carbon) for at least the next 50 to 150 years. For example, under DICE OC a $100 MTC^{-1} carbon tax is not achieved until 2055 and a $500 MTC^{-1} is not reached until 2165 (see figure 1.10).

Table 1.6 summarizes the benefit of each AC strategy. We see that AC has almost no ability to reduce the PV of abatement costs and makes only a moderate reduction in climate damages.

AC does have the benefit of removing CO_2 from the atmosphere, which is potentially less risky than SRM – assuming the CO_2 remains safely sequestered. For example, the 2205 CO_2 concentration under AC 0.5 with optimal abatement is 610 ppm, compared to 627 ppm in the OC case without AC – a reduction of 17 ppm. The 2205 reduction for AC 5.5 is 155 ppm.

Figure 1.11 displays the CO_2 concentrations differences (relative to optimal controls) for each AC and SRM scenario.

Summary of Benefits

As detailed in the previous four sections, the ability of SRM to reduce climate damages and abatement

costs appears to be dramatic. For example, a single W m^{-2} of SRM:

- Is worth over $6 trillion under OC
- Can turn an emissions control strategy of limiting temperatures to $+2°C$, which is worse than doing nothing, into a strategy better than that of DICE's OC
- Can blunt the economic damage caused by policies such as those of *The Stern Review*
- Has an economic benefit, when considering the impact of temperature changes, equivalent to capturing and sequestering over 65% of the world's annual CO_2 emissions.

Table 1.7 summarizes the benefits of SRM.

We now turn to the task of estimating the costs of the different SRM strategies.

Direct Cost Estimates of Climate Engineering

The incoming solar radiation at the top of the atmosphere (TOA) is 341 W m^{-2} (Trenberth *et al.* 2009). Of this, 102 W m^{-2} is reflected back to space corresponding to an average planetary albedo α_P of 0.299 (102/342). The change in planetary albedo

Table 1.7 Summary of SRM benefits ($ are trillion US $2005)

			OC		
SRM strategy	NC ($)	Temp < 2°C ($)	2025 start Market discount rate (5.5%) ($)	2055 start market discount rate (5.5%) ($)	Low-discount rate (2.0%) ($)
SRM 1	7.5	9.8	6.3	3.9	12.2
SRM 2	13.1	15.3	11.0	7.1	20.1
SRM 3	16.8	18.6	14.0	9.5	24.8

needed to achieve a particular change in radiative forcing ΔF is

$$\Delta\alpha_p = \frac{-\Delta F}{341}. \tag{5}$$

Thus, if one wanted to reduce the radiative forcing by 1 W m^{-2}, 2 W m^{-2}, or 3 W m^{-2} the planetary albedo would need to be increased by 0.003, 0.006, and 0.009, respectively.

In this section we consider three SRM strategies that operate at three distinctly different positions relative to the Earth's surface. The first is the enhancement of marine stratiform cloud albedo; the second is the injection of aerosols into the stratosphere. Finally, the third is a sunshade placed in orbit at the Lagrangian-1 point between the Earth and the Sun. While the direct cost estimates we describe are speculative, we will show that they are so small that it is almost certain that SRM's *direct* BCR is greater than unity.

Marine Cloud Whitening

Lenton and Vaughan (2009) develop a simple methodology to approximate the change in atmospheric albedo required to bring about a desired change in planetary albedo, the specifics of which depend upon where reflection takes place in the atmosphere. They approximate marine stratiform cloud albedo enhancement by assuming all reflection takes place just above the Earth's surface, after all atmospheric absorption. We follow Lenton and Vaughan's method, but instead of basing our estimates of global energy fluxes on the work of Kiehl and Trenberth (1997) we use the updated estimates of Trenberth *et al.* (2009). We note in passing that Kiehl and Trenberth (1997) estimated that the TOA

flux was 342 W m^{-2}, whereas the 2009 estimate (Trenberth *et al.* 2009) was 341 W m^{-2}. This 1 W m^{-2} difference parallels our SRM 1 strategy. We estimate that the required change in atmospheric albedo, when reflection occurs after absorption, $\Delta\alpha_a$, is

$$\Delta\alpha_a = 1.482\Delta\alpha_p = 1.482\frac{-\Delta F}{341}. \tag{6}$$

Thus, decreasing radiative forcing by 1 W m^{-2} would require an increase in atmospheric albedo of 0.004. To determine the required increase in low-level marine stratiform cloud albedo, we must divide by the fraction of the Earth that is covered by such clouds. Latham *et al.* (2008) estimate that the increase in marine stratiform cloud albedo, $\Delta\alpha_c$, is

$$\Delta\alpha_c = \frac{\Delta\alpha_a}{0.175\,f} = 1.482\frac{-\Delta F}{0.175 \cdot f\,0.341} \approx \frac{-\Delta F}{40 \cdot f}, \tag{7}$$

where 0.175 is the fraction of the Earth covered by marine stratiform clouds and f is the fraction of these clouds that are seeded. Further, Latham *et al.* (2008) estimate the volume of seawater (m^3 s^{-1}) that must be injected to achieve a particular increase in cloud albedo. Table 1.8 details the required increase in cloud albedo and the required rate of injection as a function of desired forcing.

The minimum fraction of clouds that must be seeded is based on Latham *et al.*'s estimate that the number of droplets within the clouds can be increased by at most ten times (Latham *et al.* 2008). The cells marked "NA" are thus technologically infeasible. The numbers in parenthesis are the percentage increases in cloud albedo based on a natural marine stratiform cloud albedo of between 0.3

Table 1.8 Required injection rate for marine stratiform cloud albedo enhancement

Negative forcing (W m^{-2})	Required change in planetary albedo, $\Delta\alpha_p$	Minimum fraction of clouds that must be seeded	Required change in cloud albedo, $\Delta\alpha_c$			Required injection rate (m^3 s^{-1})		
			$f = 0.25$	$f = 0.50$	$f = 1$	$f = 0.25$	$f = 0.50$	$f = 1$
1	0.003	0.14	0.099 (14–33%)	0.050 (7–17%)	0.025 (4–8%)	12.7	8.6	7.2
2	0.006	0.29	NA	0.099 (14–33%)	0.050 (7–17%)	NA	25.4	17.3
3	0.009	0.43	NA	0.149 (21–50%)	0.074 (11–25%)	NA	57.9	31.3

Table 1.9 BCRs for marine stratiform cloud albedo enhancement (50% seeding)

SRM strategy	Annual injected volume (km^3)	Number of vessels required	PV of costs (trillions 2005 $)		NC	Temp < 2°C	OC		
			2025 start	2055 start			2025 start market discount rate (5.5%)	2055 start market discount rate (5.5%)	2025 start low discount rate (2.0%)
SRM 1	0.27	284	0.0009	0.0003	8,531	11,227	7,188	14,580	13,925
SRM 2	0.80	830	0.0026	0.0008	5,101	5,988	4,283	9,055	7,859
SRM 3	1.83	1881	0.0058	0.0018	2,889	3,191	2,413	5,365	4,265

and 0.7 (Salter et al. 2008; Lenton and Vaughan 2009). For example, an increase in cloud albedo of 0.099 (2 W m^{-2}, $f = 0.50$) represents an increase of 14% to 33%, which seems technologically feasible (Lenton and Vaughan 2009). An injection of 25 m^3 s^{-1} equates to 0.785 km^3 (0.188 mi^3) per year, or about 4.9 billion barrels per year, which is about 17% of world oil consumption.

Salter et al. (2008) investigate a range of wind-powered vessel designs and nominally consider a design able to inject 0.03 m^3 s^{-1}. Thus, it would take 288 vessels to offset 1 W m^{-2} if 50% of the available clouds are seeded (8.6/0.03 $\cong$ 288). Salter et al. also estimate, based on vessel displacement and power requirements, that each vessel would cost between £1 million and £2 million. We take the higher estimate and assume the vessels will cost about $US 3 million each.[7] We further assume, conservatively, that this fleet must be replaced every ten years. Thus, the ten-year cost to offset 1 W m^{-2} (SRM 1) with 50% seeding would be about $860 million. The 200-year PVs of this

recurring cost, beginning in 2025, using DICE's endogenous discount rate, is $0.90 billion. Given that the benefit of SRM 1 under OC is $6.3 trillion, the direct BCR is over 7000 to 1 ($6.3/$0.00090). The BCRs for each SRM strategy and each control environment are given in table 1.9, assuming that 50% of available clouds are seeded.

In the case of a 2055 start, the costs and benefits do not begin for forty more years. The PV of SRM 1 cost is $0.27 billion, yielding a BCR of approximately 14,500 ($3.9/$0.00027). The fact that the BCR is larger for the delayed start may be surprising given that the net befits are smaller. This result is, of course, a limitation of ranking based on the ratio between benefits and costs, instead of the difference between benefits and costs.

BCRs are also quite high in the low-discount rate scenario. For example, the benefit of SRM 1 in this

---

[7] Based on an exchange rate of $1.5 per £1. Given the uncertainty in this estimate and its small magnitude, we assume these are in 2005 $.

case was $12.3 trillion. The cost is $0.90 billion, yielding a BCR of almost 14,000. Limiting temperature changes to 2°C also yields large BCRs. Clearly, the more one deviates from optimal emission reduction strategies the greater the value that should be placed on SRM.

Clearly, these BCRs are quite large. Part of the reason for this is that intervention takes place close to the Earth's surface, requiring less energy for deployment than either a sunshade or stratospheric aerosols, In addition, as Latham *et al.* (2008) point out, nature provides the energy to increase the droplet size by 4 to 5 orders of magnitude from that which enters the bottom of the cloud compared to its size at the cloud top. These results strongly suggest that marine stratiform cloud albedo enhancement should be investigated more fully.

Stratospheric Aerosol Injection

Based on the Mount Pinatubo eruption, Crutzen (2006) estimates that the radiative forcing efficiency of sulfate aerosol is -0.75 W m^{-2} per Tg S (1 trillion g = 1 billion kg = 1 million MT of sulfur). Rasch *et al.* (2008) use a coupled atmospheric model to better understand the role that aerosol particle size plays in forcing. They consider "large" particles (effective radius of 0.43 microns) that might be associated with a volcanic eruption and "small" particles (effective radius of 0.17 microns) typically seen during background conditions. Unfortunately, Rasch *et al.* do not report their forcing efficiencies, but based on their work we estimate a forcing efficiency of between -0.50 W m^{-2} and -0.60 W m^{-2} for volcanic size particles and around -0.90 W m^{-2} for the small particles. Given the uncertainty in these estimates and in the size of the particles themselves, we follow Crutzen and assume an efficiency of -0.75 W m^{-2} per Tg S. Particle residence time is another critical factor, which is also affected by particle size. Rasch *et al.* find residence times of between 2.6 and 3.0 years for the volcanic particles and between 2.4 and 2.8 years for the small particles. We assume a residence time of 2.5 years for simplicity.

In order to offset 1 W m^{-2} we require a sulfur burden of 1.3 Tg S (1/0.75). Assuming a residence time of 2.5 years, we would require yearly injections of 0.53 Tg S. To place this number in perspective, we consider two benchmarks. First, the burning of fossil fuels emits 55 Tg S per year (Stern 2005). Thus, the SRM 1 strategy requires an injection equivalent to approximately 1% of the sulfur emitted via fossil fuels. Second, Mount Pinatubo injected about 10 Tg S into the stratosphere (Crutzen 2006), which is almost twenty times larger than what is required for our SRM 1 strategy.

The mass of material that must be injected depends upon the choice of precursor. Common candidates include hydrogen sulfide (H_2S) and sulfur dioxide (SO_2). The molecular masses of H_2S and SO_2 are 34.08 g mol^{-1} (1.1 times that of S) and 64.07 g mol^{-1} (2.0 times that mass of S), respectively. The use of SO_2 would require about twice the capital as H_2S and we therefore assume the use of H_2S as a precursor. We note, however, that H_2S is both toxic and flammable. In sum, in order to offset 1 W m^{-2} we would need to inject about 0.57 Tg H_2S per year.

The NAS (1992) considered the use of 16-inch naval artillery rifles, rockets, balloons, and airplanes to inject material into the stratosphere. The cost of naval artillery and balloons were about the same, while the cost of rockets was estimated to be about five times greater. Robock *et al.* (2009) have recently revised the cost estimates for the use of airplanes. They conclude that 1 Tg of H_2S could be injected near the equator using F-15s for a yearly cost of about $4.2 billion. However, many questions remain regarding the ability of planes to continuously inject corrosive H_2S and if droplets of the correct size would be formed. Thus, in this section, we estimate direct costs based on the use of naval artillery.

The NAS assumed that each artillery shell could carry a payload of 500 kg. Therefore, our SRM 1 strategy would require 1.1 million shells per year, or the continuous firing of about 2 shells per minute. The cost of this system was estimated to be $40 per kg (2005 $), or $40 billion per Tg, to place aerosols in the stratosphere. Approximately $35 of this cost (89%) is the variable cost of the ammunition and the personnel. The remaining $5 is the capitalized cost of the equipment, which was assumed to have a forty-year lifetime. The yearly cost for SRM 1

Table 1.10 Stratospheric aerosol BCRs (naval rifles)

SRM strategy	Annual injected mass (Tg H$_2$S)	Shell firing frequency (shells min^{-1})	PV of costs (trillion '05 $) 2025 start	PV of costs (trillion '05 $) 2055 start	BCRs NC	BCRs Temp < 2°C	BCRs OC 2025 start market discount rate (5.5%)	BCRs OC 2055 start market discount rate (5.5%)	BCRs OC 2025 start low discount rate (2.0%)
SRM 1	0.57	2.2	0.23	0.07	32	43	27	56	53
SRM 2	1.13	4.3	0.46	0.14	29	34	24	51	44
SRM 3	1.70	6.5	0.68	0.21	25	27	21	46	36

would then be $22.8 billion (0.57 · 40). The 200-year PV of this yearly cost, beginning in 2025, is $230 billion, yielding a BCR under OC of 27 to 1 ($6.3/$0.23). The BCRs for each SRM strategy and emissions scenario are given in table 1.10.

The direct BCRs for stratospheric aerosols appear to be quite attractive. The use of planes instead of artillery might further improve this performance. Likewise, Teller *et al.* (2003) have suggested the development of engineered particles could reduce the cost of our SRM 3 strategy to about $1 billion per year, which is about an order of magnitude less than our current cost estimates. If true, this would result in BCRs on the order of 1000 to 1.

Space-Based Sunshade

Angel (2006) analyzes reducing the solar flux through the deployment of a large sunshade, composed of trillions of tiny ($\sim$1 g) autonomous spacecraft ("flyers"). These flyers would be placed in a one-year period orbit slightly beyond the Lagrange-1 point (L1), which is approximately 1.5 million km from the Earth. Angel optimizes his design in terms of mass, reflexivity, and distance from the Sun. As discussed below, the sheer scale of this project boggles the mind.

In order to offset 1 W m^{-2} the solar flux would need to be decreased by 1.46 W m^{-2}, taking into account the planetary albedo of 0.299 (1/(1 − 0.299)). This is a 0.43% decrease in the solar flux and based on Angel's calculations, would require a total flyer cross-section of 1.1 million km^2. Based on an individual flyer cross-section of 0.28 m^2 we find that 3.9 trillion flyers would be

required with a total mass of 4.7 million MT. Based on the current design, each launch would include 800,000 flyers and therefore approximately 5 million launches would be required to put the SRM 1 screen in place. It is enlightening to put this number into perspective. If 800,000 flyers were launched every 5 minutes it would take almost fifty years to put the sunshade in place.[8] If we wished to have the sunshade in place within one year, we would need to launch about every 6 seconds, which Angel estimates could be achieved with multiple launchers.

Angel roughly estimates that the cost of the sunshade program to offset 4.23 W m^{-2} would be on the order of $5 trillion. This is broken out as follows: electromagnetic launchers ($0.6 trillion), flyers ($1 trillion), launches/fuel ($1 trillion), and development and operations ($2.4 trillion). Of this $5 trillion, 60% is ($3 trillion) is fixed cost. These estimates are really not estimates at all, but rather cost targets. For example, when Angel considers the cost lifting 20 million MT (what is required to offset 4.23 W m^{-2}) into high-Earth orbit, he writes "for the sake of argument if we allow $1 trillion for the task, a transportation cost of $50 kg^{-1} of payload would be needed [$1 trillion/20 billion kg]." In fact, as Angel notes, the current cost to achieve high-Earth orbit is $20,000 kg^{-1}, in which case launch costs alone would be $395 trillion. Similarly, in the case of manufacturing costs, Angel writes "An aggressive target would be the same $50 per kilogram as for launch, for $1 trillion

---

[8] In fact, just keeping up with the 2 ppm yr^{-1} increase in CO$_2$ under NC would require about 138,000 launches per year, or one launch every 4 minutes.

Table 1.11 Summary of BCRs for SRM (market and low-discount-rate cases)

SRM strategy	2025 start market discount rate (5.5%)		2055 start market discount rate (5.5%)		2025 start low-discount rate (2.0%)	
	Stratospheric aerosol	Cloud albedo	Stratospheric aerosol	Cloud albedo	Stratospheric aerosol	Cloud albedo
SRM 1	27	7,188	56	14,580	53	13,925
SRM 2	24	4,283	51	9,055	44	7,859
SRM 3	21	2,413	46	5,365	36	4,265

total." The scale required to produce trillions of (tiny) spacecraft is unprecedented. The only spacecraft that have been manufactured in any "mass" quantity are the Iridium satellites, which Angel cites as costing $7000 kg^{-1}, in which case manufacturing costs for the sunshade would be around $135 trillion.

Angel's aggressive targets are based on assumptions regarding returns to scale, but the scale of this project is so far beyond anything that has ever been attempted in the space industry that we are uncomfortable using Angel's targets. Unfortunately, development of our own cost estimates is outside the scope of our current effort and we will have to leave this as an issue for further study.

BCR Summary

Both stratospheric aerosol injection and cloud albedo enhancement have attractive *direct* BCRs. In the interest of space, table 1.11 summarizes the BCRs for each SRM technology under optimal controls only, for the market discount rate and low-discount rate scenarios. BCRs decline with increasing amounts of SRM. This is related to the quadratic nature of DICE's damage equation. Policy regimes that result in significant emissions reductions in the short term (e.g. low discount rate, temperature constraint) result in higher BCRs because SRM helps to delay these costly interventions.

Possible Objections and Responses

An earlier version of this chapter received helpful comments from Anne E. Smith and Roger A. Pielke, Jr. (see Perspective papers 1.2 and

1.1, respectively). These comments raised valuable points that deserve explicit treatment. Space constraints, however, limit us to a short response within the body of this chapter rather than a rejoinder. Thus, before concluding, we address some of the main points raised by these valuable critiques.

Anne E. Smith

The main thrust of Smith's comments stresses the potential worth of applying a value of information (VOI) analysis to the issues discussed in this chapter. This suggestion is valid and we would urge that later study of this type be undertaken.

At the same time, we note that the VOI calculation requires additional assumptions and analyses such as: the direct damages that may be caused by SRM, the increase in climate damages due to more rapid warming if SRM is halted, the mitigation strategy that should be selected given that R&D on SRM is pursued, the reliability of an R&D program, etc. Smith has done an admirable job at providing these additional assumptions, and she suggests that more accurate estimates could be obtained from the DICE model. Unfortunately, doing so is outside the scope of these comments and also, as we discuss below, it is, in some cases, beyond DICE's capabilities. We regard this as an excellent area of future research.

Roger A. Pielke, Jr.

Pielke, Jr. has highlighted several important issues. He raises four objections.

A contradiction in the use of BCA   Pielke sees a contradiction in our noting that the *direct* BCRs for SRM are large, but not arguing for immediate deployment. We see no such contradiction. As we argue throughout this chapter, while the potential net benefits of SRM are large, indirect costs might still change the calculus. Only research can address this question. Our recommendation for research does not stem from a "skepticism" in our analysis, but rather from a recognition of our discussion that unknown and potentially large uncertainties remain.

The inability to accurately anticipate costs or benefits First, we begin by noting that the basis of the Copenhagen Consensus was to "present empirically based cost-benefit estimates." Thus, Pielke's criticisms of our use of BCA are outside the framework to which we all agreed.

Second, Pielke, while expressing doubts about the utility of BCA, believes that our chapter did not actually employ it. He writes that that there are "no policy recommendations [in our chapter] that result directly from the CBA." We disagree. The case for studying SRM, to begin with, rests on the evidence that in order to achieve net benefits, GHG control policies must be structured to accept substantial amounts of damage from climate change. Furthermore, our chapter uses BCA in winnowing the technologies worthy of R&D and setting priorities among them.

Third, Pielke "disagrees that CBA tells us anything meaningful about how much should be invested in research or what the potential payoffs might be." Why then does he recommend research into CE? He bases his recommendation for research into CE because it "has considerable value to advancing fundamental understandings of the global earth system." This reasoning applies to climate research in general. Why then call it "CE" research?

Finally, Pielke cites the work of Goes *et al.* (2009). He avers that "the same (or a very similar)" model could produce different results. In fact, these authors used a modified version of DICE that differed quite substantially from our own. They also assumed a fundamentally different implementation of SRM. The differences in modeling and scenarios make the results difficult to compare. Further, Goes *et al.* acknowledge that future learning might raise SRM's net benefits and expand the range of conditions in which substituting SRM for GHG control

would pass a C/B test. This observation seems to endorse precisely the kind of R&D proposed in our chapter.

Reliance on a demonstrably incorrect conceptual model of how CE influences the climate system Pielke notes that the science underlying SRM is not well understood and doubts that DICE depicts it accurately. From this note of doubt, Pielke wishes to segue to the conclusion that cost-benefit analysis (CBA) is at least useless and possibly misleading. However, as Smith argues, attempting to understand the benefits of CE, while the science is still evolving, is useful. Such an effort can help to highlight what is not known, identify critical assumptions, and focus future research. We share her views.

CE as a technological fix Pielke cites Sarewitz and Nelson (2008), as do we. Sarewitz and Nelson state three rules that, in their judgment, define where technological solutions to social problems are likely to work and where they are not. Pielke, however, claims that the Sarewitz–Nelson rules argue against research on SRM, a claim that Sarewitz and Nelson do not make. In fact, Nelson's previous work makes clear that the sort of BCA that we use to show SRM's superiority to AC is an important and valid part of the R&D selection process (Nelson and Winter 1977).

Further, Pielke's effort in this respect seems to us to stretch the Sarewitz–Nelson argument beyond its reasonable limits. Thus, rule one is merely that the solution should embody a clear cause and effect relationship. SRM does embody such a relationship. Much, although not all, of the damage caused by climate change arises from warming. SRM is designed to lessen warming. Rule two calls for clear metrics of success. SRM will either reduce warming, or it will not. Rule three suggests that odds of success are better the smaller are the required advances in science and technology. In fact, the volcanic record and existing evidence of marine-cloud formation suggest that well-established scientific knowledge underlies both stratospheric aerosols and marine cloud whitening. True, SRM will require a new knowledge of climate science. At the same time, AC faces the challenge of finding processes that can accomplish it at a cost that society is willing to pay and with less risk than they are willing to bear. Neither is certain.

Conclusion

Limitations of the Results

Any assessment of SRM and AC will be limited by the current state of knowledge, the rudimentary nature of the concepts, and the lack of prior R&D efforts. As noted in in the introduction, this analysis relies on numbers found in the existing literature and existing climate change models. These inputs to our analysis are admittedly speculative; many questions surround their validity, and many gaps exist in them. This chapter has also stressed the potential importance of transaction costs and "political market failures." Finally, many important scientific and engineering uncertainties remain. Some of these pertain to climate change itself, its pace, and its consequences. Still others are more directly relevant to SRM. How will SRM impact regional precipitation patterns and O_3 levels? To what extent can SRM be scaled to the levels considered here? What is the best method for aerosol injection? Are there other side effects that could invalidate the use of SRM? These are just a few of the questions that a well-designed research program should be designed to answer.

Principal Implications for Climate Policy

This analysis, then, can claim to be only an early and partial look at the potential benefits and costs of CE. Even so, the large scale of the estimated direct net benefits associated with the stratospheric aerosol and marine cloud whitening approaches are impressive. One might draw several preliminary conclusions from our results. These include:

- The direct BCR for stratospheric aerosol injection is on the order of 25 to 1, while the BCR for marine cloud whitening is around 5000 to 1. Net benefits are clearly large relative to plausible costs of an ambitious R&D effort. Problems could indeed surface in the course of future

research. Indirect cost issues are much more likely to preclude or severely limit the use of SRM than are direct costs. Much of the R&D effort, therefore, should seek to narrow the uncertainties that surround these issues. Nonetheless, the results of this initial BCA place the burden of proof squarely on the shoulders of those who would prevent such research or would place *ex ante* restrictions on its progress.

- The space-based sunshade appears to be an exception to this conclusion. This conjecture rests on the sunshade's far less promising BCR, the extremely high economic risks entailed by its massive fixed costs, and its large scale and high technological risk.
- The greater the degree to which the global GHG control regime falls short of optimal in the policy tools that it employs, the targets it sets, or the gaps in its participation, the greater the potential value of SRM. However, SRM yields large net benefits even with optimal controls.
- Transaction costs and failures of the policy market are likely to affect SRM just as they do all other climate policy options. These costs could greatly affect its benefits and costs. The deferred deployment scenario for SRM offers an example of the possible impact on benefits. At least in this example, net benefits fall, but remain large. In some areas, the political transaction costs of SRM may exceed those of other options. In other areas, its costs may be lower. For example, some reasons exist for hoping that SRM may be a less tempting target for pork barrel politics than are some other responses to climate change.
- Insofar as possible, the transaction costs and the effects of political market failures should be recognized as likely to affect CE, as well as all other options for dealing with climate change. Even when the effects cannot be quantified *ex ante*, they should be explored as thoroughly as possible in qualitative terms. BCA should attempt to assess these costs consistently across all climate policy options.
- SRM is more promising than AC, but the latter, despite its current high costs, merits a secondary R&D effort. It offers a particularly low-risk strategy with appealing institutional features that resemble those of SRM.

• Future research efforts should be more heavily focused on SRM. Such research should seek to remove uncertainty regarding the possible side effects that SRM may cause and the associated risks. It should also address the technical and political feasibility of aerosol injection and marine cloud whitening, and explore the degree to which these approaches can be scaled, their deployment and operational costs, and their impact on other climate change policies. Some research should also be directed towards development of engineered particles. Such particles may improve the cost and environmental profile of aerosol injection. Finally, research funding should be allocated for B/C studies so as to improve our results. This would include: the quantification of uncertainty, side effects, and the ability of CE to reduce the risk of abrupt climate change, as well as quantification of the indirect costs discussed above.

While our analysis is preliminary, we believe it makes a strong case that the potential net benefits of SRM are large; the question is whether or not the indirect costs will change the calculus. Only research can provide an answer.

Bibliography

Akbari, H., S. Menson, and A. Rosenfeld, 2009: Global cooling: increasing world-wide urban albedos to offset CO_2, *Climatic Change* **94**(3–4), 275–86

Angel, R., 2006: Feasibility of cooling the Earth with a cloud of small spacecraft near the inner Lagrange point (L1), *Proceedings of the National Academy of Sciences* **103**(46), 17184–89

Arrow, K., 2007: Global climate change: a challenge to policy, *The Economists' Voice* **4**(3), 1–5

Barrett, S., 2007a: *Why Cooperate? The Incentive to Supply Global Public Goods*, Oxford University Press, New York

  2007b: The incredible economics of geoengineering, *Environmental and Resource Economics* **39**(1), 45–54

  2009: Geoengineering's role in climate change policy, Working Paper for the AEI Geoengineering Project

Bial, J.R., D. Houser, and G.D. Libecap, 2000: Public choice issues in international collective action: global warming regulation, Working Paper 00–05, Arizona University, Tucson, AZ

Borenstein, S., 2009: Obama looking at cooling air to fight warming, *The Associated Press*, April 9

Bower, K., T. Choularton, J. Latham, J. Sahraei, and S. Salter, 2006: Computational assessment of a proposed technique for global warming mitigation via albedo-enhancement of marine stratocumulus clouds, *Atmospheric Research* **82**(1–2), 328–36

Broder, J.M. and M.L. Wald, 2009: Big science role is seen in global warming cure, *New York Times*, February 12, A24

Caldeira, K. and L. Wood, 2008: Global and Arctic climate engineering: numerical model studies, *Philosophical Transactions of the Royal Society A* **366**(1882), 4039–56

Cao, L. and K. Caldeira, 2008: Atmospheric CO_2 stabilization and ocean acidification, *Geophysical Research Letters* **35**(19), L19609

Christoffersen, P. and M.J. Hambrey, 2006: Is the Greenland Ice Sheet in a state of collapse?, *Geology Today* **22**(3), 98–103

Cohen, L.R. and R.G. Noll, with J.S. Banks, S.A. Edelman, and W.M. Pegram, 1991: *The Technology Pork Barrel*, The Brookings Institution Press, Washington, DC

Cooper, R.N., 2000: International approaches to global climate change, *World Bank Research Observer* **15**(2), 145–72

Corell, R.W., S.J. Hassol, and J. Melillo, 2008: Emerging challenges – methane from the Arctic: global warming wildcard, *UNEP Year Book 2008: An Overview of Our Changing Environment*, United Nations Environment Programme, Stevenage, UK

Crutzen, P.J., 2006: Albedo enhancement by stratospheric sulfur injections: a contribution to resolve a policy dilemma?, *Climatic Change* **77**(3–4), 211–20

Easterly, W., 2006: *The White Man's Burden: Why the West's Efforts to Aid the Rest Have Done So Much Ill and So Little Good*, Penguin Press, New York

Edgerton, D., 2007: *The Shock of the Old: Technology and Global History since 1900*, Oxford University Press, New York

Environmental Protection Agency (EPA), 2009: *Economics of Climate Change*, National Center for Environmental Economics, July 21,

http://yosemite.epa.gov/ee/epa/eed.
nsf/webpages/ClimateEconomics.html

Fleming, J.R., 2007: The climate engineers, *The Wilson Quarterly* **31**(2), 46–60

Gaskill, A., 2004: Summary of meeting with US DOE to discuss geoengineering options to prevent abrupt and long-term climate change, www.global-warming-geo-engineering.org/DOE-Meeting/DOE-Geoengineering-Climate-Change-Meeting/org.html

Global Carbon Project, 2008: *Carbon Budgets and Trends 2007*, September 26

Goes, M., K. Keller, and N. Tuana, 2009: The economics (or lack thereof) of aerosol geoengineering, submitted to *Climate Change*

Govindasamy, B. and K. Caldeira, 2000: Geoengineering Earth's radiation balance to mitigate CO_2 induced climate change, *Geophysical Research Letters* **27**(14), 2141–4

Gulledge, J., 2008: Three plausible scenarios of future climate change, in K.M. Campbell (ed.), *Climatic Cataclysm: The Foreign Policy and National Security Implications of Climate Change*, K.M. Campbell, The Brookings Institution Press, Washington, DC

Hamwey, R.M., 2007: Active amplification of the terrestrial albedo to mitigate climate change: an exploratory study, *Mitigation and Adaptation Strategies for Global Change* **12**(4), 419–39

Hu, Z.-Z., M. Latif, E. Roeckner, and L. Bengtsson, 2000: Intensified Asian summer monsoon and its variability in a coupled model forced by increasing greenhouse gas concentrations, *Geophysical Research Letters* **27**(17), 2681–4

Intergovernmental Panel on Climate Change (IPCC), 2007: *Climate Change 2007: Mitigation*, B. Metz, O. Davidson, P. Bosch, R. Dave, and L. Meyer (eds.), Cambridge University Press, New York

Jacoby, H.D., M.H. Babiker, S. Paltsev, and J.M. Reilly, 2008: Sharing the burden of GHG reductions, *MIT Joint Program on the Science and Policy of Global Change*, **Report 167**

Keith, D.W. and M. Ha-Duong, 2003: CO_2 capture from the air: technology assessment and implications for climate policy, *Greenhouse Gas Control Technologies – 6th International Conference: Proceedings of the 6th International Conference on Greenhouse Gas Control Technologies: October 1–4, 2002, Kyoto, Japan*, Pergamon, Oxford

Keith, D.W., M. Ha-Duong, and J.K. Stolaroff, 2006: Climate strategy with CO_2 capture from the air, *Climatic Change* **74**(1–3), 17–45

Kerr, R.A., 2007: Is battered Arctic sea ice down for the count?, *Science* **318**(5847), 33–4

Kiehl, J.T. and K.E. Trenberth, 1997: Earth's annual global mean energy budget, *Bulletin of the American Meteorological Society* **78**(2), 197–208

Kleypas, J.A., R.A. Feely, V.J. Fabry, C. Langdon, C.L. Sabine, and L.L. Robbins, 2006: Impacts of ocean acidification on coral reefs and other marine calcifiers: a guide for future research, Workshop report sponsored by the National Science Foundation, the National Oceanic and Atmospheric Administration, and the US Geological Survey

Kravitz, B., A. Robock, L. Oman, G. Stenchikov, and A.B. Marquardt, 2009: Sulfuric acid deposition from stratospheric geoengineering with sulfate aerosols, *Journal of Geophysical Research – Atmospheres* **114**(D14109)

Kuylenstierna, J.C.I., H. Rodhe, S. Cinderby, and K. Hicks, 2001: Acidification in developing countries: ecosystem sensitivity and the critical load approach on a global scale, *Ambio* **30**(1), 20–8

Lackner, K.S., P. Grimes, and H.-J. Ziock, 2001: Capturing carbon dioxide from air, *Proceedings of the First National Conference on Carbon Sequestration*, Washington, DC

Lal, M., U. Cubasch, R. Voss, and J. Waszkewitz, 1995: Effect of transient Increase in Greenhouse Gases and sulphate aerosols on monsoon climate, *Current Science* **69**(9), 752–63

Lane, L., K. Caldeira, R. Chatfield, and S. Langhoff, 2007: Workshop Report on Managing Solar Radiation, NASA Ames Research Center, Carnegie Institute of Washington Department of Global Ecology, **NASA/CP-2007–214558**

Lane, L. and D. Montgomery, 2008: Political institutions and greenhouse gas controls, AEI Center for Regulatory and Market Studies, Related Publication **08–09**

Latham, J., 1990: Control of global warming?, *Nature*, **347**, 339–40

Latham, J., P.J. Rasch, C.C. Chen, L. Kettles, A. Gadian, A. Gettleman, H. Morrison, K. Bower, and T. Choularton, 2008: Global temperature stabilization via controlled albedo enhancement of low-level maritime clouds, *Philosophical*

Transactions of The Royal Society A **366**(1882), 3969–87

Lenton, T.M. and N.E. Vaughan, 2009: The radiative forcing potential of different climate geoengineering options, *Atmospheric Chemistry and Physics Discussions* **9**, 2559–2608

MacCracken, M.C., 2006: Geoengineering: worthy of cautious evaluation?, *Climatic Change* **77**(3–4), 235–43

Matthews, H.D. and K. Caldeira, 2007: Transient climate – carbon simulations of planetary geoengineering, *Proceedings of the National Academy of Sciences*, **104**(24), 9949–54

May, W., 2002: Simulated changes of the Indian summer monsoon under enhanced greenhouse gas conditions in a global time-slice experiment, *Geophysical Research Letters* **29**(7), 22.1–22.4

Mearsheimer, J.J., 1994: The false promise of international institutions, *International Security* **19**(3), 5–49

Meehl, G.A. and W.M. Washington, 1993: South Asian summer monsoon variability in a model with doubled atmospheric Carbon Dioxide concentration, *Science* **260**(5111), 1101–4

National Academy of Sciences (NAS), 1992: *Policy Implications of Greenhouse Warming: Mitigation, Adaptation, and the Science Base*, National Academy Press, Washington, DC

National Research Council, 2005: *Radiative Forcing of Climate Change: Expanding the Concept and Addressing Uncertainties*, Committee on Radiative Forcing Effects on Climate, Climate Research Committe Board on Atmospheric Sciences and climate, Division on Earth and Life Studies. The National Academies Press, Washington, DC

Nelson, R.R. and S.G. Winter, 1977: In search of useful theory of innovation, *Research Policy* **6**(1), 36–76

1982: *An Evolutionary Theory of Economic Change*, Harvard University Press, Cambridge, MA

Nordhaus, W.D., 1994: *Managing the Global Commons: The Economics of Climate Change*, MIT Press, Cambridge, MA

2002: Modeling induced innovation in climate-change policy, *Technological Change and the Environment*, A. Grübler, N. Nakićenović, and W.D. Nordhaus (eds), RFF Press, Washington, DC

2008: *A Question of Balance: Weighing the Options on Global Warming Policies*, Yale University Press, New Haven, CT

Nordhaus, W.D., and J. Boyer, 2000: *Warming the World: Economics Models of Global Warming*, MIT Press, Cambridge, MA

North, D.C., 1990: *Institutions, Institutional Change, and Economic Performance*, Cambridge University Press, New York

2005: *Understanding the Process of Economic Change*, Princeton University Press, Princeton, NJ

North, D.C., J.J. Wallis, and B.R. Weingast, 2009: *Violence and Social Orders: A Conceptual Framework for Interpreting Recorded Human History*, Cambridge University Press, New York

Olson, Mancur, 1982: *The Rise and Decline of Nations: Economic Growth, Stagflation, and Social Rigidities*, Yale University Press, New Haven, CT

Pacala, S. and R. Socolow, 2004: Stabilization wedges: solving the climate problem for the next 50 years with current technologies, *Science* **305**(5686), 968–72

Pearson, J., J. Oldson, and E. Levin, 2006: Earth rings for planetary environment control, *Acta Astronautica* **58**(1), 44–57

Pielke, R.A. Jr., 2009: An idealized assessment of the economics of air capture of carbon dioxide in mitigation policy, *Environmental Science & Policy* **12**(3), 216–25

Rasch, P.J., P.J. Crutzen, and D.B. Coleman, 2008: Exploring the geoengineering of climate using stratospheric sulfate aerosols: the role of particle size, *Geophysical Research Letters* **35**(2), L02809

Rasch, P.J., S. Tilmes, R.P. Turco, A. Robock, L. Oman, C.-C. Chen, G.L. Stenchikov, and R.R. Garcia, 2008: An overview of geoengineering of climate using stratospheric sulphate aerosols. *Philosophical Transactions of The Royal Society A* **366**, 4007–37

Rau, G.H., K.G. Knauss, W.H. Langer, and K. Caldeira, 2007: Reducing energy-related CO_2 emissions using accelerated weathering of limestone, *Energy* **32**(8), 1471–77

Richels, R.G., A.S. Manne, and T.M.L. Wigley, 2004: Moving beyond concentrations: the challenge of limiting temperature change, *AEI–Brookings Joint Center for Regulatory Studies*, Working Paper **04–11**

Ridgwell, A., J.S. Singarayer, A.M. Hetherington, and P.J. Valdes, 2009: Tackling regional climate change by leaf albedo bio-geoengineering, *Current Biology* **19**(2), 146–50

Robock, A., 2008a: Whither geoengineering?, *Science* **320**, 1166–7

2008b: 20 reasons why geoengineering may be a bad idea, *Bulletin of the Atomic Scientists* **64**(2), 14–18

Robock, A. and J. Mao, 1995: The volcanic signal in surface temperature observations, *Journal of Climate* **8**(5), 1086–1103

Robock, A., A.B. Marquardt, B. Kravitz, and G. Stenchikov, 2009: The benefits, risks, and costs of stratospheric geoengineering, *Geophysical Research Letters* **36**(L19703)

Robock, A., L. Oman, and G. Stenchikov, 2008: Regional climate responses to geoengineering with tropical and Arctic SO2 injections. *Journal of Geophysical Research* **113**, D16101

Royal Society, 2005: Ocean acidification due to increasing atmospheric carbon dioxide, Policy Document **12/05**

Salter, S., G. Sortino, and J. Latham, 2008: Sea-going hardware for the cloud albedo method of reversing global warming, *Philosophical Transactions of The Royal Society A* **366**(1882), 3989–4006

Sarewitz, D. and R. Nelson, 2008: Three rules for technological fixes, *Nature* **456**, 871–2

Schelling, T.C., 2002: What makes greenhouse sense?: time to rethink the Kyoto Protocol, *Foreign Affairs* **81**(3), 2–9

Skeffington, R.A., 2006: Quantifying uncertainty in critical loads: (A) literature review, *Water, Air, and Soil Pollution* **169**(1–4), 3–24

Stern, D.I., 2005: Global sulfur emissions from 1850 to 2000, *Chemosphere* **58**(2), 163–75

Stern, N., 2007: *The Economics of Climate Change: The Stern Review*, Cambridge University Press, Cambridge

Stocker, Thomas F., 2003: Changes in the global carbon cycle and ocean circulation on the millennial time scale, *Global Climate: Current Research and Uncertainties in the Climate System*, X. Rodó and F.A. Comín (eds), Springer-Verlag, Berlin

Summers, L., 2007: Foreword, *Architectures for Agreement: Addressing Global Climate Change in the Post-Kyoto World*, J. Aldy and R. Stavins (eds), Cambridge University Press, Cambridge

Teller, E., L. Wood, and R. Hyde, 1997: Global warming and ice ages: I. Prospects for physics-based modulation of global change, University of California Lawrence Livermore National Laboratory

Teller, E., R. Hyde, and L. Wood, 2002: Active climate stabilization: practical physics-based approaches to prevention of climate change, University of California Lawrence Livermore National Laboratory

Teller, E., R. Hyde, M. Ishikawa, J. Nuckolls, and L. Wood, 2003: Active stabilization of climate: inexpensive, low-risk, near-term options for preventing global warming and ice ages via technologically varied solar radiative forcing, University of California Lawrence Livermore National Laboratory

Tetlock, P.E. and M. Oppenheimer, 2008: The boundaries of the thinkable, *Daedalus* **137**(2), 59–70

Tilmes, S., R. Müller, and R. Salawitch, 2008: The sensitivity of polar ozone depletion to proposed geoengineering schemes, *Science* **320**(5880), 1201–4

Tol, R.S.J., 2006: The Stern Review of the Economics of Climate Change: a comment, *Energy & Environment* **17**(6), 977–81

2008: The social cost of carbon: trends, outliers and catastrophes, *Economics* **2**, 2008–25

Travis, W.R., 2009: Geo-engineering the climate: an emerging technology assessment, Discussion Paper prepared for the National Research Council, Committee on America's Climate Choices, http://americasclimatechoices.org/Geoengineering_Input/attachments/Travis_geoengineering.pdf

Trenberth, K.E., J.T. Fasullo, and J.T. Kiehl, 2009: Earth's global energy budget, *Bulletin of the American Meteorological Society* **90**(2), 311–23

United Nations Framework Convention on Climate Change (UNFCCC), 2008: UNFCCC: rising industrialized countries' emissions underscore urgent need for political action on climate change at Poznan meeting, Press Release, November 17

2009: Changes in GHG emissions from 1990 to 2004 for Annex I Parties, July 21, http://unfccc.int/files/essential_background/background_publications_htmlpdf/application/pdf/ghg_table_06.pdf

United Nations Statistics Division, 2008: National Accounts, UN Statistics Division

United States Department of Energy, 2002: Response options to limit rapid or severe climate change: assessment of research needs, National Climate Change Technology Initiative

United States Energy Information Administration, 2006: *International Energy Outlook, 2006*, Energy Information Administration, Washington, DC

United States Environmental Protection Agency (EPA), 2009: Economics of climate change, US EPA National Center for Environmental Economics

Victor, D.G., M.G. Morgan, J. Apt, J. Steinbruner, and K. Ricke, 2009: The geoengineering option: a last resort against global warming?, *Foreign Affairs* **88**(2), 64–76

Weitzman, M.L., 2007: A review of The Stern Review on the Economics of Climate Change, *Journal of Economic Literature* **45**(3), 703–24

2009: On modeling and interpreting the economics of catastrophic climate change, *Review of Economics and Statistics* **91**(1), 1–19

Wigley, T.M.L., 2006: A combined mitigation/geoengineering approach to climate stabilization, *Science* **314**(5798), 452–4

Wingenter, O.W., S.M. Elliot, and D.R. Blake, 2007: New directions: enhancing the natural sulfur cycle to slow global warming, *Atmospheric Environment* **41**(34) 7373–5

Zickfeld, K., B. Knopf, V. Petoukhov, and H.J. Schellnhuber, 2005: Is the Indian summer monsoon stable against global change?, *Geophysical Research Letters* **32**, L15707

1.1 Climate Engineering

Alternative Perspective

ROGER A. PIELKE, JR.

Introduction

This Perspective paper critiques the cost-benefit analysis (CBA) of climate engineering (CE) in chapter 1 by J. Eric Bickel and Lee Lane (Bickel and Lane 2009, hereafter, BL09) in two parts. First, it argues that the analysis of solar radiation management (SRM) is, at best, arbitrary and, more critically, not grounded in a realistic set of assumptions about how the global earth system actually works. The result is an analysis that is precise but not accurate. Second, it summarizes an analysis of the potential role for air capture (AC) technologies to play in the de-carbonization of the global economy, finding the costs of AC to be directly comparable with major global assessments of the costs of conventional mitigation policies. The Perspective paper concludes, as does BL09, that there is justification for continued research into technologies of SRM, but that this judgment does not follow from a CBA. It further concludes that technologies of AC are deserving of a much greater role in mitigation policies than they have had in the past.

BL09 focuses on "climate engineering"[1] in the context of the Copenhagen Consensus exercise for climate change, where authors were tasked by the Copenhagen Consensus Center with addressing the question:

> If the global community wants to spend up to, say, $250 billion per year over the next 10 years to diminish the adverse effects of climate changes, and to do most good for the world, which solutions would yield the greatest net benefits?

More precisely, BL09 focus primarily on two technologies of CE: stratospheric aerosol injection and marine cloud whitening (which together they call SRM),[2] both of which serve to alter the radiation balance of the global earth system via changes in albedo. BL09 apply a CBA methodology to evaluate the potential value of implementation of these technologies under the assumption that "A finding that net benefits may be large suggests that we should devote some current resources to researching and developing this capacity."

This Perspective paper proceeds in two parts. The first part offers a critique of BL09's CBA methodology, arguing that the analysis is, at best, arbitrary and, more critically, not grounded in a realistic set of assumptions about how the global earth system actually works. I argue that the present understandings of the potential effects of CE are not sufficiently well developed to allow for any meaningful CBAs. Nonetheless, I agree with BL09 when they conclude that there is value in further research on CE technologies. My judgment, as apparently was the case as well for the conclusions of BL09, is not based on numbers that result from precise-looking CBAs, but rather, on the fact that our understandings are so poor. I further argue that developing informed understandings will require adopting a more scientifically realistic perspective on the role of CE in the global earth system than is reflected in the simplifications presented in BL09. I conclude that the quantitative CBA of BL09 is guilty of precision without accuracy.

The second part of the Perspective paper summarizes an analysis of the potential role for AC technologies to play in the de-carbonization of the global economy. BL09 consider AC only briefly, leaving a more detailed analysis to this paper. I

---

[1] BL09 define "climate engineering" (CE) as "the intentional modification of Earth's environment to promote habitability," and is largely synonymous with term "geoengineering" (GE).
[2] BL09 also include a brief discussion of the direct "air capture" (AC) of CO_2 from the atmosphere.

show that the costs of AC are comparable to the costs of conventional mitigation, as presented by the Intergovernmental Panel on Climate Change (IPCC) in its 2007 assessment report, as well as the widely cited *The Stern Review* report by the government of the UK (Stern 2007). Based on this conclusion I argue that AC deserves to receive a similar close scrutiny as other mitigation policies.

The Perspective paper concludes by considering more general criteria for evaluating technological fixes such as technologies of CE. I suggest that stratospheric aerosol injection and marine cloud whitening comprehensively fail these broader criteria, whereas AC does not.

A Critique of the Cost-Benefit Methodology of BL09

BL09 are to be applauded for sticking their necks out on a very complex and difficult subject. Such intellectual leadership is often followed by critical commentary, and this case is no different. A first thing to note of BL09 is that their policy recommendations do not follow from the cost-benefit analysis. Their quantitative analysis results in the following dramatic conclusions:

> The direct BCR for stratospheric aerosol injection is on the order of 25 to 1, while the BCR for marine cloud whitening is around 5,000 to 1.

One would think that with such overwhelmingly positive BCRs the authors would immediately recommend a strategy of CE as a core policy response to climate change. Instead, the authors recommend only investing in further research: "an initial investment of perhaps 0.3% ($750 million) of the global total proposed by the Copenhagen Consensus guidelines might be an appropriate average yearly expenditure for the first decade."

The authors' reluctance to recommend anything more than an initial investment in R&D reflects an appropriate degree of skepticism in their analysis, which they clearly state is preliminary and tentative. The authors are quite explicit about the limitations to their analysis:

Any assessment of SRM and AC will be limited by the current state of knowledge, the rudimentary nature of the concepts, and the lack of prior R&D efforts. As noted in . . . , this analysis relies on numbers found in the existing literature and existing climate change models. These inputs to our analysis are admittedly speculative; many questions surround their validity, and many gaps exist in them. This chapter has also stressed the potential importance of transaction costs and "political market failures." Finally, many important scientific and engineering uncertainties remain. Some of these pertain to climate change itself, its pace, and its consequences. Still others are more directly relevant to SRM. How will SRM impact regional precipitation patterns and ozone levels? To what extent can SRM be scaled to the levels considered here? What is the best method for aerosol injection? Are there other side effects that could invalidate the use of SRM?

The concerns expressed by the authors do raise a question of whether CBA is an appropriate tool to use on a subject as complex and uncertain as CE. More specifically, is it possible that the presentation of very precise-looking BCRs may do more to mislead than provide insight on the practical merits of CE?

Below I argue that the technologies of SRM and marine cloud whitening are not sufficiently developed to allow for any sort of meaningful CBA. I go further and argue that the framework used in BL09 represents a misleading simplification of how the earth system actually works, and would be unable in any case to lead to a practically meaningful assessment of the costs or benefits of even well-developed technologies of CE. Nonetheless, I fully agree with the conclusions of BL09 that CE should be the subject of continued research, perhaps proving the point that agreement on potential costs and benefits is irrelevant to deciding to lend support for additional research on the subject.

Major Issue 1: The Inability to Accurately Anticipate Costs or Benefits

It is a simple logical observation to state that to be able to conduct a meaningful CBA requires some degree of accuracy in estimates of both costs and benefits of alternative courses of action. In the cases of stratospheric aerosol injection and marine cloud whitening there are considerable uncertainties in direct costs of deployment, not least because there is "no fully worked out concept for implementing SRM." As the authors note with respect to indirect costs (i.e. impacts), there are areas of both uncertainty and fundamental ignorance where even uncertainties are not well understood.[3]

But let us assume that direct costs of the technologies (i.e. implementation) are known with some degree of accuracy, such that they pose no obstacle to conducting a meaningful CBA. It is in the areas of fundamental ignorance in estimates of indirect costs and potential benefits that are fatal to efforts to create a meaningful CBA. When a quantitative analysis of any type is operating in areas of ignorance, simplifying assumptions must be made in such a way so as to allow the calculations to occur. Such assumptions can be made in any of a number of potentially plausible ways leading to diametrically opposed conclusions. And when the outcome of an analysis rests entirely on the choice of assumptions that cannot be discriminated from one another empirically, the exercise can do more to obscure than reveal.

Consider Goes *et al.* (2009) which, as in BL09, uses a modified version of the DICE integrated assessment model (IAM) as the basis for calculating the potential indirect costs and benefits of SRM. Goes *et al.* (2009: 14) conclude the following:

> aerosol geoengineering hinges on counterbalancing the forcing effects of greenhouse gas emissions (which decay over centuries) with the forcing effects of aerosol emissions (which decay within years). Aerosol geoengineering can hence lead to abrupt climate change if the aerosol forcing is not sustained. The possibility of an intermittent aerosol geoengineering forcing as well as negative impacts of the aerosol forcing itself may cause economic damages that far exceed the benefits. Aerosol geoengineering may hence pose more than just

"minimal climate risks," contrary to the claim of Wigley (2006). Second, substituting aerosol geoengineering for CO_2 abatement fails an economic cost-benefit test in our model for arguably reasonable assumptions.

Thus, using the same (or a very similar) IAM and simply varying assumptions about "deep uncertainties" leads to results that are completely contradictory with those presented in BL09. This outcome is not because BL09 is wrong and Goes *et al.* (2009) is right, or vice versa. This outcome results because there is presently no way to discern which set of assumptions is more appropriate to use in the analysis, hence the presence of "deep uncertainty" which I have here called "ignorance."

The conclusion that should be reached from the comparison of the two studies is that while it is certainly possible that techniques of SRM can lead to very large benefits in relation to costs, it is also possible that SRM could lead to very large costs with respect to benefits. There is simply no way at this point to empirically adjudicate between these starkly different conclusions. It is this fundamental ignorance that leads to the conclusion that "more research is needed."

Underscoring the very large uncertainties present on CE, Goes *et al.* (2009: 14) cite a 1992 NRC report, finding its conclusions to still be current:

> More than a decade ago, a United States National Academies of Science committee assessing geoengineering strategies concluded that "Engineering countermeasures need to be evaluated but should not be implemented without broad understanding of the direct effects and the potential side effects, the ethical issues, and the risks" (COSEPUP[NRC] 1992). Today, we are still lacking this broad understanding.

The conclusions presented by BL09 finding BCRs of 25 to 1 and 5,000 to 1 should thus be taken with a very large dose of salt, as they reflect choices made in the analysis that, had they been made differently but also plausibly, could have resulted in very different (even opposite) conclusions. Hence, in this case the CBA leads to precision without accuracy, and risks doing more to obscure uncertainties than to clarify them.

---

[3] I do not here address the issue of political transaction costs, which are raised in BL09. I do agree with BL09 that such costs are "speculative" at this point, adding another layer of ignorance to the issue. They write: "No one can yet know how the process will distort the various options."

As a consequence, there are no policy recommendations in BL09 that result directly from the CBA. The recommendation to fund research is a matter of qualitative judgment, and the size of investment into SRM proposed by BL09 of $750 million over ten years is arbitrary. I agree with BL09 that some investment in research on CE makes sense, however, I disagree that a CBA tells us anything meaningful about how much should be invested in research or what the potential payoffs might be. Because CE research has considerable value to advancing fundamental understandings of the global Earth system, there are other justifications for its support beyond the potential development of CE technologies.

Major Issue 2: Reliance on a Demonstrably Incorrect Conceptual Model of How Climate Engineering Influences the Global Earth System

Beyond the ability to accurately assess the costs and benefits of CE, there is a more fundamental issue with the approach taken by BL09, and that is the reliance on a conceptual model of the global Earth system that is scientifically flawed. The broader complexities are discussed by Goes *et al.* (2009: 11):

> The analysis, so far, assumes that geoengineering causes environmental damages only through the effects on global mean temperatures (i.e., the value of θ was set to zero). As discussed above, the aerosol geoenginering forcing is projected to change Earth system properties such as precipitation – and surface temperature – patterns, El Niño, and polar ozone concentrations, to name [several] (Lunt *et al.* 2008; Robock 2008). A review of the current literature on the impacts of stratospheric aerosol on natural and human systems suggests that aerosol injections into the atmosphere might cause potentially sizable damages (Lunt *et al.* 2008; Robock 2008; Robock *et al.* 2008; Trenberth and Dai 2007).[4]

Specifically, BL09 approach the evaluation of costs and benefits of SRM through the framework of "radiative forcing."[5] The IPCC (2007a) notes that the concept is very useful but that "it provides a limited measure of climate change as it

does not attempt to represent the overall climate response." The IPCC (2007b) also cautions against simply summing various radiative forcing terms.[6] NRC (2005: 158) offered an even more explicit warning:

> For most policy applications, the relationship between radiative forcing and temperature is assumed to be linear, suggesting that radiative forcing from individual positive and negative forcing agents could be summed to determine a net forcing. This assumption is generally reasonable for homogenously distributed greenhouse gases, but it does not hold for all forcings. Thus, the assumed linearity of radiative forcing has been simultaneously useful and misleading for the policy community. It is important to determine the degree to which global mean TOA [top of the atmosphere] forcings are additive and whether one can expect, for example, canceling effects on climate change from changes in greenhouse gases on the one hand and changes in reflective aerosols on the other.

BL09 modifies the DICE model by using a simple additive term to represent the climatic effect of SRM, which may or may not be scientifically justifiable. Not only are there uncertainties and ignorance about the costs and benefits of CE, but there are fundamental areas of uncertainty and ignorance in how to even conceptualize those effects.

NRC (2005) presented a more complex view of radiative forcing than found in either the IPCC (or BL09) and its relationship to non-radiative forcings, indirect radiative forcings and their feedbacks,

---

[4] I note that many of the citations in this passage from Goes *et al.* (2009) are also cited in BL09.

[5] The IPCC defines "radiative forcing" as "'the change in net (down minus up) irradiance (solar plus longwave, in W m^{-2}) at the tropopause after allowing for stratospheric temperatures to readjust to radiative equilibrium, but with surface and tropospheric temperatures and state held fixed at the unperturbed values." See IPCC (2007a: 133).

[6] See the caption to figure SPM.2 in the 2007 *Summary for Policy Makers of Working Group I*, where it states, "The net anthropogenic radiative forcing and its range are also shown [in the figure]. These require summing asymmetric uncertainty estimates from the component terms, and cannot be obtained by simple addition."

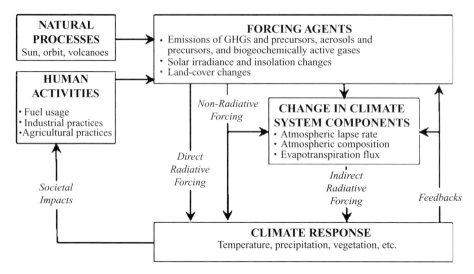

Figure 1.1.1 *Radiative forcing in context (NRC 2005), see www.nap.edu/openbook.php?record_id=11175&page=13*

as shown in figure 1.1.1. The relationship of a forcing agent, such as the injection of stratospheric aerosols or marine cloud whitening, and eventual climate impacts at global as well as regional scales manifests itself in a degree of interrelationships and feedbacks that cannot be resolved simply by adding or subtracting direct radiative forcings. Perhaps future research will show that all other relationships beyond the additive effect on direct radiative forcing can be ignored, however, current research suggests that this is not the case (see the wide range of sources cited in NRC 2005).

To summarize, the ability to conduct a CBA of CE is hindered by both uncertainties and fundamental ignorance of both costs and benefits. It is quite possible to vary assumptions in plausible ways and to arrive at diametrically opposed results. Further, the analysis in BL09 simplifies physical relationships in a manner suitable for inclusion in a simple integrated assessment model, but in the process fails to reflect that the global Earth system may

actually respond to forcing agents and changes in climate system components through direct and indirect radiative forcing, non-radiative forcings, and feedbacks among these. Consequently, I conclude that a quantitative CBA of the CE technologies of SRM is premature at best.

The Costs of Air Capture

As part of my response to BL09 I was asked to provide an overview of the costs and benefits of "air capture" technologies. "Air capture" (AC) refers to a range of methods and technologies for the direct removal of CO_2 from the ambient air, ranging from photosynthesis to chemical extraction, and has received increasing attention in recent years.[7] After removal, in order to draw down atmospheric concentrations of CO_2 the gas needs to be either sequestered or otherwise used.

AC is particularly amenable to a CBA because it directly addresses a part of the climate change issue that has been most intensively studied, the increasing accumulation of CO_2 in the atmosphere. There have been various studies of the economic benefits of limiting the accumulation of greenhouse gases (GHGs), which will not be recited here. Thus, in order to compare AC as a possible contributor to

---

[7] This section of the Perspective paper draws on Pielke (2009), which provides a more comprehensive review of AC and its economics. The focus in Pielke (2009) is on techniques of chemical extraction, however the economic analysis is a function of cost rather than specific technology and thus could be equally applied to biological or geological means of AC.

stabilizing concentrations of GHGs, it need only be compared in terms of costs to other approaches to stabilizing concentrations.[8] The fundamental question to be asked is: How does the cost of AC compare to other approaches to stabilizing concentrations of CO_2 in the atmosphere?

Estimates vary for the cost of capturing CO_2 directly from the atmosphere. Keith *et al.* (2006) suggest that using existing technology the costs could be as much as $500 per ton of carbon, and perhaps eventually under $200/ton. In 2007 Keith suggested that the cost of AC could drop below $360 per ton (Graham-Rowe 2007). Columbia University's Klaus Lackner has suggested that the costs today are less than $360 per ton of carbon, and may eventually fall beneath approximately $100 per ton. IPCC (2007a) discusses AC only in passing:

> Studies claim costs less than 75 US$/tCO$_2$ [$275/tC] and energy requirements of a minimum of 30% using a recovery cycle with Ca(OH)$_2$ as a sorbent. However, no experimental data on the complete process are yet available to demonstrate the concept, its energy use and engineering costs.[9]

In the simple exercises below I use three values for the costs of AC: (a) $500 per ton of carbon, (b) $360 per ton, and (c) $100 per ton, as described in Pielke (2009). The IPCC (2007a) estimate falls near the middle of this range.

The Costs of Stabilization via Air Capture

At 2.13 GtC equivalent to1 ppm carbon, this means that the current (idealized) costs of AC are about $1 trillion per reduced ppm of atmospheric CO_2 at a cost of AC equal to $500/tC. $1 trillion represented about 2.5% of global GDP in 2007. At $500/tC complete mitigation of net 2008 human emissions would cost about $4 trillion, or about 10% of global GDP. At $100/ton the 2007 cost would be about 2.0% of global GDP.

If the goal of AC is to limit cumulative CO_2 emissions during the remainder of the twenty-first century to less than 240 GtC (as suggested by the IPCC as being consistent with a 450 ppm target), then there are many different temporal paths over which AC might be implemented. That is, it is the cumulative emissions over the twenty-first century

Table 1.1.1a Cost of AC as a percentage of global GDP, assuming 2.9% global GDP growth to 2100 (after IPCC 2000)

	$500/GtC (%)	$360/GtC (%)	$100/GtC (%)
450 ppm Cost to 2050	2.7	1.9	0.5
550 ppm Cost to 2050	0.0	0.0	0.0
450 ppm Cost to 2100	2.1	1.5	0.4
550 ppm Cost to 2100	1.5	1.1	0.3

that matter, not the specific emissions trajectory. The further into the future one assumes deployment the lower the present value (PV) will be as a function of the discount rate chosen. The analysis below does not discount.

The analysis errs on the side of understating costs as there are no assumptions made about the economies of scale associated with a widespread deployment and likely reductions in costs of the technology (McKinsey & Co. 2008). The calculation of cost involves simply multiplying the expected capture cost per ton of carbon by the integral of the difference between projected emissions and emissions under AC. The analysis here assumes that cumulative, business-as-usual (BAU) (i.e. no AC), net CO_2 emissions will be approximately 880 GtC of carbon from 2008 to 2100, which is somewhat higher than the mid-range projection of the IPCC (see Pielke 2009 for details). Higher or lower values, which are certainly plausible, will result in corresponding changes in the cost estimates of AC.

Under these assumptions, tables 1.1.1a and 1.1.1b show the cumulative costs of AC over the periods 2008–50 and 2008–2100 for different

---

[8] Of course, all studies of the benefits of mitigation policies could be wrong, however that will affect judgments of mitigation policies in general, and not an analysis of AC specifically.

[9] *Working Group III*, Chapter 4: 286. The IPCC provides no reference or justification for its cost estimate. The IPCC's dismissal of AC in this manner is surprising, because much of the IPCC's analysis of the prospects for and costs of GHG mitigation depends upon policies and technologies whose implementation has not been proven successful in practice.

Table 1.1.1b Cost of AC as a percentage of global GDP, assuming 2.5% global GDP growth to 2100 after Stern (2007)

	$500/GtC	$360/GtC	$100/GtC
450 ppm Cost to 2050	3.0%	2.2%	0.6%
550 ppm Cost to 2050	0.0%	0.0%	0.0%
450 ppm Cost to 2100	2.7%	2.0%	0.5%
550 ppm Cost to 2100	2.0%	1.4%	0.4%

stabilization levels and different costs per ton of carbon. Table 1.1.1a assumes an annual global GDP growth rate of 2.9% following IPCC (2000), and table 1.1.1b assumes, after Stern (2007), an annual global GDP growth rate of 2.5%. Stern (2007) uses a global GDP of $35 trillion in 2005. No effort has been made here to account for the time value of money or different approaches to calculating economic growth across countries, which have been discussed elsewhere in great depth in the context of climate change, and all dollars are expressed in constant-year terms.

All of the values presented in tables 1.1.1a and 1.1.1b for the costs of stabilization at 450 ppm via AC fall within the range of those presented in Stern (2007), which suggested that stabilization at 450 ppm CO_2 would cost about 1% of global GDP to 2100 (with a range of plus/minus 3%).[10] Stern (2007: 249) explained how one might think about this value:

> if mitigation costs 1% of world GDP by 2100, relative to the hypothetical "no climate change" baseline, this is equivalent to the growth rate of annual GDP over the period dropping from 2.5% to 2.49%. GDP in 2100 would still be approximately 940% higher than today, as opposed to 950% higher if there were no climate-change to tackle.

If AC technology could be implemented at $100/ton, then the cost to stabilize emissions over

the twenty-first century would be less than the Stern median estimate. For stabilization at 550 ppm or about twice preindustrial, AC costs nothing prior to 2050.

Similarly, the ranges of costs for AC are comparable to those presented in IPCC (2007a) which estimated the costs of mitigation for 2050 at a level of 535–590 ppm CO_2 equivalent (comparable to Stern's 450 ppm CO_2) to fall within the IPCC range of $-1%$ to 5.5% of global GDP in 2050. The IPCC median value of 1.3% is less than the cost of AC at $360 cost per ton of carbon, but almost three times the cost at $100 per ton.

Making global cost estimates for any complex set of interrelated systems far into the future is a dubious enterprise. However, the analysis here shows that using very similar assumptions to the IPCC (2007c, 2007d) and Stern (2007), AC compares favorably with the cost estimates for mitigation provided in those reports. The main reason for this perhaps surprising result, given that AC has a relatively high cost per ton of carbon, is the long period for which no costs are incurred until the stabilization target is reached. Further, a factor not considered here is that the economy would likely grow at a higher rate than with early, aggressive mitigation, meaning that the costs of AC would be a smaller fraction of future GDP than comparable costs per ton of carbon requiring large costs early in the century. The cost of AC under the assumptions examined here is also less that the projected costs of unmitigated climate change over the twenty-first century, which Stern (2007) estimated to be from 5–20% of GDP annually and IPCC (2007e) estimate to be 5% of global GDP by 2050.

There are several additional factors, beyond those already discussed, which serve to overstate the cost estimates of AC found in tables 1.1.1a and 1.1.1b. Carbon dioxide emissions from power plants, representing perhaps as much as half total emissions over the twenty-first century could be captured at the source for what many believe is a cost considerably less than direct AC.[11] The technical, environmental, and societal aspects of carbon sequestration are identical for capture of CO_2 from both power plants and ambient air. To the extent that improvements in efficiency and overall emissions intensity occur, these developments would further

[10] Stern (2007) equated a 450 ppm CO_2 level with a 550 ppm CO_2 equivalent concentration, which includes other gases.
[11] For a review of the costs of carbon capture and storage (CCS), see IPCC (2005).

reduce total emissions and thus the need to rely on AC.[12] The assumptions here assume simplistically a fixed average cost of AC over time, whereas experience with technological innovation suggests declining marginal costs over time (e.g. McKinsey & Co. 2008).

Consideration of these factors could reduce the values presented in tables 1.1.1a and 1.1.1b by a significant amount, perhaps by as much as half. Uncertainties in rates of increasing emissions, economic growth, and concentrations mean that the values presented here could be more or less than under different assumptions. Because the analysis relies on the mid-range values of the IPCC for these various factors, it is unlikely that a more comprehensive treatment of uncertainties would lead to qualitatively different conclusions if one begins with assumptions underpinning and implications following from the IPCC.

To summarize, a simple approach to costing AC as a strategy of achieving CO_2 stabilization targets using 2007 technology results about the same costs as the costs estimates for stabilization at 450 ppm or 550 ppm CO_2 presented by IPCC (2007c) and Stern (2007). If the costs of AC decrease to $100 per ton of carbon, then over the twenty-first century AC would in fact cost much less than the costs estimates for stabilization presented by IPCC (2007c) and Stern (2007). This surprising result suggests, at a minimum, that AC should receive the same detailed analysis as other approaches to mitigation. To date, it has not.

Conclusion: Climate Engineering as a Technological Fix

BL09 raise important questions about how to evaluate the role of a technological fix in efforts to stabilize concentrations of GHGs (primarily CO_2) in the atmosphere. In this response I have argued that CBAs of SRM are limited in the insights they can bring to bear on highly complex systems that are incompletely understood. Writing in *Nature*, Sarewitz and Nelson (2008) offer three broader criteria by which to distinguish "problems amenable to technological fixes from those that are not." Here in conclusion I briefly apply these criteria

to the technology of CE, concluding that indirect approaches to CE such as SRM fall well short of all three of the criteria that Sarewitz–Nelson present as guidelines for when to employ a technological fix. By contrast, the technology of AC offers much greater promise.

Sarewitz–Nelson Criterion 1: The Technology must largely Embody the Cause–Effect Relationship Connecting Problem to Solution

As argued in the first part of this Perspective paper, SRM does not directly address the cause–effect relationship between emissions and increasing atmospheric concentrations of CO_2 (and other GHGs). It addresses the effects, and only in an indirect, poorly understood fashion. It is thus appropriate to consider SRM as a form of adaptation to human-caused climate change. In this instance, rather than building a levée (i.e. changing localized topography) to physically ward off rising seas, the goal of SRM is to alter the Earth system in other ways to compensate for the effects of changes in climate. Unlike levées, where cause and effect are unambiguous, SRM has unknown consequences. In contrast, AC prevents a human perturbation through the release of CO_2 into the atmosphere, and thus directly addresses the accumulation of CO_2 in the atmosphere. Thus, AC is a form of mitigation.

Sarewitz–Nelson Criterion 2: The Effects of the Technological Fix must be Assessable Using Relatively Unambiguous or Uncontroversial Criteria

As argued in the first part of this Perspective paper, the effects of CE on climate impacts of concern – including phenomena such as extreme events, global precipitation patterns, sea ice extent, biodiversity loss, food supply, and so on – would be difficult if not impossible to assess on timescales of relevance to decision makers. Research on weather

---

[12] In addition, if the allowable "carbon allocation" is understated (overstated) by the simple methodology here, then there would be less (more) need for AC and corresponding less (more) costs.

modification provides a cautionary set of lessons in this regard (cf. Travis 2009). In contrast, the technology of AC does not require developing a better understanding of the global Earth system – simply knowing that the accumulation of CO_2 poses risks worth responding to is a sufficient basis for considering deployment. In other words, if the accumulation of CO_2 in the atmosphere is judged to be a problem, then its removal logically follows as a solution.

Sarewitz–Nelson Criterion 3: R&D is most likely to Contribute Decisively to Solving a Social Problem When it Focuses on Improving a Standardized Technical Core that already Exists

CE via SRM on a planetary scale has never been attempted, and to do so would in effect be a decision to implement the technology, as we have only one Earth. Thus, its effects cannot be known, only speculated upon and researched with sophisticated scientific tools. Even so, it could easily have unpredicted or undesirable effects. By contrast AC builds upon existing (and expensive) technologies that can be deployed, evaluated, refined and improved upon with no risk to the climate system.

In short, SRM fails comprehensively with respect to the three criteria for technological fixes offered by Sarewitz and Nelson, suggesting that it offers little prospect to serve as a successful contribution to efforts to deal with increasing concentrations of CO_2. As they write, "one of the key elements of a successful technological fix is that it helps to solve the problem while allowing people to maintain the diversity of values and interests that impede other paths to effective action." Because it fails with respect to the three criteria, SRM is likely to make the politics of climate change even more complex and contested, resulting in little prospect for success. But even if SRM offers few prospects for successfully addressing the climate issue, as concluded in BL09, continued research on SRM nonetheless make sense both to keep options open and also to contribute to a further understanding of the human role in the climate system. In contrast, for reasons of a preliminary CBA as well as with respect to broader criteria of a technological fix, technologies of AC are deserving of a much greater role in mitigation policies than they have had in the past.

Bibliography

Bickel, J.E. and L. Lane, 2009: An analysis of climate change as a response to global warming, Copenhagen Consensus Center, August 7

COSEPUP [NRC] 1992. Policy implications of greenhouse warming: mitigation, adaptation, and the science base. National Academy of Science, Committee on Science Engineering and Public Policy (COSEPUP), National Academy Press. Washington, DC

Goes, M., K. Keller, and N. Tuana, 2009: The economics (or lack thereof) of aerosol geoengineering, submitted to *Climatic Change*

Graham-Rowe, D., 2007: Scientists attempt to roll back emissions, *Guardian Unlimited*, July 30, www.guardian.co.uk/technology/2007/jul/30/news.greentech

IPCC, 2000: *Special Report on Emissions Scenarios*, Intergovernmental Panel on Climate Change, Cambridge University Press, Cambridge

2005: *Special Report on Carbon Dioxide Capture and Storage*, Intergovernmental Panel on Climate Change, Cambridge University Press, Cambridge

2007a: *Working Group I – The Physical Science Basis of Climate Change*, Intergovernmental Panel on Climate Change, Cambridge University Press, Cambridge, ipcc-wg1.ucar.edu/wg1/wg1-report.html

2007b: *Summary for Policy Makers of Working Group I – The Physical Science Basis of Climate Change*, Intergovernmental Panel on Climate Change, Cambridge University Press, Cambridge, ipcc-wg1.ucar.edu/wg1/wg1-report.html

2007c: *Working Group III – Mitigation*, *Intergovernmental Panel on Climate Change*, Cambridge University Press, Cambridge, www.mnp.nl/ipcc/pages_media/AR4-chapters.html

2007d: *IPCC Fourth Assessment Report, Working Group III – Chapter 3 Issues Related to Mitigation in the Long-term Context*, *Intergovernmental Panel on Climate Change*, Cambridge University Press, Cambridge, www.mnp.nl/ipcc/pages_media/FAR4docs/chapters/Ch3_Longterm.pdf

2007e: *Summary for Policy Makers, Report of Working Group II – Impacts, Adaptation and*

Vulnerability, Intergovernmental Panel on Climate Change, Cambridge University Press, Cambridge, www.ipcc.ch/SPM040507.pdf

Keith, D.W., M. Ha-Duong, and J.K. Stolaroff, 2006: Climate strategy with CO2 capture from the air, *Climatic Change* **74**, 17–45

Lunt, D.J., A. Ridgwell, P.J. Valdes, and A. Seale, 2008: "Sunshade world": a fully coupled GCM evaluation of the climatic impacts of geoengineering, *Geophysical Research Letters* **35**(L12710),

McKinsey & Co., 2008: The carbon productivity challenge: curbing climate change and sustaining economic growth, June, www.mckinsey.com/mgi/reports/pdfs/Carbon_Productivity/MGI_carbon_productivity_full_report.pdf

National Academy of Sciences (**NAS**), 1992: *Policy Implications of Greenhouse Warming: Mitigation, Adaptation, and the Science Base*, National Academy Press, Washington, DC

National Research Council (**NRC**), 2005: *Radiative Forcing of Climate Change: Expanding the Concept and Addressing Uncertainties*, Committee on Radiative Forcing Effects on Climate, National Academies Press, Washington, DC

Pielke, R. A., Jr., 2009: "An idealized assessment of the economics of air capture of carbon dioxide in mitigation policy," *Environmental Science & Policy* **12**, 216–25

Robock, A., 2008: 20 reasons why geoengineering may be a bad idea, *Bulletin of the Atomic Scientists* **64**, 14–18

Robock, A., L. Oman, and G.L. Stenchikov, 2008: Regional climate responses to geoengineering with tropical and Arctic SO2 injections, *Journal of Geophysical Research-Atmospheres* **113**(D16101),

Sarewitz, D. and R. Nelson, 2008: Three rules for technological fixes, *Nature* **456**: 871–2

Stern, N., 2007: *The Economics of Climate Change: The Stern Review*, Cambridge University Press, Cambridge, www.hmtreasury.gov.uk./media/F/0/Chapter_9_Identifying_the_Costs_of_Mitigation.pdf; www.hm-treasury.gov.uk./media/B/7/Chapter_10_Macroeconomic_Models_of_Costs.pdf

Travis, W., 2009: Geo-engineering the climate: an emerging technology assessment, Discussion Paper prepared for the National Research Council, Committee on America's Climate Choices, americasclimatechoices.org/Geoengineering_Input/attachments/Travis_geoengineering.pdf

Trenberth, K.E. and A. Dai, 2007: Effects of Mount Pinatubo volcanic eruption on the hydrological cycle as an analog of geoengineering, *Geophysical Research Letters* **34**(L15702),

Wigley, T.M.L., 2006: A combined mitigation/geoengineering approach to climate stabilization, *Science* **314**, 452–4

1.2 Climate Engineering

Alternative Perspective

ANNE E. SMITH

Introduction

Chapter 1 by J. Eric Bickel and Lee Lane provides a thorough summary of the range of options that fall under the term "geoengineering" (GE), and some detail on the specific technological concepts for options that fall under the classification of solar radiation management (SRM). It provides less information on technological proposals for accomplishing air capture (AC), but does at least explain how SRM and AC would differ in the way they might help address risks of climate change. The chapter is clear that it focuses on providing a first-cut assessment of the potential costs and benefits of these general categories of climate engineering (CE), relying on a limited body of existing literature. It does, nevertheless, develop its own new estimates of the benefits of SRM and AC using the DICE model. Overall, this chapter is an excellent review and analysis for anyone who wishes to gain a high-level understanding of how GE methods could become an important element of long-term management of global climate change. However, what becomes apparent throughout the chapter is the large range of risks and unknowns associated with any of these methods, but most particularly with SRM. Even though the chapter's scope was necessarily limited to providing a few point estimates of benefits and costs, and the related benefit-cost ratios (BCRs), the issue of uncertainties could have been addressed more satisfactorily without resorting to a full-blown "probabilistic analysis." In this Perspective paper, I attempt to show how one can extend the initial efforts of Bickel and Lane to more directly assess the implications of these enormous uncertainties on their conclusions.

Bickel and Lane have performed a deterministic benefit-cost analysis (BCA) for SRM and AC. A range of cost estimates is noted in the discussion, some of which are treated with an abbreviated form of sensitivity analysis. Uncertainty in the benefits is, of course, also enormous, and not explored through sensitivity analysis. Rather, the deterministic analysis produces such huge BCRs, and huge net benefits as well, that the authors conclude that the very large set of unknowns and uncertainties in these estimates merit investment in a concerted program of research and development (R&D). Their proposal for that R&D budget is then developed from a bottom-up process of thinking through the steps and elements of such a program. Their recommendation is $0.75 billion dollars per year,[1] although they describe an R&D program that would start much smaller and expand rapidly as the time of full deployment approached.[2] It is useful to restate their proposed R&D budget as a present value (PV) so that it can be more readily compared to their net benefits estimates, and to the estimates of value of information (VOI) that I will be reporting below. I estimate their recommended R&D spending to have a PV of about $5–10 billion.[3] This contrasts to their deterministic net benefits estimates that range from $4 trillion to $25 trillion.[4]

---

[1] Bickel and Lane (2009: 10).

[2] "As R&D progresses, and assuming that results were favorable, spending would increase from tens of millions of dollars in early years to the low billions of dollars" (Bickel and Lane 2009: 10).

[3] I based this on an assumption that the spending that they describe will extend through 2025, given that they have suggested that 2025 would be the earliest reasonable date when the GE technologies would be ready for full deployment if R&D were to start now (Bickel and Lane 2009: 21). I applied a discount rate of 5.5%, as the authors have been using, to several spending patterns from 2010 to 2025 that have an average spending level of about $750 million per year.

[4] These are the range of benefits estimates for SRM 1 through SRM 3, under a variety of start dates, discount rates, and mitigation policies, presented in table 1.7 of Bickel and

Bickel and Lane do not advocate actually initiating SRM or AC until the R&D has proven that the uncertainties not addressed in their deterministic analysis will not eliminate the apparently very large net-benefits case for CE. They even imply that a mature SRM and AC capability might never be deployed, but only that the concepts need to be developed into real options that the world can resort to if climate changes turn out to be far worse than in their own benefits analysis.

While these arguments appear to make a good case for a spending on R&D, the failure of the chapter to address the role of the uncertainties in this decision to commit to creating a real option makes the case incomplete. In fact, the existing professional and public pressure against even a modest research program around CE is itself tied to the exceptional nature of the associated uncertainties. There are uncertainties about whether it will work, what it will cost to do, and what benefits it might produce in terms of climate damages avoided. But the most important uncertainty at play in this debate is concern about unleashing an entirely new set of environmental or economic damages in the course of trying to reduce the costs of global warming. Discussion of these risks tends to be non-specific, and lumped into a single ill-defined category called "unintended consequences." Even though they are ill defined, these risks are clearly central to the debate about whether to fund even R&D for GE, and they need to be addressed head-on in an analysis that is intended to make the case for such R&D.

Incorporating Consideration of Uncertainties with Value of Information Analysis

Most R&D investment decisions are made in the face of a similar array of uncertainties, and a standard method for assessing whether and how much to spend to obtain more information is "value of information" analysis. VOI analysis directly estimates how having better information before taking action can improve the quality of future decisions – in this case, whether to deploy CE options in 2025 or later. If research has a chance of identifying an undesirable cost of CE before full-scale deployment occurs, it can help us avoid using CE if it would reduce rather than increase net benefits. Research can therefore increase the expected benefits of CE by cutting off the downside risks that are presently associated with this option for climate risk management, while maintaining its upside potential. It does so by directly considering the extent to which the future deployment decision could be altered by resolving any of the many uncertainties before making that decision. The value of research is determined by comparing the expected net benefits of the CE decision when taken in the face of present uncertainty to the expected net benefit of the same decision when taken in the face of less uncertainty. If the latter is larger, the information has value. The difference between the two expected net benefits is viewed as the most that one should be willing to pay in order to perform the research that would provide that reduction in uncertainty. If the R&D program to obtain that information would cost less than the assessed VOI, then it is a good investment. If it costs more, then the best course of action is to make the decision under uncertainty, without the additional information-gathering. In the context of the GE assessment, if the VOI for the R&D program outlined by Bickel and Lane exceeds $10 billion, one has a case for accepting their proposal.

Bickel and Lane's initial deterministic estimates of the net benefits of SRM range from about $4 trillion to $25 trillion (PVs), with BCRs all exceeding 20. They argue that despite the significant uncertainties they did not analyze, it is difficult to imagine any situation in which an investment of $10 billion (PV) would be a bad allocation of resources. However, VOI analysis can show otherwise – all depending on initial views (assumptions) about the possibility of significant unintended consequences, and their potential magnitude. For this Perspective paper, I have developed a fairly simple structure for estimating VOI for this particular issue, and I have used it to prepare estimates of VOI based on the net benefits calculations in Bickel and Lane.

Lane (2009: 40) minus the range of respective costs in their tables 1.9 and 1.10. It includes net benefits of AC, but they estimate AC's net benefits at about 0 (i.e. a maximum net benefit ratio of 1).

Because VOI is contingent on current views about the potential size and probabilities of unintended consequences, I do not provide a single estimate of "the" VOI, but instead show how that value varies over the full range of possible viewpoints. These current viewpoints are known in technical terms as the "prior probabilities" regarding the uncertainty that is to be studied. The prior probability is a subjective assumption, and thus it can vary widely among individuals. Thus, some people may see great value in an R&D program that others would see as having no value at all, depending on their respective "priors."

In the following, I simplify the problem to consider only one crucial uncertainty, which is the possibility that SRM can lead to serious "unintended consequences" that would add new environmental costs of their own, and exacerbate climate damages as well due to sudden more rapid climate change when the program of SRM is suddenly stopped. Consideration of this single uncertainty creates some interesting patterns in the potential value of the proposed R&D program that may provide a platform for a much richer discussion of its merits. The proposed R&D program would presumably reduce uncertainties over a variety of other uncertainties such as cost, functionality, climate change damages, etc., and thus VOI from this single program might be larger or smaller than the estimates I will provide here. While I doubt they would significantly alter those patterns, a more comprehensive analysis may be desirable. I hope that the

initial structure that I present in this Perspective paper can serve as a foundation on which that more comprehensive discussion and analysis can be built.

I also wish to note that my example, like the chapter, focuses on SRM. However, the chapter does recommend research on AC in addition to SRM. This recommendation may seem a bit weak because Bickel and Lane conclude that AC may have a BCR less than 1, which is not only low, but much less than they estimate for SRM.[5] However, I would like to emphasize that the AC option not only has gross benefits as high as $5.5 trillion,[6] but it has risk management properties that could easily lend it a large positive VOI even if it does have negative net benefits when calculated deterministically as Bickel and Lane have done. Indeed, there may be some interactions between the VOI for AC and SRM when considered as a set of options rather than as individual options. While I have not extended my analysis to AC in this Perspective paper, I consider that an important next step.

Structuring the SRM Issue as a Decision Made with Uncertainty

Figure 1.2.1 illustrates the essence of the deterministic analysis in Bickel and Lane using a decision tree diagram. The numerical values on which I will base my VOI analysis are for the SRM 3 option and for the "temperature constraints" policy case.[7] The costs of each option are shown as the outcomes of each decision branch. Climate damages and mitigation costs of the option come from table 1.4 of the chapter, while costs of the SRM (on the SRM decision branch only) are from table 1.10. The total cost of each decision is the sum of these three categories, and is $4.9 trillion for the SRM option and $22.8 trillion for the No-SRM option (i.e. "SRM 0" in the chapter). Clearly the least-cost decision is SRM, and the savings of having SRM as an option in this decision tree is $17.9 trillion (i.e. 22.8 − 4.9). This result is fully consistent with the chapter's BCR of 27 for SRM 3, but this analysis is computing values as differences in total costs of two options, rather than as differences in climate

---

[5] Bickel and Lane report that the BCR from their deterministic estimates for AC are only 1 or less (2009: 37).

[6] Bickel and Lane, table 1.6 (2009: 39).

[7] I use their SRM 3 case as the only SRM option in the decision tree because it always has the largest net benefits, and any decision tree would thus always ignore the other SRM options even if they were included. I use the higher cost estimates of SRM associated with stratospheric deployment for these illustrative purposes. A complete analysis would include both forms of SRM as well as AC. I also focus on the case in which every climate management strategy must limit temperature change to 2°C, because it is easiest to use estimates of alternative mitigation costs and climate damages under different outcomes than those actually presented in the chapter without actually re-running the DICE model, which was beyond the scope of what I could do for this Perspective paper.

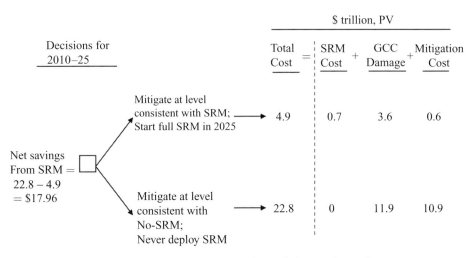

Figure 1.2.1 *Representation of deterministic analysis of chapter 1 as a decision tree*

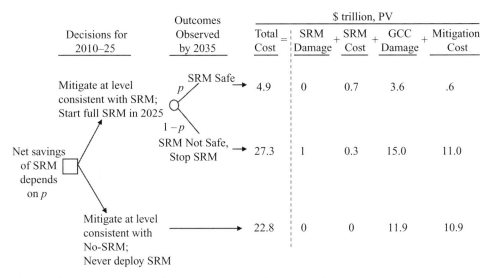

Figure 1.2.2 *Representation of SRM decision as a decision under uncertainty*

damages plus mitigation costs (i.e. $(11.9 + 10.9) - (3.6 + 0.6) = 18.6$) divided by differences in SRM costs (i.e. 0.7).

Figure 1.2.2 carries the analysis one step closer to the actual situation that characterizes the decision of whether to do SRM or not. If No-SRM is undertaken, the same outcomes occur as for that branch in figure 1.2.1. However, the decision branch for SRM now reflects the concern with "unintended consequences" that Bickel and Lane describe but do not account for in their deterministic BCA. Figure 1.2.2 presents the SRM decision as one that is taken with-

out any advance investment in research on this risk. That is, figure 1.2.2 accounts for the key uncertainty in the SRM decision, but does so as if nothing can be learned from R&D prior to making that decision. Instead, if one proceeds down the SRM decision path in 2025, there is some probability, p, that SRM will not cause any unintended consequences at all, and its costs will be just like Bickel and Lane estimated (i.e. the same as for that branch in figure 1.2.1).

However, figure 1.2.2 shows that there also is a probability, $(1 - p)$, that unintended consequences

Table 1.2.1 Expected cost savings of SRM strategy compared to a no-SRM strategy (assuming no further research on SRM risks prior to the SRM deployment decision)

		Damages from SRM if SRM not safe (PV, $ trillion)			
		1	5	10	15
	0	−4.5	−8.5	−13.5	−18.5
Prior probability that SRM is safe, "p"	0.25	1.1	−1.9	−5.6	−9.4
(subjective, varies with the person)	0.5	6.7	4.7	2.2	−0.3
	0.75	12.3	11.3	10.1	8.8
	1	17.9	17.9	17.9	17.9

will start to become apparent. At that time (e.g. 2035 in this example) the SRM will have to be stopped abruptly, costs of climate mitigation will increase to the level in the No-SRM case (or perhaps even higher because of the need to act even more rapidly on those controls than if they had been initiated much earlier), climate damages may rise substantially, due to a more rapid rate of change when accumulated CO_2 suddenly is allowed to have its full warming potential, which had been offset by the SRM for many years, *and* there would be the new costs from the bad consequences of the SRM itself. In return, the cost of the SRM would not be as high, as only its fixed costs and first ten years of operating costs would be incurred. In figure 1.2.2, one can see my approximate estimates of how these various cost elements would change relative to the costs that were formally estimated by Bickel and Lane for the other two branches of the decision tree. These are rough but reasonable estimates for purposes of the illustrative analysis. However, all but the SRM-related damages can be estimated precisely simply by running this additional scenario using the same version of DICE that Bickel and Lane have used.

When this extra branch is added, the decision of whether to undertake SRM is altered, and so too is the value of SRM. The calculations in figure 1.2.2 can be run for any number of combinations of probabilities that SRM will not produce unintended consequences (i.e. will "be safe"), and any degree of damage from any unintended consequences. Table 1.2.1 shows a set of such results, including the case that reduces figure 1.2.2 back to figure 1.2.1, which is where the prior probability that SRM will be safe, p, is 1.0. Thus the deterministic result of Bickel and Lane has the maximal SRM value of $17.9 trillion that appears on the bottom row of table 1.2.1, for $p = 1$. However, as the

prior probability p drops towards zero, the value of SRM drops. It drops more rapidly when the associated damage from the unintended consequences increases (i.e. as one moves to columns to the right in table 1.2.1). The value of SRM is negative for those who have "priors," or subjective views of the probability and the potential impact that make the *expected* cost of SRM higher than the $22.8 trillion cost of the No SRM option. The prevalent opposition to GE that exists today may be traced to people whose individual "priors" place them in the part of table 1.2.1 that has negative values. In those cases, the preferred decision in the decision tree is the No SRM option and, formally, the value of the SRM option is zero. However, the large negative values in table 1.2.1 provide an indication of the degree of concern with expected damages that may be held by some people that SRM will be deployed anyway. They may have these concerns because they think that people who will be in control of the deployment decision may have the more optimistic views that lead to a large positive subjective value for SRM.

For those who believe that there is a better than 25% probability that SRM will not have unintended consequences greater than $1 trillion (PV), the decision to act on SRM has expected benefits exceeding $1.1 trillion. For those who believe that the direct damages of SRM could be about $10 trillion (i.e., could be larger than the damages projected for climate change itself), SRM would still have a large positive value even if one assigns that outcome a 50% probability. Nevertheless, one can also see that people who assign very high probabilities to the possibility of creating "another" problem of equal proportion to climate change would certainly disagree with a planned course of action that could lead to use of SRM.

Table 1.2.1 only presents the potential expected benefits of SRM, given this single large uncertainty.

It may help clarify why there are such strong views both for and against SRM, given the current state of uncertainty. However, it does not reveal how a research program to better inform these decisions can reduce the risks of making a bad decision on SRM, while still preserving an option to use it if it proves to be as benign as assumed in the Bickel and Lane analysis. The VOI is the difference between the expected costs in the case where a research program is conducted before deciding whether to deploy SRM, and the expected costs in the case depicted in figure 1.2.2 where the best decision is made given current information.

Structuring the SRM Issue to Account for Information from an R&D Program in Advance of Any SRM Deployment Decisions

Figure 1.2.3 shows how the decision tree is altered to estimate the expected costs when a research program is conducted before making any deployment decision. This figure has the same basic elements of figure 1.2.2, except that the second stage of the decision process is preceded by new (but probably imperfect) information that would be produced by the R&D program. The decision today is now simply to choose the level of mitigation spending to engage in while the R&D on SRM is being conducted. Consistent with the previous examples, this current decision is simplified to include only whether to start with a level of mitigation that would be taken if one does *not* anticipate later having SRM as part of the management path, or with a mitigation level that would be consistent with SRM being deployed in 2025. Then, *before the decision on whether to deploy SRM is taken*, new information is assumed to result from the R&D program. This information has been simplified to be either that the research finds that SRM will be safe or not safe. (In figure 1.2.3, results of the research are identified by placing the outcome inside quotation marks. This indicates that they are merely conclusions, and not necessarily the actual state of the world.)

The decision on whether to deploy the SRM is taken after the research findings are reported, and thus is a more informed, less risky decision than in the case of figure 1.2.2; that is, that decision

is made *contingent* on the R&D findings. Thus, whatever one's *a priori* assumption about the probability of unintended consequences occurring from SRM (which decision analysts refer to as one's "prior"), if the research concludes "SRM will be safe," the probability that SRM is indeed safe will be increased. If the R&D has no risk whatsoever of coming to an erroneous conclusion, then the probability that SRM is safe becomes 1.0, regardless of any prior views, and the SRM deployment decision can be made in a risk-free manner. (This is the case used to compute the "value of perfect information.") However, most research efforts leave some chance of producing a false positive or false negative result, and the best that such research findings can hope to do is reduce but not eliminate the possibility of making an incorrect decision.

The probabilities associated with the branches of figure 1.2.3 are derived using Bayes' Rule, and depend on the subjective prior probability, p, that was used in figure 1.2.2 that unintended consequences will result from SRM. This probability is now also complemented by the probability that the R&D will produce either a false negative, q, or a false positive, s. The research produces a false negative if it concludes that "SRM is safe" even though SRM will not be found to be safe if actually deployed. Research produces a false positive when it concludes that "SRM is not safe" but it would actually be consequence-free if it were to be deployed. Each erroneous conclusion leads to a different potentially bad outcome. False negatives may lead to too much willingness to engage in SRM and false positives can lead to overly precautionary approaches that prevent society from benefiting from the cost reductions that SRM could otherwise provide.

As in the case of no R&D (i.e. figure 1.2.2), the decision one takes, and the willingness to undertake SRM, is contingent on one's prior probability of the unintended consequences, but also on one's view of the ability of the R&D to properly identify the truth about those risks. The latter are embodied in the probabilities of false negatives and positives. The VOI is the difference between the expected cost of the climate management problem as structured in figure 1.2.3 and the expected value if that same risky decision is made without the benefit of first learning the research conclusions, which is the

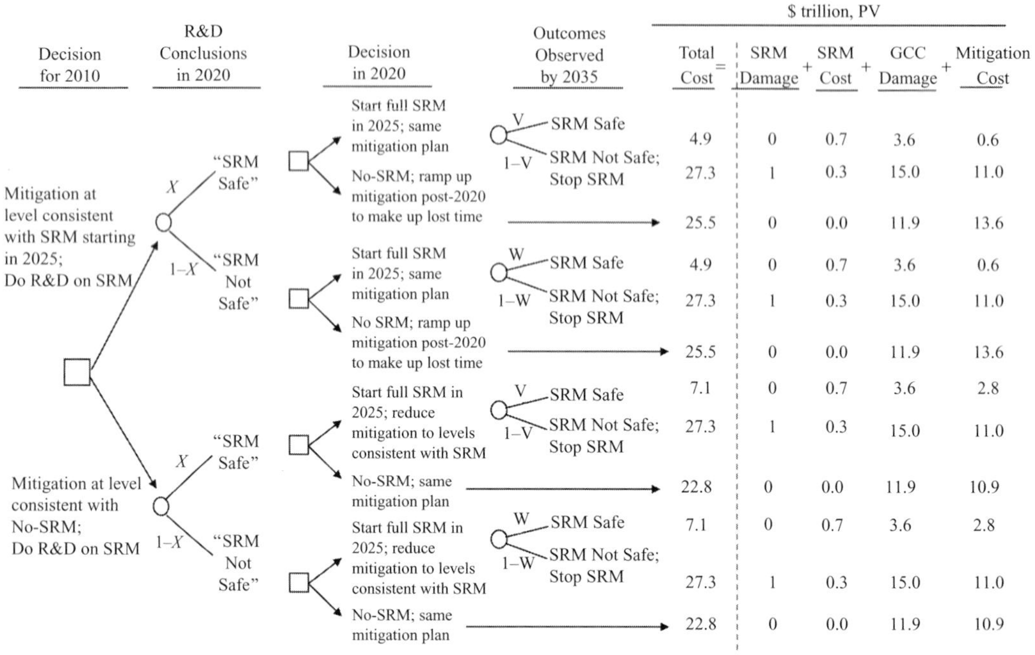

| | | | | $ trillion, PV | | | | |
Decision for 2010	R&D Conclusions in 2020	Decision in 2020	Outcomes Observed by 2035	Total Cost	SRM Damage	SRM Cost	GCC Damage	Mitigation Cost
		Start full SRM in 2025; same mitigation plan	V — SRM Safe	4.9	0	0.7	3.6	0.6
	"SRM Safe"		1−V — SRM Not Safe; Stop SRM	27.3	1	0.3	15.0	11.0
Mitigation at level consistent with SRM starting in 2025; Do R&D on SRM	X	No-SRM; ramp up mitigation post-2020 to make up lost time		25.5	0	0.0	11.9	13.6
	"SRM Not Safe" 1−X	Start full SRM in 2025; same mitigation plan	W — SRM Safe	4.9	0	0.7	3.6	0.6
			1−W — SRM Not Safe; Stop SRM	27.3	1	0.3	15.0	11.0
		No SRM; ramp up mitigation post-2020 to make up lost time		25.5	0	0.0	11.9	13.6
	"SRM Safe" X	Start full SRM in 2025; reduce mitigation to levels consistent with SRM	V — SRM Safe	7.1	0	0.7	3.6	2.8
			1−V — SRM Not Safe; Stop SRM	27.3	1	0.3	15.0	11.0
Mitigation at level consistent with No-SRM; Do R&D on SRM		No-SRM; same mitigation plan		22.8	0	0.0	11.9	10.9
	"SRM Not Safe" 1−X	Start full SRM in 2025; reduce mitigation to levels consistent with SRM	W — SRM Safe	7.1	0	0.7	3.6	2.8
			1−W — SRM Not Safe; Stop SRM	27.3	1	0.3	15.0	11.0
		No-SRM; same mitigation plan		22.8	0	0.0	11.9	10.9

Figure 1.2.3 *Decision tree for SRM decision made with improved information from R&D*

Notes: X: Probability that R&D will conclude that SRM will not produce any damages of its own – i.e. R&D outcome will be "SRM Safe." *X* includes the possibility that such a conclusion is a false negative:

$$X = p(1 - s) + (1 - p)q,$$

where:

p = prior probability that SRM will not produce damages of its own – i.e. pr(SRM Safe), as used in figure 1.2.2; s is the probability the R&D will report a false positive – i.e. pr("SRM Not Safe" | SRM Safe), and q is the probability the R&D will report a false negative – i.e. pr("SRM Safe" | SRM Not Safe).

V: Probability that SRM is safe if R&D has concluded it is safe – i.e. pr(SRM Safe | "SRM Safe"). This is a posterior value of p, conditioned on one possible R&D outcome:

$$V = \frac{p(1 - s)}{X},$$

W: Probability that SRM is safe even though R&D has concluded it is not safe – i.e. pr(SRM Safe | "SRM Not Safe"). This is another posterior value of p, conditioned on a different R&D outcome than in the case of *V*:

$$W = \frac{ps}{1 - X}.$$

Note on perfect information
R&D would produce perfect information as it has no probability of returning false positives or false negatives. This is the same as setting $q = s = 0$. If this is done, $X = p$, and $V = W = 0$. This reduces the above probability tree to a simpler tree used in calculating the value of perfect information, and this value of perfect information is also presented in the results of this Perspective paper, for cases where both q and s are zero.

expected value from the decision tree in figure 1.2.2 when using all the same parameter assumptions for p, and for the various outcome costs.

In order to perform any numerical computations using the VOI structure presented here, I needed estimates of a number of costs that I cannot obtain from the chapter, because they would require three additional DICE model runs. Because execution of those runs was not within the scope of this effort, I have made some educated guesses for those

Table 1.2.2 VOI results ($ trillion) assuming probability of false positive, $s = 0$

Prior probability SRM not safe, p (rows) vs. Probability of false negative from R&D, q (columns)

p \ q	0	0.05	0.1	0.15	0.2	0.25	0.3	0.35	0.4	0.45	0.5	0.55	0.6	0.65	0.7	0.75	0.8	0.85	0.9	0.95
0	0	0	0	0	0	0	0	0	0	0	0	0	0	0	0	0	0	0	0	0
0.05	0.8	0.6	0.4	0.2	0	0	0	0	0	0	0	0	0	0	0	0	0	0	0	0
0.1	1.6					0.6	0.4	0.2	0	0	0	0	0	0	0	0	0	0	0	0
0.15	2.4					1			0.7	0.5	0.3	0.09	0	0	0	0	0	0	0	0
0.2	3.1								1.5		1	0.8				0.5	0.3	0.2	0.1	0.07
0.25	2.8					2	1.8			1.1		1		0.6		0.3	0.3			0.06
0.3	2.5													0.4	0.4	0.3				
0.35													0.5	0.4	0.4					
0.4	1.8						1	0.9		0.6	0.5	0.5				0.3				0.05
0.45							0.7	0.6	0.6	0.5										
0.5	1.1	1		0.8		0.6	0.6				0.4					0.2				0.04
0.55	0.8	0.7																		
0.6	0.7	0.7																		
0.65																				
0.7																				
0.75	0.4					0.3	0.3				0.2					0.1				0.02
0.8																				
0.85																				
0.9	0.2					0.1	0.1				0.08					0.04				
0.95	0.02					0.01	0.01				*	*	*	*	*	*	*	*	*	*
1	0	0	0	0	0	0	0	0	0	0	0	0	0	0	0	0	0	0	0	0

Notes:
Color zones indicate the implied sequence of decisions through the decision tree:
  Darkest grey: Never do SRM no matter what R&D conclusion is.
  Mid-range grey: Plan early mitigation for no SRM, but do deploy SRM in 2025 if R&D concludes it will be safe.
  Lightest grey: Plan early mitigation for SRM, but deploy SRM only if R&D concludes it will be safe.
  No shading: Do SRM even if R&D concludes it will not be safe (in table 1.2.2, this zone only applies if prior $p = exactly$ 1.0. For 0.95 < p < 1.0, the decision is defined as for the lightest shade of grey).
* indicates when a non-zero VOI is less than $10 billion, which is the cost of the proposed R&D program.

values in figure 1.2.3. The basis for each of the cost estimate values in figure 1.2.3 is explained in an addendum at the end of this Perspective paper. The salient points to note about them here are, first, that the values of SRM cost, mitigation cost, and global climate change damages were tied as closely as possible to the estimates of the chapter. The analysis can be fine-tuned later, if desired, while still relying on the basic analytic framework that I have prepared so far. Second, the PV for potential damages from SRM if SRM proves unsafe is set at $1 trillion, consistent with the example in figure 1.2.2, and with the results in the first column of table 1.2.1. Value of information results presented below will change substantially if much higher or lower values are used for the likely magnitude of potential SRM damages, and sensitivity analysis on that assumption may be warranted and would be easy to do. However, for brevity, none are presented in this Perspective paper.

VOI Results

Table 1.2.2 shows the VOI given a range of prior probabilities, p, that SRM will be safe, and a range of probabilities that the research will produce a false negative. All of the values in table 1.2.2 assume that the probability of a false positive from the research is zero, and the costs if SRM is deployed but found to have unintended consequences are as shown in figure 1.2.2 (i.e. a PV of additional damages being $1 trillion, plus additional climate-related damages and mitigation costs to catch up).

Table 1.2.2 shows that the VOI is not always greater than zero, despite the very large net benefits and BCR in Bickel and Lane. However, the conditions that lead to zero VOI only occur if one has a very pessimistic view (i.e. less than 25% probability) that SRM will not produce unintended consequences. Further, the VOI is well above $10 billion – the proposed R&D funding level – when

Table 1.2.3 VOL results ($ trillion) assuming probability of false positive, s = 0.25

	Probability of false negative from R&D, q																			
	0	0.05	0.1	0.15	0.2	0.25	0.3	0.35	0.4	0.45	0.5	0.55	0.6	0.65	0.7	0.75	0.8	0.85	0.9	0.95
0	0	0	0	0	0	0	0	0	0	0	0	0	0	0	0	0	0	0	0	0
0.05	0.6	0.4	0.2	0	0	0	0	0	0	0	0	0	0	0	0	0	0	0	0	0
0.1	1.2			0.6	0.4	0.2	0	0	0	0	0	0	0	0	0	0	0	0	0	0
0.15	1.8						0.6		0.3	0.07	0	0	0	0	0	0	0	0	0	0
0.2	2.3										0.6	0.4	0.2	0.02	0	0	0	0	0	0
0.25	1.8					1	0.8		0.3	0.1	0	0	0	0	0	0	0	0	0	0
0.3	1.3						0.3		0.04	0	0	0	0	0	0	0	0	0	0	0
0.35	0.8	0.6		0.3		0.03	0	0	0	0	0	0	0	0	0	0	0	0	0	0
0.4	0.2	0.09	0	0	0	0	0	0	0	0	0	0	0	0	0	0	0	0	0	0
0.45	0	0	0	0	0	0	0	0	0	0	0	0	0	0	0	0	0	0	0	0
0.5	0	0	0	0	0	0	0	0	0	0	0	0	0	0	0	0	0	0	0	0
0.55	0	0	0	0	0	0	0	0	0	0	0	0	0	0	0	0	0	0	0	0
0.6	0	0	0	0	0	0	0	0	0	0	0	0	0	0	0	0	0	0	0	0
0.65	0	0	0	0	0	0	0	0	0	0	0	0	0	0	0	0	0	0	0	0
0.7	0	0	0	0	0	0	0	0	0	0	0	0	0	0	0	0	0	0	0	0
0.75	0	0	0	0	0	0	0	0	0	0	0	0	0	0	0	0	0	0	0	0
0.8	0	0	0	0	0	0	0	0	0	0	0	0	0	0	0	0	0	0	0	0
0.85	0	0	0	0	0	0	0	0	0	0	0	0	0	0	0	0	0	0	0	0
0.9	0	0	0	0	0	0	0	0	0	0	0	0	0	0	0	0	0	0	0	0
0.95	0	0	0	0	0	0	0	0	0	0	0	0	0	0	0	0	0	0	0	0
1	0	0	0	0	0	0	0	0	0	0	0	0	0	0	0	0	0	0	0	0

Row labels (left axis): Prior probability SRM not safe, p

Notes:
Color zones indicate the implied sequence of decisions through the decision tree:
  Darkest grey: Never do SRM no matter what R&D conclusion is.
  Mid-range grey: Plan early mitigation for no SRM, but do deploy SRM in 2025 if R&D concludes it will be safe.
  No shading: Do SRM even if R&D concludes it will not be safe.

it is positive. In fact, except for priors that SRM is more than 90% likely to be free of unintended consequences, the VOI is about $100 billion or more. At its peak, its value is in the range of $1 trillion to $3 trillion. (The value of perfect information is shown in the values on the leftmost column of table 1.2.2, where the probability of a false negative is zero. Since this entire table assumes that the probability of a false positive is zero, the information from research has its peak value, the "value of perfect information" in that column. That peak value is about $3 trillion.)

It is also interesting to note that the VOI is at its highest levels when one has only modest priors that SRM will be safe – where p is in the range of 10% to 50%. The most optimistic views are associated with declining VOI. One can also observe from table 1.2.2 that VOI declines as confidence in the research to identify real problems declines. However, even if the research has a 75% probability of failing to detect a real problem with SRM, it still

has value in the range of $100 billion to $500 billion (PV) if one's priors that SRM will be safe are anywhere in the range from 20% to 80%.

Table 1.2.3 shows the results for the same circumstances as in table 1.2.2 except that the probability of a false positive has been set at 25%. In this case, the VOI is very widely zero. The remaining zone of positive VOI is greatly condensed around the range where it is at its peak in table 1.2.2, and when in that range, the value still rises to about $2 trillion. However, if the research is likely to be relatively prone to concluding there will be problems with SRM-related damages when there will not actually be any, the people who would be most likely to find no value from R&D would be the optimistic individuals. The reason is that their priors that SRM will be safe are so strong that they will tend to interpret research results of potentially bad outcomes as more likely false outcomes than correct outcomes. They will be inclined to undertake SRM regardless of research results, and hence that research has relatively little chance of changing their course of actions.

When situations like that occur, research has no value. For people on the pessimistic end of the spectrum, the R&D has no value because it will not be good enough to convince people that GE is safe, even if the research does not identify any concerns. For people on the optimistic end of the spectrum, any negative findings of the R&D would only confirm what they already believe, and they may tend to interpret any positive findings (red flags) as more likely being false positives than as information that might alter the preferred course of action with respect to using SRM. Research has to have the potential to persuade people towards different courses of action depending on its findings if it is to have any actual value. So, almost counterintuitively, we find that the very people who might be most inclined to undertake SRM are also the most likely to find no value in the research at all, if the fact of doing that research presents a moderate chance of producing false alarms.

This raises a question of whether we are pursuing the correct model of VOI. It is certainly the correct one from the perspective of decision theory and practice. However, in that standard calculation, the expected value of the decisions given better information are compared to the "best" decision that the same decision maker would make if he or she were to make that decision without better information. This may be the theoretical ideal, and it is probably also the case for a corporate decision. However, with SRM one is concerned with a global public policy decision with many diffuse decision makers. The information in tables 1.2.1 through 1.2.3 have been useful because they highlight how different people may be viewing the identical decision, by showing how values vary with individual, subjective judgments on probabilities. Optimists may be prepared to promote SRM as an important option for managing climate, but they may assign little value for the R&D associated with it, if the information resulting from the R&D is not close to perfect. Why might these same individuals therefore also argue in favor of doing that R&D? One possibility is that they realize that the R&D is the only possible way to find a path under which SRM might ever be accepted. They might understand that the greater pessimism about SRM among a large portion of the other individuals contributing to this policy decision will prevent SRM from being allowed unless and until the R&D is completed. Thus, the alternative expected value that the informed case would be compared to in order to calculate the VOI in a public policy making setting might be the highly precautionary outcome of No SRM, rather than the optimized decision that uses a single set of priors. The precautionary attitude might also dominate the public policy decision if the R&D produces a "not safe" finding, such that SRM would never be allowed under that research outcome, with no consideration for the possibility that it could be a false positive. This would be a world in which those with relatively optimistic priors have to convince all of the others in society that SRM is a risk worth taking, and the VOI computed by an alternative model of choice such as this is very different than that presented above.

Table 1.2.4 shows the estimates of the VOI that one would assign to research on the potential consequences of SRM, given one's own personal prior, but now assuming that the societal decision on SRM will be driven by a majority who are inclined to act in a precautionary manner if there is no concerted additional research to help them be persuaded otherwise, and who also will act in a precautionary manner if such research identifies a "not safe" signal. The settings of probabilities in table 1.2.4 are identical to those in table 1.2.3: it fixes the probability of a false positive at 25% and reports the alternative concept of VOI in a social choice setting for the full range of combinations of prior probabilities that SRM will be safe and probabilities of false negatives from the research.

Suddenly the large zone of zero VOI associated with relatively optimistic priors on SRM's consequences that was found in table 1.2.3 is replaced by the highest VOI estimates we have estimated anywhere. In fact, these values are nearly as high as the value of SRM computed by Bickel and Lane. In a sense, in a societal decision making situation that may be dominated by precautionary attitudes, R&D may be the key that enables any consideration at all of undertaking a potentially highly valuable but also risky activity such as SRM. Thus, the R&D may have very high values to those who have

Table 1.2.4 VOL for societal decision ($ trillion) assuming probability of false positive, $s = 0.25$

Probability of false negative from R&D, q

Prior probability SRM not safe, p	0	0.05	0.1	0.15	0.2	0.25	0.3	0.35	0.4	0.45	0.5	0.55	0.6	0.65	0.7	0.75	0.8	0.85	0.9	0.95
0	0	0	0	0	0	0	0	0	0	0	0	0	0	0	0	0	0	0	0	0
0.05		0.2	0	0	0	0	0	0	0	0	0	0	0	0	0	0	0	0	0	0
0.1	1.2				0.2	0	0	0	0	0	0	0	0	0	0	0	0	0	0	0
0.15							0.06	0	0	0	0	0	0	0	0	0	0	0	0	0
0.2												0.04	0	0	0	0	0	0	0	0
0.25	2.9				2.1					1.3					0.4	0.3	0.1	0	0	
0.3																	1	0.9	0.8	0.8
0.35																1.8	2.6			
0.4															2.8	2.7	2.6			
0.45													3.7	3.6	3.5					
0.5	5.9									4.8		4.6	4.4		4.4					4.4
0.55						6.5					5.5	5.4	5.3							
0.6						6.6					6.3	6.2								
0.65						7.3		7.1			7.1									
0.7	8.3			8.1	8	8														
0.75	8.9					8.8					8.6					8.6				
0.8																				
0.85																				
0.9	10.9	10.9	10.9	10.9	10.9	10.9	10.9	10.9	10.9	10.9	10.9	10.9	10.9	10.9	10.9	10.9	10.9	10.9	10.9	10.9
0.95																				
1	12.6	12.6	12.6	12.6	12.6	12.6	12.6	12.6	12.6	12.6	12.6	12.6	12.6	12.6	12.6	12.6	12.6	12.6	12.6	12.6

Notes:
Color zones indicate the implied sequence of decisions through the decision tree:
Darkest grey: Never do SRM no matter what R&D conclusion is.
Mid-range grey: Plan early mitigation for no SRM, but do deploy SRM in 2025 if R&D concludes it will be safe.
Lightest grey: Plan early mitigation for SRM, but deploy SRM only if R&D concludes it will be safe.

strong preconceived views of the promise of a new and risky technology, even though traditional VOI theory would suggest that these same individuals ought to be the ones who would assign a relatively low VOI to the same research.

Conclusion

This Perspective paper has taken a first step to incorporate concerns about potential unintended consequences into the CBA of Bickel and Lane's chapter 1 on geoengineering, and to directly assess the value of an R&D program that would help reduce such uncertainties before any GE deployment decision might be made. It has found that the value of *perfect* information would be much higher than their proposed research budget, but it also showed that *imperfect* information may have zero value. The VOI analysis thus would appear to undercut the assertion by the chapter authors that the value of SRM is so large that uncertainties are unlikely to affect the merits of conducting the further research needed to position SRM as a real

option for potential deployment. This Perspective paper, however, has also proposed a possible alternative VOI calculation that may be more appropriate to use for societal decision making by groups of people who hold very different sets of priors. When the suggested alternative for calculating VOI in a social choice setting was applied, the value of the GE R&D program exceeded even the value of perfect information over a very wide range of priors, and regions of zero value were confined to a narrow range consistent with a very pessimistic view of both the probability that the GE will be safe and the probability that R&D will be able to identify such hazards in advance of full deployment.

This initial VOI analysis has addressed only one crucial uncertainty. It has hopefully created some interesting perspectives for discussion of the GE chapter. It has also defined an analytical structure that can be expanded into a more complete form, as needed. The proposed R&D program would presumably reduce uncertainties over a variety of other uncertainties such as cost, functionality, climate change damages, etc., and thus VOI from this single program might be larger or smaller than the estimates here. If a more complete quantitative analysis is desired, the structure developed in this Perspective paper can serve as a foundation on which to build the more comprehensive analysis.

In my opinion, one of the most important next steps for this line of analysis would be to expand it to include AC as an additional option that could be deployed with or instead of SRM. That decision would be delayed until R&D is completed and has better characterized their respective true costs and risks. Because the uncertainties and risks of AC and SRM are largely independent of each other, both may contribute value, and in different ways than would be apparent from a comparison of their respective deterministic BCRs. For example, AC might be found to have value as a backup option that could be deployed *if* SRM is used and then must be stopped suddenly.

It is also important to point out that the structuring of a decision problem is most valuable in its ability to get experts and decision makers to converse and bring new issues to the table when trying to communicate the decision problem in the highly structured format of a decision tree. That suggests that further work to expand on these foundations would probably best be done in a collaborative manner that draws in the comments, suggestions, and reactions of the range of experts, policy makers, and stakeholders engaged in the GE issue.

Addendum: basis for cost estimates used in figure 1.2.3

Category	Summary of estimates and rationale
SRM damages	$0 if full deployment of SRM is not done, or if it is done and is found to be safe. $1t if SRM is deployed and found to cause damages during it first ten years (i.e. 2025–35). This is consistent with the example in figure 1.2.2, but alternative assumptions can be explored later, as was done in table 1.2.1.
SRM cost	$0.7t in cases where SRM is deployed in 2025 and found to be safe, taken from table 1.10 of chapter 1. $0.3t in cases where SRM is deployed but stopped in 2035. (A more precise value could be provided by Bickel and Lane.) $0 if SRM never fully deployed. (Small-scale field tests of SRM prior to 2025 are not included, as these are considered to be part of the R&D budget whose value is being assessed in this analysis.)
GCC damages	$3.6t in cases where SRM is deployed and found to be safe, and $11.9t in cases where SRM is never deployed, both taken from table 1.4 of the chapter. In cases where SRM is deployed but then stopped in 2035, cost set to $15t on the assumption that the much more rapid changes in temperature that would occur post-2035 due to higher CO_2 concentrations accumulated under the SRM strategy would exacerbate climate change damages relative to the case without any SRM. (A more precise value consistent with the result of the analysis assumptions could be obtained with another DICE run.)
Mitigation cost	$0.6t when SRM is deployed and found safe and pre-2025 mitigation levels are tailored for this SRM outcome. $10.9t in cases where pre-2025 mitigation levels assume no future SRM benefits, and SRM is indeed not deployed. Both of the former values are taken from table 1.4 of the chapter. In cases where pre-2025 mitigation assumes that SRM will not occur, but SRM *is* deployed and found safe post-2025, it is assumed that 20% of the PV of the $10.9t cost will have been incurred by 2025 and this amount is added to the $0.6t associated with the SRM case, for an input value of $2.8t. (A more precise value could be estimated with another DICE run.) In cases where pre-2025 mitigation levels are tailored for an SRM deployment, but it is learned in 2020 that SRM will not be deployed, the PV of mitigation costs is assumed to be about 25% higher than if the pre-2025 mitigation levels had been consistent with no SRM, due to the need to make up for lost time starting in 2020 to still avoid exceeding 2°C. This produces a cost of $13.6t. (A more precise estimate also could be obtained from another DICE run.) In the cases where SRM *is* deployed, but must be stopped in 2035, the cost is assumed to be $11t. This does not appear to be as high as in the former case because although both cases involve rapid increases in mitigation rates to make up for lost time, the former occurs fifteen years earlier, and thus is less discounted in the PV. (A more precise estimate could be obtained from the same DICE run that could better inform the level of GCC damages for this case.)

Bibliography

Bickel, J.E. and L. Lane, 2009: An analysis of climate change as a response to global warming, Copenhagen Consensus Center, August 7

Carbon Dioxide Mitigation

RICHARD S.J. TOL

Introduction

In the Copenhagen Consensus on Climate 2009 (hereafter, CCC09), options for climate policy are evaluated and ranked. The current chapter contributes with an analysis of five alternative policies to reduce carbon dioxide (CO_2) emissions.

Ranking options is a standard tool of decision analysis (Pratt *et al.* 1995). It is important to note that options should be ranked on the basis of an internally consistent set of assumptions. That is done here for alternative ways to reduce CO_2. By the same token, the CO_2 options presented here cannot be readily compared to the other options for climate policy presented elsewhere in this forum as they are based on different assumptions with regard to (1) the future populations, economies, and emissions; (2) the working of the climate system; (3) the impact of climate change; (4) the impact of emission reduction; and (5) aggregation over space and time.

Although the chapter presents five options for CO_2 emission reduction, the options differ only in scope and intensity. It is well known that a uniform carbon tax is the cheapest way to abate emissions (Weitzman 1974; Pizer 1997; Fischer *et al.* 2003). I therefore do not consider other policy instruments, as these necessarily have a lower benefit-cost ratio (BCR) than the options analyzed here.

The chapter proceeds as follows. The next section presents a rather lengthy review of the literature on the economic impacts of climate change, as this is a crucial and controversial part of any analysis of climate policy. The chapter continues with a shorter review of the literature on the economic costs of emission reduction, and then describes the Climate Framework for Uncertainty, Negotiation, and Distribution (FUND), the integrated assessment model (IAM) used in the analysis. The fifth section presents the five policy scenarios, the sixth discusses the findings, and the seventh concludes.

Impacts of Climate Change: A Survey

Estimates of the Total Economic Effect of Climate Change

The first studies of the welfare impacts of climate change were done for the USA by (Nordhaus 1991; Cline 1992; Titus 1992; and Smith 1996). Although (Nordhaus 1991) extrapolated his US estimate to the world, and (Hohmeyer and Gaertner 1992) published some global estimates, the credit for the first serious study of the global welfare impacts of climate change goes to Fankhauser (1994b, 1995). Table 2.1 lists that study and a dozen other studies of the worldwide effects of climate change that have followed.

Any study of the economic effects of climate change begins with some assumptions on future emissions, the extent and pattern of warming, and other possible aspects of climate change, such as sea level rise and changes in rainfall and storminess. The studies must then translate from climate change to economic consequences. A range of methodological approaches are possible here. Nordhaus (1994a) interviewed a limited number of experts.

The studies by Fankhauser (1994b, 1995) (1995), (Nordhaus 1994b), and Tol (1995, 2002a, 2002b) use the *enumerative method*. In this approach, estimates of the "physical effects" of climate change are obtained one by one from natural science papers, which in turn may be based on some combination of climate models, impact models, and laboratory experiments. The physical impacts must then each be given a price, and added up. For traded goods and services, such as agricultural products,

Table 2.1 Estimates of the welfare loss due to climate change

Study	Warming (°C)	Impact (% GDP)	Worst-off region (% GDP)	Worst-off region (Name)	Best-off region (% GDP)	Best-off region (Name)
Nordhaus (1994a)	3.0	−1.3				
Nordhaus (1994b)	3.0	−4.8 (−30.0 to 0.0)				
Fankhauser (1995)	2.5	−1.4	−4.7	China	−0.7	Eastern Europe and the former Soviet Union (FSU)
Tol (1995)	2.5	−1.9	−8.7	Africa	−0.3	Eastern Europe and the FSU
Nordhaus and Yang (1996)[a]	2.5	−1.7	−2.1	Developing countries	0.9	FSU
Plamberk and Hope (1996)[a]	2.5	−2.5 (−0.5 to −11.4)	−8.6 (−0.6 to −39.5)	Asia (w/o China)	0.0 (−0.2 to 1.5)	Eastern Europe and the FSU
Mendelsohn et al. (2000a)[a,b,c]	2.5	0.0[b] 0.1[b]	−3.6[b] −0.5[b]	Africa	4.0[b] 1.7[b]	Eastern Europe and the FSU
Nordhaus and Boyer (2000)	2.5	−1.5	−3.9	Africa	0.7	Russia
Tol (2002a)	1.0	2.3 (1.0)	−4.1 (2.2)	Africa	3.7 (2.2)	Western Europe
Maddison (2003)[a,d,e]	2.5	−0.1	−14.6	South America	2.5	Western Europe
Rehdanz and Maddison (2005)[a,c]	1.0	−0.4	−23.5	Sub-Saharan Africa	12.9	South Asia
Hope (2006)[a,f]	2.5	0.9 (−0.2 to 2.7)	−2.6 (−0.4 to 10.0)	Asia (w/o China)	0.3 (−2.5 to 0.5)	Eastern Europe and the FSU
Nordhaus (2006)	2.5	−0.9 (0.1)				

Notes: Expressed as an equivalent income loss (% GDP); where available, estimates of the uncertainty are given in brackets, either as standard deviations or as 95% confidence intervals.
[a] Note that the global results were aggregated by the current author.
[b] The top estimate is for the "experimental" model, the bottom estimate for the "cross-sectional" model.
[c] Note that Mendelsohn et al. (2000a) only include market impacts.
[d] Note that the national results were aggregated to regions by the current author for reasons of comparability.
[e] Note that Maddison (2003) only considers market impacts on households.
[f] The numbers used by Hope (2006) are averages of previous estimates by Fankhauser and Tol (2005); Stern et al. (2006) adopt the work of Hope (2006).

agronomy papers are used to predict the effect of climate on crop yield, and then market prices or economic models are used to value that change in output. As another example, the impact of sea level rise constitutes coastal protection and land lost, estimates of which can be found in the engineering literature; the economic input in this case is not only the cost of dike building and the value of land, but also the decision which properties to protect. For non-traded goods and services, other methods are needed. An ideal approach might be to study how climate change affects human welfare through health and nature in each area around the world, but a series of "primary valuation" studies of this kind would be expensive and time-consuming. Thus, the monetization of non-market climate change impacts relies on "benefit transfer," in which epidemiology papers are used to estimate effects on health or the environment, and then economic values are applied from studies of the valuation of mortality risks in other contexts than climate change.

An alternative approach, exemplified in Mendelsohn's work (Mendelsohn *et al.* 2000a, 2000b) can be called the *statistical approach*. It is based on direct estimates of the welfare impacts, using observed variations (across space within a single country) in prices and expenditures to discern the effect of climate. Mendelsohn assumes that the observed variation of economic activity with climate over space holds over time as well and uses climate models to estimate the future impact of climate change. Mendelsohn's estimates are done per sector for selected countries, extrapolated to other countries, and then added up, but physical modeling is avoided. Other studies (Maddison 2003, Nordhaus 2006) use versions of the statistical approach as well. However, Nordhaus uses empirical estimates of the *aggregate* climate impact on income across the world (per grid cell), while Maddison looks at patterns of *aggregate* household consumption (per country). Like Mendelsohn, Nordhaus and Maddison rely exclusively on observations, assuming that "climate" is reflected in incomes and expenditures – and that the spatial pattern holds over time. Rehdanz and Maddison (2005) also empirically estimate the aggregate impact, using self-reported happiness for dozens of countries.

The enumerative approach has the advantage that it is based on natural science experiments, models, and data; the results are physically realistic and easily interpreted. However, the enumerative approach also raises concerns about extrapolation: economic values estimated for other issues are applied to climate change concerns; values estimated for a limited number of locations are extrapolated to the world; and values estimated for the recent past are extrapolated to the remote future. Tests of benefit transfer methods have shown time and again that errors from such extrapolations can be substantial (Brouwer and Spaninks 1999). But perhaps the main disadvantage of the enumerative approach is that the assumptions about adaptation may be unrealistic – as temperatures increase, presumably private and public sector reactions would occur to both market and non-market events.

In contrast, the statistical studies rely on uncontrolled experiments. These estimates have the advantage of being based on real-world differences in climate and income, rather than extrapolated differences. Therefore, adaptation is realistically, if often implicitly, modelled. However, statistical studies run the risk that all differences between places are attributed to climate. Furthermore, the data often allow for cross-sectional studies only; and some important aspects of climate change, particularly the direct impacts of sea level rise and CO_2 fertilization, do not have much spatial variation.

Given that the studies in table 2.1 use different methods, it is striking that the estimates are in broad agreement on a number of points – indeed, the uncertainty analysis displayed in figure 2.1 reveals that no estimate is an obvious outlier. Table 2.1 shows selected characteristics of the published estimates. The first column of table 2.1 shows the underlying assumption of long-term warming, measured as the increase in the global average surface air temperature. The assumed warming typically presumes a doubling of concentrations of greenhouse gases (GHGs) in the atmosphere. It is reasonable to think of these as the temperature increase in the second half of the twenty-first century. However, the impact studies in table 2.1 are comparative static, and they impose a future climate on today's economy. One can therefore not attach a date to these estimates. The second column of table 2.1 shows the impact on welfare at that future time, usually expressed as a percentage of income. For instance, Nordhaus (1994a, 1994b) estimates that the impact of 3°C global warming is as bad as losing 1.4% of income. In some cases, a confidence interval (usually at the 95% level) appears under the estimate; in other cases, a standard deviation is given; but the majority of studies does not report any estimate of the uncertainty. The rest of table 2.1 illustrates differential effects around the world. The third column shows the percentage decrease in annual GDP of the regions hardest-hit by climate change, and the fourth column identifies those regions. The fifth column shows the percentage change in GDP for regions that are least hurt by climate change – and in most cases would even benefit from a warmer climate – and the final column identifies those regions.

A first area of agreement between these studies is that the welfare effect of a doubling of the atmospheric concentration of GHG emissions on the current economy is relatively small – a few percentage points of GDP. This kind of loss of output

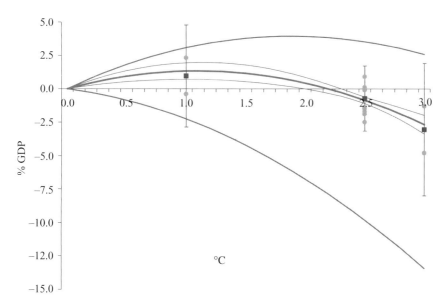

Figure 2.1 *The fourteen estimates of the global economic impact of climate change*

Note: Expressed as the welfare-equivalent income loss, as a functions of the increase in global mean temperature relative to today. The dots represent the estimates (cf. table 2.1). The squares are the sample means (for the specific global warming), and the lines are the sample means plus or minus twice the sample standard deviation. The central line is the least squares fit to the f observations: $D = 2.46$ (1.25) $T - 1.11$ (0.48) $T2$, $R2 = 0.51$, where D denotes impact and T denotes temperature; standard deviations are between brackets. The thin inner two lines are the 95% confidence interval for the central line re-estimated with one observation dropped. The thick outer two lines are the 95% confidence interval, where the standard deviation is the least squares fit to the five reported standard deviations or half-confidence intervals (cf. table 2.1): S optimistic = 0.87 (0.28) T, $R2 = 0.70$, S pessimistic = 1.79 (0.87) T, $R2 = 0.51$, where S is the standard deviation.

can look large or small, depending on the context. From one perspective, it's roughly equivalent to a year's growth in the global economy – which suggests that over a century or so, the economic loss from climate change is not all that large. On the other hand, the damage is not negligible. An environmental issue that causes a permanent reduction of welfare, lasting into the indefinite future, would certainly justify some steps to reduce such costs. Balancing these factors, cost-benefit analyses (CBAs) of climate change typically recommend only limited GHG emission reduction – for instance, Nordhaus (1993) argues that the optimal rate of emission reduction is 10–15% (relative to the scenario without climate policy) over the course of the twenty-first century. Recall that the EU calls for a 20–30% emission reduction (relative to 2005) by 2020.

A second finding is that some estimates (Mendelsohn *et al.* 2000a, 2000b; Tol 2002b; Hope 2006), point to initial benefits of a modest increase in

temperature, followed by losses as temperatures increase further. There are no estimates for a warming above 3°C, although climate change may well go beyond that (see below). All studies published after 1995 have regions with net gains and net losses due to global warming, while earlier studies only find net losses. Figure 2.1 illustrates this pattern. The horizontal axis shows the increase in average global temperature. The vertical index shows the central estimate of welfare loss. The central line shows a best-fit parabolic line from an ordinary least squares (OLS) regression. Of course, it is something of a stretch to interpret the results of these different studies as if they were a time series of how climate change will affect the economy over time, and so this graph should be interpreted more as an interesting calculation than as hard analysis. But the pattern of modest economic gains due to climate change, followed by substantial losses, appears also in the few studies that report impacts over time (Mendelsohn *et al.* 2000a, 2000b;

Nordhaus and Boyer 2000; Smith *et al.* 2001; Tol 2002b).

The initial benefits arise partly because more CO_2 in the atmosphere reduces "water stress" in plants and may make them grow faster (Long *et al.* 2006). In addition, the output of the global economy is concentrated in the temperate zone, where warming reduces heating costs and cold-related health problems. Although the world population is concentrated in the tropics, where the initial effects of climate change are probably negative, the relatively smaller size of the economy in these areas means that – at least over the interval of small increases in global temperatures – gains for the high-income areas of the world exceed losses in the low-income areas.

However, this pattern should be interpreted with care. Even if, initially, economic impacts may well be positive, it does not follow that GHG emissions should be subsidized. The climate responds rather slowly to changes in GHG emissions. The initial warming can no longer be avoided; it should be viewed as a sunk benefit. The fitted line in figure 2.1 suggests that the turning point in terms of economic benefits occurs at about 1.1°C warming (with a standard deviation of 0.7°C). Policy steps to reduce emissions of GHGs in the near future would begin to have a noticeable affect on climate some time around mid-century – which is to say, at just about the time that any medium-run economic benefits of climate change begin to decline (Tol *et al.* 2000, 2002b; Hitz and Smith 2004). In short, even though total economic effects of 1–2°C warming may be positive, incremental impacts beyond that level are likely to be negative. Moreover, if one looks further into the future, the incremental effects look even more negative.

Third, although GHG emissions per person are higher in high-income countries, relative impacts of climate change are greater in low-income countries (Yohe and Schlesinger 2002). Indeed, impact estimates for sub-Saharan Africa (SSA) go up to a welfare loss equivalent to the loss of a quarter of income (table 2.1). The estimates are higher for several reasons. Low-income countries tend to be in tropical zones closer to the equator. They are already hotter, and their output already suffers to some extent from their higher temperatures in sectors like agriculture. Moreover, low-income countries are typically less

able to adapt to climate change both because of a lack of resources and less capable institutions (Yohe and Tol 2002; Adger 2006; Alberini *et al.* 2006; Smit and Wandel 2006; Tol *et al.* 2007; Tol and Yohe 2007b; Tol 2008a).

The emissions of GHGs are predominantly from high-income countries while the negative effects of climate change are predominantly in low-income countries. This has two policy implications. First, any justification of stringent abatement for GHGs is at least in part an appeal to consider the plight of citizens of low-income countries around the world and the effects imposed on them by the citizens of high-income countries (Schelling 2000). Second, if preexisting poverty is one of the main causes for vulnerability to climate change, one may wonder whether stimulating economic growth or emission abatement is the better way to reduce the effects of climate change. Indeed, Tol and Dowlatabadi (2001) and Tol and Yohe (2006) argue that the economic growth forgone by stringent abatement of GHGs would more than offset the avoided impacts of climate change, at least in the case of malaria. Similarly, Tol (2005b) shows that development is a cheaper way of reducing climate-change-induced malaria than is emission reduction. Moreover, high-income countries may find it easier and cheaper to compensate poorer countries for the climate change damages caused, rather than to pay for reducing their own GHG emissions. Such compensation could be explicit, but would more likely take the shape of technical and financial assistance with adaptation (Paavola and Adger 2006).

Although research is scarce (O'Brien *et al.* 2004) climate change impacts would not be homogenous within countries; certainly, certain economic sectors (e.g. agriculture), regions (e.g. the coastal zone) and age groups (e.g. the elderly) are more heavily affected than others.

Fourth, estimates of the economic effects of GHG emissions have become less pessimistic over time. For the studies listed here, the estimates increase by 0.23% of GDP per year in which the study was done (with a standard deviation of 0.10% per year). There are several reasons for this change. Projections of future emissions and future climate change have become less severe over time – even though the public discourse has become shriller. The earlier studies focused on the negative impacts of climate change, whereas later studies considered the balance of positives and negatives. In addition, earlier studies tended to ignore adaptation. More recent studies – triggered by Mendelsohn *et al.* (1994) – include some provision for agents to change their behavior in response to climate change. However, more recent studies also tend to assume that agents have perfect foresight about climate change, and have the flexibility and appropriate incentives to respond. Given that forecasts are imperfect, agents are constrained in many ways, and markets are often distorted – particularly in the areas that matter most for the effects of climate change such as water, food, energy, and health – recent studies of the economic effects of climate change may be too optimistic about the possibilities of adaptation and thus tend to underestimate the economic effects of climate change.

A fifth common conclusion from studies of the economic effects of climate change is that the uncertainty is vast and right-skewed. For example, consider only the studies that are based on a benchmark warming of 2.5°C. These studies have an average estimated effect of climate change on average output of -0.7% of GDP, and a standard deviation of 1.2% of GDP. Moreover, this standard deviation is only about best estimate of the economic impacts, given the climate change estimates. It does not include uncertainty about future levels of GHG emissions, or uncertainty about how these emissions will affect temperature levels, or uncertainty about the physical consequences of these temperature changes. Moreover, it is quite possible that the estimates are not independent, as there are only a relatively small number of studies, based on similar data, by authors who know each other well.

Only five of the thirteen studies in table 2.1 report some measure of uncertainty. Two of these report a standard deviation only – which suggests symmetry in the probability distribution. Three studies report a confidence interval – of these, two studies find that the uncertainty is right-skewed, but one study finds a left-skewed distribution. Although the evidence on uncertainty here is modest and inconsistent, and I suspect less than thoroughly reliable, it seems that negative surprises should be more likely than positive surprises. While it is relatively easy to imagine a disaster scenario for climate change – for example, involving massive sea level rise or monsoon

failure that could even lead to mass migration and violent conflict – it is not at all easy to argue that climate change will be a huge boost to economic growth.

Figure 2.1 has three alternative estimates of the uncertainty around the central estimates. First, it shows the sample statistics. This may be misleading for the reasons outlined above; note that there are only two estimates each for a 1.0°C and a 3.0°C global warming. Second, I re-estimated the parabola fourteen times with one observation omitted. This exercise shows that the shape of the curve in figure 2.1 does not depend on any single observation. At the same time, the four estimates for a 1.0°C or 3.0°C warming each have a substantial (but not significant) effect on the parameters of the parabola. Third, five studies report standard deviations or confidence intervals. Confidence intervals imply standard deviations, but because the reported intervals are asymmetric I derived two standard deviations, one for negative deviations from the mean, and one for positive deviations. I assumed that the standard deviation grows linearly with the temperature, and fitted a line to each of the two sets of five "observed" standard deviations. The result is the asymmetric confidence interval shown in figure 2.1. This probably best reflects the considerable uncertainty about the economic impact of climate change, and that negative surprises are more likely than positive ones.

In other words, the level of uncertainty here is large, and probably understated – especially in terms of failing to capture downside risks. The policy implication is that reduction of GHG emissions should err on the ambitious side.

The kinds of studies presented in table 2.1 can be improved in numerous ways, some of which have been mentioned already. In all of these studies, economic losses are approximated with direct costs, ignoring general equilibrium and even partial equilibrium effects.[1]

---

[1] General equilibrium studies of the effect of climate change on agriculture have a long history (Darwin 2004; Kane *et al.* 1992). These papers show that markets matter, and may even reverse the sign of the initial impact estimate (Yates and Strzepek 1998). Bosello *et al.* (2007) and Darwin and Tol (2001) show that sea level rise would change production and consumption in countries that are not directly affected, primarily through the food market (as agriculture is affected

In the enumerative studies, effects are usually assessed independently of one another, even if there is an obvious overlap – for example, losses in water resources and losses in agriculture may actually represent the same loss. Estimates are often based on extrapolation from a few detailed case studies, and extrapolation is to climate and levels of development that are very different from the original case study. Little effort has been put into validating the underlying models against independent data – even though the findings of the first empirical estimate of the impact of climate change on agriculture by Mendelsohn *et al.* (1994) were in stark contrast to earlier results like those of Parry (1990), which suggests that this issue may be important. Realistic modeling of adaptation is problematic, and studies typically either assume no adaptation or perfect adaptation. Many effects are unquantified, and some of these may be large (see below). The uncertainties of the estimates are largely unknown. These problems are gradually being addressed, but progress is slow. Indeed, the list of warnings given here is similar to those in Fankhauser and Tol (1996, 1997).

A deeper conceptual issue arises with putting value on environmental services. Empirical studies have shown that the willingness-to-pay (WTP) for improved environmental services may be substantially lower than the willingness to accept compensation (WTAC) for diminished environmental services (Horowith and McConnell 2002). The difference between WTP and WTAC goes beyond income effects, and may even hint at loss aversion and agency effects, particularly around involuntary risks. A reduction in the risk of mortality due to GHG emission abatement is viewed differently than an increase in the risk of mortality due to the emissions of a previous generation in a distant

most by sea level rise through land loss and salt water intrusion) and the capital market (as sea walls are expensive to build). Ignoring the general equilibrium effects probably leads to only a small negative bias in the global welfare loss, but differences in regional welfare losses are much greater. Similarly, Bosello *et al.* (2006) show that the direct costs are biased towards zero for health – that is, countries that would see their labor productivity fall (rise) because of climate change would also lose (gain) competitiveness. Berrittella *et al.* (2006) also emphasize the redistribution of impacts on tourism through markets.

country. The studies listed in table 2.1 all use WTP as the basis for valuation of environmental services, as recommended by Arrow *et al.* (1993). Implicitly, the policy problem is phrased as: "How much are we willing to pay to buy an improved climate for our children?" Alternatively, the policy problem could be phrased as "How much compensation should we pay our children for worsening their climate?" This is a different question, and the answer would be different if the current policy makers assume that future generations would differentiate between WTP and WTAC much like the present generation does. The marginal avoided compensation would be larger than the marginal benefit, so that the tax on GHG emission would be higher.

Estimates of the Marginal Cost of Greenhouse Gas Emissions

The marginal damage cost of CO_2, also known as the "social cost of carbon," is defined as the net present value (NPV) of the incremental damage due to a small increase in CO_2 emissions. For policy purposes, the marginal damage cost (if estimated along the optimal emission trajectory) would be equal to the Pigouvian tax that could be placed on carbon, thus internalizing the externality and restoring the market to the efficient solution.

A quick glance at the literature suggests that there are many more studies of the marginal cost of carbon than of the total cost of climate change. Table 2.1 has 13 studies and 14 estimates; in contrast, Tol (2009) reports forty-seven studies with 232 estimates. Some of the total cost estimates (Mendelsohn *et al.* 2000a, 2000b; Maddison 2003; Rehdanz and Maddison 2005; Nordhaus 2006) have yet to be used for marginal cost estimation. Therefore, the 200-plus estimates of the social cost of carbon are based on nine estimates of the total impact of climate change. The empirical basis for the size of an optimal carbon tax is much smaller than is suggested by the number of estimates.

How can nine studies of total economic cost of climate change yield well over 200 estimates of marginal cost? Remember that the total cost studies are comparative static, and measure the economic cost of climate change in terms of a reduction in welfare below its reference level. This approach to describing total costs can be translated into marginal costs of current emissions in a number of ways. The rate at which future benefits (and costs) are discounted is probably the most important source of variation in the estimates of the social cost of carbon. The large effect of different assumptions about discount rates is not surprising, given that the bulk of the avoidable effects of climate change is in the distant future. Differences in discount rates arise not only from varying assumptions about the rate of pure time preference, the growth rate of *per capita* consumption, and the elasticity of marginal utility of consumption.[2] Some more recent studies have also analyzed variants of hyperbolic discounting, where the rate of discount falls over time.

However, there are other reasons why two studies with identical estimates of the total economic costs of climate change, expressed as a percentage of GDP at some future date, can lead to very different estimates of marginal cost. Studies of the marginal damage costs of CO_2 emissions can be based on different projections of CO_2 emissions, different representations of the carbon cycle, different estimates of the rate of warming, and so on. Alternative population and economic scenarios also yield different estimates, particularly if vulnerability to climate change is assumed to change with a country or region's development.

For example, the estimate of Nordhaus (1991) of the total welfare loss of a 3.0°C warming is 1.3% of GDP. In order to derive a marginal damage cost estimate from this, you would need to assume when in the future 3.0°C would occur, and whether damages are linear or quadratic or some other function of temperature (and precipitation etc.). And then the future stream of incremental damages due to today's emissions needs to be discounted back to today's value.

Marginal cost estimates further vary with the way in which uncertainty is treated (if it is recognized at all). Marginal cost estimates also differ with how regional effects of climate change are aggregated. Most studies add monetary effects for certain regions of the world, which roughly reflects the assumption that emitters of GHGs will compensate the victims of climate change. Other studies add utility-equivalent effects – essentially assuming a social planner and a global welfare function. In these studies, different assumptions about the

shape of the global welfare function can imply widely different estimates of the social cost of carbon (Fankhauser *et al.* 1997, 1998; Anthoff *et al.* 2009).

Table 2.2 shows some characteristics of a meta-analysis of the published estimates of the social cost of carbon. Columns (2)–(5) show the sample statistics of the 232 published estimates. One key issue in attempting to summarize this work is that just looking at the distribution of the medians or modes of these studies is inadequate, because it does not give a fair sense of the uncertainty surrounding these estimates – it is particularly hard to discern the right tail of the distribution which may dominate the policy analysis (Tol 2003; Tol and Yohe 2007a; Weitzman 2009). Because there are many estimates of the social cost of carbon, this can be done reasonably objectively. (The same would not be the case for the total economic impact estimates.) Thus, the idea here is to use one parameter from each published estimate (the mode) and the standard deviation of the entire sample – and then to build up an overall distribution of the estimates and their surrounding uncertainty on this basis using the methodology in (Tol 2008b).[3] The results are shown in columns (6)–(8) of table 2.2.

---

[2] The elasticity of marginal utility with respect to consumption plays several roles. It serves as a measure of risk aversion. It plays an important role in the discount rate (Ramsey 1928), as it also partly governs the substitution of future and present consumption. Furthermore, this parameter drives the tradeoffs between differential impacts across the income distribution, both within and between countries. Although conceptually distinct, all climate policy analyses that I am aware of use a single numerical value (Saelen *et al.* 2008; Atkinson *et al.* 2009). The reason is simply that although these distinctions are well recognized, welfare theorists have yet to find welfare and utility functions that make the necessary distinctions and can be used in applied work.

[3] I fitted a Fisher–Tippett distribution to each published estimate using the estimate as the mode and the *sample* standard deviation. The Fisher–Tippett distribution is the only two-parameter, fat-tailed distribution that is defined on the real line. A few published estimates are negative, and given the uncertainties about risk, fat-tailed distributions seem appropriate (Tol 2003; Weitzman 2009). The joint probability density function follows from addition, using weights that reflect the age and quality of the study as well as the importance that the authors attach to the estimate – some estimates are presented as central estimates, others as sensitivity analyses or upper and lower bounds. See www.fnu.zmaw. de/Social-cost-of-carbon-meta-analy.6308.0.html

Table 2.2 The SCC of carbon ($/tC)

	Sample (unweighted)				Fitted distribution (weighted)			
	All	Pure rate of time preference			All	Pure rate of time preference		
		0%	1%	3%		0%	1%	3%
Mean	105	232	85	18	151	147	120	50
Stdev	243	434	142	20	271	155	148	61
Mode	13	–	–	–	41	81	49	25
33%	16	58	24	8	38	67	45	20
Median	29	85	46	14	87	116	91	36
67%	67	170	69	21	148	173	142	55
90%	243	500	145	40	345	339	272	112
95%	360	590	268	45	536	487	410	205
99%	1500	–	–	–	1687	667	675	270
N	232	38	50	66	–	–	–	

Note: Sample statistics and characteristics of the Fisher–Tippett distribution fitted to 232 published estimates, and to three subsets of these estimates based on the pure rate of time preference.

Table 2.2 reaffirms that the uncertainty about the social costs of climate change is very large. The mean estimate in these studies is a marginal cost of carbon of $105 per MT of carbon, but the modal estimate is only $13/tC. Of course, this divergence suggests that the mean estimate is driven by some very large estimates – and indeed, the estimated social cost at the 95th percentile is $360/tC and the estimate at the 99th percentile is $1500/tC. The fitted distribution suggests that the sample statistics underestimate the marginal costs: the mode is $41/tC, the mean $151/tC, and the 99th percentile $1687/tC.

This large divergence is partly explained by the use of different pure rates of time preference in these studies. The columns for sample statistics and for the fitted distribution of table 2.2 divide up the studies into three subsamples which use the same pure rate of time preference. A higher rate of time preference means that the costs of climate change incurred in the future have a lower present value (PV), and so for example, the sample mean social cost of carbon for the studies with a 3% rate of time preference is $18/tC, while it is $232/tC for studies that choose a 0% rate of time preference. But these columns also show that even when the same discount rate is used, the variation in estimates is large. For the fitted distribution, the means are roughly double the modes – showing that the means are being pulled higher by some studies with very high estimated social costs.[4] Table 2.2 shows that the estimates for the whole sample are dominated by the estimates based on lower discount rates.

The sample and distribution characteristics of table 2.2 also allow us to identify outliers. On the low side, the results of Tol (2005b) stand out with a social cost of carbon of −$6.6/tC for a 3% pure rate of time preference and $19.9/tC for a 0% rate. The reason is that Tol's model was the first used for marginal cost estimation that had initial benefits from climate change. In later work by the same author, the early benefits are less pronounced. On the high side, the results of Ceronsky *et al.* (2006) stand out, with a social cost estimate of $2400/tC

---

[4] Some readers may wonder why the estimates with a discount rate of 0% don't look all that substantially higher than the estimates with a discount rate of 1%. The main reason is that most estimates are (inappropriately) based on a finite time horizon. With an infinite time horizon, the SCC would still be finite, because fossil fuel reserves are finite and the economy would eventually equilibrate with the new climate, but the effect of the 0% discount rate would be more substantial. For the record, there is even one estimate (Hohmeyer and Gaertner 1992) based on a 0 consumption discount rate (Davidson 2006, 2008) and thus a *negative* pure rate of time preference.

for a 0% pure rate of time preference and $120/tC for a 3% rate. The reason is that Ceronsky *et al.* (2006) consider extreme scenarios only – while they acknowledge that such scenarios are unlikely, they do not specify a probability. At a 1% pure rate of time preference, the $815/tC estimate of Hope (2008) stands out. Again, this is the result of a sensitivity analysis in which Hope sets risk aversion to 0 so that the consumption discount rate equals 1% as well.

Although table 2.2 reveals a large estimated uncertainty about the social cost of carbon, there is reason to believe that the actual uncertainty is larger still. First of all, the social cost of carbon derives from the total economic impact estimates – and I argue above that their uncertainty is underestimated, too. Second, the estimates only contain those impacts that have been quantified and valued – and I argue below that some of the missing impacts have yet to be assessed because they are so difficult to handle and hence very uncertain. Third, although the number of researchers who published marginal damage cost estimates is larger than the number of researchers who published total impact estimates, it is still a reasonably small and close-knit community who may be subject to group-think, peer pressure and self-censoring.

To place these estimated costs of carbon in context, a carbon tax in the range of $50–$100 per MT of carbon would mean that new electricity generation capacity would be carbon-free, be it wind or solar power or coal with carbon capture and storage (Weyant *et al.* 2006). In contrast, it would take a much higher carbon tax to de-carbonize transport, as biofuels, batteries and fuel cells are very expensive still (Schaefer and Jacoby 2005, 2006). Substantial reduction of carbon emissions thus requires a carbon tax of at least $50/tC – which is just barely justifiable at the mean estimate for a pure rate of time preference of 3%.

Missing Impacts

The effects of climate change that have been quantified and monetized include the impacts on agriculture and forestry, water resources, coastal zones, energy consumption, air quality, and human health.

Obviously, this list is incomplete. Even within each category, the assessment is incomplete. I cannot offer quantitative estimates of these missing impacts, but a qualitative and speculative assessment of their relative importance follows. For more detail, see Tol (2008c).

Many of the omissions seem likely to be relatively small in the context of those items that have been quantified. Among the negative effects, for example, studies of the effect of sea level rise on coastal zones typically omit costs of saltwater intrusion in groundwater (Nicholls and Tol 2006). Increasing water temperatures would increase the costs of cooling power plants (Szolnoky *et al.* 1997). Redesigning urban water management systems, be it for more or less water, would be costly (Ashley *et al.* 2005), as would implementing safeguards against increased uncertainty about future circumstances. Extratropical storms may increase, leading to greater damage and higher building standards (Dorland *et al.* 1999). Tropical storms do more damage, but it is not known how climate change would alter the frequency, intensity, and spread of tropical storms (McDonald *et al.* 2005). Ocean acidification may harm fisheries (Kikkawa *et al.* 2004).

The list of relatively small missing effects would also include effects that are probably positive. Higher wind speeds in the mid-latitudes would decrease the costs of wind and wave energy (Breslow and Sailor 2002). Less sea ice would improve the accessibility of Arctic harbors, would reduce the costs of exploitation of oil and minerals in the Arctic, and might even open up new transport routes between Europe and East Asia (Wilson *et al.* 2004). Warmer weather would reduce expenditures on clothing and food, and traffic disruptions due to snow and ice (Carmichael *et al.* 2004).

Some missing effects are mixed. Tourism is an example. Climate change may well drive summer tourists towards the poles and up the mountains, which amounts to a redistribution of tourist revenue (Berrittella *et al.* 2006). Other effects are simply not known. Some rivers may see an increase in flooding, and others a decrease (Kundzewicz *et al.* 2005; Svensson *et al.* 2005).

These small unknowns, and doubtless others not identified here, are worth some additional research, but they pale in comparison to the big unknowns: extreme climate scenarios, the very long term, bio-diversity loss, the possible effects of climate change on economic development and even political violence.

Examples of extreme climate scenarios include an alteration of ocean circulation patterns – such as the Gulf Stream that brings water north from the equator up through the Atlantic Ocean (Marotzke 2000). This may lead to a sharp drop in temperature in and around the North Atlantic. Another example is the collapse of the West Antarctic Ice Sheet (Vaughan and Spouge 2002; Vaughan 2008), which would lead to sea level rise of 5–6 m in a matter of centuries. A third example is the massive release of methane (CH_4) from melting permafrost (Harvey and Zhen 1995), which would lead to rapid warm-ing worldwide. Exactly what would cause these sorts of changes, or what effects they would have, are not at all well understood, although the chance of any one of them happening seems low. But they do have the potential to happen relatively quickly, and if they did, the costs could be substantial. Only a few studies of climate change have examined these issues. Nicholls *et al.* (2008) find that the impacts of sea level rise increase ten-fold should the West Antarctic Ice Sheet collapse, but the work of Olsthoorn *et al.* (2008) suggests that this may be too optimistic as Nicholls *et al.* (2008) may have overestimated the speed with which coastal protec-tion can be built. Link and Tol (2004) estimate the effects of a shutdown of the Thermohaline Circu-lation (THC). They find that the resulting regional cooling offsets but does not reverse warming, at least over land. As a consequence, the net economic effect of this particular change in ocean circulation is *positive*.

Another big unknown is the effect of climate change in the very long term. Most static analy-ses examine the effects of doubling the concentra-tion of atmospheric CO_2; most studies looking at effects of climate change over time stop at 2100. Of course, climate change will not suddenly halt in 2100. In fact, most estimates suggest that the nega-tive effects of climate change are growing, and even accelerating, in the years up to 2100 (cf. figure 2.1). It may be that some of the most substantial bene-fits of addressing climate change occur after 2100, but studies of climate change have not looked seri-ously at possible patterns of emissions and atmo-spheric concentrations of carbon after 2100, the potential physical effects on climate, nor the mon-etary value of those impacts. One may argue that impacts beyond 2100 are irrelevant because of time discounting, but this argument would not hold if the impacts grow faster than the discount rate – because of the large uncertainty, this cannot be excluded.

Climate change could have a profound impact on biodiversity (Gitay *et al.* 2001), not only through changes in temperature and precipitation, but in the ways climate change might affect land use and nutrient cycles, ocean acidification, and the prospects for invasion of alien species into new habitats. Economists have a difficult time analyz-ing this issue. For starters, there are few quanti-tative studies of the effects of climate change on eco-systems and biodiversity. Moreover, valuation of eco-system change is difficult, although some methods are being developed (Champ *et al.* 2003). These methods are useful for marginal changes to nature, but may fail for the systematic impact of climate change. That said, valuation studies have consistently shown that, although people are will-ing to pay something to preserve or improve nature, most studies put the total WTP for nature con-servation at substantially less than 1% of income (Pearce and Moran 1994). Unless scientists and economists develop a rationale for placing a sub-stantially higher cost on biodiversity, it will not fundamentally alter the estimates of total costs of climate change.

A cross-sectional analysis of *per capita* income and temperature may suggest that people are poor because of the climate (Gallup *et al.* 1999; Ace-moglu *et al.* 2001; Masters and McMillan 2001; van Kooten 2004; Nordhaus 2006), although oth-ers would argue that institutions are more important than geography (Acemoglu *et al.* 2002; Easterly and Levine 2003). There is an open question about the possible effects of climate change on annual rates of economic growth. For example, one pos-sible scenario is that low-income countries, which are already poor to some extent because of climate, will suffer more from rising temperatures and have less ability to adapt, thus dragging their economies down further. (Fankhauser and Tol 2005) argue that only very extreme parameter choices would imply such a scenario. In contrast, Dell *et al.* (2008) find that climate change would slow the *annual* growth rate of poor countries by 0.6 to 2.9 percentage points. Accumulated over a century, this effect would dominate all earlier estimates of the economic effects of climate change. However, Dell *et al.* (2008) have only a few explanatory variables in their regression, so their estimate may suffer from specification or missing variable bias; they may also have confused weather variability with climate change. One can also imagine a scenario in which climate change affects health, particularly the prevalence of malaria and diarrhea, in a way that affects long-term economic growth (Galor and Weil 1999); or in which climate change-induced resource scarcity intensifies violent conflict (Zhang *et al.* 2006; Tol and Wagner 2008) and affect long-term growth rates through that mechanism (Butkiewicz and Yanikkaya 2005). These potential channels have not been modeled in a useful way. But the key point here is that if climate change affects annual rates of growth for a sustained period of time, such effects may dominate what was calculated in the total effects studies shown earlier in table 2.1.

Besides the known unknowns described above, there are probably unknown unknowns too. For example, the direct impact of climate change on labor productivity has never featured on any list of "missing impacts," but Kjellstrom *et al.* (2008) show that it may well be substantial.

The "missing impacts" are a reason for concern and further emphasize that climate change may spring nasty surprises. This justifies GHG emission reduction beyond that recommended by a cost-benefit analysis (CBA) under quantified risk. The size of the "uncertainty premium" is a political decision. However, one should keep in mind that there is a history of exaggeration in the study of climate change impacts. Early research pointed to massive sea level rise (Schneider and Chen 1980), millions dying from infectious diseases (Haines and Fuchs 1991) and widespread starvation (Hohmeyer and

Gaertner 1992). Later, more careful research has dispelled these fears.

Impacts of Emission Reduction: A Survey

Options

Carbon dioxide emissions are driven by the Kaya–Bauer identity:

$$M = P \frac{Y}{P} \frac{E}{Y} \frac{C}{E} \frac{M}{C} \tag{1}$$

where M is emissions, P is population, Y is income, E is emissions, and C is CO_2 generated. That is, (1) has that emissions are equal to the number of people times their *per capita* income, times the energy intensity of the economy, times the carbon intensity of the economy, times the fraction of emissions that is vented to the atmosphere.

Although it is an accounting identity, (1) provides insight into how emissions can be abated. One may reduce the number of people. This is generally not considered to be a policy option, but a few governments are actively pursuing this (albeit not for reasons of climate change). One may also reduce economic growth, or induce economic shrinkage. Again, this is not typically seen as an option for climate policy, but the economic downturn that followed the collapse of the Soviet Union and the current depression have reduced emissions considerably.

The three right-most terms of (1) are seriously considered for climate policy. First, one may increase the overall energy efficiency of the economy – that is, deliver the same economic value using less energy. Second, one may decrease the overall carbon intensity of the energy system – that is, deliver the same amount of energy emitting less carbon. Third, one may prevent CO_2 from entering the atmosphere.

None of these options is free or easy. Energy is a cost to businesses and households. The market therefore pushes for increased energy efficiency. When energy is cheap, this often means that more services are delivered for the same amount of energy input. When energy is dear, the same services are typically delivered with less energy.

Historically, the rate of energy efficiency improvements has ranged between 0.5% and 1.5% per year (Lindmark 2002; Tol et al. 2009). This is quite an achievement considering that this rate is maintained over the long term and applies to often mature technologies.

Suppose for the sake of argument that, in the absence of climate policy, energy efficiency improves by 1% per year, that the economy grows by 2%, and that the carbon intensity is constant. Then, emissions grow by 1% per year. In order to stabilize emissions, the rate of energy efficiency improvement has to double from 1% to 2%. Because of decreasing returns to scale, doubling the rate of technological progress means that the effort that is being put into improving energy efficiency has to be more than doubled. This is easy to do for a specific technology, but hard across the entire economy. Furthermore, only a fraction of appliances, vehicles, and machines are replaced each year. That is, technological progress applies to a fraction of the capital stock only. Premature retirement of capital is very expensive.

Similar arguments apply to de-carbonization. Energy supply has shifted dramatically before climate policy. In the early stages of industrial development, biomass was replaced by coal as the main source of energy, leading to a rapid rise of CO_2 emissions. Later, oil and gas started to replace coal, reducing the carbon intensity of the energy supply, but not sufficiently so to reduce emissions (Tol et al. 2009). In times of high energy prices, alternative energy sources have established niche applications but never captured the market. At present, non-fossil energy is too expensive for commercial application in the absence of government support.

There are a number of alternative, carbon-free energy sources: biomass, wind, water, wave, tidal, solar, geothermal, and nuclear power. Water and nuclear power have low costs, but are politically constrained. Wave, tidal, and geothermal power are experimental technologies, with a few niche applications. Wind power has expanded rapidly on the back of generous subsidies, but its unpredictable nature prevents it from even attaining a dominant position in the market. Biomass energy and solar power are currently very expensive still, but rapid progress is being made piggy-backing on advances in biotechnology and materials science.

Finally, there is the option to capture CO_2 just before it would be released into the atmosphere. Carbon capture, transport, and storage are all proven technologies, but have never been applied at the scale needed to reduce emissions. Cost is a major issue with carbon capture. The process significantly increases the capital invested in a power plant, while a substantial part of the energy generated is used to capture carbon. Reliability and safety are main issues with carbon storage. Leaky storage postpones rather than prevents emissions, and accidental releases of a large amount of CO_2 may kill animals and humans. At present, there are various plans to build demonstration plants for carbon capture and storage (CCS).

Costs

The IPCC[5] periodically surveys the costs of emission abatement (Hourcade et al. 1996, 2001; Barker et al. 2007); there are the EMF[6] overview papers (Weyant 1993, 1998, 2004; Weyant and Hill 1999; Weyant et al. 2006), and there a few recent meta-analyses as well (Repetto and Austin 1997; Barker et al. 2002; Fischer and Morgenstern 2006; Kuik et al. 2009). There are two equally important messages from this literature. First, a well-designed, gradual policy can substantially reduce emissions at low cost to society. Second, ill-designed policies, or policies that seek to do too much too soon can be orders of magnitude more expensive. While the academic literature has focused on the former, policy makers have opted for the latter.

Figure 2.2 illustrates the costs of emission reduction, here represented as the average reduction in gross world product over eight models participating in the Energy Modeling Forum (EMF22) for three alternative scenarios.[7] Stabilizing the atmospheric concentrations of all GHGs in the atmosphere at a level of 650 ppm CO_{2eq} may cost only 2.6% of GDP over a century. This is roughly equal to losing

[5] IPCC, www.ipcc.ch/.
[6] Energy Modeling Forum, http://emf.stanford.edu/.
[7] Note that the most stringent target is infeasible according to half of the models.

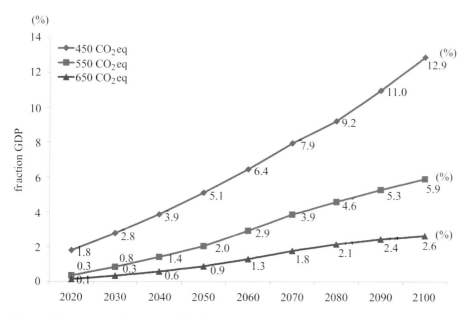

Figure 2.2 *Average reduction in GDP due to climate policy aiming at three alternative stabilization targets for atmospheric GHGs*

Note: The models used are FUND, GTEM, two versions of MERGE, MESSAGE, MiniCAM, SGM, and WITCH.

one year of growth in a hundred years. If the target is 550 ppm CO_{2eq}, costs go up to 5.9% of GDP. The costs are twice as high, but still small compared to economic growth. Half of the models cannot meet a target of 450 ppm CO_{2eq}. The other half report an average cost of 12.9%.

The cost estimates of figure 2.2 were achieved under the assumption that all GHG emissions from all sources in all countries are taxed by the same amount, and that the tax rate increases with the discount rate. That is, the stabilization target is met at the lowest possible cost. Figure 2.3 shows the estimates of the carbon tax, averaged for the eight (four) models. In order to achieve stabilization at 650 ppm CO_{2eq}, an $8/tCO_{2eq}$ carbon tax in 2020 rising to $320/tCO_{2eq}$ may be enough. However, for 450 ppm CO_{2eq}, the carbon tax would need to start at $100/tCO_{2eq}$ and rise to $4,000/tCO_{2eq}$ – keeping in mind that half of the models suggest that this target cannot be reached. Stabilizing at 450 ppm CO_{2eq} is needed to have a decent chance of keeping temperatures below 2°C above preindustrial levels.

The costs of emission reduction increase, and the feasibility of meeting a particular target decreases if:

- Different countries, sectors, or emissions face different explicit or implicit carbon prices (Manne and Richels 2001; Boehringer *et al.* 2006a, 2006b, 2008; Reilly *et al.* 2006).
- The carbon prices rise faster or more slowly than the effective discount rate (Wigley *et al.* 1996; Manne and Richels 1998, 2004).
- Climate policy is used to further other, non-climate policy goals (Burtraw *et al.* 2003).
- Climate policy adversely interacts with pre-existing policy distortions (Babiker *et al.* 2003).

Unfortunately, each of these four conditions is likely to be violated in reality. For instance, only selected countries have adopted emissions targets. Energy-intensity sectors that compete on the world market typically face the prospect of lower carbon prices than do other sectors. Climate policy often targets CO_2 but omits CH_4 and nitrous oxide (N_2O). Emission trading systems have a provision

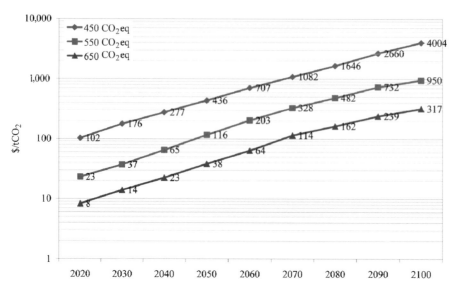

Figure 2.3 *Average carbon tax needed for three alternative stabilization targets for atmospheric GHGs*

Note: The models used are FUND, GTEM, two versions of MERGE, MESSAGE, MiniCAM, SGM, and WITCH.

for banking permits for future use, but not for borrowing permits from future periods. Climate policy is used to enhance energy security and create jobs. Climate policy is superimposed on energy and transport regulation and taxation.

The costs of emission reduction would also increase if emissions grow faster, if the price of fossil fuels is lower, or if the rate of technological progress in alternative fuels is slower than anticipated. This risk is two-sided. Emissions may grow more slowly, the price of fossil energy may be higher, and the alternative fuels may progress faster than expected.

The Model

I use Version 2.9e of the Climate Framework for Uncertainty, Negotiation, and Distribution (FUND). Version 2.9 of FUND has the same basic structure as previous versions (Tol 1999, 2005a, 2006). The source code and a complete description of the model can be found at www.fund-model. org/. A more succinct description is in the appendix (p. 96).

Essentially, FUND is a model that calculates impacts of climate change and climate policy

for sixteen regions of the world by making use of exogenous scenarios of socioeconomic variables. The scenarios comprise projected temporal profiles of population growth, economic growth, autonomous energy efficiency improvements and carbon efficiency improvements (de-carbonization), emissions of CO_2 from land use change, and emissions of CH_4 and of N_2O. CO_2 emissions from fossil fuel combustion are computed endogenously on the basis of the Kaya identity. The calculated impacts of climate change perturb the default paths of population and economic outputs corresponding to the exogenous scenarios. The model runs from 1950 to 2300 in time steps of a year, though the outputs for the 1950–2000 period is only used for calibration, and the years beyond 2100 are ignored in this chapter. The scenario up to the year 2100 is based on the EMF14 Standardized Scenario, which lies somewhere in between IS92a and IS92f (Leggett *et al.* 1992) and is somewhat similar to the SRES A2 scenario (Nakicenovic and Swart 2001). For the years from 2100 onward, the values are extrapolated from the pre-2100 scenarios. Radiative forcing is based on Forster *et al.* (2007). The global mean temperature is governed by a geometric buildup to its equilibrium (determined by the radiative forcing) with a half-life of fifty years. In the base case, the global mean temperature increases by 2.5°C in equilibrium for a doubling of CO_2 equivalents.

The climate impact module (Tol 2002a, 2002b) includes the following categories: agriculture, forestry, sea level rise, cardiovascular and respiratory disorders related to cold and heat stress, malaria, dengue fever, schistosomiasis, diarrhea, energy consumption, water resources, unmanaged ecosystems, and tropical and extra tropical storms. The last two are new additions (Narita *et al.* 2008, 2009). Climate change-related damages can be attributed to either the rate of change (benchmarked at 0.04°C/year) or the level of change (benchmarked at 1.0°C). Damages from the rate of temperature change slowly fade, reflecting adaptation (Tol 2002b).

People can die prematurely due to climate change, or they can migrate because of sea level rise. Like all impacts of climate change, these effects are monetized. The value of a statistical life is set to be 200 times the annual *per capita* income. The resulting value of a statistical life (VSL) lies in the middle of the observed range of values in the literature (Cline 1992). The value of emigration is set to be three times the *per capita* income (Tol 1995), the value of immigration is 40% of the *per capita* income in the host region (Cline 1992). Losses of dryland and wetlands due to sea level rise are modeled explicitly. The monetary value of a loss of 1 km^2 of dryland was on average $4 million in OECD countries in 1990 (Fankhauser 1994a). Dryland value is assumed to be proportional to GDP per km^2. Wetland losses are valued at $2 million per km^2 on average in the OECD in 1990 (Fankhauser 1994a). The wetland value is assumed to have a logistic relation to *per capita* income. Coastal protection is based on a CBA, including the value of additional wetland lost due to the construction of dikes and subsequent coastal squeeze.

Other impact categories, such as agriculture, forestry, energy, water, storm damage, and ecosystems, are directly expressed in monetary values without an intermediate layer of impacts measured in their "natural" units (Tol 2002a). Impacts of climate change on energy consumption, agriculture, and cardiovascular and respiratory diseases explicitly recognize that there is a climatic opti-

mum, which is determined by a variety of factors, including plant physiology and the behavior of farmers. Impacts are positive or negative depending on whether the actual climate conditions are moving closer to or away from that optimum climate. Impacts are larger if the initial climate conditions are further away from the optimum climate. The optimum climate is of importance with regard to the potential impacts. The actual impacts lag behind the potential impacts, depending on the speed of adaptation. The impacts of not being fully adapted to new climate conditions are always negative (Tol 2002b).

The impacts of climate change on coastal zones, forestry, tropical and extratropical storm damage, unmanaged ecosystems, water resources, diarrhea, malaria, dengue fever, and schistosomiasis are modelled as simple power functions. Impacts are either negative or positive, and they do not change sign (Tol 2002b).

Vulnerability to climate change changes with population growth, economic growth, and technological progress. Some systems are expected to become more vulnerable, such as water resources (with population growth), heat-related disorders (with urbanization), and ecosystems and health (with higher *per capita* incomes). Other systems such as energy consumption (with technological progress), agriculture (with economic growth) and vector- and water-borne diseases (with improved health care) are projected to become less vulnerable, at least over the long term (Tol 2002b). The income elasticities (Tol 2002b) are estimated from cross-sectional data or taken from the literature.

We estimated the social cost of carbon (SCC) by computing the total, monetized impact of climate change along a business-as-usual (BAU) path and along a path with slightly higher emissions between 2005 and 2014.[8] Differences in impacts were calculated, discounted back to the current year, and normalized by the difference in emissions.[9] The

---

[8] The SCC of emissions in future or past periods is beyond the scope of this chapter.

[9] We abstained from levelizing the incremental impacts within the period 2005–14 because the numerical effect of this correction is minimal while it is hard to explain.

SCC is thereby expressed in dollars per ton of carbon at a point in time – the standard measure of how much future damage would be avoided if today's emissions were reduced by 1 ton.[10] That is,

SCC_r

$$= \sum_{t=2005}^{3000} \frac{I_{t,r}\left(\sum_{s=1950}^{t-1} E_s + \delta_s\right) - I_{t,r}\left(\sum_{s=1950}^{t-1} E_s\right)}{\prod_{s=2005}^{t} 1 + \rho + \eta g_{s,r}}$$

$$\Bigg/ \sum_{t=2005}^{2014} \delta_t \qquad (2)$$

where

- SCC_r is the regional SCC (in US $ per ton of carbon)
- r denotes region
- t and s denote time (in years)
- I are monetized impacts (in US $ per year)
- E are emissions (in MT of carbon)
- δ are additional emissions (in MT of carbon)
- ρ is the pure rate of time preference (in fraction per year)
- η is the elasticity of marginal utility with respect to consumption
- g is the growth rate of *per capita* consumption (in fraction per year).

This chapter only considers emission reduction of CO_2. Initially, marginal abatement costs rise more than proportionally with abatement effort, but marginal costs become linear above $100/tC. There are mild intertemporal spillovers between and within regions that reduce costs (Tol 2005a; Clarke *et al.* 2008; Gillingham *et al.* 2008). An instantaneous emission reduction of 1% from the baseline would cost roughly 0.01% of consumption, and a 10% reduction would cost 1%. CH_4 (NO_2) emission reduction is two (four) orders of magnitude cheaper, but only CO_2 emission reduction can contain climate change. CO_2 emissions are

---

[10] Full documentation of the FUND model, including the assumptions in the Monte Carlo analysis, is available at www.fund-model.org.

strictly positive in FUND. FUND's cost estimates are well in line with other models (Kuik *et al.* 2009).

Scenarios

CCC09 hypothetically dispenses $250 billion per year on climate policy for a period of ten years. In this chapter, climate policy is restricted to abatement of CO_2 emissions from industrial processes (largely cement production) and from fossil fuel combustion.

There are many ways to reduce CO_2 emissions. I here restricted the analysis to a carbon tax/cap-and-trade scheme with auctioned permits. As there is no stochasticity in the model, these two options are equivalent. I omit other options (e.g. direct regulation; subsidies) because of the excess costs of such measures.

I consider five scenarios. In the first scenario, the countries of the OECD implement a uniform carbon tax such that the net present value (NPV) of the abatement cost equals $2 trillion, the NPV of $250 billion per year for ten years. The discount rate is 5% per year. Costs are discounted to 2009. This is achieved by a carbon tax of $700/tC, starting in 2010 and rising with the discount rate. The carbon tax is 0 from 2020 onwards.

In the second scenario, all countries implement a carbon tax of $250/tC in 2010, rising with the discount rate, but returning to 0 in 2020. This also leads to an abatement cost of $2 trillion.

In the third scenario, I assume that climate policy after 2020 will continue as before. That is, the carbon tax keeps rising with the discount rate between 2020 and 2100.

The fourth scenario is different. For ten years, $250 billion is invested in a trust fund. This trust fund finances a century-long programme of emission abatement such that the NPV of the abatement cost *over the century* equals $2 trillion. This is achieved by a uniform carbon tax for all countries, which starts at $12/tC in 2010 and rises with the discount rate.

The fifth scenario is different again. Only a part of the $250 billion is invested. The carbon tax in 2010 is set equal to the Pigou tax ($2/tC), also known as the marginal damage costs of CO_2 emissions and

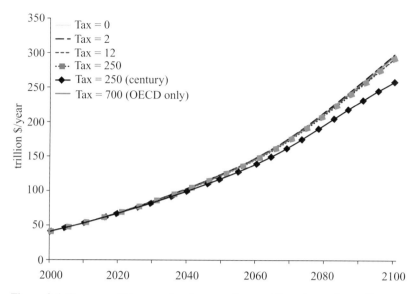

Figure 2.4 *Gross world income for the no-policy and five alternative policy scenarios*

the SCC. The carbon tax is applied worldwide, and equal for all countries.

Results

Figure 2.4 shows the gross world income for the no-abatement case and the five alternative policy scenarios. To cite Thomas Schelling, if these lines were drawn with a thick pencil, you would not see the difference. This is a recurrent theme in the climate economics literature. Given time and a clever policy design, substantial emission abatement can be achieved at acceptable cost. Even the most drastic policy considered – a worldwide carbon tax of $250/tC in 2010 rising with the rate of discount to over $20000/tC in 2100 – leads to a reduction of income of only 13% in 2100 (while income increases more than seven-fold in the no-policy scenario).

Figure 2.5 shows just how substantial emission cuts can be. Figure 2.5 also demonstrates the importance of long-lived climate policy – that is, a climate policy that is in line with the slow turnover of capital and the gradual progress of technology. The two policy scenarios that concentrate effort in the first decade are less effective than the scenario that spends the same amount of money over the cen-

tury. Even the $2/tC century-long policy is about as effective in the long run as the $250/tC decade-long policy, and at a fraction of the cost. If the $250/tC initial carbon tax is maintained over the century, CO_2 emissions fall by more than 90% in 2050 and by almost 100% in 2100 compared to the baseline;[11] 2050 emissions are some 20% of 2000 emissions in this scenario.

Figure 2.6 shows the impact of the five policy scenarios on the ambient concentration of CO_2. Figure 2.6 highlights that climate change is a stock problem. Emissions respond only slowly to policy, and concentrations respond even more tardily. A $12/tC initial carbon tax would almost stabilize the CO_2 concentration at around 680 ppm. An initial carbon tax of $250/tC would keep the concentration below 450 ppm; as other GHG are uncontrolled, the temperature continues to rise to 2.4°C above pre-industrial in 2100.[12]

Figure 2.7 depicts the economic impacts of climate change. As argued above, moderate warming

[11] Note that this is an artefact of the model which was never designed for such an aggressive policy. At such a high carbon tax, it would make economic sense to remove carbon from the atmosphere. FUND does not allow for that.
[12] Compare this to the vapid announcement of the G8 in 2009, which called for a 50% emission reduction in 2050 in order to keep the global mean temperature below 2°C.

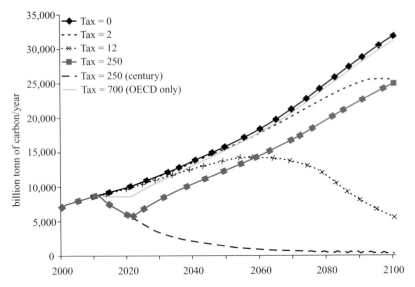

Figure 2.5 *Global CO$_2$ emissions from fossil fuel combustion and industrial processes for the no-policy and five alternative policy scenarios*

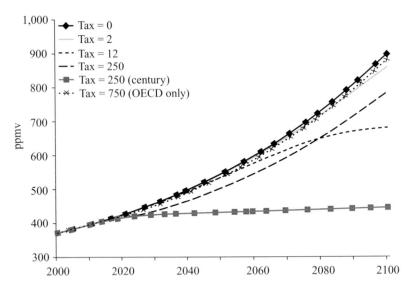

Figure 2.6 *Atmospheric concentration of CO$_2$ for the no-policy and five alternative policy scenarios*

has a positive effect. However, these are sunk benefits, hardly affected by emission abatement. In the longer term, the impacts of climate change are decidedly negative and rapidly accelerate in the absence of policy. That said, the policy scenarios considered here only slow the negative impacts of climate change, with the exception of the $250/tC

century-long policy which has net positive impacts of climate change throughout the century.

Table 2.3 shows the net present costs and benefits as well as the benefit-cost ratios (BCRs) of the five alternative policy scenarios. The five scenarios are ordered in the intensity of climate policy in the OECD in the coming decade. Starting at the bottom,

Table 2.3 NPV of abatement costs and benefits for the five scenarios

Initial carbon tax/Period	NPV cost		NPV benefit	BCR
	2010–20	2010–2100	2010–2100	2010–2100
World: 2 $/tC (century)	$0.2 10^9	$0.1 10^{12}	$0.1 10^{12}	1.51
World: 12 $/tC (century)	$5.6 10^9	$2.0 10^{12}	$0.5 10^{12}	0.26
World: 250 $/tC (decade)	$2.0 10^{12}	$17.8 10^{12}	$0.2 10^{12}	0.01
World: 250 $/tC (century)	$2.0 10^{12}	$46.7 10^{12}	$1.1 10^{12}	0.02
OECD: 700 $/tC (decade)	$2.0 10^{12}	$13.3 10^{12}	$0.0 10^{12}	0.00

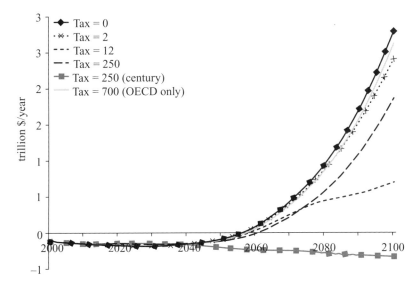

Figure 2.7 *Monetized impact of climate change for the no-policy and five alternative policy scenarios*

spending a lot of money on CO_2 emission reduction in the near term in the OECD does not pay off. A much greater benefit can be achieved if the same money is used to finance worldwide abatement – essentially because emission reduction is cheaper in poorer countries – but the BCR is about 1 to 100. If the same program is repeated decade after decade, abatement costs go up considerably but benefits rise even faster. Still, the BCR is only 1 to 50.

The BCR improves considerably if the $250 billion is spent over the century rather than over the decade. A BCR is of 1 to 4 is the result. This policy – a worldwide carbon tax of $12/tC in 2010, rising at 5% per year – does not improve global welfare, but recall that the model ignores the substantial concerns about equity and uncertainty. An

equity and risk premium of 400% on the benefits would not be outrageous.

If the initial carbon tax is set equal to the estimated marginal damage cost, the BCR unsurprisingly exceeds unity: 3 to 2 (cf. table 2.3). Over the century, this policy spends only one-twentieth of the funds (hypothetically) available to the Copenhagen Consensus.

Figure 2.8 shows the benefits as a percentage of gross world product (GWP) over time. Figure 2.9 shows the costs. Figure 2.10 shows the BCR per year. As climate change is initially beneficial, emission reduction brings damages at first. There are benefits only after 2040, and the benefits rise rapidly. Costs rise, too. In case climate policy only lasts for a decade, costs are roughly constant

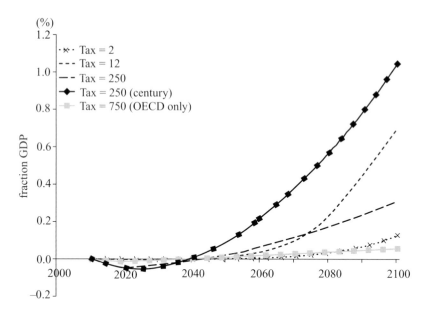

Figure 2.8 *Monetized and normalized benefit of the five alternative policy scenarios*

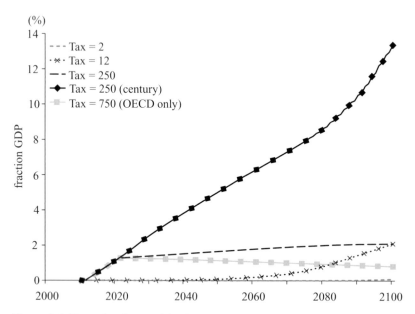

Figure 2.9 *Normalized cost of the five alternative policy scenarios*

as a fraction of GDP (and thus fall as a fraction of world GDP if abatement only applies to the countries of the OECD). The BCR is thus negative until 2040 (not shown). After that, the BCR rises over time but does not exceed unity. The $2/tC initial carbon tax scenario is the exception. The BCR exceeds unity after 2055. However, it reaches a maximum in 2078, after which current costs rise faster than current benefits. This suggests that the carbon tax rises too fast. This implies that this policy is not optimal, and that there is a policy with a higher BCR still.

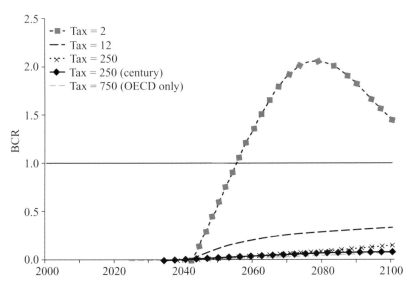

Figure 2.10 *BCR of the five alternative policy scenarios*

Discussion and Conclusion

This chapter consists of two parts: a literature review and policy analysis. In the first half of the literature review, I argue that the impacts of climate change are very uncertain. On the basis of our current knowledge, climate change seems to be a real problem but not the biggest problem in the world, or even the biggest environmental problem. However, as the marginal impacts are negative, this externality should be regulated. In the second half of the literature review, I argue that emission reduction is feasible and as cheap as policy is clever. Putting the two halves of the literature review together, one may wonder what all the fuss is about. Such speculation would be beyond this chapter.

In the second part of the chapter, I consider five alternative policy options for CO_2 emission reduction. In the first scenario, all funds ($2.5 trillion) of the CCC09 are spent on emission reduction in the OECD between 2010 and 2020. This is a rather silly thing to do. The BCR is below 1/100. In the second scenario, the same amount of money is spent on emission reduction worldwide, but policy still ceases in 2020. This is a much better plan than the first scenario, because non-OECD abatement is cheaper, but the BCR is still only 1/100. In the third scenario, I assume that there will be CCC in

2019, 2029, and so. That is, CO_2 abatement continues after 2020. Emission reduction costs are much larger, obviously, but the consequent abatement is so stringent that most of the negative impacts of climate change would be avoided altogether. Nonetheless, this policy does not pay off. The BCR is 1/50. Although the benefits are substantial, the costs are larger still. In the fourth scenario, I do not assume that there will be future CCC. Instead, the $2.5 trillion available to the CCC09 is invested in a trust fund. The trust fund finances emission reduction across the world such that it runs out of money in 2100. This policy has a BCR of 1/4. Finally, in the fifth scenario, 1/20 of the money available to the CCC09 is put into the trust fund. In this case, the BCR is 3/2. The fifth scenario is the only project worth funding.

These results are based on many assumptions, none of which is tested in a sensitivity analysis. Given the large uncertainties and the large inequities of climate change, one may justifiably argue that the "right" policy is more stringent than the "optimal" policy shown here. One may similarly argue that the discount rate used here is too high. That said, a cursory look at aid and trade policies do not suggest great care for the welfare of people in faraway lands; and pension policies suggest that the future is not a high priority. Therefore, one may go beyond the policy with a global carbon tax of \$2/tC in 2010, rising with the rate of discount. However, the analysis presented here also omits suboptimal policy design. Carbon price differentiation and direct regulation may well increase abatement costs by a substantial margin. Therefore, one should perhaps not go too far beyond the optimal policy outlined here.

Appendix: Model Description

Carbon dioxide emissions are calculated on the basis of the Kaya identity:

$$M_{t,r} = \frac{M_{t,r}}{E_{t,r}} \frac{E_{t,r}}{Y_{t,r}} \frac{Y_{t,r}}{P_{t,r}} P_{t,r} =: \psi_{t,r} \varphi_{t,r} Y_{t,r} \tag{A1.1}$$

where M denotes emissions, E denotes energy use, Y denotes GDP, and P denotes population; t is the index for time, r for region. The carbon intensity of energy use, and the energy intensity of production follow from:

$$\psi_{t,r} = g^{\psi}_{t-1,r} \psi_{t-1,r} - \alpha_{t-1,r} \tau^{\psi}_{t-1,r} \tag{A1.2}$$

and

$$\varphi_{t,r} = g^{\varphi}_{t-1,r} \varphi_{t-1,r} - \alpha_{t-1,r} \tau^{\varphi}_{t-1,r} \tag{A1.3}$$

where τ is policy intervention and α is a parameter. The exogenous growth rates g are referred to as the Autonomous Energy Efficiency Improvement (AEEI) and the Autonomous Carbon Efficiency Improvement (ACEI). The values are specified at www.fund-model.org/. Policy also affects emissions via

$$M_{t,r} = \left(\psi_{t,r} - \chi^{\psi}_{t,r} \right) \left(\varphi_{t,r} - \chi^{\varphi}_{t,r} \right) Y_{t,r}, \tag{A1.1'}$$

$$\chi^{\psi}_{t,r} = \kappa_{\psi} \chi_{t-1,r} + (1 - \alpha_{t-1,r}) \tau^{\psi}_{t-1,r}, \tag{A1.4}$$

and

$$\chi^{\varphi}_{t,r} = \kappa_{\varphi} \chi_{t-1,r} + (1 - \alpha_{t-1,r}) \tau^{\varphi}_{t-1,r}. \tag{A1.5}$$

Thus, the variable $0 < \alpha < 1$ governs which part of emission reduction is *permanent* (reducing carbon and energy intensities at all future times) and which part of emission reduction is *temporary* (reducing current energy consumptions and carbon emissions), fading at a rate of $0 < \kappa < 1$. In the base

case, $\kappa_{\psi} = \kappa_{\phi} = 0.9$ and

$$\alpha_{t,r} = 1 - \frac{\tau_{t,r}/100}{1 + \tau_{t,r}/100} \tag{A1.6}$$

So that $\alpha = 0.5$ if $\tau = \$100/tC$. One may interpret the difference between permanent and temporary emission reduction as affecting commercial technologies and capital stocks, respectively. The emission reduction module is a reduced form way of modelling that part of the emission reduction fades away after the policy intervention is reversed, but that another part remains through technological lock-in. Learning effects are described below. The parameters of the model are chosen so that FUND roughly resembles the behavior of other models, particularly those of the EMF (Weyant 2004; Weyant et al. 2006).

The costs of emission reduction C are given by

$$\frac{C_{t,r}}{Y_{t,r}} = \frac{\beta_{t,r} \tau^2_{t,r}}{H_{t,r} H^g_t}. \tag{A1.7}$$

H denotes the stock of knowledge. Equation (A1.6) gives the costs of emission reduction in a particular year for emission reduction in that year. In combination with (A1.2)–(A1.5), emission reduction is cheaper if spread out over a longer time period. The parameter β follows from

$$\beta_{t,r} = 0.784 - 0.084 \sqrt{\frac{M_{t,r}}{Y_{t,r}} - \min_s \frac{M_{t,s}}{Y_{t,s}}}. \tag{A1.8}$$

That is, emission reduction is relatively expensive for the region that has the lowest emission intensity. The calibration is such that a 10% emission reduction cut in 2003 would cost 1.57% (1.38%) of GDP of the least (most) carbon-intensive region; this is calibrated to Hourcade et al. (1996, 2001). An 80% (85%) emission reduction would completely ruin the economy. Later emission reductions are cheaper by (A1.7) and (A1.8). Emission reduction is relatively cheap for regions with high emission intensities. The thought is that emission reduction is cheap in countries that use a lot of energy and rely heavily on fossil fuels, while other countries use less energy and less fossil fuels and are therefore closer to the technological frontier of

Table A2.1   Parameters of the CH$_4$ and N$_2$O emission reduction cost curve; the 67% confidence interval is given in brackets

	Methane			Nitrous oxide		
USA	5.74E-04	(4.15E-04	7.90E-04)	2.14E-05	(1.91E-05	2.39E-05)
CAN	1.20E-03	(8.70E-04	1.64E-03)	6.92E-05	(6.29E-05	7.60E-05)
WEU	3.71E-04	(2.34E-04	5.80E-04)	7.26E-06	(6.60E-06	7.98E-06)
JPK	1.27E-04	(8.75E-05	1.84E-04)	5.32E-07	(3.21E-07	8.57E-07)
ANZ	4.12E-03	(3.03E-03	5.57E-03)	2.08E-04	(1.89E-04	2.29E-04)
EEU	3.90E-03	(2.81E-03	5.38E-03)	9.39E-05	(8.89E-05	9.93E-05)
FSU	8.87E-03	(7.49E-03	1.05E-02)	1.05E-05	(1.00E-05	1.10E-05)
MDE	6.32E-03	(4.86E-03	8.19E-03)	1.05E-05	(1.00E-05	1.10E-05)
CAM	3.65E-03	(2.87E-03	4.62E-03)	2.35E-04	(2.19E-04	2.53E-04)
SAM	2.75E-02	(1.81E-02	4.14E-02)	1.05E-05	(1.00E-05	1.10E-05)
SAS	3.16E-02	(2.43E-02	4.08E-02)	5.64E-04	(5.29E-04	6.01E-04)
SEA	1.43E-02	(1.06E-02	1.91E-02)	2.55E-15	(2.16E-15	3.01E-15)
CHI	1.26E-02	(9.50E-03	1.67E-02)	2.16E-05	(2.02E-05	2.30E-05)
NAF	1.43E-02	(1.06E-02	1.91E-02)	1.05E-05	(1.00E-05	1.10E-05)
SSA	1.43E-02	(1.06E-02	1.91E-02)	1.05E-05	(1.00E-05	1.10E-05)
SIS	1.43E-02	(1.06E-02	1.91E-02)	1.05E-05	(1.00E-05	1.10E-05)

emission abatement. For relatively small emission reduction, the costs in FUND correspond closely to those reported by other top-down models, but for higher emission reduction FUND finds higher costs, because FUND does not include backstop technologies, that is, a carbon-free energy supply that is available in unlimited quantities at fixed average costs.

The regional and global knowledge stocks follow from

$$H_{t,r} = H_{t-1,r}\sqrt{1 + \gamma_R \tau_{t-1,r}} \qquad (A1.9)$$

and

$$H_t^G = H_{t-1}^G\sqrt{1 + \gamma_G \tau_{t,r}}. \qquad A1.10$$

Knowledge accumulates with emission abatement. More knowledge implies lower emission reduction costs. The parameters γ determine which part of the knowledge is kept within the region, and which part spills over to other regions as well. In the base case, $\gamma_R = 0.9$ and $\gamma_G = 0.1$. The model is similar in structure and numbers to that of Goulder and Schneider (1999) and Goulder and Mathai (2000).

Emissions from land use change and deforestation are exogenous as specified at www.fund-model.org/ and cannot be mitigated.

CH$_4$ emissions are exogenous, specified at www.fund-model.org/. There is a single scenario only, based on IS92a (Leggett et al. 1992). The costs of emission reduction are quadratic. Table A2.1 specifies the parameters, which are calibrated to Tol (2006) and USEPA (2003).

N$_2$O oxide emissions are exogenous, specified at www.fund-model.org/. There is a single scenario only, based on IS92a (Leggett et al. 1992). The costs of emission reduction are quadratic. Table A2.1 specifies the parameters, which are calibrated to Tol (2006) and USEPA (2003).

SF$_6$ emissions are linear in GDP and GDP per capita. Table A2.2 gives the parameters. The numbers for 1990 and 1995 are estimated from IEA data (http://data.iea.org/ieastore/product.asp?dept_id=101&pf_id=305). There is no option to reduce SF$_6$ emissions.

Sulfur dioxide (SO$_2$) emissions grow with population (elasticity 0.33), fall with per capita income (elasticity 0.45), and fall with the sum of energy

Table A2.2   Determinants of SF$_6$ emissions

	C	GDP	GDP/capita
1990	1.6722E-01	5.0931E-06	−5.7537E-05
	(1.9297E-01)	(2.3482E-07)	(1.8505E-05)
1995	1.6255E-01	5.7234E-06	−6.0384E-05
	(2.1143E-01)	(2.3082E-07)	(1.8727E-05)
Used	1.6489E-01	5.4083E-06	−5.8961E-05
	(1.4312E-01)	(1.6464E-07)	(1.3164E-05)

Note: SF$_6$ emissions are in million MT of CO_2 equivalent. GDP is in million dollars (1995, MEX). GDP/*capita* is in dollars (1995, MEX).

Table A2.3   Parameters of (A1.11)

Gas	α^a	β^b	Preindustrial concentration
Methane (CH$_4$)	0.3597	1/8.6	790 ppb
Nitrous oxide (N$_2$O)	0.2079	1/120	285 ppb
Sulfur hexafluoride (SF$_6$)	0.0398	1/3200	0.04 ppt

Notes:
[a] The parameter α translates emissions (in million MT) into concentrations (in parts per billion or trillion by volume).
[b] The parameter β determines how fast concentrations return to their pre-industrial (and assumedly equilibrium) concentrations; $1/\beta$ is the atmospheric life-time (in years) of the gases.

efficiency improvements and decarbonization (elasticity 1.02). The parameters are estimated on the IMAGE scenarios (IMAGE Team 2001). There is no option to reduce SO$_2$ emissions.

CH$_4$, N$_2$O, and Sulfur hexafluoride (SF$_6$) are taken up in the atmosphere, and then geometrically depleted:

$$C_t = C_{t-1} + \alpha E_t - \beta(C_{t-1} - C_{pre}),$$
$$\text{(A1.11)}$$

where C denotes concentration, E emissions, t year, and *pre* preindustrial. Table A2.3 displays the parameters α and β for all gases. Parameters are taken from Schimel *et al.* (1996).

The atmospheric concentration of CO_2 follows from a five-box model:

$$Box_{i,t} = \rho_i Box_{i,t} + 0.000471\alpha_i E_t \quad \text{(A1.12a)}$$

with

$$C_t = \sum_{i=1}^{5} \alpha_i Box_{i,t}, \quad \text{(A1.12b)}$$

where α_i denotes the fraction of emissions E (in million MT of carbon) that is allocated to $Box\ i$

(0.13, 0.20, 0.32, 0.25, and 0.10, respectively) and ρ the decay-rate of the boxes ($\rho = \exp(-1/\text{lifetime})$, with lifetimes infinity, 363, 74, 17, and 2 years, respectively). The model is due to Maier-Reimer and Hasselmann (1987), its parameters are due to Hammitt *et al.* (1992). Thus, 13% of total emissions remains forever in the atmosphere, while 10% is – on average – removed in two years. CO_2 concentrations are measured in ppm by volume.

There is a feedback from climate change on the amount of CO_2 that is stored and emitted by the terrestrial biosphere. Instead of modeling the full dynamics, I keep the uptake by the terrestrial biosphere as it is – that is, (A1.12) is not affected – and add emissions from the terrestrial biosphere, primarily due to forest dieback. Emissions from the terrestrial biosphere follow:

$$E_t^B = \beta (T_t - T_{2000}) \frac{B_t}{B_{\max}} \quad \text{(A1.13a)}$$

with

$$B_t = B_{t-1} - E_{t-1}^B, \quad \text{(A1.13b)}$$

where E^B are emissions (in million MT of carbon); t denotes time; T is the global mean temperature (in °C); B_t is the remaining stock of potential emissions (in million MT of carbon; $B_{\max}$ is the total stock of potential emissions; $B_{\max} = 1,900$ GtC; β is a parameter; $\beta = 2.6$ GtC/°C, with a lower and upper bound of 0.6 and 7.5 GtC/°C. The model is calibrated to Denman *et al.* (2007).

Radiative forcing is specified as follows:

$$
\begin{aligned}
RF_t ={}& 5.35 \ln\left(\frac{CO_2}{275}\right) + 0.036\left(\sqrt{CH_4} - \sqrt{790}\right) \\
&+ 0.12\left(\sqrt{N_2O} - \sqrt{285}\right) \\
&- 0.47 \ln\left(1 + 2.01 \cdot 10^{-5} CH_4^{0.75} 285^{0.75}\right. \\
&\left. + 5.31 \cdot 10^{-15} CH_4^{2.52} 285^{1.52}\right) \\
&+ 0.47 \ln\left(1 + 2.01 \cdot 10^{-5} 790^{0.75} N_2O^{0.75}\right. \\
&\left. + 5.31 \cdot 10^{-15} 790^{2.52} N_2O^{1.52}\right) \\
&+ 0.00052(SF_6 - 0.04) - 0.03\frac{SO_2}{14.6} \\
&- 0.08 \frac{\ln\left(1 + \dfrac{SO_2}{34.4}\right)}{\ln\left(1 + \dfrac{14.6}{34.4}\right)} \quad \text{(A1.14)}
\end{aligned}
$$

Parameters are taken from Forster *et al.* (2007).

Bibliography

Acemoglu, D., S. Johnson, and J.A. Robinson, 2001: The colonial origins of comparative development: an empirical investigation, *American Economic Review* **91**(4), 1369–1401
2002: Reversal of fortune: geography and institutions in the making of the modern world income distribution, *Quarterly Journal of Economics* **117**(4), 1231–94

Adger, W.N., 2006: Vulnerability, *Global Environmental Change* **16**(3), 268–81

Alberini, A., A. Chiabai, and L. Muehlenbachs, 2006: Using expert judgement to assess adaptive capacity to climate change: evidence from a conjoint choice survey, *Global Environmental Change* **16**(2), 123–44

Anthoff, D., C.J. Hepburn, and R.S.J. Tol, 2009: Equity weighting and the marginal damage costs of climate change, *Ecological Economics* **68**(3), 836–49

Arrow, K.J., R.M. Solow, P.R. Portney, E.E. Leamer, R. Radner, and H. Schuman, 1993: Report of the NOAA Panel on Contingent Valuation, *Federal Register* **58**(10), 4016–64

Ashley, R.M., D.J. Balmforth, A.J. Saul, and J.D. Blanksby, 2005: Flooding in the future – predicting climate change, risks and responses in urban areas, *Water Science and Technology* **52**(5), 265–73

Atkinson, G.D., S. Dietz, J. Helgeson, C.J. Hepburn, and H. Saelen, 2009: Siblings, not triplets: social preferences for risk, inequality and time in discounting climate change, Economics Discussion Papers **2009–14**, Kiel Institute of the World Economy, Kiel

Babiker, M.H., G.E. Metcalf, and J.M. Reilly, 2003: Tax distortions and global climate policy, *Journal of Environmental Economics and Management* **46**, 269–87

Barker, T., I. Bashmakov, A. Alharthi, M. Amann, L. Cifuentes, J. Drexhage, M. Duan, O. Edenhofer, B.P. Flannery, M.J. Grubb, M. Hoogwijk, F.I. Ibitoye, C.J. Jepma, W.A. Pizer, and K. Yamaji, 2007: Mitigation from a cross-sectoral perspective, in *Climate Change 2007: Mitigation – Contribution of Working Group III to the Fourth Assessment Report of the Intergovernmental Panel on Climate Change*, B. Metz, R. Davidson, P.R. Bosch, R. Dave, and C.A. Meyer (eds.), Cambridge University Press, Cambridge, 619–90

Barker, T., J. Koehler, and M. Villena, 2002: The costs of greenhouse gas abatement: a meta-analysis of post-SRES mitigation scenarios, *Environmental Economics and Policy Studies* **5**, 135–66

Berrittella, M., A. Bigano, R. Roson, and R.S.J. Tol, 2006: A general equilibrium analysis of climate change impacts on tourism, *Tourism Management* **27**(5), 913–24

Boehringer, C., T. Hoffmann, and C. Manrique-de-Lara-Penate, 2006a: The efficiency costs of separating carbon markets under the EU emissions trading scheme: a quantitative assessment for Germany, *Energy Economics* **28**(1), 44–61

Boehringer, C., H. Koschel, and U. Moslener, 2008: Efficiency losses from overlapping regulation of EU carbon emissions, *Journal of Regulatory Economics* **33**(3), 299–317

Boehringer, C., A. Loeschel, and T.F. Rutherford 2006b: Efficiency gains from "what"-flexibility in climate policy: an integrated CGE assessment, *Energy Journal* (Multi-Greenhouse Gas Mitigation and Climate Policy Special Issue **3**), 405–24

Bosello, F., R. Roson, and R.S.J. Tol, 2006: Economy-wide estimates of the implications of climate change: human health, *Ecological Economics* **58**(3), 579–91
2007: Economy-wide estimates of the implications of climate change: sea level rise, *Environmental and Resource Economics* **37**(3), 549–71

Breslow, P.B. and D.J. Sailor, 2002: Vulnerability of wind power resources to climate change in the continental United States, *Renewable Energy* **27**(4), 585–98

Brouwer, R. and F.A. Spaninks, 1999: The validity of environmental benefits transfer: further empirical testing, *Environmental and Resource Economics* **14**(1), 95–117

Burtraw, D., A. Krupnick, K. Palmer, A. Paul, M. Toman, and C. Bloyd, 2003: Ancillary benefits of reduced air pollution in the US from moderate greenhouse gas mitigation policies in the electricity sector, *Journal of Environmental Economics and Management* **45**, 650–73

Butkiewicz, J.L. and H. Yanikkaya, 2005: The impact of sociopolitical instability on economic growth: analysis and implications, *Journal of Policy Modeling* **27**(5), 629–45

Carmichael, C.G., W.A. Gallus, Jr., B.R. Temeyer, and M.K. Bryden, 2004: A winter weather index for estimating winter roadway maintenance costs in the Midwest, *Journal of Applied Meteorology* **43**(11), 1783–90

Ceronsky, M., D. Anthoff, C.J. Hepburn, and R.S.J. Tol, 2006: Checking the Price Tag on Catastrophe: The Social Cost of Carbon under Non-linear Climate Response, Working Paper **87**, Research unit Sustainability and Global Change, Hamburg University and Centre for Marine and Atmospheric Science, Hamburg

Champ, P.A., K.J. Boyle, and T.C. Brown (eds.) 2003: *A Primer on Nonmarket Valuation*, Kluwer Academic Publishers, Dordrecht/ Boston, and London

Clarke, L.E., J.P. Weyant, and J. Edmonds, 2008: On the sources of technological change: what do the models assume?, *Energy Economics* **30**(2), 409–24

Cline, W.R., 1992: The Economics of Global Warming, Institute for International Economics, Washington, DC

Darwin, R.F., 2004: Effects of greenhouse gas emissions on world agriculture, food consumption, and economic welfare, *Climatic Change* **66**(1–2), 191–238

Darwin, R.F. and R.S.J. Tol, 2001: Estimates of the economic effects of sea level rise, *Environmental and Resource Economics* **19**(2), 113–29

Davidson, M.D., 2006: A social discount rate for climate damage to future generations based on regulatory law, *Climatic Change* **76**(1–2), 55–72

2008: Wrongful harm to future generations: the case of climate change, *Environmental Values* **17**, 471–88

Dell, M., B.F. Jones, and B.A. Olken, 2008: Climate Change and Economic Growth: Evidence from the Last Half Century, Working Paper **14132**, National Bureau of Economic Research, Washington DC

Denman, K.L., G.P. Brasseur, A. Chidthaisong, P. Ciais, P.M. Cox, R.E. Dickinson, D. Hauglustaine, C. Heinze, E. Hollad, D. Jacob, U. Lohmann, S. Ramachandran, P.L. de Silva Dias, S.C. Wofsy, and X. Zhang, 2007: Couplings between changes in the climate system and biogeochemistry', in *Climate Change 2007: The Physical Science Basis – Contribution of Working Group I to the Fourth Assessment Report of the Intergovernmental Panel on Climate Change*, S. Solomon, D. Qin, M. Manning, Z. Chen, M. Marquis, K.B. Averyt, M. Tignor, and H.L. Miller (eds.), Cambridge University Press, Cambridge, 499–587

Dorland, C., R.S.J. Tol, and J.P. Palutikof, 1999: Vulnerability of the Netherlands and Northwest Europe to storm damage under climate change, *Climatic Change* **43**(3), 513–35

Easterly, W. and R. Levine, 2003: Tropics, germs, and crops: how endowments influence economic development, *Journal of Monetary Economics* **50**(1), 3–39

Fankhauser, S., 1994a: Protection vs. retreat – the economic costs of sea level rise, *Environment and Planning A* **27**, 299–319

1994b: The economic costs of global warming damage: a survey, *Global Environmental Change* **4**(4), 301–9

1995: *Valuing Climate Change – The Economics of the Greenhouse*, 1st edn., EarthScan, London

Fankhauser, S. and R.S.J. Tol, 1996: Climate change costs – recent advancements in the economic assessment, *Energy Policy* **24**(7), 665–73

1997: The social costs of climate change: the IPCC Second Assessment Report and beyond, *Mitigation and Adaptation Strategies for Global Change* **1**(4), 385–403

2005: On climate change and economic growth, *Resource and Energy Economics* **27**(1), 1–17

Fankhauser, S., R.S.J. Tol, and D.W. Pearce, 1997: The aggregation of climate change damages: a welfare theoretic approach, *Environmental and Resource Economics* **10**(3), 249–66

1998: Extensions and alternatives to climate change impact valuation: on the critique of IPCC Working Group III's Impact Estimates, *Environment and Development Economics* **3**, 59–81

Fischer, C. and R.D. Morgenstern, 2006: Carbon abatement costs: why the wide range of estimates?, *Energy Journal* **272**, 73–86

Fischer, C., I.W.H. Parry, and W.A. Pizer, 2003: Instrument choice for environmental protection when technological innovation is endogenous, *Journal of Environmental Economics and Management* **45**, 523–45

Forster, P., V. Ramaswamy, P. Artaxo, T.K. Berntsen, R.A. Betts, D.W. Fahey, J. Haywood, J. Lean, D.C. Lowe, G. Myhre, J. Nganga, R. Prinn, G. Raga, M. Schulz, and R. van Dorland, 2007: Changes in atmospheric constituents and in radiative forcing, in *Climate Change 2007:*

The Physical Science Basis – Contribution of Working Group I to the Fourth Assessment Report of the Intergovernmental Panel on Climate Change, S. Solomon, D. Qin, M. Manning, Z. Chen, M. Marquis, K.B. Averyt, M. Tignor, and H.L. Miller (eds.), Cambridge University Press, Cambridge, 129–234

Gallup, J.L., J.D. Sachs, and A.D. Mellinger, 1999: Geography and economic development, CAER II Discussion Papers **39**, Harvard Institute for International Development, Cambridge, MA

Galor, O. and D.N. Weil, 1999: From Malthusian stagnation to modern growth, *American Economic Review* **89**(2), 150–4

Gillingham, K., R.G. Newell, and W.A. Pizer, 2008: Modeling endogenous technological change for climate policy analysis, *Energy Economics* **30**(6), 2734–53

Gitay, H., S. Brown, W.E. Easterling, III, B.P. Jallow, J.M. Antle, M. Apps, R.J. Beamish, F.S. Chapin, W. Cramer, J. Frangi, J.K. Laine, E. Lin, J.J. Magnuson, I.R. Noble, J. Price, T.D. Prowse, T.L. Root, E.-D. Schulze, O.D. Sirotenko, B.L. Sohngen, and J.-F. Soussana 2001: Ecosystems and their goods and services, in *Climate Change 2001: Impacts, Adaptation and Vulnerability – Contribution of Working Group II to the Third Assessment Report of the Intergovernmental Panel on Climate Change*, J.J. McCarthy, O.F. Canziani, N.A. Leary, D.J. Dokken, and K.S. White (eds.), Cambridge University Press, Cambridge, 235–342

Goulder, L.H. and K. Mathai, 2000: Optimal CO_2 abatement in the presence of induced technological change, *Journal of Environmental Economics and Management* **39**, 1–38

Goulder, L.H. and S.H. Schneider, 1999: Induced technological change and the attractiveness of CO_2 abatement policies, *Resource and Energy Economics* **21**, 211–53

Haines, A. and C. Fuchs, 1991: Potential impacts on health of atmospheric change, *Journal of Public Health Medicine* **13**(2), 69–80

Hammitt, J.K., R.J. Lempert, and M.E. Schlesinger, 1992: A sequential-decision strategy for abating climate change, *Nature* **357**, 315–18

Harvey, L.D.D. and H. Zhen, 1995: Evaluation of the potential impact of methane clathrate destabilization on future global warming, *Journal of Geophysical Research* **100**(D2), 2905–26

Hitz, S. and J.B. Smith, 2004: Estimating global impacts from climate change, *Global Environmental Change* **14**(3), 201–18

Hohmeyer, O. and M. Gaertner, 1992: *The Costs of Climate Change – A Rough Estimate of Orders of Magnitude*, Fraunhofer-Institut für Systemtechnik und Innovationsforschung, Karlsruhe

Hope, C.W., 2006: The marginal impact of CO_2 from PAGE2002: an integrated assessment model incorporating the IPCC's five reasons for concern, *Integrated Assessment Journal* **6**(1), 19–56

2008: Optimal carbon emissions and the social cost of carbon over time under uncertainty, *Integrated Assessment Journal* **8**(1), 107–22

Horowith, J.K. and K.E. McConnell, 2002: A review of WTA/WTP studies, *Journal of Environmental Economics and Management* **44**(3), 426–47

Hourcade, J.-C., K. Halsneas, M. Jaccard, W.D. Montgomery, R.G. Richels, J. Robinson, P.R. Shukla, and P. Sturm, 1996: A review of mitigation cost studies, in *Climate Change 1995: Economic and Social Dimensions – Contribution of Working Group III to the Second Assessment Report of the Intergovernmental Panel on Climate Change*, J.P. Bruce, H. Lee, and E.F. Haites (eds.), Cambridge University Press, Cambridge, 297–366

Hourcade, J.-C., P.R. Shukla, L. Cifuentes, D. Davis, J.A. Edmonds, B.S. Fisher, E. Fortin, A. Golub, O. Hohmeyer, A. Krupnick, S. Kverndokk, R. Loulou, R.G. Richels, H. Segenovic, and K. Yamaji, 2001: Global, regional and national costs and ancillary benefits of mitigation, in *Climate Change 2001: Mitigation – Contribution of Working Group III to the Third Assessment Report of the Intergovernmental Panel on Climate Change*, O.R. Davidson and B. Metz (eds.), Cambridge University Press, Cambridge, 499–559

IMAGE Team, 2001: *The IMAGE 2.2 Implementation of the SRES Scenarios: A Comprehensive Analysis of Emissions, Climate Change, and Impacts in the 21st Century*, RIVM CD-ROM Publication **481508018**, National Institute for Public Health and the Environment, Bilthoven

Kane, S., J.M. Reilly, and J. Tobey, 1992: An empirical study of the economic effects of climate change on world agriculture, *Climatic Change* **21**(1), 17–35

Kikkawa, T., J. Kita, and A. Ishimatsu, 2004:
    Comparison of the lethal effect of CO_2 and
    acidification on red sea bream (Pagrus major)
    during the early developmental stages, *Marine
    Pollution Bulletin* **48**(1–2), 108–10

Kjellstrom, T., R.S. Kovats, S.L. Lloyd, M.T. Holt,
    and R.S.J. Tol, 2008: The direct impact of
    climate change on regional labour productivity,
    Working Paper **260**, Economic and Social
    Research Institute, Dublin

Kuik, O., L. Brander, and R.S.J. Tol, 2009: Marginal
    abatement costs of greenhouse gas emissions: a
    meta-analysis, *Energy Policy* **37**(4), 1395–1403

Kundzewicz, Z.W., D. Graczyk, T. Maurer, I.
    Pinkswar, M. Radziejeswki, C. Svensson, and
    M. Szwed, 2005: Trend detection in river flow
    series: 1. annual maximum flow, *Hydrological
    Sciences Journal* **50**(5), 797–810

Leggett, J., W.J. Pepper, and R.J. Swart, 1992:
    Emissions scenarios for the IPCC: an update, in
    *Climate Change 1992: The Supplementary
    Report to the IPCC Scientific Assessment*, 1st
    edn., vol. 1, J.T. Houghton, B.A. Callander, and
    S.K. Varney (eds.), Cambridge University Press,
    Cambridge, 71–95

Lindmark, M., 2002: An EKC-pattern in historical
    perspective: carbon dioxide emissions,
    technology, fuel prices and growth in Sweden
    1870–1997, *Ecological Economics* **42**,
    333–47

Link, P.M. and R.S.J. Tol, 2004: Possible economic
    impacts of a shutdown of the thermohaline
    circulation: an application of FUND,
    Portuguese Economic Journal **3**(2),
    99–114

Long, S.P., E.A. Ainsworth, A.D.B. Leakey, J.
    Noesberger, and D.R. Ort, 2006: Food for
    thought: lower-than-expected crop yield
    stimulation with rising CO_2 concentrations,
    Science **312**(5811), 1918–21

Maddison, D.J., 2003: The amenity value of the
    climate: the household production function
    approach, *Resource and Energy Economics*
    25(2), 155–75

Maier-Reimer, E. and K. Hasselmann, 1987:
    Transport and storage of carbon dioxide in the
    ocean: an inorganic ocean circulation carbon
    cycle model, *Climate Dynamics* **2**, 63–90

Manne, A.S. and R.G. Richels, 1998: On stabilizing
    CO_2 concentrations – cost-effective emission
    reduction strategies, *Environmental Modeling
    and Assessment* **2**, 251–65

2001: An alternative approach to establishing
    trade-offs among greenhouse gases, *Nature* **410**,
    675–7
2004: US rejection of the Kyoto Protocol: the
    impact on compliance costs and CO_2 emissions,
    Energy Policy **32**, 447–54

Marotzke, J., 2000: Abrupt climate change and
    thermohaline circulation: mechanisms and
    predictability, *Proceedings of the National
    Academy of Science* **97**(4), 1347–50

Masters, W.A. and M.S. McMillan, 2001: Climate
    and scale in economic growth, *Journal of
    Economic Growth* **6**(3), 167–86

McDonald, R.E., D.G. Bleaken, D.R. Cresswell,
    V.D. Pope, and C.A. Senior, 2005: Tropical
    storms: representation and diagnosis in climate
    models and the impacts of climate change,
    Climate Dynamics **25**(1), 19–36

Mendelsohn, R.O., W.D. Nordhaus, and D. Shaw,
    1994: The impact of climate on agriculture: a
    Ricardian analysis, *American Economic Review*
    84(4), 753–71

Mendelsohn, R.O., W.N. Morrison, M.E.
    Schlesinger, and N.G. Andronova, 2000a:
    Country-specific market impacts of climate
    change, *Climatic Change* **45**(3–4), 553–69

Mendelsohn, R.O., M.E. Schlesinger, and L.J.
    Williams, 2000b: Comparing impacts across
    climate models, *Integrated Assessment* **1**(1),
    37–48

Narita, D., D. Anthoff, and R.S.J. Tol, 2008:
    Damage costs of climate change through
    intensification of tropical cyclone activities: an
    application of FUND, Working Paper **259**,
    Economic and Social Research Institute,
    Dublin

Nakicenovic, N. and R.J. Swart (eds.), 2001: *IPCC
    Special Report on Emissions Scenarios*,
    Cambridge University Press, Cambridge
2009: Economic costs of extratropical storms
    under climate change: an application of FUND,
    Working Paper **274**, Economic and Social
    Research Institute, Dublin

Nicholls, R.J. and R.S.J. Tol, 2006: Impacts and
    responses to sea level rise: a global analysis of
    the SRES scenarios over the twenty-first
    century, *Philosophical Transactions of the
    Royal Society A* **364**(1849), 1073–95

Nicholls, R.J., R.S.J. Tol, and A.T. Vafeidis, 2008:
    Global estimates of the impact of a collapse of
    the West Antarctic ice sheet: an application of
    FUND, *Climatic Change* **91**(1–2), 171–91

Nordhaus, W.D., 1991: To slow or not to slow: the economics of the greenhouse effect, *Economic Journal* **101**(444), 920–37

1993: Rolling the 'DICE': an optimal transition path for controlling greenhouse gases, *Resource and Energy Economics* **15**(1), 27–50

1994a: Expert opinion on climate change, *American Scientist* **82**(1), 45–51

1994b: *Managing the Global Commons: The Economics of Climate Change* , MIT Press, Cambridge, MA

2006: Geography and macroeconomics: new data and new findings, *Proceedings of the National Academy of Science* **103**(10), 3510–17

Nordhaus, W.D. and J.G. Boyer, 2000: *Warming the World: Economic Models of Global Warming*, MIT Press, Cambridge, MA and London, UK

Nordhaus, W.D. and Z. Yang, 1996: RICE: a regional dynamic general equilibrium model of optimal climate-change policy, *American Economic Review* **86**(4), 741–65

O'Brien, K.L., L. Sygna, and J.E. Haugen, 2004: Vulnerable or resilient? A multi-scale assessment of climate impacts and vulnerability in Norway, *Climatic Change* **64**(1–2), 193–225

Olsthoorn, A.A., P.E. Van Der Werff, L.M. Bouwer, and D. Huitema, 2008: Neo-Atlantis: the Netherlands under a 5-m sea level rise, *Climatic Change* **91**(1–2), 103–22

Paavola, J. and W.N. Adger, 2006: Fair adaptation to climate change, *Ecological Economics* **56**(4), 594–609

Parry, M.L., 1990: *Climate Change and World Agriculture*, EarthScan, London

Pearce, D.W. and D. Moran, 1994: *The Economic Value of Biodiversity*, EarthScan, London

Pizer, W.A., 1997: Prices vs. quantities revisited: the case of climate change, Discussion Paper **98–02**, Resources for the Future, Washington, DC

Plamberk, E.L. and C.W. Hope, 1996: PAGE95 – an updated valuation of the impacts of global warming, *Energy Policy* **24**(9), 783–93

Pratt, J.W., H. Raiffa, and R. Schlaifer, 1995: *Introduction to Statistical Decision Theory*, 1st edn., MIT Press, Cambridge, MA

Ramsey, F., 1928: A mathematical theory of saving, *Economic Journal* **38**, 543–9

Rehdanz, K. and D.J. Maddison, 2005: Climate and happiness, *Ecological Economics* **52**(1), 111–25

Reilly, J.M., M. Sarofim, S. Paltsev, and R. Prinn, 2006: The role of non-CO_2 GHGs in climate policy: analysis using the MIT IGSM, *Energy Journal* (Multi-Greenhouse Gas Mitigation and Climate Policy Special Issue **3**), 503–20

Repetto, R. and D. Austin, 1997: *The Costs of Climate Protection: A Guide for the Perplexed*, World Resources Institute, Washington, DC

Saelen, H., G.D. Atkinson, S. Dietz, J. Helgeson, and C.J. Hepburn, 2008: Risk, inequality and time in the welfare economics of climate change: is the workhorse model underspecified?, Discussion Paper **400**, Department of Economics, Oxford University, Oxford

Schaefer, A. and H.D. Jacoby, 2005: Technology detail in a multisector CGE model: transport under climate policy, *Energy Economics* **27**(1), 1–24

2006: Vehicle technology under CO_2 constraint: a general equilibrium analysis, *Energy Policy* **34**(9), 975–85

Schelling, T.C., 2000: Intergenerational and international discounting, *Risk Analysis* **20**(6), 833–7

Schimel, D., D. Alves, I. Enting, M. Heimann, F. Joos, M. Raynaud, R. Derwent, D. Ehhalt, P. Fraser, E. Sanhueza, X. Zhou, P. Jonas, R. Charlson, H. Rodhe, S. Sadasivan, K.P. Shine, Y. Fouquart, V. Ramaswamy, S. Solomon, J. Srinivasan, D.L. Albritton, I.S.A. Isaksen, M. Lal, and D.J. Wuebbles, 1996: Radiative forcing of climate change, in *Climate Change 1995: The Science of Climate Change – Contribution of Working Group I to the Second Assessment Report of the Intergovernmental Panel on Climate Change*, 1st edn., J.T. Houghton, (eds.), Cambridge University Press, Cambridge, 65–131

Schneider, S.H. and R.S. Chen, 1980: Carbon dioxide warming and coastline flooding: physical factors and climatic impact, *Annual Review of Energy* **5**, 107–40

Smit, B. and J. Wandel, 2006: Adaptation, adaptive capacity and vulnerability, *Global Environmental Change* **16**(3), 282–92

Smith, J.B., 1996: Standardized estimates of climate change damages for the United States, *Climatic Change* **32**(3), 313–26

Smith, J.B., H.-J. Schellnhuber, M.Q. Mirza, S. Fankhauser, R. Leemans, L. Erda, L. Ogallo, A.B. Pittock, R.G. Richels, C. Rosenzweig, U. Safriel, R.S.J. Tol, J.P. Weyant, and G.W. Yohe, 2001: Vulnerability to climate change and reasons for concern: a synthesis, in *Climate Change 2001: Impacts, Adaptation, and*

Vulnerability – Contribution of Working Group II to the Third Assessment Report of the Intergovernmental Panel on Climate Change, J.J. McCarthy, O.F. Canziani, N.A. Leary, D.J. Dokken, and K.S. White (eds.), University of Cambridge, Cambridge, 913–67

Stern, N. *et al.*, 2006:

Svensson, C., Z.W. Kundzewicz, and T. Maurer, 2005: Trend detection in river flow series: 2. Flood and low-flow index series, *Hydrological Sciences Journal* **50**(5), 811–24

Szolnoky, C., K. Buzas, and A. Clement, 1997: Impacts of the climate change on the operation of a freshwater cooled electric power plant, *Periodica Polytechnica: Civil Engineering* **41**(2), 71–94

Titus, J.G., 1992: The costs of climate change to the United States, in *Global Climate Change: Implications, Challenges and Mitigation Measures*, S.K. Majumdar *et al.* (eds.), Pennsylvania Academy of Science, Easton, 384–409

Tol, R.S.J., 1995: The damage costs of climate change toward more comprehensive calculations, *Environmental and Resource Economics* **5**(4), 353–74

1999: Spatial and temporal efficiency in climate change: applications of FUND, *Environmental and Resource Economics* **14**(1), 33–49

2002a: Estimates of the damage costs of climate change – part 1: benchmark estimates, *Environmental and Resource Economics* **21**(1), 47–73

2002b: Estimates of the damage costs of climate change – part II: dynamic estimates, *Environmental and Resource Economics* **21**(2), 135–60

2003: Is the uncertainty about climate change too large for expected cost-benefit analysis?, *Climatic Change* **56**(3), 265–89

2005a: An emission intensity protocol for climate change: an application of FUND, *Climate Policy* **4**, 269–87

2005b: Emission abatement versus development as strategies to reduce vulnerability to climate change: an application of FUND, *Environment and Development Economics* **10**(5), 615–29

2006: Multi-gas emission reduction for climate change policy: an application of FUND, *Energy Journal* (Multi-Greenhouse Gas Mitigation and Climate Policy Special Issue 3), 235–50

2008a: Climate, development and malaria: an application of FUND, *Climatic Change* **88**(1), 21–34

2008b: The social cost of carbon: trends, outliers and catastrophes, *Economics – the Open-Access, Open-Assessment E-Journal* **2**(25), 1–24

2008c: Why worry about climate change? A research agenda, *Environmental Values* **17**(4), 437–70

2009: Why worry about climate change?, *Research Bulletin* **09**(1), 2–7

Tol, R.S.J. and H. Dowlatabadi, 2001: Vector-borne diseases, development and climate change, *Integrated Assessment* **2**(4), 173–81

Tol, R.S.J., K.L. Ebi, and G.W. Yohe, 2007: Infectious disease, development, and climate change: a scenario analysis, *Environment and Development Economics* **12**, 687–706

Tol, R.S.J., S. Fankhauser, R.G. Richels, and J.B. Smith, 2000: How much damage will climate change do?, *World Economics* **1**(4), 179–206

Tol, R.S.J., S.W. Pacala, and R.H. Socolow, 2009: Understanding long-term energy use and carbon dioxide emissions in the USA, *Journal of Policy Modeling* **31**(3), 425–45

Tol, R.S.J. and S. Wagner, 2008: Climate change and violent conflict in Europe over the last millennium, Working Paper **154**, Research unit Sustainability and Global Change, Hamburg University and Centre for Marine and Atmospheric Science, Hamburg

Tol, R.S.J. and G.W. Yohe, 2006: Of dangerous climate change and dangerous emission reduction, in *Avoiding Dangerous Climate Change*, H.-J. Schellnhuber *et al.* (eds.), Cambridge University Press, Cambridge, 291–8

2007a: Infinite uncertainty, forgotten feedbacks, and cost-benefit analysis of climate change, *Climatic Change* **83**(4), 429–42

2007b: The weakest link hypothesis for adaptive capacity: an empirical test, *Global Environmental Change* **17**(2), 218–27

USEPA, 2003: *International Analysis of Methane and Nitrous Oxide Abatement Opportunities: Report to Energy Modeling Forum, Working Group 21*, United States Environmental Protection Agency, Washington, DC

van Kooten, G.C., 2004: *Climate Change Economics – Why International Accords Fail*, Edward Elgar, UK and Northampton, MA

Vaughan, D.G., 2008: West Antarctic ice sheet collapse – the fall and rise of a paradigm, *Climatic Change* **91**(1–2), 65–79

Vaughan, D.G. and J.R. Spouge, 2002: Risk estimation of collapse of the West Antarctic sheet, *Climatic Change* **52**(1–2), 65–91

Weitzman, M.L., 1974: Prices vs. quantities, *Review of Economic Studies* **41**(4), 477–91

2009: On modelling and interpreting the economics of catastrophic climate change, *Review of Economics and Statistics* **91**(1), 1–19

Weyant, J.P., 1993: Costs of reducing global carbon emissions, *Journal of Economic Perspectives* **7**(4), 27–46

1998: The costs of greenhouse gas abatement', in *Economics and Policy Issues in Climate Change*, W.D. Nordhaus (ed.), Resources for the Future, Washington, DC, 191–214

2004: Introduction and overview, *Energy Economics* **26**, 501–15

Weyant, J.P., F.C. de la Chesnaye, and G.J. Blanford, 2006: Overview of EMF-21: multigas mitigation and climate policy, *The Energy Journal: Multi-Greenhouse Gas Mitigation and Climate Policy* (Special Issue **3**), 1–32

Weyant, J.P. and J.N. Hill, 1999: Introduction and overview of the special issue, *Energy Journal Special Issue on the Costs of the Kyoto Protocol: A Multi-Model Evaluation*, **vii–xliv**

Wigley, T.M.L., R.G. Richels, and J.A. Edmonds, 1996: Economic and environmental choices in the stabilization of atmospheric CO_2 concentrations, *Nature* **379**, 240–3

Wilson, K.J., J. Falkingham, H. Melling, and R.A. de Abreu, 2004: Shipping in the Canadian Arctic: other possible climate change scenarios, *International Geoscience and Remote Sensing Symposium* **3**, 1853–6

Yates, D.N. and K.M. Strzepek, 1998: An assessment of integrated climate change impacts on the agricultural economy of Egypt, *Climatic Change* **38**(3), 261–87

Yohe, G.W. and M.E. Schlesinger, 2002: The economic geography of the impacts of climate change, *Journal of Economic Geography* **2**(3), 311–41

Yohe, G.W. and R.S.J. Tol, 2002: Indicators for social and economic coping capacity – moving towards a working definition of adaptive capacity, *Global Environmental Change* **12**(1), 25–40

Zhang, D.D., C.Y. Jim, G.C.S. Lin, Y.-Q. He, J.J. Wang, and H.F. Lee, 2006: Climatic change, wars and dynastic cycles in China over the last millennium, *Climatic Change* **76**, 459–77

Zhang, D.D., J. Zhang, H.F. Lee, and Y.-Q. He, 2007: Climate change and war frequency in Eastern China over the last millennium, *Human Ecology* **35**(4), 403–14

Carbon Dioxide Mitigation
Alternative Perspective

ONNO KUIK

Introduction

Chapter 2 on Carbon Dioxide Mitigation by Richard S.J. Tol includes a survey of assessments of the economic impacts (damages) of climate change, a survey of assessments of the economic impacts (costs) of greenhouse gas (GHG) emissions mitigation measures, a description of the integrated assessment model (IAM) FUND, that was used to compute the damages and costs, and, finally, a report of benefit-cost (B/C) estimates of the FUND model for a number of Copenhagen Consensus CO_2 mitigation scenarios.

The survey of impact assessments is divided into three sections: (1) a survey of assessments of the total economic impacts of climate change, (2) a survey of the assessments of marginal economic impacts, and (3) a discussion of the impacts that are missing in the surveyed assessments. The survey and discussion are very clear and competent, as may be expected from an author who has such a formidable track record in this area. The purpose of the present Perspective paper is merely to add a few observations to this excellent survey.

This Perspective paper will also make a few remarks on the estimated benefit-cost ratios (BCRs) that were computed with the FUND model, with a view to highlighting some of the assumptions that lie behind the reported ratios and to help with their interpretation.

Survey on the Economic Impacts of Climate Change

The number of global assessments of the economic impacts of climate change can be counted on the fingers of two or three hands. Given the contin-

ued public interest in climate change and climate change policies over a period of two decades, and the potentially large social values at stake, this is a remarkable fact. Most of the estimates of the social cost of carbon (SCC), including the celebrated estimate of Sir Nicholas Stern in *The Stern Review*, are variations on a remarkably small set of original studies. Tol (2008:4) has poignantly assessed the deplorable situation of current economic research in this area:

> There are a dozen studies. The number of authors is lower, and can be grouped into a UCL group and a Yale one. Most fields are dominated by a few people and fewer schools, but dominance in this field is for want of challengers. The impact of this is unknown, but this insider argues below that the field suffers from tunnel-vision. This situation is worrying. Politicians proclaim that climate change is the greatest challenge of this century. Billions of dollars have been spent on studying the problem and its solutions, and hundreds of billions may be spent on emission reduction (e.g. Weyant *et al.* 2006). Yet, the economics profession has essentially closed its eyes to the question whether this expenditure is justified.

This is a serious complaint and should be kept in mind in the discussion that follows. While the Copenhagen Consensus is bravely attempting to address the "closed eyes" part of this complaint, it does of course not have the means to address the "want of challengers" part of it.

In his survey of impact assessments, Tol distinguishes between *enumerative* and *statistical* assessment methods. The former method enumerates all physical effects, quantifies them in natural (mostly physical) dimensions, and then attaches economic values to the quantified effects. By contrast, the statistical method makes use of observed variations in expenditures and prices of the same activities in different climatic zones to discern the effect of climate on the economy. Both methods have a longer history of application in other areas of environmental concern (e.g. air pollution) and have known strengths and weaknesses. Tol does not mention a third method, which could be called the *subjective* method. This method directly examines agents' revealed or stated preferences for the mitigation of climate change. An example of this method can, for example, be found in Brouwer *et al.* (2008) who used a survey instrument to elicit the willingness-to-pay (WTP) of air travellers for a tax on their air travel to offset their CO_2 emissions. Brouwer *et al.* (2008) found that 75% of the passengers were willing to pay €25/tCO_2-eq on average. There is a small body of research that takes this subjective approach (see, e.g. Kuik *et al.* 2008) and although it has not resulted yet in a robust assessment of the social costs of climate change, it represents an interesting addition to the more "objective" statistical and enumerative approaches. In particular, it directly addresses the disutility of (perceived) risk, something that the more objective methods have difficulties of coping with. In the words of Brouwer *et al.* (2008:310):

> subjective or perceived risk of climate change is an important additional motivation for tackling climate change. In our survey, people generally dislike being at risk and are willing to pay to reduce their exposure to risks associated with climate change. This reduced disamenity through mitigating climate change is an important economic benefit of action.

Tol writes in the interesting "missing impacts" section of the chapter that the size of an "uncertainty premium" as a benefit of climate change mitigation would have to be based on a political decision. What I try to argue is that economic research can to some extent help the political process to establish such uncertainty premiums in the context of climate change.

Scenarios

Tol develops five different mitigation scenarios. In the first two scenarios, a carbon tax is implemented for a period of ten years only. In the first scenario

the carbon tax is in OECD countries only and in the second a global carbon tax is implemented. Because of the short-term nature of the tax policy the benefits in terms of avoided climate change damage are very small compared to the costs.

The third scenario has a global tax for the entire century, starting at $250/tC ($68/tCO_2) in 2010 and rising with the discount rate (5% per year) to more than $20,000/tC ($5,500/tCO_2) in 2100. The FUND model predicts that this tax would drive CO_2 emissions to zero in the second half of this century. But the costs still outweigh the benefits.

The fourth scenario has a global tax for the entire century starting at $12/tC ($3/tCO_2) in 2010 and rising with the discount rate (5% per year) to almost $1000/tC ($260/tCO_2) in 2100. The BCR is 1 to 4.

The fifth scenario has a smaller global tax of 2$/tC ($0.5/tCO_2) in 2010 and rising with the discount rate (5% per year) to $161/tC ($44/tCO_2) in 2100. The BCR is 3 to 2.

These scenarios basically say that (1) climate change is a long-term problem that requires a long-term policy, (2) an optimal mitigation policy is a global policy that starts with a relatively low tax that increases with the discount rate (compare the Hotelling rule for optimal depletion). This is a conventional and reasonable view on an optimal CO_2 mitigation policy.

There are some issues with the numbers, however.

One can, of course, always quarrel over specific assumptions – like, for example, the height of the discount rate, the marginal utility of income (or equity-weighting), the economic value of impacts on biodiversity, the risk premium for uncertainty, the possible impact of climate change on economic growth, and so forth. These assumptions are ably discussed in Tol's chapter, but it should be noted that the treatment of the assumptions in the calculations leads to very conservative BCRs: i.e. a relatively high discount rate, no equity weighting, zero value for impacts on biodiversity, no risk premium, no effects on economic growth itself.

Further, even for the scenarios that have a policy during the entire century, the cut-off date of the year 2100 has a serious negative effect on the BCRs. Due to the relatively long lag times of climate change (represented in the FUND model), the rather high carbon taxes in, say, the last quarter of this century will reduce emissions that will mitigate climate change mainly after the year 2100. Thus, in the BCRs as presented in the chapter, a significant cost is made in the latter part of this century whose benefit is not accounted for. If the mitigation policy were terminated in say, 2080, we would save more than 20% of the costs (in Net Present Value, NPV), whereas the benefits (in this century!) would likely be little affected.

If the cost side of the equation is taken for granted, I would argue that the presented benefits of mitigation and the BCRs are a bit on the low side. First, as Tol also argues in his chapter, because the ratios do not reflect the substantial concerns about equity and uncertainty; and, second, because a substantial part of the benefits (after the year 2100) is not accounted for.

By Way of Conclusion

Cost-benefit analysis (CBA) has come a long way from appraising public investment projects such as the construction of a water reservoir to the appraisal of policies to mitigate climate change.

At the 2009 Conference of the European Association of Environmental and Resource Economists in Amsterdam, Cameron Hepburn of Oxford University asked himself how far CBA could be stretched before it would break. In contrast to small-scale public investment projects, the climate change problem is:

- international
- intergenerational
- uncertain and ambiguous
- non-marginal
- (partly) irreversible and non-linear.

Cameron Hepburn answered his own question in the affirmative, but there were others in the audience who were more skeptical: if CBA would not break in the appraisal of the mitigation of GHG emissions, then where *would* it break?

The Copenhagen Consensus project offers an interesting opportunity to reflect further on this question.

Bibliography

Brouwer, R., L. Brander, and Van P. Beukering, 2008: "A convenient truth": air travel passengers' willingness to pay to offset their CO_2 emissions, *Climatic Change* **90**, 299–313

Kuik, O.J., B. Buchner, M. Catenacci, A. Goria, E. Karakaya, and R.S.J. Tol, 2008: Methodological aspects of recent climate change damage cost studies. *The Integrated Assessment Journal* **8**, 19–40

Tol, R.S.J., 2008: The economic impact of climate change, ESRI Working Paper, **255**, Dublin

Weyant, J.P., F.C. de la, Chesnaye, and G.J. Blanford, 2006: Overview of EMF-21: multigas mitigation and climate policy, *The Energy Journal: Multi-Greenhouse Gas Mitigation and Climate Policy* (Special Issue **3**), 1–32

2.2 Carbon Dioxide Mitigation

Alternative Perspective

ROBERTO ROSON

Introduction

The purpose of this Perspective paper is to critically review Richard Tol's chapter 2 on Carbon Dioxide Mitigation, and provide a counterbalance to it, thus ensuring the Expert Panel a comprehensive presentation of climate change and its viable solution.

Chapter 2 has three parts. The first part is a general survey on economic modeling of climate change impacts, whose content is basically the same as a recently published paper (Tol 2009). This is a very useful overview which is, however, only indirectly related to mitigation policy. The second part provides some information about the structure and characteristics of the FUND model, which is used to simulate a number of alternative mitigation policies, all based on carbon taxation or emissions trading. Much of the material here is drawn from the technical documentation of the model, available at www.fnu.zmaw.de/FUND.5679.0.html. The third part illustrates the simulation exercises and comments on the results. The working hypothesis here is the availability of a budget of $250 billion per year for a period of ten years, to be spent on climate change mitigation. The FUND model is used to explore a number of policy options, ranking them in order of benefit-cost ratios (BCRs).

Since the FUND model is central to this analysis, it is important to understand its structure, potential, and limitations. I will address this issue in the second section of this Perspective paper. Only after knowing something more about the model's capabilities shall we understand how much we should trust the results, and how they could possibly be interpreted.

In a third section I shall focus on the numerical exercise and output of the model. Finally, a concluding section provides some overall evaluation of chapter 2, proposing also some general thoughts about climate change mitigation policies.

The FUND Model

In the FUND web page, one can read:

> It is the developer's firm belief that most researchers should be locked away in an ivory tower. Models are often quite useless in unexperienced hands, and sometimes misleading. No one is smart enough to master in a short period what took someone else years to develop. Not-understood models are irrelevant, half-understood models treacherous, and mis-understood models dangerous.

This is true. However, I would add that results cannot be trusted, especially for policy guidance, without a certain degree of understanding of model characteristics. Models should never be black boxes. Furthermore, models are based on a number of assumptions and simplifications, which must be recognized when assessing the output of numerical simulations. This is precisely why I am starting this Perspective paper by looking at the structure and features of the FUND model.

The most recent technical description of the model is by Anthoff and Tol (Anthoff and Tol 2008). Unfortunately, this document does not completely describe the model structure, but deals primarily with the climatic and impacts modules. That is, how emissions translate into temperature changes and how a number of climate change impacts are valued. Nothing is said about the number of sectors considered, substitution possibilities, trade, income, and capital flows.

Still, one can understand a number of key features. Perhaps the most relevant one is that: population and *per capita* income follow exogenous scenarios. This should mean that there is no fully fledged (dynamic) economic model inside. Also, neither climate change impacts nor policy (mitigation or adaptation) affect (potential) economic growth. Actual growth is influenced by a number of impacts, modeled by specific equations, in which one impact (often, but not always, valued in monetary terms) is a function of temperature (level and change). The ad hoc equations appear to be reduced form relationships derived from sectoral micro-studies. The reliability of these relationships therefore depends on the quality of the underlying studies, which seems to be variable. For example, the most recent version of FUND includes an extreme weather module, expressing the economic damage due to an increase in the intensity of tropical storms. I was quite surprised to find such a function in the model, as I know there is no consensus among climatologists about how and whether climate change affects the number, location, and intensity of storms.

Another rather obscure point is the link between impacts and national income, on the one hand, and between costs and GDP, on the other hand. National income is a flow variable, accounting for market transactions, but impacts relate also to stock variables (e.g. water resources) and non-marketed goods and services (e.g. ecosystems).

Mitigation and adaptation costs are not always macroeconomic costs, because what is cost for one individual may be income for another. For example, suppose that dikes are built to protect the seacoast, or expenditure on health services prevents the spreading of diseases, related to climate change. These are monetary transactions between agents within the same economic system. There is no loss of primary resources; therefore only second-order effects (whose sign is *a priori* ambiguous) affect GDP and income.

How are these costs considered in the FUND framework, and how is income influenced? Do these costs enter in the BCRs? We do not really know.

Carbon dioxide (CO_2) emissions are calculated on the basis of the Kaya identity, and therefore depend on economic growth as well as on energy and carbon efficiency. Energy and carbon efficiency may be affected by policy intervention, but it is not clear how. Methane (CH_4) and nitrous oxide (N_2O) emissions are exogenous. Sulfur hexafluoride (SF_6) emissions are linear in GDP and GDP *per capita* (exogenously given). Sulfur dioxide (SO_2) emissions follow the growth of population (elasticity 0.33), fall with *per capita* income (elasticity 0.45), and fall with the sum of energy efficiency improvements and de-carbonization (elasticity 1.02). There is no option to reduce SO_2 emissions.

The FUND model does not (explicitly) account for technological progress (except through trends in efficiency improvements, and other ad hoc formulations of abatement costs). It is not possible to simulate policies aimed at fostering climate-friendly technologies. There are no backstop technologies. All these characteristics suggest that mitigation and adaptation costs may be overestimated.

What policy instruments are available in FUND, and what policies can be simulated? Again, this is not well explained in the technical documentation, but the simulation exercise illustrated in chapter 2 relies on a pure carbon tax which, ideally, would be equivalent to a perfect emissions trading regime (with no exemptions, no transaction costs, no uncertainty and – I think – no banking). Still, even if we are sure that carbon taxation can be simulated, it is not clear how this brings about a reduction of emissions in the model and how tax revenue is redistributed (or, equivalently, how emissions rights are allocated). What is known from other papers and models is that the way rights are allocated does make a huge difference in the results. Property rights or redistribution schemes can be cleverly designed to realize a system of incentives which could dramatically lower the costs of mitigation and adaptation policies, especially when these are linked to technological improvements, knowledge, and research.

Simulation Scenarios

The Copenhagen Consensus on Climate 2009 (hereafter, CCC09) hypothetically dispenses $250 billion per year on climate policy for a period of ten years. Five scenarios are considered in chapter 2.

In the first scenario, the countries of the OECD implement a uniform carbon tax such that the net present value (NPV) of the abatement cost equals $2 trillion, the NPV of $250 billion per year for ten years. The discount rate is 5% per year (which is a lot!). Costs are discounted to 2009. This is achieved by a carbon tax of $700/tC, starting in 2010 and rising with the discount rate. The carbon tax is 0 from 2020 onwards.

In the second scenario, all countries implement a carbon tax of $250/tC in 2010, rising with the discount rate, but returning to 0 in 2020. This also leads to an abatement cost of $2 trillion.

In the third scenario, it is assumed that climate policy after 2020 will continue as before. That is, the carbon tax keeps rising with the discount rate between 2020 and 2100.

In the fourth scenario, for ten years, $250 billion is invested in a trust fund. This trust fund finances a century-long programme of emission abatement such that the NPV of the abatement cost over the century equals $2 trillion. This is achieved by a uniform carbon tax for all countries, which starts at $12/tC in 2010 and rises with the discount rate.

In the fifth scenario, only a part of the $250 billion is invested. The carbon tax in 2010 is set equal to the Pigou tax ($2/tC), also known as the marginal damage costs of CO_2 emissions and the social cost of carbon. The carbon tax is applied worldwide, and is equal for all countries.

Before looking at the results, let me comment on the characteristics of the five scenarios, and what we should expect. Using Tol's own words (Tol 2009: 29):

Climate change is the mother of all externalities: larger, more complex, and more uncertain than any other environmental problem. The sources of GHG emissions are more diffuse than any other environmental problem. Every company, every farm, every household emits some GHGs. The effects are similarly pervasive ... Climate change is also a long-term problem. Some GHGs have an atmospheric lifetime measured in tens of thousands of years.

Therefore, if the climate change externality is "so much global" (in terms of space and time) we can expect that policies which affect a limited number of regions for a limited number of years should be quite ineffective, whereas policies involving all countries for long time should be preferred. We do not need a model to understand this.

Second, in which sense is a carbon tax "a cost"? From basic public and welfare economics we know that any tax generates revenue and revenue should be accounted for in the total welfare, as well as in the gross domestic product (GDP). Costs associated with taxation are only due to price distortions (e.g. deadweight losses) and dynamic inefficiency (if one can prove that the economy grows less because of taxation). We do not know if and how the carbon tax revenue has been redistributed in the model. As mentioned above, this little detail makes a lot of difference in the real world.

A similar kind of reasoning applies to the third and fourth scenarios, where money goes to emissions abatement. Let me first say that forcing the economy to spend the money in emissions abatement is like adding a constraint, so results cannot be better than the case where a price for carbon is introduced, but consumers and firms are free to choose whether to reduce emissions, or to abate, etc. Anyway, who gets the money spent on carbon abatement? This is left unexplained.

The last scenario departs from the CCC09 prescriptions, and simply applies a Pigouvian tax. Surprisingly, the author assumes, as something obvious, that the marginal damage cost of carbon emissions is known and equal to only $2/tC. But this is not at all obvious. There is no consensus on this and many alternative estimates are available in the literature. Furthermore, I think that the same model used for the simulation exercises (FUND) has also been used to estimate the marginal damage cost! Now, in a perfect world with perfectly competitive markets, the introduction of a perfect Pigouvian tax is a first-best policy, which must bring about a total welfare improvement, unless the tax is paid to the Mars economy.

Having said all this, now let us look at the results and see whether we can find any surprises.

The two policy scenarios that concentrate effort in the first decade are less effective than the scenario that spends the same amount of money over the century. This does not come as a surprise. Even the $2/tC century-long policy is about as effective in the long run as the $250/tC decade-long policy, and at a fraction of the cost. If the $250/tC initial carbon tax is maintained over the century, CO_2 emissions fall by more than 90% in 2050 and by almost 100% in 2100 compared to the baseline; 2050 emissions are some 20% of 2000 emissions in this scenario. No surprise here.

Emissions respond only slowly to policy, and concentrations respond even more tardily. A $12/tC initial carbon tax would almost stabilize the CO_2 concentration at around 680 ppm. An initial carbon tax of $250/tC would keep the concentration below 450 ppm; as other greenhouse gases (GHGs) are uncontrolled, the temperature continues to rise to 2.4°C above preindustrial in 2100.

Table 2.3 in the chapter shows the net present costs and benefits as well as the BCRs of the five alternative policy scenarios. The five scenarios are ordered in the intensity of climate policy in the OECD in the coming decade. Starting at the bottom, spending a lot of money on CO_2 emission reduction in the near term in the OECD does not pay off, as expected. A much greater benefit can be achieved if the same money is used to finance worldwide abatement – essentially because emission reduction is cheaper in poorer countries – but the BCR is about 1 to 100. If the same programme is repeated decade after decade, abatement costs go up considerably but benefits rise even faster. Still, the BCR is only 1 to 50. The BCR improves considerably if the $250 billion is spent over the century rather than over the decade, of course. A BCR of 1 to 4 is the result. If the initial carbon tax is set equal to the estimated marginal damage cost, the BCR unsurprisingly exceeds unity: 3 to 2. Over the century, this policy spends only 1/20th of the funds (hypothetically) available to CCC09.

Conclusion

FUND is an interesting model whose main advantage is the integration, often in the form of reduced form relationships, of many sectoral micro-models,

coming from different scientific areas. This is very important in a field like climate change science, which is intrinsically multidisciplinary. This advantage, however, becomes a disadvantage to the extent that heterogeneity in the model components creates internal inconsistency. How severe a problem like this can be in the FUND model is difficult to say, as its technical documentation is not very informative.

Richard Tol's chapter 2 is largely based on the FUND model and the results of a set of simulation exercises, where a number of policy options are explored and assessed. In this Perspective paper, I have noted a series of limitations of the FUND model, as well as some other points which remain quite obscure and limit the interpretation of the results. However, when considering the simulation scenarios, I also made some general remarks, which are confirmed by the model results and bring me to think that we could have got about the same findings with a different model (or possibly without any model at all!). In other words, we can trust the results even if we do not trust the model.

Furthermore, the considerations above cause me to think about the usefulness of assessing mitigation policy through numerical simulation, in which hypothetical but simple carbon taxation or an emissions trading scheme (ETS) is implemented. Mitigation policy in the real world is much more complicated. Think about the European ETS, as an example. There are industries exempted, there are relevant transaction costs, informational asymmetries, uncertainty. In short, there are many implementation details that cannot be easily captured by a stylized model but ultimately may make a difference between a successful and an unsuccessful scheme. In this sense, I think the keyword should be "incentives." Successful mitigation policies should aim at creating systems of incentives (often implying positive externalities), especially in the presence of technological and organizational innovation.

I am a modeler myself and I am naturally sympathetic to all efforts aimed at using quantitative tools for policy assessment. There is no perfect model and all models imply simplifications of some kind. The important thing is not to hide them under the carpet which, in this context, would take the form of a complicated set of equations, unexplained assumptions, etc. This kind of danger is much more present in a multidisciplinary IAM, like FUND, because, for example, it would be very difficult for an economist like me to notice that an equation in the climatic submodel was inconsistent with another one, say, about impacts on water resources.

Bibliography

Anthoff, D. and R.S.J. Tol, 2008: The climate framework for uncertainty, negotiation and distribution (FUND), Technical Description, Version 3.3, www.fnu.zmaw.de/FUND.5679.0. html

Tol, R.S.J., 2009: The economic effects of climate change, *Journal of Economic Perspectives* **23**(2), 29–51

Forestry Carbon Sequestration

BRENT SOHNGEN*

Introduction

There is widespread belief now that forests can be used to reduce the costs of slowing climate change. While the role of forests in the global carbon cycle has long been acknowledged, recent discussions within the context of the United Nations Framework Convention on Climate Change (FCCC), as well as efforts to write climate change legislation in the USA, have emphasized the role forests might play. The most recent policy efforts have focused on near-term actions to reduce deforestation in tropical countries.

The rationale for considering forests at all in the policy mix stems partly from the physical components of the issue. The world's forest estate is exceedingly large: It contains roughly 3.9 billion ha of forestland and 1 trillion tons[1] of CO_2 (UN Food and Agricultural Organization 2006). Current estimates indicate that roughly 11 million ha each year are lost in tropical regions due to deforestation and conversion of land to agriculture (Houghton 1999, 2003). These losses cause emissions of about 3.6–4.5 billion tons of CO_2, so that deforestation accounts for around 17% of global carbon emissions. Countries like Indonesia and Brazil are near the top of total emissions when estimated by country if deforestation is included in carbon emission calculations. Efforts to slow these emissions, of course, could have enormous benefits for society.

In contrast to the story in the tropics, the area of forests in temperate zones is fairly stable. Carbon stocks are increasing in most temperate regions as forests continue to age (Smith *et al.* 2003, Sohngen *et al.* 2005), although perturbations around natural cycles can cause large emissions (Kurz and Apps 1999). Current estimates from the Intergovernmental Panel on Climate Change (IPCC) suggest that northern forests presently may sequester 3.2 billion tons of CO_2 per year currently (IPCC 2007). Growth in northern forests may offset much of the loss in tropical zones. Efforts to increase these carbon stocks by changing management, increasing forest area, or shifting species, could also help reduce net emissions of greenhouse gases (GHGs) and could benefit society.

Can avoiding deforestation in tropical areas, reforesting old agricultural lands in temperate and tropical regions, or changing management practices increase the total uptake of carbon into the forests? Several studies so far suggest that forest actions can cost effectively provide roughly 30% of the total global effort needed in all sectors to meet climate mitigation strategies (Sohngen and Mendelsohn 2003; Tavoni *et al.* 2007). This chapter examines these and other results in the literature, and argues that the evidence clearly indicates that forests should be an important part of any national or global strategy aimed at avoiding climate change. If society is both serious about climate mitigation and serious about containing costs, there is little choice but to develop programs that increase the stock of carbon in forests.

Of course, developing a program that fundamentally alters future land use by valuing carbon stored on the landscape will not be easy, or cheap. It will require that countries agree to manage their forest resources in different ways

* The author very much appreciates the comments of several reviewers, including Sabine Fuss. He would also like to acknowledge the generous funding of the US Environmental Protection Agency, Climate Change Division, and the US Department of Energy, office of Biological and Environmental Research, for development of the forestry modeling tools employed in this analysis.
[1] Tons in this paper are metric tons (MT), or 1,000 kg.

(e.g. to value maintenance of the stock over conversion to agriculture). It will require the development and innovation of new systems for measuring monitoring and verifying the carbon gains that are made, whether these systems are accomplished with satellites or the proverbial boots on the ground. It will require new types of services that can assemble carbon and deliver it to an emerging carbon market place. None of this will in fact be easy, but if incentives are large enough then there is no reason to believe that carbon in forests will not become an important, valued commodity across the landscape.

To examine the potential for carbon sequestration in forests, this chapter begins with a brief discussion of the categories of costs that are important to include in an analysis of forestry options. Then, several of the forestry options that have been widely discussed in the literature are presented. The technical components of these options are briefly described to provide readers with some general background. The chapter then presents estimates of the potential costs of a large-scale carbon sequestration program, considering which options appear most cost effective and manageable, and which options may be more difficult. The chapter then examines new calculations of the benefits of forest carbon sequestration options derived from integrating a forestry and land use model with a global integrated assessment model. Finally, the chapter describes some of the limitations associated with implementing a large-scale forest carbon sequestration program.

Cost Categories

It is perhaps useful to begin with a discussion about the categories of costs that should be considered when addressing the economics of forest carbon sequestration. The most important category for land-based activities like forestry is *opportunity cost*. Opportunity costs are the costs of holding land in forests. Opportunity costs arise because land has other potential uses and those uses would also provide value, thus opportunity costs are defined as the value of the next-best alternative use of the land. If one converts cropland to forests to sequester carbon, the opportunity costs are the value of the forgone returns to agriculture.

A second cost category is the *implementation and management cost*, which includes all the direct costs of installing or implementing a practise and maintaining and managing it over time. Implementation costs include those costs that can be directly attributed to the action. For example, the costs of buying seedlings to plant and the costs of the labor to plant the seedlings are implementation costs. The costs of herbicide or nutrient treatments used to increase the value of the stand over time would also be included in this category. In addition, any costs of thinning or ultimately harvesting a stand would be considered here as well, although one must also be careful to include the benefits of thinning and harvesting operations in the analysis, as discussed below.

A third category is *measurement, monitoring, and verification costs (MMV)*. These costs include the costs of measuring the carbon in areas that have undergone afforestation, or improved management. They also include the costs of monitoring and verifying stands to ensure that the carbon under contract actually is there. While these costs will be very important to consider in forest carbon cost analyses, they often are assumed to be programmatic in nature, and they are ignored. That is, authors typically measure the opportunity, implementation, and management costs, but assume that MMV will be undertaken programmatically so that the costs are not borne by the individual actors. The studies on which this chapter reports by and large do not account for MMV costs, but an analysis is conducted below to assess the potential implications of these costs on benefits and costs.

A fourth category is *other transaction costs*. The modifier "other" is used here because some authors include MMV costs in transaction costs. Other transaction costs are any other unaccounted-for costs associated with developing and implementing contracts for carbon sequestration on the landscape. These could include the time costs of learning about the biology of carbon sequestration, the costs of hiring lawyers to draft contracts, the costs third parties impose to bring together buyers and sellers, etc. There are many potential categories of these types of costs, some of which may be borne directly by buyers and sellers, and some of which may be more programmatic in nature. Most of the existing literature on carbon sequestration costs does not include these costs, and there is actually very little literature on what the extent of these costs may be. The estimates provided in this study do not include them but, as with MMV costs, an analysis is conducted below to consider them.

A final category of costs is called *system-wide adjustment costs*. These costs arise specifically from the design of the sequestration program. One example of this type of cost is leakage, which occurs when an incomplete program is developed. Such a program may provide incentives only for some forest options or some regions of the world. Subsequent adjustments in timber prices in the market may cause shifts in other regions that offset the sequestered carbon. Another important secondary effect of forest carbon sequestration may occur in land markets and land prices. For instance, if reducing deforestation reduces the area of productive agricultural land, then crop or livestock prices could rise. These rising prices would be expected to increase the opportunity costs of land. These secondary effects could have important implications for estimation of carbon sequestration costs. Some studies do in fact model both forest and agriculture, and thus capture these secondary effects,[2] while most studies do not. Neither of these issues is explicitly addressed in the cost estimates provided in this study, although the study does discuss the potential implications of systemwide adjustment costs below.

In summary, the cost estimates discussed in this section and with the forestry model focus on opportunity costs and implementation and management costs. The cost estimates also account for any timber market benefits that may accrue to the activities through timber harvesting. These benefits may be particularly important for afforestation and forest management activities. MMV and other transaction costs, while important, are not considered in this section of the chapter, but will be addressed later. In addition, discussions about leakage and secondary market effects are saved to later in the chapter.

Options for Carbon Sequestration in Forests

Afforestation

Afforestation has been the most widely recognized and studied option for mitigation using forests to date. Afforestation refers to taking agricultural land and converting it into forests. Because agricultural land stores very little carbon in aboveground biomass, converting the land to trees, and allowing those trees to grow, will remove carbon from the atmosphere. A forest that is growing can remove 5–11 tons of CO_2 per ha per year, depending on location and productivity. A large proportion of the world's crop and grazing lands are rain-fed, indicating that they also can support trees. As a result, there are many opportunities to sequester carbon by converting this agricultural land into forests.

Of course, converting land from agriculture to forests comes with a cost. Afforestation requires implementation and management costs, as well as opportunity costs. Depending on the region, tree species, labor costs, site quality and other factors, planting and managing trees may cost $700 to more than $3000 per ha in present value (PV) terms. As noted above, opportunity costs associated with converting land from agricultural uses to forest will also be important, and they will depend on the value of the land in agricultural production. The costs are important, but it is also important to recognize that there may be future benefits to planting trees. That is, because afforestation ultimately leads to standing forest stocks with potentially valuable timber, there may be some future benefits that can reduce the costs. When measuring the net costs of afforestation, all of these categories (planting, management, opportunity costs, and benefits) must be included.

The main reason why afforestation is so widely acknowledged as having large potential throughout the world relates to the rather substantial value of the carbon embodied in forests. Consider a southern upland hardwood forest in the USA, which may

---

[2] The only model we report on that captures the full range of price effects across markets is the work of Murray *et al.* (2005).

typically be harvested at age 50. A stand like this may contain 257 tons of CO_2 per ha in aboveground carbon (Sohngen *et al.* 2009). If there is no value to carbon sequestration, under current timber prices, such a stand would have a typical return of $30–$40 per ha per year. If, however, carbon prices are $14 per ton of CO_2, then annual returns (inclusive of timber harvests) would be $75–$80 per ha per year, and if they rise to $28 per ton of CO_2, then annual returns increase to $130–$140 per ha per year. The increase in returns to planting forests when the embodied carbon is valued is substantial, for higher carbon prices it quickly makes forest competitive with some crop and grazing land.

Many estimates of the sequestration potential for afforestation have been made over the years. Sedjo (1989) presented the first economic analysis of the potential, finding that forest plantations could sequester up to 10.7 billion tons of CO_2 per year for less than $2 per ton of CO_2. That study assumed that crop and grazing land was very cheap and could readily be converted to forests. Subsequent analysis suggests that these estimates may be too optimistic – at least with respect to the costs. For example, a global land use model by Sohngen and Mendelsohn (2003, 2007) suggested that 0.7–2.2 billion tons of CO_2 can be sequestered globally per year for $8–$30 per ton of CO_2. Richards and Stokes (2004), in one of the most thorough reviews of the literature to date, find that 7.0 billion tons of CO_2 per year may be sequestered globally, but that the costs could be as much as $41 per ton of CO_2. All of the estimates discussed in this section account for opportunity costs, and installation and management costs, but not for MMV costs, other transaction costs, or systemwide costs.

Reductions in Deforestation

Since 2005, much attention has focused on the idea that reductions in deforestation could reduce emissions of carbon dioxide (CO_2) into the atmosphere, and also be a relatively low-cost option for mitigation. Of course, deforestation has always been an important contributor to carbon emissions, so it is surprising it took the policy makers so long to get engaged in the issue. Given the scale of deforestation globally, interests among some develop-

ing countries to achieve larger reductions in net emissions sooner rather than later, and the interests of environmental non-governmental organizations (NGOs), avoided deforestation is now widely recognized as a vital ingredient for international climate negotiations.

Deforestation causes about 5 billion tons of CO_2 emissions per year, or around 17% of total global emissions (IPCC 2007). From a technical standpoint, avoiding deforestation makes great sense. Standing tropical forests may contain 300–400 tons of CO_2 per ha in biomass (Kindermann *et al.* 2008). If these standing forests are converted to agriculture, some wood may make its way into markets, but the vast majority of it will be burned on site when the land is converted. Other wood will decompose over time. Either way, when standing tropical forests are converted to agriculture there is a relatively quick emission of carbon into the atmosphere. Holding that carbon on the landscape in trees can substantially alter net global emissions each year.

The value of holding this carbon on the landscape is exceedingly large. If carbon prices are $14 per ton of CO_2, the annual rental value of the carbon embodied in a standing tropical forest with 350 tons of CO_2 in measureable aboveground carbon is $245 per hectare per year.[3] If carbon prices double, to roughly $30 per ton of CO_2, then rental values would be $525 per ha per year. Values this high would compete with agricultural production in some of the world's most productive regions. They are sure to compete with agricultural production in the tropics at the forest–agricultural margin where, by definition, opportunity costs are low. Unlike afforestation, there are no upfront costs associated with planting and managing these forests. One needs to arrange to pay a rental fee to maintain the stock (i.e. to cover the opportunity costs), but these fees do not need to include large-scale outlays to plant and manage timber. The estimates of costs of avoided deforestation presented in this chapter thus

---

[3] Prices are assumed constant for these estimates of rents, and under those circumstances, the annual rental value is calculated as $r^*P_C^*(tCO_2$ per ha), where r is the interest rate (assumed to be 5% in this case) and P_C is the price of carbon dioxide.

include only opportunity costs, and losses associated with not harvesting timber. It is widely recognized, however, that there may be some institutional difficulties associated with accomplishing deforestation reductions in developing countries, and thus there may be some other transaction costs that are important. These costs will alter the quantity of carbon obtained, as discussed in the analysis on transaction costs below.

Recent estimates indicate that slowing or stopping this deforestation could have important consequences for the atmosphere. Estimates by Kindermann et al. (2008) suggest that for $30 per ton of CO_2, around 2.8 billion tons CO_2 emissions per year could be reduced in tropical regions by avoiding deforestation. These estimates in Kindermann et al. (2008) are derived from global land use models and they tend to be higher than many other estimates that have so far been done. See also Murray et al. (2009). However, even these estimates imply that there is great hope that avoided deforestation can be a low-cost option that is meaningfully applied to climate policy. As noted in their paper, the estimates in Kindermann et al. (2008) do not account for MMV costs, other transaction costs, or systemwide adjustment costs.

Forest Management

The third mitigation option considered in this chapter involves forest management. There is a surprisingly wide range of options available to increase carbon through such management. Some of the options would provide carbon benefits in the near term, while others would provide longer-term benefits. In managed forests, the quickest way to increase carbon on the landscape is to increase the forest rotation age (Sohngen and Brown 2008). Even small increases in forest rotations, when implemented over large areas with millions of ha, could produce measurable increases in carbon stock on the landscape. Given that many of the world's intensively managed plantation forests are managed in rotations, with timber outputs in mind, these landowners could be persuaded to extend their rotation if the carbon price were high enough. Sohngen and Mendelsohn (2003), Murray et al. (2005), and

Sohngen and Sedjo (2006) all suggest that increases in the rotation age could be an important component of any carbon policy that values carbon stored on the landscape.

Over the longer run, many additional management strategies can be undertaken to increase total carbon in the forest. For instance, it is always possible to bring new forests under management. Planting forests rather than relying on natural regeneration after harvest, or forest fire, or other disturbance can increase the rate of carbon accumulation in early years and increase the overall quantity of carbon on the site in the long run (Hoehn and Solberg 1994). Alternatively, shifting forests from one type to another can increase total carbon sequestration across the landscape (Sohngen and Brown 2006).

Summary Estimates of the Costs of Carbon Sequestration Options

A marginal costs (MC) curve for carbon sequestration in global forests, including estimates for temperate/developed regions and tropical regions separately, is shown in figure 3.1, using data derived from the IPCC (2007). The MC in figure 3.1 are derived from three global land-use models, where the models are run under differing assumptions about current and future carbon prices. The results are summarized for the year 2030 only. Estimates shown in figure 3.1 indicate that up to 13 billion tons of CO_2 per year may be sequestered in the world's forests in 2030 for $100 per ton of CO_2 (figure 3.1). For low carbon prices (e.g. $0–$20 per ton of CO_2), most of the carbon is derived from activities undertaken in tropical countries. As prices rise, developed/temperate countries become a larger share of the total, but they do not exceed tropical potential over this range of CO_2 prices for the year 2030.

These results can be disaggregated by considering a single carbon price ($30 per ton of CO_2), and calculating the carbon sequestered by different activities (afforestation, forest management, and reduced deforestation) in different regions of the world. For this disaggregation, only one of the models used to calculate the marginal cost curves in figure 3.1 is used, namely the global timber model of Sohngen and Mendelsohn (2003, 2007), but the results are supplemented with estimates of costs

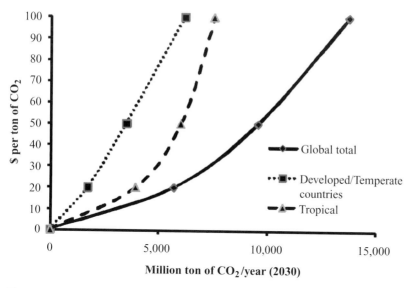

Figure 3.1 *Marginal cost functions for carbon sequestration (2030)*

Notes: Cost estimates include opportunity costs and implementation and management costs, but not MMV and other transaction costs.

Source: Data from IPCC (2007). Cost estimates include opportunity costs and implementation and management costs, but not MMV and other transaction costs.

from regional studies where such studies have been conducted (table 3.1). The annual sequestration potential is averaged for the period 2020–50.

Table 3.1 illustrates that most of the carbon potential over the 2020–50 time period results from avoided deforestation (reducing emissions from deforestation in developing countries, REDD) in tropical countries, followed by forest management in temperature and boreal regions, and finally by afforestation. At \$30 per ton of CO_2, 6.8 billion tons of CO_2, or about 15% of the total emission of CO_2 and of CO_2 equivalents currently can be sequestered. What is perhaps most surprising is that the economic estimates presented in table 3.1 indicate that the largest share of carbon potential is derived from REDD and forest management. The focus of policy over the past ten–fifteen years has been afforestation, and while afforestation is important, it represents the smallest potential share of carbon. The results in table 3.1 are largely consistent with other compilations of results that have been conducted over the years (e.g. Sedjo *et al.* 1995; Richards and Stokes 2004; van Kooten *et al.* 2004). Estimates for REDD are based on Kindermann

et al. (2008), and the estimates in that study are substantially more expensive than other recent estimates, such as Blaser and Robledo (2008), Eliasch (2008), and Grieg-Gran (2008), for example.

Forestry Program Implications

To assess the implications of the forestry program for the overall control of greenhouse gases (GHGs), it is useful to combine these results with an IAM. Sohngen and Mendelsohn (2003) conducted the first such analysis by linking their global forestry model with the DICE model of Nordhaus and Boyer (2000). They found that forestry could efficiently accomplish 30% of the total abatement across the century (e.g. from the present to 2100). A subsequent analysis by Tavoni *et al.* (2007) utilized an updated version of the same land-use model, but a different integrated assessment model, considered how forestry would affect the costs of meeting a 550 ppmv (parts per million by volume) CO_2 concentration target. That study found that forestry would also be about 30% of the total mitigation

Table 3.1 Average annual potential net emissions reductions through forestry for the period 2020–50

	Afforestation	REDD[a]	Management	Total
	Million tons of CO_2 per year for 2020–50			
Temperate				
USA	471 (325–2,267)[c]	0	291 (268–314)[c]	762
Canada	87[b]	0	148[b]	234
Europe	32[b]	0	132[b]	164
Russia	25[b]	0	414[b]	439
China	104[b]	0	348[b]	451
Japan	34[b]	0	25[b]	59
Oceania	24[b]	0	21[b]	45
Total temperate	777	0	1,378	2,155
Tropics				
South & Central America	356[b]	1,209 (800–1600)[d]	0	1,565
SE Asia	288[b]	402 (141–1153)[d]	696[b]	1,387
Africa	258[b]	1,216 (884–1407)[d]	0	1,474
India	168[b]	0	2[b]	170
Total tropics	1,070	2,827	698	4,595
Total all	**1,848**	**2,827**	**2,076**	**6,751**

Notes: Compilation from various studies. Cost estimates include opportunity costs, and implementation and management costs, but not MMV and other transaction costs. Carbon price assumed to be constant at $30 per ton of CO_2.
[a] REDD = Reductions in emissions from deforestation and forest degradation.
[b] Global Timber Model (Sohngen and Mendelsohn 2003, 2007).
[c] Range from Adams *et al.* (1994), Plantinga *et al.* (1999), Stavins (1999), Murray *et al.* (2005), Lubowski *et al.* (2006), Sohngen and Mendelsohn (2007).
[d] Kindermann *et al.* (2008).

effort over the century, but that it could reduce the costs of meeting the fairly strict carbon cap by around 40%.

For this chapter, the earlier analysis of Sohngen and Mendelsohn (2003) is updated using a new version of the land-use model described in Sohngen and Mendelsohn (2007), and the new version of the DICE model described in Nordhaus (2009). Two scenarios are conducted. First, an "optimal" scenario is considered, in which the original optimal policy scenario from Nordhaus (2009) is adjusted to account for land-based sequestration. Because land-based sequestration is fairly large, the models are iterated until the prices and quantities of sequestration in the two models are the same.[4] Second, a scenario that limits the overall temperature increase

to 2°C above preindustrial levels is examined. The resulting carbon prices, carbon sequestration, and temperature change over the coming century are shown in table 3.2.

In this analysis, estimates of three potential benefits associated with including forestry in the greenhouse gas (GHG) control program are calculated for each scenario: changes in consumption, changes in damages, and changes in energy abatement costs. Note that in some cases, changes in consumption could be negative, resulting in losses rather than gains. In all cases, the value of the benefits presented in the chapter are PV calculations, using the internal interest rates (rate of return on capital, IRR) calculated by the DICE model. These interest rates start around 5.6% and fall over time to about 5.0% by 2100. The costs of the forestry program are calculated as the quantity of the carbon sequestration

[4] See Sohngen and Mendelsohn (2003) for the methods used.

Table 3.2 Carbon sequestration pathways for combined forestry and DICE model for the optimal scenario and a maximum 2°C temperature change 2010–2100

	NC	Optimal scenario			2° C Limit		
	△° C	Sequestration ton of CO₂/yr	$/t CO₂	△° C	Sequestration ton of CO₂/yr	$/t CO₂	△° C
2010	0.85	5,727	7.23	0.83	5,471	8.51	0.83
2020	1.09	5,552	11.07	1.02	5,952	13.16	1.02
2030	1.34	6,076	14.09	1.21	6,508	17.50	1.20
2040	1.59	5,998	17.54	1.39	7,134	23.10	1.37
2050	1.84	6,114	21.45	1.56	6,877	30.50	1.53
2060	2.08	4,679	25.87	1.74	7,299	40.56	1.67
2070	2.33	3,951	30.85	1.91	8,817	54.45	1.80
2080	2.57	3,658	36.43	2.08	10,580	73.68	1.90
2090	2.81	4,762	42.69	2.24	14,234	99.49	1.96
2100	3.05	5,078	49.68	2.39	15,192	130.83	1.99

provided times the current carbon market price. For forestry program costs, PV calculations are also made using the IRRs calculated by DICE.

Under the optimal scenario, the introduction of forestry amounts to a cumulative 516 billion tons of CO_2 sequestered in forests (additional to the baseline) or an increase of about 17% relative to the baseline, and an increase in about 900 million ha of forestland. As in Sohngen and Mendelsohn (2003), carbon prices fall only modestly in the optimal scenario, by around 2–3% over the decade. With this small decrease in carbon prices, energy abatement costs decline by only 7%, or $66 billion, relative to the baseline case. Forestry has important implications for the temperature change experienced over the century. The increase in total abatement effort reduces the temperature change by the end of the century by about 0.2°C (2.39°C vs. 2.59°C in the optimal case with energy only). This leads to a reduction in damages of $1,042 billion (table 3.3). While the reduction in damages ordinarily would increase consumption over time, to get these benefits, society must spend money on the forestry program. In net, consumption declines modestly, by $29 billion. The sum of these benefit categories is therefore $1,081 billion ($1042 + $68 − $29). The forestry program costs $1,062 billion, suggesting a benefit-cost ratio (BCR) of around 1.02 if all of these benefits are considered.

Table 3.3 B/C estimates for the optimal scenario and the scenario that limits temperature increase to 2°C with interest rate (r) = 5%; and the same two scenarios with lower interest rates (r = 3%), both assuming no transaction costs

	r = 5%		r = 3%	
	Optimal	2°C limit	Optimal	2°C limit
	Billion US$ (PV)			
Benefits				
Consumption gain	(29)	832	164	4,970
Reduction in damage	1,042	496	10,247	2,436
Reduction in energy costs	68	2,679	2,191	13,294
Total benefit	1,081	4,007	12,602	20,701
Forest cost	1,062	2,297	11,651	11,918
BCRs				
All benefits	**1.02**	**1.74**	**1.08**	**1.74**

In the 2°C limiting scenario, carbon prices are substantially higher and the forestry program is substantially bigger, particularly towards the end of the century when the 2°C temperature limit ultimately becomes binding (table 3.2). The size and scope of the forestry program in the 2°C limiting scenario is very similar to the optimal scenario for

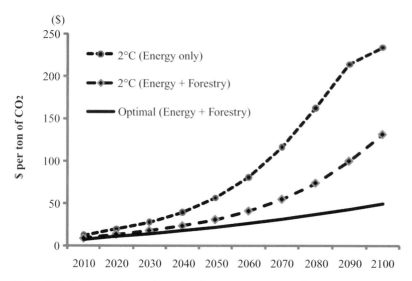

Figure 3.2 *Carbon price paths under three scenarios (r = 5%)*

the first forty years; however, it diverges after that as carbon constraints on the economy become more binding. From an economic perspective, the most important implication of the forestry program is that it reduces overall costs of meeting this very stringent temperature limitation dramatically. For example, when forestry is included as a control, carbon prices fall by over 50% over the century (figure 3.2).

Due to the relatively large reduction in carbon prices, energy abatement costs fall substantially more than the optimal scenario, by 56%, or $2,679 billion in PV terms. Because both the scenarios with and without forestry have similar temperature profiles (due to the ultimate 2°C limit), the reduction in damages when comparing this case with and without the forestry program included is only $496 billion in this scenario. The change in consumption, however, is positive, and it amounts to an increase of $832 billion. The sum of these three benefits is $4,007 billion. The cost of the forestry program is $2,297 billion, suggesting a BCR of 1.74 (table 3.3).

Implications for Policy Design

These results have important implications for policy design. First, the results indicate that if poli-

cies are designed to incorporate forestry, the three primary forest actions can have strong effects on carbon prices when strict limits on emissions are in place. In the case of the 2°C limiting scenario, carbon prices fall by more than 50% when forestry is included in the global control strategy. A reduction in carbon prices by such a large amount would have enormous benefits for society by directly reducing compliance costs, and freeing resources for other productive investments. If society undertakes a much more modest control strategy closer to the optimal strategy in Nordhaus (2009), the market benefits are not as great, although forestry still provides benefits greater than the costs.

Second, forestry is not just a bridge to the future – it should be an important part of any control strategy across the entire century. The cumulative abatement required and the proportion accomplished by forestry and the energy sectors over the century is shown in table 3.4. The pattern is similar under both control strategies (optimal and 2°C limiting). Forestry accomplishes roughly 64% of the total control by 2030, 50% by mid-century, and 34% by the end of the century. While the results do show how important it is to integrate a forest strategy into climate policy right away, the results also show that any forest policy should be enduring – that is, it should be something that lasts for an entire

Table 3.4 Cumulative abatement and proportion from forests and energy sectors under the two scenarios

	2030	2050	2100
		Optimal	
Cumulative (Gt CO_2)	225	515	1616
% Forest	65%	52%	30%
% Energy	35%	48%	70%
		2 deg.	
Cumulative (Gt CO_2)	238	575	2410
% Forest	63%	50%	34%
% Energy	37%	50%	66%

century. Building and maintaining carbon stocks in forests will be important for long-term climate stabilization.

Third, reductions in emissions from deforestation are the largest source of abatement in the first twenty–thirty years of the program (table 3.5). REDD in tropical countries amounts to 52–54%

of total abatement in the two scenarios in 2020. By 2050, REDD is still important, but it represents only about 15–25% of total abatement effort, and by the end of the century it is a very small part of the total abatement effort. In contrast, afforestation grows in importance over time, rising from around 20% of the effort initially to over 50% of the effort by the end of the century.

Summary of B/C Estimates and Interest Rate Sensitivity

A summary of the estimates of benefits and costs under a "5%" interest rate assumption are shown in table 3.3 for the scenarios described in the preceding section. One of the tricky issues associated with calculating the PV of benefits and costs in this study is that interest rates are an endogenous variable in the DICE model, and they change over time, while they are an exogenous variable in the forestry model, and they are assumed to be fixed over time. The baseline assumptions for the DICE

Table 3.5 Method of sequestration in temperature and tropical forests under the two policies

	Optimal			2° C limiting		
	Temperate	Tropics	Total	Temperate	Tropics	Total
			Million tons of CO_2 per year			
2020						
Afforestation	404	832	1,236	453	946	1,400
REDD	0	3,030	3,030	0	3,123	3,123
Management	1,273	13	1,286	1,429	1	1,430
Total	**1,677**	**3,874**	**5,552**	**1,883**	**4,070**	**5,952**
2050						
Afforestation	689	1,380	2,069	1,026	1,825	2,851
REDD	0	1,680	1,680	0	1,343	1,343
Management	632	1,732	2,364	964	1,720	2,684
Total	**1,321**	**4,793**	**6,114**	**1,990**	**4,887**	**6,877**
2100						
Afforestation	1,009	1,186	2,195	3,158	4,851	8,009
REDD	0	489	489	0	734	734
Management	1,497	897	2,393	5,404	1,043	6,448
Total	**2,506**	**2,572**	**5,078**	**8,563**	**6,629**	**15,192**

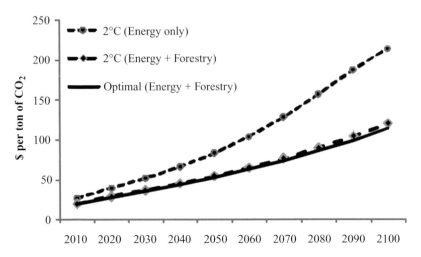

Figure 3.3 *Carbon price paths under three scenarios (r = 3%)*

model resolve interest rates at about 5.6% initially, declining to about 5% by the end of the century. The forestry model uses a 5% interest rate in the baseline. The analysis uses the original assumptions on interest rates used in Nordhaus (2009), and a constant 5% interest rate for the forestry model for the baseline case. This scenario represents the "$r = 5\%$" assumption in table 3.3.

To analyze a lower interest rate, several important parameters in the DICE model are changed. First, I assumed that the pure rate of time preference is 0.1%, compared to the baseline assumption of 1.5%. Second, I changed the elasticity parameter on the utility function to 1.8, compared to a baseline level of 2.0. Under these assumptions, a new baseline is calculated with the DICE model. Interest rates are endogenous in this alternative scenario and they initially are at about the same level as the original baseline, 5.6%, but they fall more rapidly to a lower level of around 2.7% by 2100. They average about 3.0% over the first 150 years. In order to reflect lower interest rates in the forestry model, I shifted the forestry model interest rate to 3% and held that level constant (as before) in the forestry model. These alternative assumptions are used to calculate alternative scenarios, called the "$r = 3\%$" assumption. As before, interest rates are not set strictly at 3% in all periods in the DICE model, but they are lower in all periods than in the previous

set of scenarios. The optimal policy and a 2°C limiting policy are both examined with and without forestry in the control, and the results are shown in table 3.3.

It is not surprising that the value of the benefits and the value of costs are greater under the lower-discount rate assumption than the higher-discount rate assumption. Despite the change, the BCRs are similar to the 5% case. If forestry makes sense under the higher discount rate, it also makes sense under the lower discount rate. The BCR in the 2°C limiting case is slightly greater under the higher discount rate because forestry provides its most important benefits in the near term, when the benefits of reducing expenditures on energy abatement are greatest.

Nordhaus (2009) discusses a number of important implications of the lower-interest rate assumptions. Lower interest rates lead to more savings and lower productivity growth in the future than historically. Lower interest rates also lead to more climate control – e.g. carbon prices are higher and forestry carbon sequestration is greater. One of the more interesting results of the lower interest rate scenario is that when forestry is included, the optimal scenario and the 2°C limiting scenario result in similar temperature trajectories, and similar carbon price paths under the lower-interest rate assumption (figure 3.3).

Implementing Forestry Sequestration Programs

These results illustrate that forestry can be a cost-effective component of international climate policy. Any policy that tackles climate change should also tackle forestry and land-use change. At a carbon price of $30 per ton of CO_2, forestry could provide up to 6.7 billion tons of CO_2 of annual net emission reductions globally. Around 42% of this carbon would be derived efficiently from avoided deforestation in the next thirty–fifty years, an additional 31% from forest management adjustments, and the rest from afforestation. The analysis of optimal policy design indicates that forestry has an important role to play whether the policy follows the optimal policy of Nordhaus (2009), who suggests an initial carbon tax of about $7.50 per ton of CO_2, rising at around 2–3% per year, or whether the policy attempts to place strict limits on temperature increases or CO_2 quantities in the atmosphere.

The surprising importance of forestry raises several "inevitable" questions about whether or not these estimates are even realistic. The results in this chapter imply that society could sequester up to 151 billion tons of CO_2 in forests by 2030 by shifting management, and by converting an additional 376 million ha of land that would otherwise be used for crops into forests. Changes of this scale imply changing land use on 18–19 million ha per year, or stopping 11 million ha per year of tropical deforestation, and afforesting in the temperate zone by 7–8 million ha per year. While we do not have much experience with government programs this large that have been successful, the experience of stopping and reversing deforestation in North America over the past century does suggest that markets can play an important role. In that case, reversion of croplands to forests in the Northeast, Southern, and Midwestern USA resulted mainly from economic forces that lowered crop prices over time and increased opportunity costs of land in other uses (e.g. houses on woodlots). If market forces can be harnessed in the case of carbon, it is possible that a large land-use-change program could achieve the large-scale changes needed.

Assuming that a program of this, or smaller, scale is undertaken, how does one design such a program to actually obtain carbon? Will not problems like MMV ultimately become too expensive? Does society run the risk that other transaction costs will emerge that will ultimately raise costs to an unsustainable level? Will leakage, additionality, and permanence problems lead to large-scale inefficiencies? The discussion below addresses these questions in turn, but the chapter recognizes that many of them remain unanswered. Answering these questions actually represents the frontier of research on carbon sequestration through forests and forestry.

Measuring, Monitoring, and Verification

As discussed in the first section of the chapter, a forestry carbon program can only work if a valid system of MMV for carbon credits on the landscape can be developed and implemented cost-effectively. While much is made of this issue, it actually seems to be fairly straightforward. Estimates of the costs of measuring carbon in biological systems are around $1–2 per ton of CO_2 (Antle et al. 2003; Antinori and Sathaye 2007). These are important, but if carbon prices really are going to rise to the levels described in the scenarios above, MMV costs will represent only a small proportion of the total value of carbon in forests. Further, one would expect these costs to decline over time as new methods are developed to measure and monitor carbon. It is likely that the actual costs of MMV for carbon will be no more than $1 per ton of CO_2 over the long run.

Transaction Costs

Transaction costs encompass other issues than MMV. Consider the following example. Given the sheer number of actors in the land-using sectors, aggregators are likely to emerge. These aggregators will work with individual landowners to create carbon assets, and the aggregators will then bundle the carbon assets of individuals with the carbon assets of other individuals. The aggregators will then sell the bundles to people who value the carbon. The activity of bundling is technical and administrative in nature, but it will use resources that will reduce the net value of the carbon asset to the landowner.

It is not yet clear how large or important the cost of this bundling activity will be. At first blush, one imagines that it could in fact be fairly large. The aggregators need to be fairly well trained, for instance, to know how to organize a measurement system of their own and implement it (or to evaluate some external measurement system). They need to have a working knowledge of accounting. They will need to be able to negotiate. Hiring individuals with all these talents could take real resources. Unfortunately, there simply are not many examples of programs that do what a carbon sequestration program is supposed to do, so it is hard to determine how extensive these activities will be. There are few studies so far that have examined how large transaction costs may be. One of them is Cacho and Lipper (2007), which suggests that for projects involving many small landholders in developing countries, transaction costs for the buyers alone could be $5-$7 per ton of CO_2, including MMV costs.

There is some information available from existing government programs in developed countries. For example, in the USA, the Conservation Reserve Program (CRP) is widely acknowledged as a successful government-run program that has changed land-use on over 12 million ha in the USA since the early 1980s. Sohngen (2008) estimates the costs of running the CRP in the USA, and suggests that the transaction costs of that program, ignoring MMV costs, would amount to less than $2 per ton of CO_2. In the case of the CRP, the transaction costs include the costs of the government office workers and engineers who do the work that aggregators do. This program likely presents a close analogy to the case of a government-run program for carbon sequestration.

The USA, as a developed country, may be an optimal place to try a large-scale land-use-change program. Other regions of the world may be less suitable. Much has been made of the lack of tenure and secure property rights in many frontier regions where deforestation is occurring. If society is unable to secure the rights to maintain forests in those regions, or if large sums of money are squandered unsuccessfully in trying to do so, then society will not be able to rely on forestry to help mitigate climate change as projected above.

It is clearly useful to acknowledge that these difficulties could affect our ability to implement a large land-use-change program in frontier regions where property rights are not well established, but it is also important to not oversell these concerns. For relatively modest returns to grazing or growing marginal crops, landowners and others seem all too willing in these regions to convert land from one use to another. Imagine if there was a real market for standing forest stocks and those funds could make their way to the same decision makers. Carbon markets with carbon prices shown in table 3.2 would generate land rents for forested land in tropical regions of greater than $400 per ha per year in many regions. Such payments would provide exceedingly strong incentives to change land use, particularly when the marginal activity is grazing or some other currently low-value use. The key likely lies less in designing government programs to pay for land use and more in figuring out clever ways to link the payments from demanders to those who actually control the land.

The main effect of MMV and transaction costs in markets will be to raise costs, but raising costs does not mean that forestry projects should not go forward. Accounting for the potential effect of transaction costs on the carbon sequestration programs can be done by shifting the cost functions. To the extent that transaction costs, including MMV costs, affect the market, they will insert a wedge between the market price and the price sellers receive. The DICE model resolves the market carbon price, and the forestry and land-use model pays the carbon price to landowners. The price in the forestry model is thus the "seller price." Transaction costs will be eaten up by other institutions.

To account for these costs, the optimal and 2°C limiting scenarios are re-calculated with transaction costs included. The simplifying assumption that transaction costs represent 20% of the value of carbon on the market is made – e.g. the marginal cost curves are shifted upwards by 20%, as shown in figure 3.4. Thus, the price determined by the DICE model is reduced by 20% to determine the

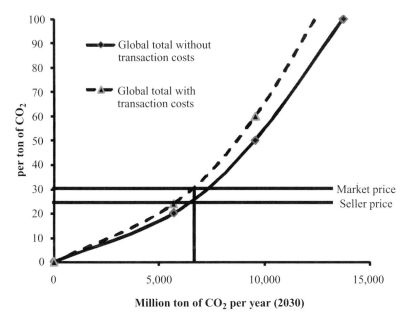

Figure 3.4 *Global marginal cost curve (2030) with and without transaction costs*

Notes: Transaction costs in this case are assumed to be 20% of the total costs. The without-transaction costs marginal cost function shown here is the same as in figure 3.1.

seller price, and this price is used in the forestry and land-use model.

The results of the transaction costs scenarios are shown in columns A and B of table 3.6 for the optimal and 2°C limiting scenario under interest rates of 5% only. Transaction costs do reduce total forest carbon sequestration in both scenarios. In the optimal scenario, forest carbon sequestration declines by 14% when the transaction costs are included, but forest carbon sequestration still amounts to 27% of the total (energy and forestry) abatement by the end of the century. In the 2°C limiting scenario, total forest carbon sequestration declines by 10% when transaction costs are included. Despite the transaction costs, benefits still outweigh costs in both the optimal and 2°C limiting scenario, and the BCRs are not substantially different than those shown in table 3.3. This is perhaps surprising, at first glance, but note that if transaction costs are present, they reduce both the benefits and the costs. Costs are smaller because less forestry carbon sequestration is obtained when transaction costs are present, and because the program is smaller, benefits decline. Because transaction costs reduce the benefits that

landowners obtain by reducing the price they actually see for carbon (i.e. it is 20% less than the market price), transaction costs cause landowners to eliminate the highest-cost (least "efficient") projects from the overall carbon sequestration portfolio. Thus, while transaction costs do have important implications – they reduce total carbon sequestration potential – they do not invalidate the use of forestry sequestration.

Additionality and Leakage

Much is made of additionality. Additionality is a problem because it is virtually impossible to determine, or know, what actions landowners will undertake with their land before the fact. We can perfectly well observe what they did with their land after the fact, but not before. However, the carbon we are actually interested in saving on the landscape is the carbon that someone actually will release into the atmosphere. Paying individuals who would not otherwise have released carbon to hold it raises the costs of a carbon sequestration program.

Table 3.6 B/C estimates for the cases with transaction costs equaling 20% of the total cost of abatement

	A	B	C	D
	Forest and energy compared to energy only		Forestry only compared to NC	
	Transaction cost = 20% r = 5%		Transaction cost = 20%	
	Optimal	2°C limit	r = 5%	r = 3%
		Billion US$ (PV)		
Benefits				
Consumption gain	($24)	$684	($52)	$1,128
Reduction in damage	$901	$462	$946	$9,372
Reduction in energy costs	$66	$2,543	$0	$0
Total benefit	$943	$3,690	$894	$10,501
Forest cost	$917	$2,186	$917	$11,387
BCRs				
All benefits	1.03	1.69	0.97	0.92

Notes: Columns A and B compare a scenario with energy abatement only to a scenario with energy abatement and forestry sequestration. Columns C and D compare an NC scenario to a scenario that includes only forestry abatement

All of the cost estimates above assume that society is able to determine perfectly which carbon is truly additional. Obviously, there could be some additional costs associated with assessing the baseline for each person who enters a carbon contract. Such estimates have been undertaken for a number of different carbon projects (Sohngen and Brown 2004; Antinori and Sathaye 2007), so it is clearly plausible to determine baselines and use these estimates for contracting purposes. Estimating baselines and additionality would be considered to be part of the transaction costs discussed above, and thus already assessed in the chapter.

Leakage is a far more important problem for carbon sequestration because it is unlikely that all countries will enter into a global climate treaty at the same time. Furthermore, many countries will have trouble developing systemwide carbon sequestration programs, so they will experience leakage within their boundaries. Because some countries remain outside the scope of the regulatory regime, and because some countries will develop programs that are geographically limited in scope, leakage will occur. Empirical estimates of leakage illustrate the seriousness of the problem. Estimates

from the project level indicate that leakage could range from 10–90% (Murray *et al.* 2007). A recent paper by Sun and Sohngen (2009) using the same forestry model as used in this chapter illustrates that leakage could be nearly 100% in the near term under a global policy that seeks to set-aside forests with high carbon potential.

It is not possible to fully account for leakage in the CBA framework used above because it is unclear which countries will and will not enter into a global climate treaty (and further, which countries will engage in carbon payments for trees). The optimal scenario and the optimization over the 2°C limiting scenario assume that all regions participate. This is an admittedly strong assumption, but it is maintained for all sectors in the analysis. The leakage problem in forestry, however, illustrates the problem with less than full action for carbon sequestration programs. All indications based on the current empirical analysis suggest that if countries do not participate in a global carbon sequestration program, significant inefficiencies could arise. These inefficiencies could be large enough to reduce BCRs to less than 1.0. It is thus important for policy makers to ensure that the largest possible number of countries is involved in the carbon sequestration program in order to avoid or reduce the scope for leakage.

Biofuels

Other policies will interact with forest-based carbon sequestration in important ways. One important policy relates to biofuels. Current US and European legislation mandate increases in biofuel consumption. When considering these current policies, however, they are unlikely to have long-term consequences that are important for climate stabilization policy. Searchinger *et al.* (2008), for example, indicate that about 3.7 billion tons of additional CO_2 would be emitted as a result of the US and European biofuel mandates. While any increase in emissions is potentially bad, this is less than a year's worth of total emissions from deforestation. The results of Searchinger *et al.* (2008) certainly tell us that current biofuel policy is inefficient, but they do not suggest that biofuel policy as it is currently constructed will substantially raise the costs of climate mitigation via forests. On the contrary, if biofuel policies can be promoted so that each ha of biofuel land provides an equal benefit to the atmosphere as a ha of forests, then biofuel will be just as efficient a means to achieve climate policy as sequestration policies (or some combination of the two may be most efficient). Analyzing such policy is beyond the scope of this chapter, although this will be important to examine in the future.

Forests Alone?

One interesting question remains: how much could forestry do alone? As a "stand-alone" option against climate change, can forestry substantially reduce climate change? To address this question, a final analysis is conducted in which the DICE model is run without any controls for 250 years, and it is then run with just forestry options and no energy abatement. Forestry is paid at the social cost of carbon calculated by the DICE model. The analysis is conducted for 5% and 3% interest rate cases, and in both cases transaction costs of 20% are included in the analysis. The 2°C limiting case is not con-

sidered under these circumstances because it is not possible to use forestry alone to meet the 2°C limit.

The results of this analysis are presented in column C and D of table 3.6. It turns out that if society decides to use just forestry, the BCR is less than 1 under both interest rate assumptions. The BCR is less than 1 in this case because there are no gains associated with avoiding or reducing energy costs (energy abatement is considered neither in the base, nor in the forest sequestration case). When the interest rate is 5%, the reduction in damages is at least greater than the costs of the forestry program, but consumption declines relative to the no-control case, and the BCR is only 0.97. The forestry program is the same financial size as under the optimal scenario with transaction costs (shown in column A of table 3.6) because the social cost of carbon is roughly the same in both scenarios.

These results show that, as a stand-alone policy for climate change, a large-scale forestry sequestration program is close to, but it does not pass a B/C test. Forestry should be a complement with energy policy, but alone forestry actions do not have a large enough effect on temperatures to substantially reduce damages. Under the optimal scenarios with only energy abatement, the temperature increase in 2100 is 2.59°C, and with forestry and energy together (assuming 20% transaction costs), the temperature change is 2.42°C in 2100. By the end of the century, when forestry is a complement to energy abatement, it has a 0.17°C effect on temperatures. Under the no-control for 250 years case, the temperature change in 2100 is 3.05°C, but when forestry is included, the temperature change in 2100 is 2.90 °C. When forestry acts alone, it has a smaller effect on temperatures.

Conclusion

This chapter examines the potential for establishing a global forest carbon sequestration policy as part of a global effort to combat global warming. The chapter begins by describing important categories to consider when measuring the costs of forestry carbon sequestration, including opportunity costs, implementation and management costs, MMV costs, other transaction costs and systemwide costs (e.g. leakage or impacts in other markets). Then, the chapter briefly examines three forestry options – afforestation, forest management, and reductions in emissions from deforestation – and describes current cost estimates available in the literature. The economics of these options have been widely explored in the literature, and the chapter describes the results of a number of studies considering the costs and quantity of carbon that may be sequestered in each of these activities. One issue noted in the chapter, however, is that the current studies have focused largely on the opportunity costs and implementation and management costs. Other categories of costs have not been addressed as completely to date.

The current research indicates that the three key forestry activities can sequester large quantities of carbon. Over the next thirty years, the upper limit of potential sequestration could be as much as 15 billion tons of CO_2 at more than $100 per ton of CO_2. At $30 per ton of CO_2, it may be possible to obtain around 6.7 billion tons of CO_2 per year, with around 40% of this arising from avoided deforestation, 31% from forest management activities, and the rest from afforestation. A global sequestration program really is global, with potential contributions from virtually every region of the world. The largest share of sequestration is derived from avoided deforestation, but the roles of forest management and afforestation cannot be discounted.

To conduct a CBA, the chapter combines a global forestry and land-use model (which estimates carbon sequestration potential), with an integrated assessment model (which estimates the implications of this sequestration on carbon markets). Two global climate policies are considered in a "with" and "without" forest carbon sequestration comparison. One policy is the optimal policy suggested by Nordhaus (2009) and the other limits global carbon emissions such that global temperatures remain below 2°C over all time. Forestry turns out to be about 30% of the global abatement effort in each case. When carbon emissions are adjusted in a way to meet the 2°C limitation, the market implications are astounding. The inclusion of forestry in the control of GHGs reduces carbon prices by 50%. This large reduction in compliance costs allows society

to invest in other productive activities and provides a strong benefit in terms of increased consumption. The BCR is 1.0 under the optimal scenario and 1.7 under the 2°C limiting scenario. The baseline interest rate is 5%, and when sensitivity analysis is conducted over a lower interest rate, roughly 3%, the qualitative results are the same. The benefit-cost ratio in the optimal scenario is a bit bigger in the 3% case as in the 5% case, but in the 2°C limiting scenario, the BCR is about the same (1.7) in the 3% case.

The results establish the importance of including forestry and land use, but they also show the importance of thinking long-term about these options. Forestry can provide carbon mitigation services in the near term through reductions in deforestation and increased forest management, but with the right incentives (e.g. rising carbon prices) it continues to provide mitigation services throughout the century. There is no evidence that the role of forestry saturates – in fact, in the 2°C limiting case, forestry becomes a larger and larger program throughout the century as the carbon constraint becomes more binding. The type of actions undertaken over the century will change, but forestry remains important.

Of course, it is also important to acknowledge that there will be difficulties associated with starting and running carbon sequestration programs. Measuring monitoring and verification protocols must be established and implemented. People will have to learn how to sell their carbon credits onto a market. There could be some inefficiencies associated with leakage and additionality. Based on the current literature, the known MMV costs plus other transaction costs appear to be less than $3 per ton of CO_2. One of the problems with current cost estimates of forestry options is that these additional costs are often ignored. While it is beyond the scope of this chapter to actually estimate these, it presents additional scenario analysis taking potential transaction costs into account. The results of this analysis show that transaction costs will reduce the size of the forestry program, and some of the benefits will accrue to bureaucratic functions, but at the levels considered in the chapter, transaction costs do not negate the central conclusions about the importance of forestry in a global climate policy.

The chapter does not address questions related to leakage, particularly in the CBA framework, although a literature review on this subject shows that it clearly will be a problem if climate policy is incomplete (e.g. some regions are left out of the control). This may be a bigger issue for forestry and land use than other abatement options because of the international nature of markets for end products. Given the scale of potential leakage, with some analyses suggesting that it could be as much as 90%, this represents one of the most important uncertainties to resolve. One related issue to leakage relates to impacts in other markets. For example, if forestland area increases over time and cropland area decreases, crop prices may rise, causing welfare losses in the food sector. This chapter implicitly accounts for these impacts by using opportunity costs, although the full range of potential price impacts in the agricultural sector is not calculated here.

The chapter also does not address the other benefits that would accrue with a forestry carbon sequestration program. Most of the forestry programs imply an increase in overall global forestland of up to 1.0 billion additional ha over the century. From an ecological perspective, these forests would provide habitat for countless species, including many species that are presently endangered. Further, forest cover could help moderate water flows in large drainage basins and provide other hydrologic benefits. These benefits have not been quantified and addressed in this chapter, although they certainly would be important.

Bibliography

Antinori, C. and J. Sathaye, 2007: *Assessing Transaction Costs of Project-Based Greenhouse Gas Emissions Trading*, Lawrence Berkeley National Laboratory Report **LBNL-57315**

Antle, J.M., S.M. Capalbo, S. Mooney, E.T. Elliot, and K.H. Paustian, 2003: Spatial heterogeneity, contract design, and the efficiency of carbon sequestration policies for agriculture, *Journal of Environmental Economics and Management* **46**, 231–50

Blaser, J. and C. Robledo, 2008: Initial analysis on the mitigation potential in the forestry sector, Update of a background paper prepared for the UNFCCC Secretariat in August 2007, presented to the International Expert Meeting on Addressing Climate Change through Sustainable Management of Tropical Forests, Yokohama, Japan, April 30–May 2

Cacho, O. and L. Lipper, 2007: Abatement and transaction costs of carbon-sink projects involving smallholders, FEEM Working Paper **27.2007**, Fundacion Eni Enrico Mattei, Milan

Eliasch, J., 2008: *Climate Change: Financing Global Forests*, Office of Climate Change, UK Government, London

Grieg-Gran, M., 2008: The cost of avoiding deforestation, Update of the Report prepared for The Stern Review of the Economics of Climate Change, Working Paper, International Institute for Environment and Development, Cambridge

Hoehn, H.F. and B. Solberg, 1994: Potential and economic efficiency of carbon sequestration in forest biomass through silvicultural management, *Forest Science* **40**, 429–51

Houghton, R.A., 1999: The annual net flux of carbon to the atmosphere from changes in land use 1850–1990, *Tellus*, **51B**, 298–313

  2003: Revised estimates of the annual net flux of carbon to the atmosphere from changes in land use and land management 1850–2000, *Tellus*, **55b**, 378–90

Intergovernmental Panel on Climate Change (IPCC), 2007: *Mitigation of Climate Change – Report of Working Group III, Intergovernmental Panel on Climate Change*, Cambridge University Press, Cambridge

Kurz, W.A. and M.J. Apps, 1999: A 70-year retrospective analysis of carbon fluxes in the Canadian forest sector, *Ecological Applications* **9**(2), 526–47

Murray, B., R. Lubowski, and B. Sohngen, 2009: Including international forest carbon incentives in a United States climate policy: understanding the economics, Working Paper, The Nicholas Institute for Environmental Policy Solutions, Duke University, Durham, NC

Murray, B.C., B. Sohngen, and M.T. Ross, 2007: Economic consequences of consideration of permanence, leakage and additionality for soil carbon sequestration projects, *Climatic Change*, **80**(1–2), 127–43

Nordhaus, W., 2009: *A Question of Balance: Weighing the Options on Global Warming Policies*, Yale University Press, New Haven, CT

Nordhaus, W. and J. Boyer, 2000: *Warming the World: Economic Models of Global Warming*, MIT Press, Cambridge, MA

Richards, K. and C. Stokes, 2004: A review of forest carbon sequestration cost studies: a dozen years of research, *Climatic Change* **63**, 1–48

Searchinger, T., R. Heimlich, R.A. Houghton, F. Dong, A. Elobeid, J. Fabiosa, S. Tokgoz, D. Hayes, and T-H Yu, 2008: Use of US croplands for biofuels increases greenhouse gases through emissions from land-use change, *Science* **319**, 1238–40

Sedjo, R.A., 1989: Forests: a tool to moderate global warming?, *Environment* **31**(1), 14–20

Sedjo, R.A., J. Wisniewski, A.V. Sample, and J.D. Kinsman, 1995: The economics of managing carbon via forestry: assessment of existing studies, *Environmental and Resource Economics* **6**, 139–65

Smith, W.B., P.D. Miles, J.S. Vissage, and S.A. Pugh, 2003: Forest resources of the United States, 2002, *General Technical Report* **NC-241**, US Department of Agriculture, Forest Service, North Central Research Station, St Paul, MN

Sohngen, B., 2008: Paying for avoided deforestation – should we do it?, *Choices* **23**(1), www.choicemagazine.org/2008–1/theme/2008–1-08.htm>

Sohngen, B., K. Andrasko, M. Gytarsky, G. Korovin, L. Laestadius, B. Murray, A. Utkin, and D. Zamolodchikov, 2005: *Stocks and Flows: Carbon Inventory and Mitigation Potential of the Russian Forest and Land Base*, World Resources Institute, Washington, DC

Sohngen, B., A. Golub, and T. Hertel, 2009: The role of forestry in carbon sequestration in general equilibrium models, chapter 11 in T. Hertel,

S. Rose, and R. Tol. (eds.), *Economic Analysis of Land Use in Global Climate Change Policy*, Routledge, New York

Sohngen, B. and S. Brown, 2004: Measuring leakage from carbon projects in open economies: a stop timber harvesting project as a case study, *Canadian Journal of Forest Research* **34**, 829–39

2006: The influence of conversion of forest types on carbon sequestration and other ecosystem services in the South Central United States, *Ecological Economics* **57**, 698–708

2008: Extending timber rotations: carbon and cost implications, *Climate Policy* **8**, 435–51

Sohngen, B. and R. Sedjo, 2006: Carbon sequestration in global forests under different carbon price regimes, *Energy Journal* **27**, 109–26

Sun, B. and B. Sohngen, 2009: Set-asides for carbon sequestration: implications for permanence and leakage, *Climatic Change*, published online, July 23, **DOI: 10.1007/s10584–009-9628–9**

Tavoni, M., B. Sohngen, and V. Bosetti, 2007: Forestry and the carbon market response to stabilize climate, *Energy Policy* **35**(11), 5346–53

UN Food and Agricultural Organization (FAO), 2006: Global forest resources assessment 2005: progress towards sustainable forest management, FAO Forestry Paper 147, United Nations Food and Agricultural Organization, Rome, www.fao.org

Van Kooten, G., Cornelis, Alison J. Eagle, James Manley, and Tara Smolak, 2004: How costly are carbon offsets? A meta-analysis of carbon forest sinks, *Environmental Science & Policy* **7**, 239–51

Forestry Carbon Sequestration
Alternative Perspective

S A B I N E   F U S S*

Introduction

I find myself in broad agreement with Brent Sohn-gen's chapter 3 on the costs and benefits associated with the climate change mitigation options offered by forestry carbon sequestration, which includes afforestation, reductions in deforestation (REDD), and forest management. This Perspective paper will provide a review of Sohngen's approach and conclusions, highlight the most important findings, identify gaps and their implications for the calculations, and thus put the results into perspective. The areas dealt with concern competition for land and its potential impact on opportunity costs, the role of various types of uncertainty and their implications for implementing forestry carbon sequestration programs, the effect of accounting for eco-systems services and biodiversity on benefit assessments, and the relevance of option values in considering REDD strategies. The conclusion drawn from the analysis coincides with Sohngen's findings; he claims that forest carbon will be needed as part of a strategy to mitigate climate change. Solutions can thus not arise exclusively from the technosphere, especially in the face of time and other resource constraints. This will be discussed in the final paragraph with the purpose of putting the forestry option in perspective also with respect to the other mitigation solutions.

Messages from Chapter 3

Chapter 3 provides an exploration of the climate change mitigation possibilities associated with enhancing forest carbon stocks. It introduces and explains the rationale of forest carbon as a component of the policy mix for combating cli-mate change and then presents the most important options of forestry carbon sequestration: afforestation, REDD, and forest management.[1] The estimates of the costs for these carbon sequestration options are summarized in Sohngen's figure 3.1, which depicts the marginal costs of sequestering carbon, where a distinction is also made between regions (temperate/developed vs. tropical). Disaggregating the results for a single carbon price of $30 per ton of CO_2 and calculating the amount of carbon sequestered per activity, the author finds that the largest reduction potential over the next forty years comes from REDD, followed by forest management and finally by afforestation, which has been the focus of policy over the past decade.

Combining these results with an integrated assessment model (DICE), Sohngen produces estimates for two scenarios, where the first is called "optimal" and is the optimal policy scenario adjusted from Nordhaus (2009), while the second scenario is one where policy aims at limiting the temperature increase to 2°C above preindustrial levels. Benefits include increases in consumption and reductions in damages and energy costs. In the

* The author wants to thank Mykola Gusti, Petr Havlik, and Hannes Böttcher at IIASA for sharing their latest work on forestry carbon sequestration, Brent Sohngen for the fruitful collaboration on this project and the Copenhagen Consensus Center and particularly Kasper Thede Anderskov for the possibility to participate in the project and support during it. The IIASA research mentioned in this Perspective paper has been conducted in the framework of the following projects: Climate Change – Terrestrial Adaptation and Mitigation in Europe (CC-TAME, EC FP7), Full Cost Climate Change (FCCC, EC FP7), and Development of forward-looking REDD scenarios in the Congo Basin (World Bank). The author further acknowledges support by Environmental Defense Fund (EDF), which supports work on uncertainty issues revolving around REDD.

[1] The sum of these items is often referred to as REDD+.

optimal scenario, the benefits accrue mainly from a reduction in damages; (carbon prices fall only modestly). Present value (PV) benefits roughly equal the costs of the program in this scenario, suggesting a benefit-cost ratio (BCR) of 1. In the 2°C scenario the ultimate target of limiting the temperature level leads to similar temperature profiles for both the case with and the case without forestry, so there is only a very marginal reduction in damages due to including forestry. Through the reduction in mitigation costs, however, consumption increases substantially, implying a BCR of 1.7. These results are robust across interest rates (5% and 3% are tested).

Sohngen's table 3.4 shows that forestry features significantly in terms of cumulative abatement when compared to the energy sector. This is especially true in the near term. However, with a rising carbon price, the forestry proportion continues to play an important role also in the longer run. In the 2°C limiting scenario, the proportion even increases over time.

Following this analysis, Sohngen discusses the policy implications and implementation issues. Furthermore, there are several problem areas that are of importance and have repeatedly been raised in the debate: first, there will be costs associated with measuring, monitoring, and verifying (MMV) carbon credits, for the analysis of which the author conducts additional scenario analysis. Even though the results indicate that the size of the forestry program will shrink in response to including MMV costs, the importance of forestry as a mitigation option is still vital. Second, there are transaction costs, which might negatively affect the ability to implement large land-use-change programs in frontier regions, where property rights are not well defined; however, the author warns not to overestimate these concerns, especially not for land where the marginal activity is of low-value use. Third, additionality and leakage are mentioned as problems, where the former is deemed to be less grave, since a baseline for each person entering a carbon

contract could readily be determined; the impact of leakage on large-scale policies, however, cannot be determined as there are no estimates of international leakage potential. Fourth, a forestry program might interact with a biofuel policy, which in itself might not be efficient, but which does not necessarily raise the costs of mitigation through forestry, especially if the biofuel policy aims at equalizing the benefits to the atmosphere from 1 ha of biofuel land to those arising from 1 ha of forest.

Finally, Sohngen presents an analysis of forestry as a stand-alone policy, where the DICE model is run without energy-abatement options. In particular, he compares a run without any controls to one where only forestry is possible. The BCR in that case is less than 1, indicating that forestry is not sufficient as a stand-alone strategy to mitigate climate change. As a complement to energy-related abatement it is an indispensable ingredient, however. I will come back to this point in the final paragraph.

Gap Analysis

Even though I agree with the messages outlined in the previous section, there are some issues which Sohngen does not deal with in detail. In this section, I am trying to identify these issues and fill the gaps, where needed, to put the option of forest carbon better into perspective.

Competition for Land

Sohngen's assessment has one very important implication: the forestland will have to be expanded by a substantial area under the proposed programs. The question arises whether there will not be more competition for such a large amount of land in the long run: in fact, growing food demand and other trends reinforcing other land uses could lead to quite some tension in the realization of large-scale forest programs. In the case of food demand, crop yields would need to increase tremendously in order to keep competition for land within its confines.[2]

---

[2] Increases in crop yields would, however, raise land values and therefore also opportunity costs. Other exogenous factors, which could have such an impact on land value and thus opportunity cost, are, e.g., interest rate changes or increases in timber prices.

In addition, policies such as those concerning biofuels mentioned in chapter 3 might interfere with goals of expanding forest area for the sake of using it as a carbon sink. As a result, the opportunity costs should be adjusted as larger and larger areas of land need to be reserved for forests. It is clear that this gap needs to be filled by the modeling community in order to offer a full account of the costs involved in REDD (and REDD+).

Actually, some of the current estimates already do account for this effect. A framework combining the Global Forest Model (G4M) and the Global Biomass Optimization Model (GLOBIOM)[3] accounting for the effect of competition over land is currently being developed at the International Institute for Applied Systems Analysis (IIASA). The Eliasch Review (Eliasch 2008), for example, already uses the REDD potential estimates of an earlier version of this model cluster (see Gusti et al., 2008).[4]

Currently, a new version of G4M is being used for determining optimal CO_2 prices for reducing deforestation and raising afforestation. GLOBIOM's predictions for land and forest product price changes are used in G4M to determine afforestation and deforestation patterns in geographic space. First findings show that the leading countries by potential for sequestration of additional carbon and cost competitiveness are Brazil, Zaire, Indonesia, Bolivia, and Tanzania. Most importantly, changes in agricultural and forestry markets are found to influence the competition for land. This will have a large impact on economic incentives for carbon sequestration (Gusti et al. 2009).

With respect to interactions with biofuel policy mentioned in the chapter, a new study by Wise et al. (2009) employs an integrated assessment model to look at the implications of emissions reductions for land use and land-use change. They find that the costs of meeting targets decreases. However, unmanaged eco-systems and forests expand and food crop and live stock prices increase. This result applies when there is a carbon tax on both land-use change and energy and industrial emissions. If only the latter are taxed, then energy crops require larger and larger amounts of land, and achieving climate goals becomes more expensive. These find-

ings underline the importance of valuing terrestrial carbon.

Eco-systems Services and Biodiversity

Whereas the previous point suggests that the costs of a forestry program might be larger than estimated by (partially) ignoring issues of competition over land, there is also an underestimation in the benefits of REDD. In fact, avoided deforestation has a wide variety of ancillary benefits – most importantly the preservation of biodiversity, natural habitats, and other ecosystems services such as the regulation of water balance and flow of the river, the adjustment of regional climate and weather patterns, and the moderation of the spread of infectious diseases (see, e.g. Foley et al. 2007).

While these benefits are admittedly difficult if not impossible to quantify and monetize, the BCRs of the assessment should be considered in the light of these additional advantages when comparing to other options, as Sohngen also suggests in the conclusion to the chapter.

The Role of Uncertainty for REDD

Chapter 3 does not go into detail about some of the problems relating to implementation of a large-scale forestry program and the role that uncertainty plays in this context. As can be concluded

---

[3] G4M provides spatially explicit estimates of annual above-ground wood increment, development of aboveground forest biomass, and costs of forestry options such as forest management, afforestation, and deforestation by comparing the income of alternative land uses. GLOBIOM is a global partial equilibrium model integrating the agricultural, bioenergy, and forestry sectors with the aim of giving policy advice on global issues concerning land use competition between the major land-based production sectors.

[4] This new version of G4M combined with GLOBIOM is planned to be used in the GAINS (Greenhouse Gas and Air Pollution Interactions and Synergies) model developed at IIASA's Atmospheric Pollution and Economic Development Program (see Böttcher et al. 2008, and information on the latest GAINS workshop at http://gains.iiasa.ac.at/index. php/home-page). In addition, there is OSIRIS, which is an open-source spreadsheet tool, designed to support UNFCCC negotiations on REDD using results from G4M for their simulations (see Busch et al. 2009).

from the extent and diversity of debates surrounding the implementation of REDD and other forest carbon sequestration programs, these points cannot be neglected in a thorough discussion and when making comparisons to other options. I will here highlight just a few points in order to put chapter 3's results into perspective.

A key problem featuring among the uncertainties affecting REDD is the definition of the "true" baseline. Most of the proposals to date still suggest using historical baselines, which might not be reliable due to lack of high-quality data. In choosing the right method to determine the business-as-usual (BAU) baseline, according to which REDD will be measured, it is important to note that there are significant financial incentives at stake: tropical countries will want to maximize the compensation they receive for reducing deforestation below the baseline by having a higher baseline to begin with. On the other hand, developed countries will have to offer sufficient compensation in order to get forest countries engaged in REDD. This is also linked to the issue of additionality Sohngen discusses.

Furthermore, much of the uncertainty surrounding the implementation of a forestry program comes from climate policy uncertainty itself: uncertainty about climate policy emanates from ambiguity about the stabilization target sufficient to achieve an acceptable increase in temperature. More precisely, limited knowledge about climate sensitivity and feedbacks make it difficult to determine the acceptable degree of warming and relate that to a concentration level. Recent findings by Hansen *et al.* (2008), for instance, explain that paleoclimate evidence and ongoing climate change suggest that carbon will need to be reduced to much lower levels than we might have been prepared for. They claim that "the largest uncertainty in the target arises from possible changes of non-CO_2 forcing." Whether REDD will be needed to mitigate climate change and to what extent is thus unclear, so a reduction in deforestation rates might be postponed, which will make this option more costly in the long run.

In addition, uncertainty about the future opportunity cost of forest land to supplier countries also complicates agreements. Those countries may have different expectations and assumptions concerning the development of commodity prices and thus look at a larger range of future opportunity costs. In this respect, more research is needed to determine the value of different future portfolios of land uses that forest nations consider. Related to this, there is uncertainty about the amount of funding that could be raised in order to finance REDD. Voluntary funds might not be sufficient to cover the expenses for implementing REDD, whereas compliance markets promise a bigger potential.[5]

Uncertainties associated with the need for MMV of carbon credits on the landscape (see Angelsen *et al.* (2009) for an overview of the issues involved in monitoring) and leakage have already been mentioned in the chapter and might lead to higher costs in the calculation of the BCRs. Again, the estimation of these costs would be very difficult, especially in the case of cross-border leakage. Murray (2008) finds empirical evidence indicating that leakage from avoided deforestation policies could be substantial and claims that this needs to be addressed by policy design (e.g. include discounts to reduce the number of REDD credits issued, broadening of the policy scope). In addition, permanence problems could raise the costs of a REDD program significantly.

It is beyond the scope of this Perspective paper to precisely estimate the costs associated with MMV, but Sohngen's assessment provides a very useful discussion and additional scenario analysis, which points to the impact such costs could have on the BCRs. The results actually seem to be robust to shifts in the cost curve, but the forestry program will of course be smaller than without these extra costs.

Finally, it is important to make a distinction between the cost uncertainties that will feature most significantly in the near term compared to those that have more significant implications in the long run. The most important source of uncertainty in the near future is probably the one relating to the estimation of opportunity costs, since good opportunity cost estimates require good estimates of the value of land. However, there are differences in the

[5] See, e.g., Murray *et al.* (2009) for an overview of financing structures currently being considered.

estimation of agricultural suitability and land values (FAO 2000; Ramankutty *et al.* 2002; Benitez *et al.* 2004; Naidoo and Iwamura, 2007; van Velthuizen *et al.* 2007). In addition, ignoring the carbon stock of alternative land uses can significantly overestimate the unit costs of actual net emission reductions (Pagiola and Bosquet 2009).

In the long run, another source of uncertainty will gain importance: climate change itself will have an influence on the suitability of land for agriculture and it will also affect forestry. As some regions become drier, for example, land value and therefore opportunity costs might be affected negatively, but forestry will also suffer from the changed conditions. It is thus inherently difficult to determine long-run costs.[6] In addition, it is important to note that not only existing forest carbon stocks, but also those that will be created will be threatened by climate change. "Sustainable" forestry carbon sequestration strategies therefore also need to involve adaptation to increase forest resilience towards new disturbances such as increased risk of wild fires, storms, etc. It is difficult to determine the magnitude of these uncertainties associated with future climate change and the resulting potential disturbances, so any forestry carbon sequestration strategy has to be seen in the context of other complementing actions – a point to which I will come back in the discussion at the end of this Perspective paper.

The Option Value Behind Forest Carbon Sequestration

Relating to the point about climate policy uncertainty made in the previous section, if more ambitious goals will need to be achieved than previously assumed, an *option* on REDD could potentially serve as a kind of insurance for meeting the target. This is an example of a "real" option, where relative irreversibility (the forest can only be re-grown at a relatively slow rate) and uncertainty (it is unclear which concentration level will ultimately be needed) imply that there is a value to waiting and keeping the option of using the forest to meet the target open.[7]

From the perspective of the market, the general fear of "market flooding", claiming that cheap REDD offsets might drive the carbon price down and thus deter investment in cleaner technologies and research and development (R&D) can also be reduced by thinking in an options framework: recent work by Golub *et al.* (2009) adopts a real-options approach to show that this does not necessarily have to be the case – if REDD credits were linked to carbon markets as options and only a limited amount of these options would be available, for instance. Pricing these REDD options as a derivative of the CO_2 permit price would ensure that it was high enough, so as not to drive down prices in the carbon market. Firms which have bought REDD options could then exercise them at the initially negotiated strike price, which would enable them to avoid spikes in the permit price. The results show that firms do not experience changes in their average profitability. However, they can smooth out some of the variability arising from permit trading by buying REDD options. An option contract on REDD-backed offsets could therefore be an attractive alternative to direct offsets.

Concerning the potential threat to R&D, it is important to realize that technological progress is an inherently uncertain process as well: whether and when a cost- or emission-reducing innovation will be made is largely unknown. Major advances will probably take longer than policy makers want to wait and REDD can offer the possibility of "bridging" the time it takes to transform the energy system. R&D can at the same time be regarded as having an option value by firms: it offers them some flexibility to respond to emission reduction demands with more efficient and less expensive technology if they move early. In particular, if REDD were linked to the global carbon market, it could be a powerful risk-management tool at the firm level. Investment into new technologies always carries certain risks, but could still be encouraged if REDD offsets provide firms with an affordable

---

[6] On the other hand, the benefits from avoiding climate change through forestry programs in the near term might be much larger as a result of these considerations.

[7] The theory of real options is formalized in Dixit and Pindyck (1994).

alternative to fill the gap in their "carbon budget" if deployment of new technologies were delayed for technical or other reasons. Without such flexibility, a firm might be even reluctant to engage in R&D targeted at carbon-saving innovations, since in case of a strong policy it would have to invest before having time for the technology to develop. Tavoni *et al.* (2007), for example, look at the impact of REDD on energy sector innovation and find larger effects than Bosetti *et al.* (2009). The difference from Bosetti *et al.* (2009) is that Tavoni *et al.* (2007) consider a less stringent target and encompass all forestry options – i.e. not only REDD (Murray *et al.* 2009).

The crucial idea behind such options thinking is the economic value associated with being flexible in responding to the outcomes of uncertain processes. Another way of capturing some of this valuable flexibility that has been suggested in the context of REDD credits is banking (e.g. Dinan and Orszag 2008). If banking is allowed, then it has been shown that firms – in anticipation of a tightening cap – will buy credits prematurely in order to comply with their reduction obligations at a cheaper price later on. Acceleration in abatement may then lead to an additional benefit of reducing the amount of greenhouse gas (GHG) persisting in the atmosphere (Murray *et al.* 2009).

Conclusion

While chapter 3 by Sohngen gives a good estimate of the costs and benefits involved in a potential forest carbon sequestration program, the multiple uncertainties and unresolved issues outlined in the chapter should remind us to be cautious and puts the numbers into perspective for any assessment and comparison across other options.

On the one hand, costs might be larger as future modeling efforts lead to adjustment of opportunity costs under competition for land. Furthermore, problems relating to permanence and leakage will add to total costs of REDD, afforestation, and forest management projects. On the other hand, taking into account ancillary benefits in terms of ecosystems services and preservation of biodiversity will lead to an upward adjustment of the benefit numbers, which currently rely on the change in consumption only. In addition, the benefits of avoiding higher degrees of climate change are of great importance as well.

Most importantly, however, one has to make a distinction between the uncertainties relating to opportunity cost estimates relying on inaccurate land values, which matter most significantly in the near term, and the uncertainty concerning the impacts of climate change on agriculture and forestry. The latter will certainly affect both costs in the long run and actually also increase the benefits of mitigation in the near future.

Further to the issue of uncertainty, current research by Gusti *et al.* (2009) sheds more light on the problems that will be raised as competition for land increases, which will exert further pressure on costs. A new version of G4M using results from the land-use model GLOBIOM as inputs finds that prices of forestry products and agricultural land could increase substantially if commodity market effects were taken into account. Future research will have to be expanded to provide even more accurate B/C estimates.

Another issue raised in this Perspective paper relates to the option values implicit in forestry carbon sequestration and particularly in REDD: continuing deforestation at the current pace will disbar us from using the full forest as a means to meet stabilization targets – forests can only be re-grown at a relatively slower rate. Given this irreversibility, there is an option value to holding on to forests and using them as a carbon sink later in the face of uncertainties about the amount of GHG reductions needed for stabilization.

Finally, it is of great importance to emphasize that Sohngen's chapter 3 and this Perspective paper both point in the same direction: it should by now be clear that we can ultimately not rely exclusively on solutions emanating from the technosphere to tackle global warming. The biosphere has large potential to help comply with our targets and can also serve as a bridge, while cleaner technologies are being further developed. Even though it has been the task of chapter 3 to present and evaluate the option of forestry carbon sequestration, the latter should be understood as a necessary and cost-reducing, but not sufficient component of the overall strategy, which must eventually comprise mitigation, R&D, and other options as well. This will be discussed in more detail in the final section.

Discussion: A Portfolio of Options

As has been mentioned, both Sohngen's chapter 3 and this Perspective paper point to the fact that forestry carbon sequestration cannot be seen as a stand-alone solution. In this final section I want to highlight two issues, which make it therefore important to consider a portfolio of mitigation options rather than singling out those with the most attractive BCRs.

The first issue is about the risks, which are inherent in each of the options evaluated in this Perspective paper. It is clear that these differ substantially across options: in order to keep risks at a minimal level, only a mix of strategies can lead to meeting our objective at minimal cost. In addition, some of the options considered bear significant risks in their tail. For example, large-scale stratospheric aerosol insertion could have substantial side effects, which might be difficult to quantify and monetize (see also Perspective paper 1.2 by Anne E. Smith, who underlines the importance of considering potential unintended consequences from geoengineering (GE)). Relying merely on cost-benefit analysis (CBA) might therefore not be the best policy guidance if it relies on the subjective choice of probability assumptions. The optimal portfolio of options would thus try to *hedge* against tail risk. Furthermore, due to the uncertainty about the "right" target in terms of atmospheric GHG stabilization and the associated degree of warming, we might not be in a position to disregard options merely on the basis of (slightly) higher costs: adjusted for the risks they bear, these might be quite different at the second glance.

The second issue concerns the dynamics underlying climate change and any strategy trying to tackle warming. It might be both costly and difficult (if not impossible) to rely on solutions focusing exclusively on the technosphere. Carbon-neutral or carbon-saving technologies will take a while to be developed and diffused; biomass-fired electricity generation with carbon capture and storage is one example. While the technology option is thus still an important element in the strategy mix and R&D can actually buy an option on such technology for the medium to long run, it is clear that in the near term other solutions have to be considered to be able to reach the stabilization target (which might turn out to be a lower one than anticipated, also in the light of potential threshold effects). Forestry carbon sequestration is one such example, which would feature more importantly in the near-term portfolio.

In conclusion, we should be looking for a portfolio of options to hedge the risks associated with each of the options considered. At the same time, we should be aware that this is not a static optimization, but that we are in fact choosing an optimal portfolio *across time*, where we should take into account the underlying dynamics of both the solutions suggested and the uncertainties associated with the climate system itself.

Bibliography

Adams, D.M., R.J. Alig, B.A. McCarl, J.M. Callaway, and S.M. Winnett, 1999: Minimum cost strategies for sequestering carbon in forests, *Land Economics* **75**, 360–74

Angelsen, A., S. Brown, C. Loisel, L. Peskett, C. Streck, and D. Zarin, 2009: Reducing emissions from deforestation and forest degradation (REDD): an options assessment report, Report prepared for the government of Norway, www.redd-oar.org/links/REDD-OAR_en.pdf

Benitez, P., I. McCallum, M. Obersteiner, and Y. Yamagata, 2004: Global supply for carbon sequestration: identifying least-cost afforestation sites under country risk consideration, Interim Report **IR-04–022**, International Institute for Applied System Analysis, Laxenburg, Austria

Bosetti, V., R. Lubowski, A. Golub, and A. Markandya, 2009: Linking reduced deforestation and a global carbon market: impacts on costs, financial flows, and technological innovation, FEEM Nota di Lavoro **56.2009**, feem.it/NR/rdonlyres/396DBB3D-C57B-4878–9AE6-E493471DDC89/2930/5609.pdf

Böttcher, H., K. Aoki, S. De Cara, M. Gusti, P. Havlik, G. Kindermann, U.A. Schneider, and M. Obersteiner, 2008: GHG mitigation potentials and costs from land-use, land-use change and forestry (LULUCF) in Annex 1 countries, GAINS Methodology Report; International Institute for Applied System Analysis, Laxenburg, Austria

Busch, J., B. Strassburg, A. Cattaneo, R. Lubowski, F. Bolz, R. Ashton, A. Bruner, and D. Rice, 2009: Open source impacts of REDD incentives spreadsheet (OSIRIS v2.4), collaborative modeling initiative on REDD economics, www.conservation.org/osiris/Pages/overviews.aspx

Dinan, T. and P. Orszag, 2008: *Policy Options for Reducing CO2 Emissions*, Congressional Budget Office, US Congress, Washington, DC, February, www.cbo.gov/ftpdocs/89xx/doc8934/02-12-Carbon.pdf

Dixit, A.K. and R.S. Pindyck, 1994: *Investment under Uncertainty* (Princeton University Press, Princeton, NJ

Eliasch, J., 2008: *Climate Change: Financing Global Forests*, Office of Climate Change, London, UK

Food and Agriculture Organization of the United Nations (FAO), 2000: The global outlook for future wood supply from forest plantations, Working Paper **GFPOS/WP/03**, FAO, Rome

Foley, J.A., G.P. Asner, M. Heil Costa, M.T. Coe, R. DeFries, H.K. Gibbs, E.A. Howard, S. Olson, J. Patz, N. Ramankutty, and P. Snyder, 2007: Amazonia revealed: forest degradation and loss of ecosystem goods and services in the Amazon Basin, *Frontiers in Ecology* **5**, 25–32

Golub, A., S. Fuss, J. Szolgayova, and M. Obersteiner, 2009: Effects of low-cost offsets on energy investment: new perspectives on REDD, FEEM Nota di Lavoro 17.2009, http://papers.ssrn.com/sol3/papers.cfm?abstract_id=1397144 or http://www.feem.it/NR/rdonlyres/80EAC8FD-07E5-4135-8642-5D959AE95C25/2832/1709.pdf

Gusti, M., P. Havlik, and M. Obersteiner, 2008: Technical description of the IIASA model cluster, The Eliasch Review, Office of Climate Change, London, UK, www.occ.gov.uk/activities/eliasch/Gusti_IIASA_model_cluster.pdf

2009: How much additional carbon can be stored in forests if economic measures are used and how much could it cost?, Research Reports of the National University of Bioresources and Nature Management of Ukraine, www.nbuv.gov.ua/Portal/chem_biol/nvau/2009_135/9m.pdf

Hansen, J., Mki. Sato, P. Kharecha, D. Beerling, R. Berner, V. Masson-Delmotte, M. Pagani, M. Raymo, D.L. Royer, and J.C. Zachos, 2008: Target atmospheric CO2: where should humanity aim?, *Open Atmospheric Science Journal* **2**, 217–31

Kindermann, G., M. Obersteiner, B. Sohngen, J. Sathaye, K. Andrasko, E. Rametsteiner, B. Schlamadinger, S. Wunder, and R. Beach, 2008: Global cost estimates of reducing carbon emissions through avoided deforestation, *Proceedings of the National Academy of Sciences* **105**(30), 10302–7

Lubowski, R., A. Plantinga, and R. Stavins, 2006: Land-use change and carbon sinks: econometric estimation of the carbon sequestration supply function, *Journal of Environmental Economics and Management* **51**, 135–52

Murray, B., 2008: Leakage from an avoided deforestation compensation policy: concepts, empirical evidence, and corrective policy options, Working Paper **08–02**, Nicholas Institute for the Environmental Policy Solutions, Duke University, www.nicholas.duke.edu/institute/wp-leakage.pdf

Murray, B., R. Lubowski, and B. Sohngen, 2009: Including international forest carbon incentives in a United States climate policy: understanding the economics, Working Paper, Nicholas Institute for the Environmental Policy Solutions, Duke University, Durham, NC

Murray, B.C., B.L. Sohngen, A.J. Sommer, B.M. Depro, K.M. Jones, B.A. McCarl, D. Gillig, B. DeAngelo, and K. Andrasko, 2005: EPA-R-05-006, Greenhouse gas mitigation potential in US forestry and agriculture, US Environmental Protection Agency, Office of Atmospheric Programs, Washington, DC

Naidoo, R. and T. Iwamura, 2007: Global-scale mapping of economic benefits from agricultural lands: implications for conservation priorities," *Biological Conservation* **140**, 40–9

Nordhaus, W., *A Question of Balance: Weighing the Options on Global Warming Policies*, Yale University Press, New Haven, CT

Pagiola, S. and B. Bosquet, 2009: Estimating the costs of REDD at the country level, Forest Carbon Partnership Facility, www.forestcarbonpartnership.org/fcp/sites/forestcarbonpartnership.org/files/The%20Costs%20of%20REDD%2004-09-09.pdf

Plantinga, A.J., T. Mauldin, and D.J. Miller, 1999: An econometric analysis of the costs of sequestering carbon in forests, *American Journal of Agricultural Economics* **81**, 812–24

Ramankutty, N., J.A. Foley, J. Norman, and K. McSweeney, 2002: The global distribution of cultivable lands: current patterns and sensitivity to possible climate change, *Global Ecology & Biogeography* **11**, 377–92

Sohngen, B. and R. Mendelsohn, 2003: An optimal control model of forest carbon sequestration, *American Journal of Agricultural Economics*. **85**(2), 448–57

2007: A sensitivity analysis of carbon sequestration, chapter 19 in M. Schlesinger, H.S. Kheshgi, J. Smith, F.C. de la Chesnaye, J.M. Reilly, T. Wilson, and C. Kolstad (eds.), *Human-Induced Climate Change: An Interdisciplinary Assessment*, Cambridge University Press, Cambridge

Stavins, R., 1999: The costs of carbon sequestration: a revealed preference approach, *American Economic Review* **89**, 994–1009

Tavoni, M., B. Sohngen, and V. Bosetti, 2007: Forestry and the carbon market response to stabilize climate, *Energy Policy* **35**, 5346–53

Van Velthuizen, H. *et al.*, 2007: Mapping biophysical factors that influence agricultural production and rural vulnerability, Environment and Natural Resources Series XI, Food and Agriculture Organization of the United Nations and International Institute for Applied Systems Analysis, Rome

Wise, M. *et al.*, 2009: Implications of Limiting CO_2 concentrations for land use and energy, *Science* **324**, 1183–6

Black Carbon Mitigation

ROBERT E. BARON, W. DAVID MONTGOMERY, AND
SUGANDHA D. TULADHAR

Introduction

Much attention has been given to mitigation policies designed to limit the emissions of carbon dioxide (CO_2) and other greenhouse gases (GHGs) that contribute to atmospheric warming. However, it is generally agreed that as much as 40% of current net warming (10–20% of gross warming) is attributable to black carbon (Jacobsen 2007: 3).[1] Because of its large effect on radiative forcing and relatively short residence time in the atmosphere, black carbon presents some unique opportunities for postponing the effects of climate change.[2] Whereas CO_2 has a lifetime of up to about forty years, black carbon remains in the atmosphere for as little as several weeks.[3] As such, reducing emissions of black carbon can have an immediate near-term impact on atmospheric warming. Furthermore, since black carbon is considered responsible for about 30% of the Arctic melting, black carbon emission reductions can rapidly reduce the rate at which Arctic ice is melting and avert associated consequences. Black carbon reduction policies can also result in large health benefits, especially to citizens of developing countries.

Black carbon emissions originate in both industrial countries (mostly from diesel emissions) and developing countries (from residential activities, crop management, and diesel emissions). Over the past century, technological advances have significantly mitigated black carbon emissions in industrial countries. Developing countries, on the other hand, have often been unable to afford the same technological advances and, in turn, as their populations have grown, so have black carbon emissions. Moreover, the extensive practice of closed-area cooking and heating has exacerbated the black carbon problem in these developing countries. This practice can result in negative health effects on those living in close quarters, often women and children. This chapter argues that controlling black carbon emissions in developing countries is a potentially cost-effective means of postponing the effects of global warming, while at the same time improving the health and quality of life (QALY) of those living in those countries. Black carbon can be controlled in developing countries through the implementation of cleaner fuels, new cooking technologies, and changing crop management practices. This chapter also presents potential ways to implement these policies.

Climate Change and the Role of Quick-Acting Solutions

Climate change is driven by the concentration of GHGs in the atmosphere, making climate change what economists call a "stock externality."[4] Concentrations of GHGs do not respond to changes in emissions over short periods of time, but depend mostly on cumulative emissions over long time periods.[5] Black carbon is different. Its short residence time in the atmosphere means that measures

[1] "Gross warming" is defined as the aggregate warming effect of GHG emissions, black carbon, and urban heat island effects. "Net warming" is defined as the aggregate warming effect of GHG emissions, black carbon, and urban heat island effects less the opposing effect of cooling particles.

[2] Jacobson (2007: 3).

[3] With "lifetime defined" as the time required for concentration to decline to 37% of its original value.

[4] Newell and Pizer (2003).

[5] GHGs include: carbon dioxide (CO_2), methane (CH_4), nitrous oxides (N_2O), chlorofluorocarbons (CFCs), hydrochlorofluorocarbons (HFCs) and perfluorinated carbon compounds.

to reduce its emissions would result in more immediate impacts on its concentration.

Bringing black carbon into climate strategy expands the choices of pathways for emissions over time that can lead to identical outcomes in global average temperatures.[6] Wide-scale requirements for GHG reductions in advance of the availability of cost-effective technology can be unnecessarily costly. If research and development (R&D) can reduce the cost of technologies that replace fossil fuels and other sources of GHG emissions with alternatives having lower or zero GHG emissions, then timing emission reductions in order to take advantage of these new lower-cost innovations can reduce the overall cost of meeting climate goals. Reducing the cost of moving to a low- or zero-carbon economy can also make a worldwide agreement to reduce GHG emissions more likely. Since reductions in black carbon emissions can delay warming for a matter of decades, black carbon policies can buy time for R&D to achieve these kinds of cost reductions.[7]

Likewise, action by industrial countries alone to reduce their emissions to minimal levels by mid-century cannot be cost-effective because during this century a majority of GHG emissions will originate from developing countries.[8] However, developing countries have been unwilling to adopt carbon limits that would in any way interfere with their aspirations for economic growth. It is particularly important to identify actions that can reduce their contribution to warming and at the same time improve growth prospects or reduce poverty.

However, in many developing countries where market, property, and legal institutions have not developed sufficiently to support market-based policies, cost-effective changes in energy use or fuel choice may be difficult to achieve. Institutional reforms that increase economic freedom can simultaneously improve growth prospects, lead to more efficient energy use (Montgomery and Bate 2005 and Montgomery and Tuladhar 2006), and facilitate the use of market forces. However, institutional change is not a rapid or easily influenced process.

With the current state of climate science it is not possible to make robust calculations of the expected benefits of reducing GHG emissions. The uncertainty in the most fundamental relationship between

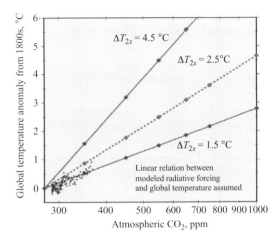

Figure 4.1 *Uncertainty in temperature sensitivity*

Source: Kheshgi (2008).

GHG concentrations and temperature has not been reduced by decades of research (figure 4.1). Even the benchmark range for a doubling of CO_2 concentrations from preindustrial levels represents a huge range of potential risk.

Assessing temperature sensitivity as a function of atmospheric GHG concentrations is the relatively easy part, but even if it is possible to put some kind of probability distribution on temperature change, it is not possible to assign probabilities to the rest of the steps in the causal chain that must be followed to calculate avoided damages.[9] From temperature change, it is necessary to project a wide variety of potential climate change effects, including sea level rise, changes in precipitation, polar melting, and probability of severe weather events. It is also necessary to understand how these physical phenomena will affect conditions of human life and ecosystems, including health, agricultural productivity, property loss, changes in heating and cooling costs, damage to unmanaged eco-systems, loss of recreational benefits, species decrease/increase or adaptation. The uncertainty about these effects is profound, with no objective basis for the assignment of probabilities. Thus calculation of expected

[6] Kheshgi (2007) and Wigley *et al.* (2007).
[7] Ramanathan (2006).
[8] Jacoby *et al.* (2008).
[9] Sokolov *et al.* (2007).

benefits is not a scientific possibility and any estimates of such benefits are highly speculative.

Nevertheless, many studies have attempted this. The most current survey of the social cost of carbon emissions finds that the mean global value taken from studies that employed reasonable discount rate and value of life assumptions is less than $25 per ton of CO_2.[10] This suggests that policies with a marginal abatement cost greater than this $25 per ton of CO_2 cut-off are not worthwhile.

Concerns about potential catastrophic change have been cited as a reason for reducing emissions much more rapidly than any calculation of expected costs and benefits would support. These concerns are based on admittedly speculative calculations of catastrophic climate change that hypothesize that a "tipping point" may be reached within a matter of years, putting the world on an inevitable course toward catastrophe thereafter.[11]

What is not sufficiently recognized is that there may be no way to avoid such tipping points, if they are indeed close enough that they could be triggered by small additional increases in GHG concentrations. Factors limiting the effectiveness and feasibility of short-term, stringent GHG reductions include:

- the slow removal of GHGs in the atmosphere – with CO_2 having a life of up to about forty years
- the unwillingness of developing countries to consider absolute reductions in their emissions for decades
- the slow turnover of the capital stock in countries that do act.

If proposed policy solutions focus exclusively on abating GHG emissions, then it is very likely that any short-term tipping points associated with catastrophic climate change will not be averted.

---

[10] Tol (2007).
[11] Barrett (2007), Gulledge (2008), and Weitzman (2008).
[12] "Carbon dioxide cycles between the atmosphere, oceans and land biosphere. Its removal from the atmosphere involves a range of processes with different time scales. About 50% of a CO_2 increase will be removed from the atmosphere within 30 years, and a further 30% will be removed within a few centuries. The remaining 20% may stay in the atmosphere for many thousands of years" (IPCC 2007).
[13] Ramanathan (2007b: 2).

If the only reason for immediate action to reduce GHG emissions is to avoid a catastrophe that will be inevitable if concentrations increase above current levels, then taking no action to reduce traditional GHG emissions is the only logical course. If catastrophe is inevitable, it must be prepared for, and expenditures on measures that are useless to prevent the catastrophe are wasted.

Whereas changes in GHG emissions occur so slowly as to be ineffective in dealing with possible catastrophic change, black carbon mitigation is capable of quickly bringing about changes in radiative forcing. Black carbon emitted today will by and large leave the atmosphere in a month or less, while CO_2 emitted today will linger in the atmosphere for decades.[12] If tipping points or catastrophic change can be predicted even a few years in advance, then black carbon provides a potential method of avoiding what would otherwise be unavoidable effects. This is a particularly important role when it is impossible to assign a probability to catastrophe and justify longer-term policy interventions on a benefit-cost (B/C) basis.

Black carbon removal may also be justified because of its ability to delay warming for several decades.[13] It would provide time for the R&D required to develop game-changing technologies that could significantly reduce the cost of emission reduction. In this case, the benefit of black carbon reduction would become the avoided cost of premature GHG emission reductions.

Moreover, black carbon from indoor heating and cooking in developing countries has severe adverse health effects, and is also thought to have important regional effects on floods and drought. Thus, policies to reduce black carbon emissions would also have near term co-benefits.

The remainder of this chapter discusses the contribution of black carbon to warming, the sources of black carbon emissions and methods of control, and the cost-effectiveness of black carbon reductions in reducing near-term warming potential. The final section discusses cost-effective policies that can be devised that result in emission reductions. Since most of the cost-effective opportunities for black carbon reduction occur in developing countries, we search for policy measures that are also consistent with poverty reduction and economic growth.

Black Carbon: Science and Estimated Contributions to Warming

Incomplete combustion entails only partial burning of a fuel, such as coal or biomass. This may be due to a lack of oxygen or low temperature.

Black carbon is a term to describe the elemental carbon substance formed during incomplete combustion. Other by-products are also formed including organic carbon and brown carbon.[14] These three substances are differentiated in one regard by their light-absorbing properties. Black carbon absorbs all wave lengths, organic carbon scatters light, and brown carbon has light-absorbing properties between black carbon and organic carbon.[15]

Black carbon emissions originate from a variety of sources: open biomass burning is estimated to be responsible for 42%, residential burning 24%, transportation 24%, and industry/power 10% (with large regional differences).[16]

Black carbon emissions can remain in the atmosphere for several weeks to a month or more. While at altitude, black carbon is transported from its point of origin to other parts of the globe. The extent and direction of travel depends on a number of factors including point of origin and wind currents. With time, black carbon will fall back to earth by settling and by precipitation.

Black carbon can affect climate both while in the atmosphere and after it returns to the earth's surface. Black carbon aerosols absorb all wave lengths of solar radiation from the short-wave ultraviolet light waves to longer infrared waves.[17] Furthermore, black carbon converts ultraviolet to infrared which it re-radiates to the atmosphere around the particle. This warms the atmosphere at higher altitudes and creates a dimming effect on the surface which can result in lower temperatures in the lower atmosphere.

As an aerosol, black carbon also interacts with other compounds in the atmosphere which can potentially enhance its impact on climate. Black carbon particles can grow with time due to coalescence and possibly impact its ability to absorb solar radiation. Interactions between aerosols and clouds are not well understood.

Black carbon also causes climatic effects at the regional level. Black carbon is thought to affect precipitation patterns through its effect on atmospheric temperature profiles. Rainfall levels and patterns can be affected in areas where black carbon is emitted in large quantities. Some associate the change in rainfall patterns in sub-Saharan Africa with the increasing burning of biomass for cooking and land-clearing by the local population.[18] Both processes can produce black carbon in large quantities.

When black carbon settles back to the Earth's surface it can impact on melting. Even with concentrations in the snow so low as not to be discernable to the naked eye, black carbon particles reduce snow and ice reflectivity.[19] This results in accelerated rates of melting. As the melting exposes land and open water an even greater portion of the solar radiation is absorbed due to its relatively darker surfaces. This creates a positive feedback which further accelerates melting.

Black carbon is also unique because the timing of the emissions also contributes to the extent of Arctic melting. Preparing fields for planting in the late winter or early spring often involves burning the residual plant materials. The fields are often left to smolder. The resulting black carbon is carried to the Arctic region where it settles on the snow and

---

[14] Many of the carbonaceous sources of black carbon and organic carbon contain sulfur. The sulfur oxides that are produced during combustion also travel into the atmosphere and are further oxidized to form sulfates in various compounds. In upper levels of the atmosphere, both the organic carbon and sulfates will scatter solar radiation, preventing it from reaching the Earth's surface.

[15] Traditionally research has focused only on the atmospheric concentrations of black carbon and organic carbon and their estimated effects on global warming. More recent research, for instance by Andreae and Geleneséi (2006), indicates that brown carbon is an important contributor to atmospheric warming and that estimates of the warming effect based upon black carbon and organic carbon alone may be biased.

[16] Bond (2007: 2).

[17] Black carbon is different from GHGs which absorb infrared energy emitted from the Earth's surface thus "trapping" energy in the atmosphere and causing warming.

[18] Ramanathan (2007a).

[19] Stegeman (2008) and www.polarfoundation.org/www_sciencepoles/index.php?/articles_interviews/black_carbon_playing_a_major_role_in_arctic_climate_change/&uid=1253&pg=8.

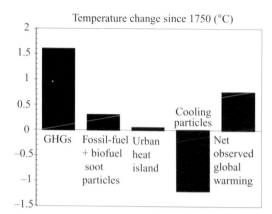

Figure 4.2 *Contribution of black carbon to increases in average global temperature*

Source: Jacobsen (2007).

ice just as the Arctic warming season begins. The result is a compounding of the forces that promote melting.[20]

The use of primitive indoor cooking stoves produces black carbon in confined quarters. When this black carbon is inhaled by those in the immediate area (often women and children) adverse health effects, including pulmonary disease and respiratory infections can result.[21]

It is generally agreed that as much as 40% of net warming (10–20% of gross warming) is attributable to black carbon.[22] Figure 4.2 illustrates the relative contribution of the components to the change in average global temperature. The global-warming potential (GWP) for black carbon for a 100-year time interval can range from 210 to 1,500 with a central value of 680 over a 100-year time period.[23]

---

[20] Pettus (2009: 28).
[21] Larsen *et al.* (2008).
[22] Jacobson (2007: 3).
[23] GWP is a metric that measures how much a substance contributes to global warming compared to the same mass of CO_2 over a specific time period. A value of 680 means that a kg of black carbon would have the same effect as 680 kg of CO_2 over a time interval of 100 years.
[24] Bond and Sun (2005).
[25] Boucher and Reddy (2008).
[26] Bond *et al.* (2007).
[27] Bond *et al.* (2004) and Cofala *et al.* (2006).
[28] Cofala *et al.* (2006).
[29] Bond (2007).

For a twenty-year time interval, the GWP can vary from 690 to 4700.[24] An estimate for black carbon GWP emitted in Europe is about 374 while for Africa it is 677.[25]

Sources of Black Carbon Emissions

Global emissions from black carbon have almost doubled during the past century rising from 2,300 giga-grams (Gg) to about 4,600 Gg.[26] During the same period, the growth in black carbon emissions has shifted from industrial countries to developing countries, with emission growth coming primarily from China and India. Developing countries in regions such as South Asia, East Asia, South East Asia, and Africa continue to generate black carbon emissions at high growth rates, while OECD regions have witnessed continued decline in emissions. Over the past decade, black carbon emissions in China (East Asia) have grown by an annual average growth rate of 3.5%, while for India (South Asia) black carbon emissions are growing at an annual rate of about 2% (figure 4.3). Overall, from 1990 to 2000, black carbon emissions from Asia accounted for about 60% of total global black carbon emissions.[27]

The sources of black carbon emissions vary by region depending on lifestyle and the types of fuel used (figure 4.4 and figure 4.5). Globally, small-scale combustion accounts for 65%, while transportation related activities emit about 24% of the global black carbon emissions (figure 4.4).[28] At a more detailed breakdown, in 2000, 42% of energy-related black carbon emissions came from open biomass burning (including open forest and savanna burning for land-clearing, anthropogenic or otherwise), 18% from residential biofuel, 14% from on road transportation, 10% from industry and power generation, and 6% from residential use of coal and other fuels.[29]

In the developing countries, black carbon is primarily emitted from "domestic use." "Domestic use" refers to small-scale combustion sources such as heating and cooking using mainly biomass fuel. A large part of the developing countries' population lives below the poverty line and hence are unable to afford cooking methods or technologies that result

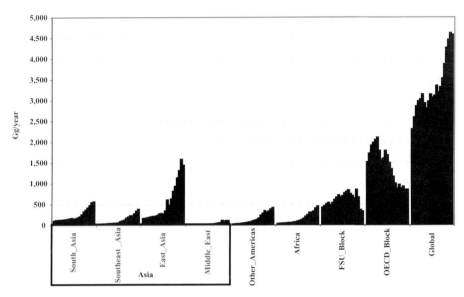

Figure 4.3 *Black carbon emissions (1900–2000) (Gg/year)*

Source: Compiled from Bond *et al.* (2004) data.

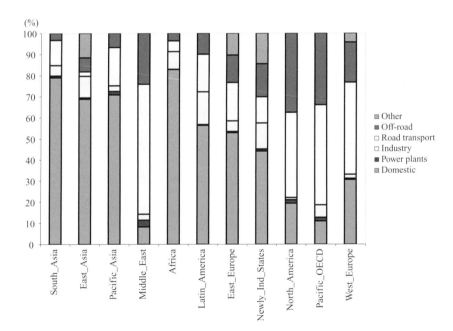

Figure 4.4 *Regional share of black carbon emissions (2000)*

Source: Compiled from IIASA (2007).

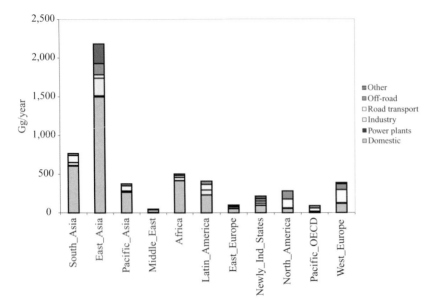

Figure 4.5 *Black carbon emissions by region (2000) (Gg/year)*

Source: Compiled from IIASA (2007)

in complete combustion of the fuel. These subsistence methods of living and the associated combustion practices contribute heavily to expanding the black carbon emissions in the developing countries. Figure 4.4 shows that in the South Asia (India) and East Asia (China) regions, black carbon from "Domestic" sources alone exceed 70%. The share of black carbon from domestic use is also large in regions such as Africa, Latin America, and Eastern Europe (in excess of 50%). However, for the industrial countries, black carbon originates mainly from the combustion of diesel fuel in the transportation sector. In North America, "Road Transportation" and "Off Road" emissions account for 80% and in Western Europe more than 70% of black carbon emissions.[30]

In much of the developing world, cooking entails open burning at low temperatures. This means of

cooking often uses low-cost/low-quality fuels such as wood, coal, and dung. These fuels, which often are wet and burned in chunks, produce black carbon due to their poor burning characteristics.[31] Localized health effects may result from the inhalation of the smoke, particularly by women cooking and or by children nearby. Low-tech solutions to improve combustion are commercially available today. These technologies improve the burning characteristics by pulverizing the fuel prior to burning (i.e. increasing the fuel's surface-to-volume ratio) and using a fan to improve air circulation in the stove (i.e. improving oxygen fuel contact).[32] More sophisticated approaches have also been tried, such as harnessing solar energy. A switch away from traditionally low-cost fuels to cleaner-burning fuels such as kerosene would also reduce black carbon formation, but may be cost-prohibitive. Traditional cooking using open burning does result in some of the taste inherent in the food. A move towards other means of cooking would need to overcome any cultural barriers.

Crop burns represent an important source of black carbon generation. It is common practice in some areas of the world to prepare fields for

---

[30] Cofala *et al.* (2006).

[31] Combustion rates accelerate when the fuel and air are well mixed and the fuel has a high surface-to-volume ratio. This results in higher burn temperatures which also promote greater oxidation of the carbonaceous fuel, resulting in improved combustion efficiency.

[32] Rosenthal (2009).

planting by burning the residual crop materials from the previous year. In China, it had been common practice to gather residual crop materials to use as animal fodder and for fuel. However, as China improved its crop yields and became more prosperous, supply of residual crop materials outpaced demand and farmers increasingly adopted crop fires as an expeditious means of crop management. Between 2001 and 2003, 30–40% of fires in China occurred in croplands.[33] The importance of crop burns is enhanced because of their timing. Crop burns occur during the late winter and early spring. Thus resulting black carbon can be transported and deposited in the Arctic region during the spring melting period, enhancing any ice melting phenomena.

According to 2000 inventory, biomass burning which includes crop burns and forest fires, contributed 42% of the black carbon emitted to the atmosphere.[34]

Costs and Benefits of Black Carbon Reduction

The uncertainties associated with quantifying the effect of black carbon on climate change make it difficult to quantify the benefits of black carbon reduction as a climate change strategy, and the dispersed nature of black carbon emissions makes it even harder to incorporate it in an overarching policy architecture such as emission trading. In particular, the effects of black carbon on climate are not well represented by the current metric of GWP metric, and there is little consensus on the level of past, current and future emissions of black carbon. Estimates of black carbon show a wide range of variation because of different assumptions between studies of emission factors (see table 4.1). Regional differences in atmospheric black carbon concentrations, and hence the warming effect of black carbon, depends upon the regional climate, radiation properties, and deposition pathways.[35]

The short atmospheric residence time of black carbon means that the black carbon forcing will be different than that for GHGs. The normal measure is to compare climate effects of different long-lived gases such as the Kyoto gases (CH_4, N_2O,

Table 4.1 Comparison of global emission estimates of black carbon (1980–2000)

Source	Year	Black carbon
Penner et al. (1993)	1980	12610
Cooke and Wilson (1996)	1984	7970[a]
Cooke et al. (1999)	1984	51009[a]
Bond et al. (2004)[b]	1996	9478
Bond et al. (2004)[c]	1996	4954 (3296–11019)
Cofala et al. (2006)	1995/2000	5551–5342

Notes:
[a] Emissions from fossil fuel use.
[b] Activity as in Bond et al. (2004); emission factors from Cooke et al. (1999). Includes open burning of crop residues
[c] Totals adjusted to include open burning of crop residue for which ranges estimated from ranges given for total open burning.
Source: Cofala et al. (2006: table 1).

and hydrofluorocarbons, HFCs). The GWP is the ratio of the time-integrated radiative forcing from a pulse emission of 1 kg of a substance, relative to that of 1 kg of CO_2 over a fixed horizon period (IPCC 2001). GWP can also be used to compute short-lived gases using a short time period.

However, these computations are controversial.[36] The general measure of GWP computed for long-life GHGs over a 100-year horizon is inappropriate for black carbon particles that have an atmospheric residence time of only weeks because it obscures the potential for large near-term reductions in radiative forcing through black carbon reductions. An additional shortcoming of the GWP measure in characterizing the climate effects of black carbon lies in its incomplete treatment of Arctic deposition. Indeed, even if two regions share the same value of GWP, it may have different climate impact due to regional effects on Arctic deposition.[37] Shine et al. (2005) introduced a Global Temperature Potential (GTP) concept to address this criticism. In contrast to the GWP measure, GTP takes into account the time profile of radiative forcing which is not the case in GWP

---

[33] Pettus (2009).
[34] Bond (2007).
[35] Rypdal et al. (2009).
[36] Rypdal et al. (2009).
[37] Hansen et al. (2005) and Rypdal et al. (2009).

Table 4.2 Level of black carbon under two different scenarios (Tg/year)

Black carbon Region	Current legislation (CLE)					Maximum feasible reductions		
	1990	2000	2010	2020	2030	2010	2020	2030
OECD90:	0.83	0.76	0.58	0.46	0.41	0.34	0.26	0.23
North America	0.32	0.28	0.24	0.19	0.18	0.08	0.09	0.09
Western Europe	0.41	0.39	0.27	0.21	0.18	0.24	0.16	0.12
Pacific OECD	0.09	0.09	0.06	0.06	0.06	0.02	0.02	0.02
REF:	0.67	0.31	0.33	0.3	0.32	0.17	0.15	0.14
Central and Eastern Europe	0.13	0.1	0.07	0.05	0.05	0.05	0.03	0.02
Russia and NIS	0.54	0.21	0.26	0.25	0.27	0.12	0.11	0.12
Asia:	3.09	3.32	3.52	3.13	2.89	2.46	2.16	1.92
Centrally Planned Asia	2.03	2.18	2.43	2.16	1.88	1.7	1.49	1.26
Other Pacific Asia	0.32	0.37	0.36	0.31	0.34	0.25	0.21	0.24
South Asia	0.74	0.77	0.73	0.67	0.67	0.51	0.45	0.43
ALM:	0.89	0.95	0.95	0.93	0.8	0.68	0.66	0.48
Latin America	0.5	0.41	0.35	0.28	0.22	0.2	0.15	0.08
Middle East	0.04	0.05	0.03	0.04	0.04	0.01	0.01	0.01
Africa	0.35	0.5	0.57	0.61	0.53	0.48	0.5	0.39
World Total	5.48	5.34	5.38	4.82	4.41	3.65	3.23	2.78

Source: Cofala *et al.* (2006).

measure. GTP is defined as the global change in surface temperature at a time horizon induced by a pulse emission.[38] For substances with short atmospheric residence time, the difference in the measure of GWP and GTP will be large.

Developing a cost-benefit analysis (CBA) for black carbon has been and still is challenging given significant data limitations. The limited extent of such analysis points to this poor state of data on black carbon. Nevertheless, researchers in the recent past have attempted to quantify reduction potentials of black carbon under various scenarios and related impacts on the radioactive forcing.[39] Rypdal *et al.* (2009) developed global scenarios with regional differences in climate impacts, regional marginal cost of abatement (MAC), and ability to pay, as well as the direct and indirect (snow-albedo) climate impact of black carbon.

Cofala *et al.* (2006) estimate future levels of black carbon under two scenarios: (i) each country continues its current environmental legislation based on current national economic development assumptions; and (ii) the adoption of currently available advanced emission control technologies. They estimate that black carbon emissions will decrease by 20–30% by 2030, with the majority of the reduction coming from phasing-out of solid fuels for domestic use, improving residential cooking technologies, and enacting legislation to control emissions from mobile sources. Table 4.2 shows the range of emissions from black carbon under the two different scenarios.

Rypdal *et al.* (2009) describe a scenario in which the regional allocation of black carbon reductions is based on minimizing global cost. Under the cost effectiveness rule, 70% of the global abatement would come from China and India. The flatness of the marginal abatement cost curve at lower prices clearly indicates that a large amount can be abated at a relatively low price (figure 4.6).

[38] Boucher and Reddy (2008).
[39] Bond (2007), Cofala (2006), and Boucher and Reddy (2008).

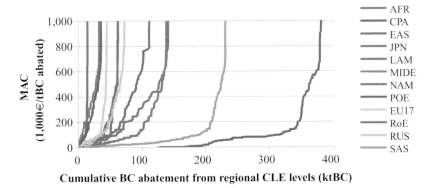

Figure 4.6 *MACs for black carbon*

Source: Rypdal *et al.* (2009).

These results support the conclusions of other studies that black carbon reductions from Annex I countries are highly expensive in the short run compared to reduction in other GHGs.[40] A more effective global strategy would be to address the reduction of black carbon from developing countries with targeted policies that support these countries' efforts to mitigate black carbon.

One proposed policy for reducing black carbon emissions in industrial countries is to require particle traps on existing vehicles (the most recent round of regulations that reduced particulate emissions on new vehicles in the USA were phased in during the last few years).[41] This a relatively expensive means of black carbon emission reduction (table 4.3). Retrofitting existing diesel powered vehicles not originally designed for particle-trap technologies has potential implications for the operation and maintenance of these vehicles beyond the cost of the particle trap itself.

Proposed Solutions for Black Carbon Mitigation

Different stories emerge from calculating costs and benefits based on a hypothetical reduction in black carbon emissions vs. those based on specific policies with predictable effects on black carbon emissions. Hypothesized reductions in diesel emissions and residential heating/cooking have

been estimated to delay warming from other GHGs by several decades.[42] However, realistic and achievable policies, though still cost-effective, are likely to be able to achieve much less in a short time.

Finding effective policy approaches to achieve these hypothetical benefits is much more difficult. Due to slow fleet turnover, reduced diesel emissions from new sources will only penetrate the fleet fully in one–two decades, and retrofitting particle traps on diesel engines is not cost-effective. Replacing traditional cooking and heating with more modern technology in Asia and Africa could achieve large health, quality of life (QALY) and temperature benefits. However, for poverty-reduction programs with similar goals and methods, institutional and implementation barriers have resulted in limited acceptance. This leaves coal and diesel emissions in China and India as attractive targets, but again the institutional capacity to enforce a large-scale change in the existing fleet is dubious.

Policy Solutions

The challenge of designing realistic policies to reduce black carbon emissions is thus very similar to the challenge of developing effective poverty

---

[40] Bond and Sun (2005).
[41] www.dieselnet.com/standards/us/hd.php.
[42] Ramanathan (2007b).

Table 4.3 Costs of various black carbon control technologies

Emitting technology	Abatement technology	EF-black carbon (g/kg)	Fuel (kg/yr)	Life-time (yr)	Life-time black carbon (kg)	CO$_2$ eq (t) 100-yr	CO$_2$ eq (t) 20-yr	Cost ($-t CO$_2$ eq) 100-yr	Cost ($-t CO$_2$ eq) 20-yr
Diesel engines Current light vehicle	Particle trap ($250–500)	0.9	1500	10	14	10	31	25–50	8–16
Superemitting light	Repair ($500–1000+);	3	1500	5	23	15	50	30–130	10–40
Vehicle	Vehicle turnover (several $K)								
Pre-regulation truck	Particle trap ($5K–10K)	2	10000	10	200	140	440	36–71	11–23
Residential solid fuel Wood cookstove	Cleaner stoves, fuel switching ($3–100)	0,7	2000	3	4,2	2,9	9,2	1–34	0.3–11
Coal cookstove	Same as wood stove	8	1000	3	24	16	53	0.2–6	0.1–2
Other transport Gasoline: 2-stroke engine	Education, engine switching	1	300	5	1,5	1,1	3,3	n.a.	n.a.
Industry & power Coal: low-tech brick kiln	Switch kiln type	5	500000	1	2500	1750	5500	18–35	5.5–11

Note: n.a. = Not available.
Source: Bond (2007).

reduction or technology modernization programs. Black carbon mitigation should be an attractive option for developing countries such as India.[43] From the point of view of a country's negotiating position, a developing country attempting to reduce black carbon emissions would signal to industrial countries that it was serious about addressing climate change. More importantly, reduction in black carbon emissions from cooking would improve the quality of life of many poor people by averting poor air quality. More than 400,000 people die each year in India due to smoke inhalation from indoor cooking using rudimentary cooking methods.[44] Thus, black carbon reductions in developing countries may be justified on a health benefit basis alone without consideration of climate benefits. For industrial countries, achieving potential climate benefits in a cost-effective way provides a reason to devote resources to support black carbon reduction in developing countries.

Incentives for developing countries to undertake black carbon emission reductions should be different from an abatement regime that addresses long-lived gases. Even if it were not difficult to measure black carbon reductions to a quality demanded by financial trading systems, the 100-year GWP metric does not provide the incentives for developing countries to participate in programs to reduce black carbon emissions from all black carbon sources. Short time-horizon GWP or other metrics, such as GTP, may be more appropriate for properly rewarding black carbon abatement. However, they introduce even greater complications in an emission trading system.

Reducing black carbon emissions in developing countries relies upon the actions of millions of people that live in poverty. Changing behavioral patterns of such a vast number of people requires policy approaches that are realistic, affordable,

[43] Carl (2009).
[44] Carl (2009).

and sustainable. Often, too many programs – and black carbon reduction programs will be much like poverty reduction programs – are launched without such consideration, leading to failed policy prescriptions.

One promising approach is a new cook stove initiative (Project Surya) launched in India that focuses on understanding and overcoming local sensitivities in order to provide a sustainable solution.[45] The Project Surya initiative provides inexpensive solar and other energy-efficient cookers (biogas plants) in rural India with local participation. An innovative aspect of this project engages village children in data collection to enhance awareness of its benefits in an effort to make it long-lasting. Project Surya is a people-centered approach which recognizes that the individual can provide accessible and relevant information for motivating the inhabitants, empower the community with skills and the ability to act on the information and sustain the changes, and create an enabling environment to support and facilitate change.[46]

The success of these micro-level interventions across large populations requires effective partnership. Financing is an important part of the solution to make it sustainable for a long period of time. Many such programs fail because these one-time substitutions eventually break down and poor people often do not have the disposable income to replace it or repair it, so they retreat back to traditional methods. Effective micro-finance services can make a contribution towards this end. Micro-finance can either lend directly to households so that they can buy or repair these cook stoves, or develop a network of supporting businesses to provide finance, maintenance, and repair. The key advantage of well-managed micro-finance institutions is strength in distribution channels, clientele, linkages, credibility, and efficiency that allows them to reach to millions of poor people in an effective way.[47]

Five specific recommendations for policies and incentives that would be a component of an overall program to encourage developing countries to reduce carbon black emissions are:

- **Improve the black carbon emissions inventory** Emissions data for black carbon sources are subject to a high level of uncertainty. Any policy intervention in reducing black carbon will require evaluation of baseline emissions. Currently, data on black carbon emissions in some developing countries is not sufficiently complete or reliable to permit determination of a quality baseline. Technical and financial support should be provided to developing countries to allow collection and dissemination of black carbon emission data.

- **Make use of better technology for current combustion practice** Small-scale programs such as Project Surya and others are taking root in developing countries. They provide cook stoves to improve indoor air quality, alternative energy sources such as biogas plants to supplant poor combustion practices of burning raw coal, and solar technology for domestic use. These programs promote the substitution of clean energy for traditional sources. These small-scale projects are critical to reducing black carbon emissions and must be financially supported on a long term basis.

- **Provide financing for these commercially available technologies on a sustainable basis** Micro-finance institutions (MFIs) can play a major role in developing countries to engage local people, enterprises, and local governments to make small-scale operations a success. MFIs are especially suitable for reducing black carbon emissions in developing countries because they operate at the grass roots level and have access to the people that most need capital investment. They provide or subsidize the purchase of equipment by local entrepreneurs and offers incentives for entrepreneurs to create demand and provide service. Use of MFIs creates entrepreneurial local businesses that are motivated to provide training and repair services. MFIs allow small-scale operations to be replicated at a large number of localities. Table 4.4 illustrates how MFIs can contribute at local, institutional and board policy levels.

---

[45] Ramanathan and Balakrishnan (2007).
[46] Ramanathan and Balakrishnan (2007).
[47] CGAP (2009).

Table 4.4 Areas where MFIs can respond to climate change

Customer level (actions that affect micro-finance clients directly at the household and micro-business levels)	• Clean energy products • Lightning • Cooking • Forestation, avoided deforestation • Biofuels • Low-carbon agriculture • Community-level projects • Crop choices and farming practices • Financial products to help clients manage risk
Institutional level (actions that affect the function and finance of MFIs)	• Reduced emissions • Carbon finance and aggregation
Systemic (actions at national and international levels)	• Monitoring and using information about climate change • Smart subsidies • Advocacy and contribution to policy debate

Source: CGAP (2009).

Table 4.5 Portfolio of black carbon reduction programs

	2000 Gg black carbon
Total black carbon emissions	8,000
• Implementation of improved stoves in China and India using for residential cooking and heating	220
• Coal to briquette use programs for domestic cooking and heating	91
• Community forestry programs and resource property right awareness programs to control savannah and open burning	1,190
Savings in black carbon	1,502
Savings share (%)	19

• **Experiment with different approaches** All options should be pursued in order to reduce black carbon emissions in developing countries. The risk of small programs will be minimal and more easily adaptable to the village setting. Small-scale programs include stove replacement at village level, distributed generation and small-grid development, and distribution systems for petroleum-based fuels to replace wood and biomass. These are but a few of the candidate programs that could contribute to an overall effort to improve air quality as well as reduce black carbon emissions.

• **Provide incentives for developing countries that reward reduction of black carbon emissions** A parallel climate regime that includes participation by developing countries and provides rewards for reduction accomplishments would expedite the effort to mitigate black carbon emissions and improve air quality. The Asia Pacific Partnership (APP) and similar partnerships offer an opportunity to define an approach to black carbon policy that can reconcile objectives of air quality/environmental improvement with other competing interests of developing countries. Answers to four key questions would provide a basis on which these part-

nerships could move forward on an agenda of addressing black carbon emission reduction:

– Identification of local benefits and obstacles
– Technical training
– Design of appropriate technology
– Identify institutional and governance failures that need correction

Benefit-Cost Analysis

Our benefit-cost analysis (BCA) is based on a policy scenario that could achieve a reduction of 19% in black carbon emissions from a portfolio of programs consisting of improved stoves using solar energy, increased biofuels or briquette use in India and China, and mitigation of open burning in Africa and South America (table 4.5). Given the short-lived nature of the black carbon emission, the program is implemented over a single model year in the DICE model (Nordhaus 2008). Hence, the discount factor in the model does not impact the benefit-cost comparison. Because of the data quality concerns and uncertainties outlined in this chapter, these estimates represent our best judgment, but remain highly uncertain.

Black carbon emissions in 2000 were estimated to be 8,000 Gg. Based on the experience of cook stove replacement in China and India, we assume that 50% of existing stoves are replaced with these improved stoves by 2020, implying yearly net increases of 52 million stoves for China and India. We further assume that 30% of the new installations are wood-based, 10% use agriculture waste, 30% are animal waste-based, and 29% use briquettes or charcoal.[48] We assume that 5% of the installed improved stoves must be replaced every year. This gives a program cost of $359 million based on the assumption that a single cook stove cost $12.[49]

Each existing stove consumes 5 kg of fuel every day (Indian data).[50] We use a black carbon emission factor for wood as 0.85 g/kg of dry matter, 1 g/kg for agriculture waste, 0.53 g/kg for animal waste, 1.0 g/kg for charcoal, and 0.15 g/kg for briquettes.[51]

We assume that the improved stoves provide savings of 35% in fuel use.[52] Wood, agriculture waste, and animal waste-based improved stoves reduce black carbon by 220 Gg, while savings from using briquette reduces 91 Gg of black carbon. Programs to reduce open burning in Africa and South America provides an additional reduction of 1,190 Gg of black carbon. We assume that these programs have no cost, as their basis would be the institution of meaningful property rights and improved agricultural technology. Using the above assumptions, the black carbon emissions saving are 1,502 Gg, which is 19% of the total black carbon emissions.

In order to estimate benefits, we use the DICE model.[53] The policies to reduce black carbon resulted in a reduction in black carbon emissions of 19%. We assumed in the DICE model a linear relationship between black carbon emission reductions and radiative forcing from other sources. We keep CO_2 emissions constant by holding the emission control rate at the baseline level, so as to estimate the *ceteris paribus* costs and benefits of a black carbon program. DICE default assumptions were used for all other parameters. We calculated a climate benefit of $375 dollars. This provides a benefit-cost ratio (BCR) of 1.04, when only taking into account the climate benefit.

We estimated the health benefits of improved indoor air quality. It is estimated that 400,000 deaths per year occur in India (Carl 2009) due to indoor smoke inhalation and that there are 240 million existing stoves. By replacing half the existing stoves with the improved stoves, we estimated a

Table 4.6 Overview table of costs and benefits

B/C (million USD)			
	Benefit	Costs	BCR
Value of DALY Low	1,275	359	3.6
High	4,875	359	13.6

reduction of 200,000 deaths per year. We used two values for disability adjusted life years (DALYs) of $1,000 and $5,000. We assumed indoor smoke inhalation cause death to be premature by 0.5 years, a very conservative assumption. The operating life for the improved stove is three years, so that each installation provides three years of climate and health benefits. Based upon these assumptions, we estimated a health benefit of between $0.9 billion and $4.5 billion. Using a health benefit of $0.9 billion and climate benefit of $375 million, we get an overall BCR of 3.6.

Conclusions

The relatively short atmospheric residence time of black carbon emissions and the availability of existing commercially available reduction technologies create unique opportunities. A reduction in black carbon emissions represents a potential near-term opportunity to postpone the effects of rising GHG levels on the global climate. The delay in global warming offered by reducing black carbon emissions creates a window of opportunity for the research and development of new technologies that lower or eliminate GHG emissions at a cost far less than that of current technological options.

Today, developing countries represent the lowest-cost option for implementing black carbon

---

[48] Shares were calculated from estimates of black carbon emissions from residential fuel sources (Bond *et al.* 2004: 28).

[49] Estimates for the cost of an improved stove vary from $3 to $20. We used a mid-range value for our calculations. See Rosenthal (2009).

[50] Ramanathan and Balakrishnan (2007).

[51] Bond *et al.* (2004: 17).

[52] Hedon (2007).

[53] DICE Delta Version 8 was used for this analysis (Nordhaus 2008).

reduction programs. Such programs would have potential co-benefits for improving air quality and reducing health issues for those who live in these areas. However, the lack of institutions and cultural resistance to change necessitate the need for creative, yet practical approaches to the implementation of such policies.

Bibliography

Andreae, M.O. and A. Geleneser, 2006: Black carbon or brown carbon? The nature of light absorbing carbonaceous aerosols, *Atmospheric Chemistry and Physics* **6**, 3131–48

Bahner, M., 2007: *Use of Black Carbon and Organic Carbon Inventories for Projections and Mitigation Analysis*, EPA Emissions Inventory Conference, Research Triangle Park, NC

Barrett, S., 2007: *Why Cooperate? The Incentive to Supply Global Public Goods*, Oxford University Press, New York

Bate, R. and W.D. Montgomery, 2005: Beyond Kyoto: real solutions to greenhouse emissions from developing countries, in *Towards a Liberal Utopia?*, Philip Booth (ed.), The Institute of Economic Affairs, London

Bond, T., 2006: *Bottom-Up Methods to Improve Global BC/OC Inventory*, Energy Modeling Forum, BC/OC Subgroup, Washington, DC
   2007: Testimony for the hearing on black carbon and climate change, House Committee on Oversight and Government Reform, October 18

Bond, T.C., E. Bhardwaj, R. Dong, R. Jogani, S. Jung, C. Roden, D.G. Streets, S. Fernandes, and N. Trautmann, 2004: Historical emissions of black and organic carbon aerosol from energy-related combustion, 1850–2000, *Global Biogeochemical Cycles* **21**: GB2018

Bond, T., D. Streets, K. Yarber, S. Nelson, J. Woo, and Z. Klimont, 2004: A technology-based global inventory of black and organic carbon emissions from combustion, *Journal of Geophysical Research* **109**(D14203)

Bond, T.C., D.G. Streets, K. F. Yarber, S. M. Nelson, J.-H. Woo, and Z. Klimont, 2005: A technology-based global inventory of black and organic carbon emissions from combustion, *Journal of Geophysical Research* **109**(D14)

Bond, T. and H. Sun, 2005: Can reducing black carbon emissions counteract global warming?, *Environmental Science Technology* **39**, 5921–6

Boucher, O. and M.S. Reddy, 2008: Climate trade-off between black carbon and carbon dioxide emissions, *Energy Policy* **36**, 193–200

Carl, J., 2009: *Cookstoves, Carbon, and Climate*, Indian Express, April

CGAP, 2009: *Microfinance and Climate Change: Threats and Opportunities*, Focus Note, CGAP, Washington, DC, February

Chandler, K., K. Vertin, T. Alleman and N. Clark, 2003: *Ralph's Grocery EC–Diesel^{TM} Truck Fleet: Final Results*, US Department of Energy, Washington, DC

Cofala, J., M. Amann, Z. Klimont, K. Kupiainen, and L. Hoglund-Isaksson, 2006. *Scenario of Global Anthropogenic Emissions of Air Pollutants and Methane Until 2030*, Final Report for WP3, Atmospheric Pollution and Economic Development Program, International Institute for Applied Systems Analysis (IIASA), Laxenburg, Austria

Cooke, W.F. and J.J.N. Wilson, 1996: A global black carbon aerosol model, *Journal of Geophysical Research* **101**(D14), 19,395–19,409

Cooke, W.F., C. Liousse, H. Cachier and J. Feichter, 1999: Construction of a 1° × 1° fossil fuel emission data set for carbonaceous aerosol and implementation and radiative impact in the ECHAM4 model, *Journal of Geophysical Research* **104**, 22,137–62

Gulledge, Jay, 2008: Three plausible scenarios of future climate change, in *Climatic Cataclysm*, K.M. Campbell (ed.), The Brookings Institution, Washington, DC

Hanson, J., M. Sato, R. Ruedy, L. Nazarenko, A. Lacis, G.A. Schmidt, G. Russell, I. Aleinov, M. Bauer, S. Bauer, N. Bell, B. Cairns, V. Canuto, M. Chandler, Y. Cheng, A. Del Genio, G. Faluvegi, E. Fleming, A. Friend, T. Hall, C. Jackman, M. Kelley, N. Niang, D. Koch, J. Lean, J. Lerner, K. Lo, J. Muller, P. Minnis, S. Menon, T. Novavov, V. Oinas, Ja. Perlwitz, Ju. Perlwitz, D. Rind, A. Romanou, D. Shindell, P. Stone, S. Sun, N. Tansner, D. Thresher, B. Wrelick, T. Wong, M.Yao, and S. Zhang, 2005: *Geophysical Research* **110**(D18),

Hedon, 2007: Household Energy Network, *Improved Cookstove in Africa – An Analysis of the Difficulties*, September

IGSD/INECE, 2008: Reducing black carbon may be fastest strategy for slowing climate change, Institute for Governance and Sustainable Development (IGSD) and International

Network for Environmental Compliance and Enforcement (INECE), 12

IPCC, 2001: *Climate Change 2001: The Scientific Basis – Contribution of Working Group I to the Third Assessment Report of the Intergovernmental Panel on Climate Change*, J.T. Hongton, Y. Ding, D.J. Griggs, M. Noguer, P.J. van der Linden, and D. Xiaosu (eds.), Cambridge University Press, Cambridge

2007: *Climate Change 2007: The Physical Science Basis – Contribution of Working Group I to the Fourth Assessment Report of the IPCC*, Cambridge University Press, Cambridge

Jacobson, M., 2007: Testimony for the Hearing on Black Carbon and Arctic, House Committee on Oversight and Government Reform, October 18

Jacoby, H.D., B.H. Babiker, S. Paltsev, and J.M. Reilly, 2008: Sharing the burden of GHG reductions, MIT Joint Program on the Science and Policy of Global Change, Report **167**, November

Kheshgi, H.S., 2007: Probabilistic estimates of climate change: methods, assumptions and examples, chapter 4 in M. Schlesinger, H. Kheshgi, J. Smith, F. de la Chesnaye, J.M. Reilly, T. Wilson, and C. Kolstad (eds.), *Human-Induced Climate Change: An Interdisciplinary Assessment*, Cambridge University Press, Cambridge

2008. R&D as key to climate policy success, Conference on Technology Policy and Climate Change, Palo Alto, California, October

Larsen, B., G. Hutton, and N. Khanna, 2008: Air pollution, Copenhagen Consensus 2008 Challenge Paper, April

Montgomery, W.D. and S.D. Tuladhar, 2006: The Asia Pacific Partnership: its role in promoting a positive climate for investment, economic growth and greenhouse gas reductions, International Council for Capital Formation, Brussels

Montgomery, W.D. and R. Bate, 2005: A (mostly) painless path forward: reducing greenhouse gases through economic freedom, in *Climate Change Policy and Economic Growth: A Way Forward to Ensure Both*, M. Thorning and A. Illarionov (eds.), International Council for Capital Formation, Brussels

Newell, R.G. and William A. Pizer, 2003: Regulating stock externalities under uncertainty, *Journal of Environmental Economics and Management* **45**(2), Supplement 1, March, 416–32

Nordhaus, W., 2008: *A Question of Balance, Weighing the Options of Global Warming Policies*, Yale University Press, New Haven, CT

Penner, J.E., H. Eddleman, and T. Novakov, 1993: Towards the development of a global inventory of carbon emissions, *Atmospheric Environment*, A **27**, 1277–95

Pettus, A., 2009: Agricultural fires and Arctic climate change, Clean Air Task Force, May

Ramanathan, V., 2007a: Testimony for the Hearing on the Role of Black Carbon on Global and Regional Climate Change, House Committee on Oversight and Government Reform, October 18

2007b: Global warming & dangerous climate change: buying time with black carbon reductions, Proposal to King Abdullah University of Science and Technology, November 15

Ramanathan, V. and K. Balakrishnan, 2007: Project Surya, reduction of air pollution and global warming by cooking with renewable sources: a controlled and practical experiment in rural India, Project Surya, www.projectsurya.org/ storage/Surya-WhitePaper.pdf

Rosenthal, E., 2009: Third-world stove soot is target in climate fight, *New York Times*, April 16

Rypdal, K., N. Rive, T. Berntsen, Z. Klimont, T. Mideksa, G. Myhre, and Ragnhild Skeie, 2009: *Costs and global impacts of black carbon abatement strategies, Tellus Series B Chemical and Physical Meteorology* **4**, 625–41

Shine, K.P., J.S. Fuglestvedt, A.K. Hailemariam, and N. Stuber, 2005: Alternatives to the global warming potential for comparing climate impacts of emissions of greenhouse gases, *Climatic Change* **68**, 281–302

Sokolov, A.P., P.H. Stone, C.E. Forest, R. Prinn, M.C. Sarofim, M. Webster, S. Paltsev, C.A. Schlosser, D. Kicklighter, S. Dutkiewicz, J. Reilly, C. Wang, B. Felzer, and H.D. Jacoby, 2009: Probabilistic forecast for 21st century climate based on uncertainties in emissions (without policy) and climate parameters, *Journal of Climate* **22**(19), 5175–204

Stegeman, R., 2008: Black carbon: playing a major role in arctic climate change, www. polarfoundation.org/www_sciencepoles/index. php?/articles_interviews/black_carbon_playing_a_ major_role_in_arctic_climate_change/&uid= 1253&pg=8

Tol, R.S.J., 2008: The social cost of carbon: trends, outliers, and catastrophes, economics, the open-access, *Open-Assessment E-Journal* **2**(25), 1–24

Weitzman, M.L., 2008: On modeling and interpreting the economics of catastrophic climate change, *Review of Economics and Statistics* **91**(1), 1–19

Wigley, T.M.L., R. Richels, and J.A. Edmonds, 2007: Overshoot pathways to CO_2 stabilization in a multi-gas context, in *Human Induced Climate Change: An Interdisciplinary Assessment*, M. Schlesinger, F.C. de la Chesnaye, H. Kheshgi, C.D. Kolstad, J. Reilly, J.B. Smith, and T. Wilson (eds.), Cambridge University Press, Cambridge, 84–92

Black Carbon Mitigation

Alternative Perspective

MILIND KANDLIKAR, CONOR C.O. REYNOLDS, AND ANDREW P. GRIESHOP

Introduction

Black carbon aerosols have recently been identified as important contributors to positive radiative forcing on the global climate. Black carbon emission reductions – especially those focusing on contained combustion sources – can provide a win–win opportunity with both health and climate benefits. Air pollution emissions from contained combustion, including black carbon, are associated with a total of about 2 million deaths annually in the developing world. There is also a compelling case for the inclusion of black carbon in a climate emission reductions regime; reducing black carbon emissions offers a large abatement potential ($\sim$15% of current excess radiative forcing), as well as a rapid impact mitigation approach ("buying" a delay of around a decade as part of a climate mitigation strategy). However, it is important to recognize that black carbon reductions are not a substitute for reductions in emissions of carbon dioxide (CO_2). The two approaches must be applied together to stabilize atmospheric concentrations of CO_2 to acceptable levels of risk. In this Perspective paper we offer a comprehensive critique of chapter 4 by Robert E. Baron, W. David Montgomery, and Sugandha D. Tuladhar (2009, hereafter, BMT). We then assess benefit-cost ratios (BCRs) for five different options to reduce black carbon emissions.

CO_2 is unavoidably emitted whenever a carbon-based fuel is burned, but other pollutants that are unintended by-products of non-ideal combustion are also released. These pollutants may include carbon monoxide (CO), oxides of sulfur and nitrogen (SO_X and NO_X), volatile organic compounds (VOCs), and PM (PM or aerosols). Until recently, these pollutants were regarded as having primarily *local* health and environmental impacts. There is

increasing scientific consensus, however, that some of these species can also impact the global climate system by changing the Earth's radiative balance. In particular, scientists have turned their attention to aerosols that can have both cooling and heating effects. Sulfate and other reflecting aerosol particles tend to cool the Earth's atmosphere by reflecting incident solar radiation. On the other hand, some carbonaceous aerosols, and black carbon in particular, absorb incident solar radiation leading to warming. The "win–win" prospect of reducing local air pollution while addressing climate change has led to calls for the inclusion of black carbon as a greenhouse agent under the United Nations Framework Convention on Climate Change (UNFCCC) (Grieshop *et al.* 2009).

This Perspective paper addresses the potential for climate mitigation from a reduction in black carbon emissions. We begin with a short scientific overview of black carbon and its climatic and health effects. Next we analyze the arguments used in favor of black carbon reductions in BMT. While the case for black carbon reductions is strong, we argue that BMT overstate the case in four ways. First, they overestimate the potential for black carbon reductions. Second, they overestimate the potential for black carbon mitigation efforts to delay the need for global CO_2 mitigation efforts. Third, the case they make for black carbon reductions as a way to cope with catastrophic climate change is far-fetched and not supported by current scientific evidence. Finally, they ignore the role of organic carbon that is co-emitted by a number of black carbon source types. This final point is a serious error that grossly overstates the role that black carbon emissions reductions – particularly those from open burning – might play in a climate control regime. This shortcoming is addressed in the section on "Assessing Black Carbon Mitigation Options", in which we propose and analyze a set of additional options for reducing black carbon emissions. The following section, "Benefit-Cost Ratios," discusses the cost-benefit analyses (CBAs) of the proposed options. We recognize that quantifying the benefits of carbon reductions is a matter of much debate, and so we use a "Stern" perspective (Stern 2007) and a "Nordhaus" perspective (Nordhaus 2007) to capture two polar views on the "true" value of marginal damage per ton of emitted CO_2-equivalents. We conclude by briefly highlighting the institutional challenges associated with these proposed interventions that are not captured by the BCRs.

The Climate and Local Health Impacts of Black Carbon Emissions

While exact definitions of light-absorbing carbonaceous aerosols vary (Andreae and Gelencsér 2006), there is uniform agreement that black carbon is a strong absorber of solar radiation; it absorbs approximately 1 million times more solar energy than CO_2 per unit mass (Bond and Sun 2005). Black carbon is thought to be the second-largest contributor to global excess radiative forcing after CO_2 (Ramanathan and Carmichael 2008). During its week-long residence in the atmosphere a black carbon particle may directly absorb radiation from below (Ramanathan et al. 2007), within (Jacobson 2006; Koren et al. 2008), and above (Haywood and Ramaswamy 1998) clouds. When deposited on bright ice and snow surfaces such as glaciers or in polar regions, black carbon particles may cause several more months of warming by reducing the reflection of light – this latter effect helps make black carbon an especially effective warming agent that is responsible for approximately 15% of global excess radiative forcing (Forster et al. 2007). Because of its short lifetime and varied atmospheric interactions, the climate impacts of black carbon vary with source location. For example, relative to those emitted near the equator, black carbon aerosols from sources in northern latitudes ($>40°N$) have a lower direct absorption effect because of the reduced solar irradiance in

their location, but are more likely to have strong *indirect* effects due to proximity to Arctic ice and snow sheets (Shindell and Faluvegi 2009).

The relative contribution of black carbon to the total aerosol from combustion sources varies considerably with source type (Bond et al. 2004). Black carbon emissions from some sources are accompanied by emissions of organic carbon compounds and sulfate, which have a cooling effect via direct light scattering and interactions with clouds (Ramanathan and Carmichael 2008). Though the net climate impact of black carbon and non-black carbon particles emitted by many source types is uncertain, analysis that includes most relevant uncertainties suggests that black carbon-dominated sources, such as residential combustion of solid fuels and high-emitting diesel engines, have a net warming impact (Bond 2007). Open biomass burning is also a large source of black carbon, but its emissions are generally dominated by organic carbon (Bond et al. 2004) and thus likely have a net cooling effect.

Locally, black carbon and other products of incomplete combustion are among the largest contributors to ambient air pollution. Extensive use of solid fuels in the poorest developing countries makes indoor exposure to emissions the fourth largest contributor to their disease burden (Ezzati et al. 2002); the World Health Organization (WHO) estimates that over 1.8 million people, mostly women and children, die from exposure to indoor smoke from solid fuels annually (Ezzati et al. 2006). Older vehicles, dirty industries, and an array of other black carbon sources contribute to outdoor urban PM air pollution levels that are up to ten or more times higher than those in developed nations' cities (Molina and Molina 2004).

Assessing BMT's Arguments

BMT make several arguments in support of black carbon emissions reductions, which fall into four main categories:

1 Those related to the *magnitude* of reductions in radiative forcing of climate from black carbon emissions reductions relative to mitigation of CO_2 emissions.

2 Those related to the *timing* of emissions reductions. In particular, BMT claim that black carbon emissions reductions can "delay warming for a matter of decades" and "black carbon policies can buy time for R&D" to achieve reductions in the cost of CO_2 reduction.

3 Those related to black carbon reductions' potential for coping with *catastrophic* climate change.

4 Black carbon reductions as a way to bring *developing countries* on board the UNFCCC, while improving health and poverty-related illnesses.

In what follows, we will examine these four arguments in turn.

The Magnitude of Black Carbon Reductions

BMT note that "40% of current net warming (10–20% of gross warming)" is related to black carbon. Since use of net warming is scientifically inaccurate and inflates the role that black carbon plays,[1] the attribution of between 10 and 20% of excess warming is more appropriate and consistent with other studies. Specifically, of the 2°C rise in global mean temperature since preindustrial times (from c. 1760), black carbon's contribution is approximately 0.3°C (Jacobson 2004; Bice *et al.* 2009). To the first order, eliminating current black carbon sources should reduce excess temperature forcing by about 15%.

BMT also assert that the benefits of reducing CO_2 are uncertain and "highly speculative" and "the calculation of expected benefits [from CO_2 reductions] is not a scientific possibility."[2] In other words, BMT imply that climate impacts (e.g. such as those assessed and summarized by the IPCC) are simply conjectures that are based on speculation and not on rigorous scientific assessment. Further, despite the uncertainties they highlight, BMT also support an upper bound for the price of carbon of $25 per ton of CO_2; in fact this is a "low" value consistent with the belief that climate change is unlikely to be a significant problem in the future. Overall, BMT appear to be arguing that CO_2 emissions reductions have speculative benefits which are only realized over the longer term, while black carbon reductions have more certain and immediate benefits. Consequently, they claim that climate mitigation policies should focus on black carbon reductions rather than on CO_2 mitigation. We examine the implications of this contention in more detail below.

In a recent paper (Grieshop *et al.* 2009), we showed that black carbon emissions could add up to a carbon emissions reduction "wedge," as described in an influential article by Pacala and Socolow's (2004). Pacala and Socolow (2004) identify fifteen greenhouse gas (GHG) reduction strategies each of which represents a "wedge" equivalent to 1 billion tons of carbon (1 GtC) mitigated over a fifty-year period (2005–55). The paper states that implementing a cluster of seven such wedges would help stabilize atmospheric concentrations of CO_2 to below double preindustrial levels (500 ppmv). Examples of Pacala and Socolow's wedges include a doubling of the fuel economy of all cars in 50 years, tripling existing nuclear power capacity by 2055, and increasing solar photovoltaic power generation by a factor of 700.[3] If we take the climate problem seriously (rather than simply writing off its impacts as "speculative"), then meeting a target of limiting CO_2 concentrations in the year 2100 to twice preindustrial levels represents a pragmatic goal that seeks to balance emissions reduction costs and impacts.[4] Meeting such a target requires seven times the carbon-equivalent reduction that black carbon reductions can alone provide. In other words, black carbon reductions are not a

---

[1] BMT define "net warming" as the aggregate warming effect of GHG emissions, black carbon, and the urban heat island effect less the opposing effect of cooling particles. "Gross warming" is defined as the aggregate warming effect of GHG emissions, black carbon, and urban heat island effect. If one were to similarly attribute net warming to CO_2, CH_4, tropospheric ozone (O_3) and N_2O as BMT do black carbon, the corresponding percentages would be 75%, 28%, 20% and 8%, respectively, and the aggregate attributions would well exceed 150%!

[2] By doing so, they write off in one fell swoop the entire literature on decision making with regards to climate change including the recent Stern–Nordhaus–Weitzman debate on long-term costs of climate change.

[3] While technologically feasible, each wedge represents a substantial commitment of resources the implementation of which is an impressively daunting task.

[4] In fact, many argue that doubled CO_2 concentrations represent a risk that is well beyond the threshold for "dangerous" climate change.

substitute for CO_2 reductions, as BMT would have us believe, but instead are one among a range of strategies needed to meet climate change mitigation targets. Black carbon reductions represent an opportunity that is complementary to CO_2 reductions, and not a substitute for them. To treat them as substitutes to CO_2 emission reductions is tantamount to denying the seriousness of the global threat of climate change.

The Timing Impact of Black Carbon Reductions

BMT find black carbon emission reductions attractive for two reasons. First, on a per-mass basis black carbon is a much more potent warmer than CO_2, with a 100-year global warming potential of approximately 450 times that of CO_2. Second, they are short-lived so their removal can have a rapid impact on global temperature and thus represents an important short-term strategy. Consequently, BMT argue that "black carbon policies can buy time for R&D" needed to achieve cost reductions in CO_2 mitigation. BMT believe that "reductions in black carbon emissions can delay warming for a matter of decades."[5] We have little disagreement with BMT on the broad strokes of these claims. However, BMT do not provide any quantitative assessment of the extent of the CO_2 mitigation delays that may be possible if black carbon reductions are introduced. We are only left with a vague sense that black carbon represents a very immediate and large opportunity to delay emissions of the order of "decades."

A coherent discussion of delays is possible only when targets and timetables for future CO_2 concentrations are in place. Comparing an emissions path that achieves a known CO_2 concentration target (e.g. 500 ppm) at a known future time (2100 AD) with and without black carbon reduction can help evaluate the extent of delay in CO_2 emissions reductions that black carbon reductions can facilitate. While BMT do not present such calculations,

other researchers have done so (Bice *et al.* 2009). They show that phasing out *all* black carbon emissions from fossil fuel use can help delay CO_2 reductions by about fifteen years when a stringent concentration target of 450 ppm is to be met by 2100. A higher concentration target (550 ppmv), consistent with BMT's low "cut-off" price for carbon, reduces the urgency of immediate CO_2 reductions. In this latter case, black carbon reductions can help meet strict CO_2 concentration targets with greater flexibility, but the amount of delay they can "buy" is smaller (on the order of a decade) than BMT's exposition might have us believe.

Black Carbon Reductions and Catastrophic Climate Change

BMT present black carbon reductions as a solution to possible catastrophic impacts of climate change. This view is inconsistent with the other arguments made for black carbon reductions, which present them as a more immediate *near-term* reduction option. Catastrophic impacts are not on the near-term horizon, and most experts agree that we are nowhere close to a climatic "tipping point." The most commonly discussed (and the most well studied) catastrophic impact is the shutting down of the North Atlantic Thermohaline Circulation (THC). The current probability of a THC shutdown is very small and becomes a worry only with a *quadrupling* of preindustrial CO_2 levels or a 4°C increase in global mean temperature (Zickfeld *et al.* 2007). While catastrophic climate change is clearly a concern, it is one that exists mainly on multidecadal and century-long time-scales. Further, when catastrophe is imminent, the reduction of black carbon emissions (such as those from cook stoves in developing countries) is not likely to be the chosen rapid-response option. Resource-constrained governments that currently lack adequate institutions to address air pollution concerns are likely to find it difficult to change the way that people live in a short period of time. Rather, policy makers in the North are likely to adopt aggressive geoengineering (GE) approaches – e.g. using nanoparticles to increase the Earth's albedo (Wigley 2006, and see chapter 1 in this volume),

---

[5] BMT cite Ramanathan and Carmichael (2008) to back up their claim that black carbon reductions can delay emissions reduction by a few decades. BMT do not mention, however, that the original reference states that delays are only possible "in tandem" with CO_2 reductions. This is a crucial omission that overstates the climate mitigation role of black carbon while downplaying the role of CO_2.

which can be deployed rapidly and at relatively low cost on a mass scale. Consequently, we find BMT's argument for using black carbon reductions to stave off climate catastrophe far-fetched. The good news is that one does not need a climate catastrophe justification to engage in black carbon reductions.

Black Carbon Emissions and the Developing World

The final justification for black carbon reductions presented by BMT is that they provide a double dividend. Black carbon reductions reduce health impacts (largely in the developing world) and climate warming at the same time. This is a sound justification for black carbon reductions, and one that both developing and developed countries can get behind (Grieshop et al. 2009). However, the case for black carbon reductions in the developing world is more nuanced than that presented by BMT. Black carbon emissions can be classified into open burning sources (e.g. crop residue, forests, savannah) and contained combustion sources (e.g. household cook stoves, diesel vehicles, coal-powered industry). Organic carbon (OC) tends to dominate open burning aerosol emissions by mass, while many contained combustion sources are dominated by black carbon. In the case of open burning, OC aerosols have a cooling effect that more than compensates for warming from the black carbon component. Consequently, open burning emissions are thought to lead to *net cooling* of the Earth's atmosphere (Bond et al. 2004; Bice et al. 2009). In reality, the picture is even more complex. Open burning emissions likely have global cooling effects, but by contributing to Atmospheric Brown Clouds (ABCs) they have serious regional climate impacts (Ramanathan and Carmichael 2008). For example, ABCs are implicated in reduced monsoon precipitation over the Indian sub-continent and reduced agricultural yields (Auffhammer et al. 2006). Black carbon-dominated contained combustion sources are linked much more strongly to warming, and are also linked directly to the deaths of 1.5 million individuals (mostly women and children) in the developing world, primarily through exposure to indoor air pollution from household combustion sources.

In their chapter 4, BMT ignore the effect of co-emitted OC species altogether. This leads them to make incorrect conclusions about the magnitude of the responsibility of developing countries for black carbon-related warming. Consequently, they vastly overestimate the potential to reduce warming by mitigating open burning sources in the developing world. In fact, the single largest black carbon reduction proposed by BMT, biomass burning in Africa and South America, makes up 80% of their proposed reductions goal (19% of global black carbon emissions). However, because of co-emitted OC species – the OC to EC mass ratio for biomass burning of forests and savannah is 9 and 7, respectively (Bice et al. 2009) – this source contributes to net cooling.

In summary, black carbon reductions – especially those related to contained combustion – can provide a win–win opportunity. Closed combustion sources are responsible for a total of about 2 million premature deaths annually in the developing world. There is also a compelling case for their inclusion in a climate emission reductions regime – because of their large abatement potential (~15% of current radiative forcing), as well as short-term timing considerations ("buying" a delay of around a decade). However, their potential should not be overestimated and black carbon reductions are not a substitute for CO_2 reduction.

Assessing Black Carbon Mitigation Options: Cook Stoves and Diesel Vehicles

BMT present various approaches to mitigating black carbon from key sources, namely indoor solid fuel use in developing countries, diesel transportation, and open burning. For the former two sources, we have performed a cost-effectiveness analysis of several realistic interventions that use proven technologies and have relatively well-understood costs. We take into account the co-emitted cooling aerosol species, in particular organic carbon, and GHG emissions in estimating both climate change and health impacts. We have not analyzed interventions that promise to mitigate aerosol emissions

Table 4.1.1 Intervention cost ranges

Intervention to reduce black carbon	Cost	
	Low (US$)	High (US$)
Household solid-fuel use 3-Stone fire to gasifier stove	30	100
Coal stove to LPG stove[a]	30 + 30/year	100 + 100/year
Heavy duty diesel vehicles (HDDV) Diesel to CNG[b]	2,000[b]	10,000
Retrofit with particle traps	6,000	12,000
Repair super-emitters	1,000	5,000

Notes:
[a] LPG stove intervention assumes a similar stove cost to a biomass-gasifier stove but includes a yearly expense due to the additional fuel cost relative to the base fuel (coal) associated with LPG use.
[b] The lower estimate of costs for conversion to CNG is reduced because natural gas costs are estimated to be significantly less than diesel, so some of the cost of conversion could pay for itself in the first year. Over the remaining lifetime of the vehicle, there may be a net economic benefit to the vehicle owner due to fuel savings.

from open burning because, as discussed in the previous section, such emissions are dominated by cooling species. The two approaches to reduce aerosols from cook stoves analyzed here are: (1) improved biomass-gasifier stoves replacing traditional stoves in India, and (2) liquefied petroleum gas (LPG) stoves replacing household coal use in China. Three interventions are evaluated for reducing black carbon from diesel-fueled buses and trucks in the urban areas of India and China: (1) converting urban heavy-duty diesel vehicles (HDDV) to compressed natural gas (CNG) fuel, (2) retrofitting them with after-treatment devices (particle traps), and (3) repairing "superemitting" diesel vehicles, which have excessive emissions due to their poor condition. Table 4.1.1 gives an overview of cost estimates per unit for each intervention.

Indoor Solid Fuel Use in Developing Countries

Cooking and heating fires using solid fuels are a major global emission source of black carbon and other aerosol species (Bond *et al.* 2004). Due to elevated indoor exposures to these emissions, such indoor fires represent one of the largest sources of premature mortality and illness in developing countries, especially among women and children (Ezzati *et al.* 2002). In the past several decades, interventions to replace such primitive cooking and heating fires with "improved" stoves or those using cleaner-burning fuels have been implemented, studied, and advocated as an efficient means to protect global public health (Mehta and Shahpar 2004). More recently, such interventions have also been suggested as cost-effective means to mitigate GHG (Smith and Haigler 2008) and black carbon (Bond and Sun 2005) emissions. Here we present cost-effectiveness analyses of two hypothetical stove interventions, building upon prior work, which demonstrate the potential efficiency of such interventions and the possibility for significant co-benefits associated with climate mitigation.

The simple analyses presented here examine the health and climate cost-effectiveness of two distinct stove interventions: (1) replacing simple "unimproved" stove or 3-stone biomass-fire cooking methods with improved biomass-gasifier stoves in India and, (2) the replacement of household coal-burning stoves in China with those burning LPG. The "scoping" analysis of Smith and Haigler (2008) presents calculations of the cost-effectiveness of these options in reducing the loss of Disability Adjusted Life Years (DALYs) and emitted GHGs (in tons CO_2-equivalent or tCO_2eq) from domestic fuel use during a ten-year stove-program implementation. Here, we adopt their major assumptions for the implementation of a stove program (see the footnote to table 4 in Smith and Haigler 2008), but expand their analyses to include the climate effects of the reduced black carbon and OC emissions that accompany these interventions.

In brief, 100,000 stoves are distributed over a ten-year period, with the assumption that 10% of stoves in place must be replaced on a given year and that stove-program effectiveness is 50% (half of distributed stoves are in active use). Here, both stove interventions are assumed to cost the same as that in Smith and Haigler's analysis ($60/stove, including 50% of total cost for program expenses), but an additional cost of LPG fuel-use is added in the Coal-to-LPG stove-use cost calculation, based on estimates of fuel cost and annual usage for each fuel

Table 4.1.2 Health impact and climate mitigation cost-effectiveness values for two potential stove interventions in developing Asia

Intervention	Health cost-effectiveness ($/DALY)	Climate cost-effectiveness ($/tCO$_2$eq)	Population affected[d] (million)
3-stone fire to gasifier stove in India[a]	600 (260–1400)[b]	6 (4–7)	850
Household coal to LPG stoves in China[c]	1400 (900–2000)	45 (10–60)	430

Notes:
[a] Efficiency increases of an improved stove of ∼×2 are assumed to be offset by the 50% "effectiveness" of stove programs assumed by Smith and Haigler (2008) in calculating improved stove fuel usage.
[b] The range in health cost-effectiveness was estimated based on ranges presented by Mehta and Shahpar (2004).
[c] Fuel cost is assumed to increase due to purchase of LPG: central values for fuel price and usage are assumed to be $0.25/kg and 1000 kg/stove/year for coal stoves and $1.50/kg and 200 kg/stove/year for LPG stoves.
[d] Values estimated from Mehta and Shahpar (2004).

type (table 4.1.1). Health cost-effectiveness values ($/DALY) can be directly taken from Smith and Haigler's work while the climate cost effectiveness ($/tCO$_2$eq) is a combination of their calculations for GHGs and ours for particles. Black carbon, OC, and SO$_2$ emission factors for the three stove types (Zhang *et al.* 2000; Bond *et al.* 2004; MacCarty *et al.* 2008) characterize emissions from fuel-use before and after the stove program. Global warming potentials (GWPs) are used to convert particle (and particle precursor, in the case of SO$_2$) emissions to CO$_2$-eq (Reynolds and Kandlikar 2008).

Biomass-Gasifier Program

Installing biomass-gasifier stoves to replace traditional household cooking methods in India is a cost-effective method to improve health (Smith and Haigler 2008) with estimates for price per DALY ranging between $260 and $1400 (table 4.1.2). Such an intervention is also an effective way to reduce GHG emissions at a cost of $7/tCO$_2$eq. Including the climate impacts of particles potentially improves the cost-effectiveness of climate mitigation to as low as $4/tCO$_2$eq. Therefore

the climate impacts associated with black carbon particles play a substantial but not overwhelming role in the climate mitigation effects of this intervention.

LPG Stove Program

The cost-effectiveness of switching from coal- to LPG-fired stoves in China is lower than for the biomass-gasifier stove, due to higher fuel costs for similar levels of CO$_2$eq emission reduction. The health cost-effectiveness of LPG replacement is slightly worse than the gasifier case with estimates for price per DALY ranging between $900 and $2000 (table 4.1.2). LPG is a non-renewable fuel and thus is a net-CO$_2$-emitter, a fact that leads to substantially lower climate cost-effectiveness based strictly on GHG emissions. Burning LPG emits very low levels of black carbon relative to coal combustion in simple stoves. Therefore including black carbon emissions in CO$_2$eq calculations can more than double the cost-effectiveness of this intervention (with up to 80% of the total benefits coming from the aerosol impacts), highlighting the importance of considering *both* GHGs and aerosols when determining the potential climate impacts of such programs.

Diesel Vehicles in Urban Areas of Developing Countries

HDDV in cities are mostly goods vehicles and buses, and they play a critical role in the functioning of an economy. Worldwide, outdoor air pollution is estimated to result in at least 800,000 excess deaths annually (Rodgers *et al.* 2002). The greater health burden falls on urban populations in less-developed countries: about 60% of these excess deaths occur in Asia alone (Anderson *et al.* 2004). HDDVs in developing countries are in general older and more poorly maintained than in OECD countries. In addition, emissions controls are less stringent for new vehicles and emissions are not effectively monitored and regulated for in-use vehicles. Therefore particulate matter (PM) emissions are very high on a per-fuel-use basis. PM from diesel vehicles is mostly black carbon (50–80%) (Bond *et al.* 2004), which makes it an ideal candidate for the implementation of control interventions.

We calculate climate mitigation cost effectiveness ($/tCO$_2$eq) according to the method introduced by Bond and Sun (2005), but extend their analysis to account for the cooling effect of co-emitted organic carbon and SO$_2$ (a precursor to reflective sulfate aerosol) and warming GHG emissions.[6] We take into account the change in fuel efficiency (and hence direct CO$_2$ emissions) and increased methane (CH$_4$) emissions (CH$_4$ has a 100-year GWP of 23) in the case of the diesel-to-CNG conversion. Health impact calculations apply the results of epidemiology studies, which show correlation between the concentration of ambient fine PM in urban areas (which includes black carbon, organic carbon, sulfates, and other species) and adverse health outcomes such as cardiovascular disease, lung cancer, and acute lower respiratory infection. Our approach follows that of Smith and Haigler (2008): for the cost-effectiveness assessment of health impacts, we evaluate the intervention cost (in US$1,000) per DALY reduced. The potential for ambient PM reduction is based on the three emission-control scenarios applied in the urban regions of India and China.

The average heavy-duty vehicle fleet is assumed to be composed of 85% "normal" trucks and buses with average PM emission factors of 2.2 g/kg fuel, while the remainder are "super-emitters" with PM emission factors of 8.4 g/kg, or almost four times higher than the regular vehicles (Subramanian *et al.* 2009). HDDVs are assumed to travel an average of 75,000 km per year, and have an average fuel consumption of 3.0 km/kg diesel; vehicles converted to burn CNG have a lower fuel efficiency of 2.5 km/kg natural gas. Intervention cost ranges are given in table 4.1.1, and include the one-time capital cost as well as an estimate of lifetime fuel savings and change in maintenance costs; note that the latter costs are highly discounted to reflect the barriers to capital investment in resource-constrained economies.

[6] The estimated 100-year GWP for diesel PM is 350 ± 200 based on aerosol component GWPs and emission fractions (Reynolds and Kandlikar 2008).

Diesel to Compressed Natural Gas

Conversion of heavy-duty vehicles from diesel to CNG fuel is a proven means of reducing PM emissions. For example in Delhi, India, all public transport vehicles were converted following regulation in 2001 (Reynolds and Kandlikar 2008). The major barrier to conversion is the supply of natural gas, since it is not available in all cities. Therefore only vehicles that operate within range of a CNG refueling infrastructure would be eligible for conversion (we assume half of the urban fleet in Asia). In our cost-effectiveness analysis we do not include the cost of refueling infrastructure because this is assumed to be revenue-neutral for the private companies or governments who undertake the endeavor. If such infrastructure were taken into account, health and climate cost-effectiveness would obviously be reduced. Switching to CNG is assumed to result in a 90% reduction of PM emissions. Although this PM reduction has a substantial climate benefit (–250 tCO$_2$eq over the lifetime of the vehicle), almost three-quarters of the benefit is offset by increased CH$_4$ emissions (+110 tCO$_2$eq) and reduced fuel efficiency (+75tCO$_2$eq). Therefore the climate cost-effectiveness is considerably less than for the cook stove options, at around $100 per tCO$_2$eq (table 4.1.3). The health cost effectiveness is $0.94M per DALY (i.e. 1,000 times more expensive than the cook stove interventions) because exposure to outdoor transportation emissions is much lower than exposure to indoor smoke.

Retrofit of In-Use Diesel Vehicle with Particle Trap

Older model vehicles can be retrofitted with exhaust particulate traps that reduce PM emissions significantly. In our calculations we assume that they are 70% effective over their lifetime, which is about eight years. Maintenance costs are assumed to increase and the fuel efficiency of retrofitted vehicles is reduced slightly, leading to both higher fuel costs and direct CO$_2$ emissions. The lifetime climate benefit of installing a particulate trap on a diesel vehicle is a reduction of approximately –80 tCO$_2$eq, taking into account the increased CO$_2$ emissions due to the fuel efficiency penalty. The

Table 4.1.3 Health impact and climate mitigation cost effectiveness values for three potential interventions to reduce black carbon from heavy-duty transport in China and India

Intervention	Health cost-effectiveness (1,000$/DALY)[a]	Climate cost-effectiveness ($/tCO$_2$e)
Diesel to CNG	940 (310–1570)	100 (35–165)
Retrofit with particulate trap	2,320 (1550–3090)	115 (75–155)
Repair super-emitters	470 (160–790)	15 (5–25)

Notes: The exposed population is the combined urban population of the two nations, which is estimated to be 1,100 million people, however the diesel–CNG option may not be viable in all cities due to infrastructure requirements.

[a] Health cost-effectiveness values indicate that diesel vehicle interventions are 1,000 times more expensive than cook stove interventions because of the much lower exposure levels (and consequently less impact on health outcomes) and the higher cost of intervention.

technology is effective at reducing PM but it is expensive, so climate cost-effectiveness is similar to that for the diesel–CNG fuel switch ($115 per tCO$_2$eq), and health cost effectiveness is more than three times less than for the fuel-switching option ($2.3M per DALY).

Repair Super-Emitters

Although super-emitters by definition make up a relatively small proportion of the vehicle fleet (around 15%), they represent a great opportunity for emissions reduction because their PM emissions are many times higher than the bulk of the fleet. We assume that a viable inspection and maintenance program would identify (and ensure successful repair of) half of the super-emitters, and that the result of the repair would be a 50% reduction in PM emissions. Setting up a good inspection and maintenance system is not a trivial task, but it can be done so that it is a revenue-neutral endeavor that does not put an overly large burden on vehicle operators or the government (Hausker 2004). This analysis suggests that identification and repair of super-emitters has a high climate impact per vehicle (approximately –200 tCO$_2$e). This approach could be one of the most cost-effective means of reducing both climate and health effects of PM

from transport, at $15 per tCO$_2$eq and $0.47M per DALY (table 4.1.3). However, since the proportion of super-emitters is small the scope of this intervention is limited.

Benefit-Cost Ratios

From the preceding analyses it is clear that indoor cooking interventions dominate the explored possibilities based on the health cost-effectiveness metric; the $/DALY metric for indoor cooking is lower than that from vehicular traffic controls by three orders of magnitude. From a climate cost-effectiveness perspective most proposed interventions are within an order of magnitude of each other. Conversion to biomass gasifier stoves and controlling super-emitting diesel vehicles are most attractive (<$15/tCO$_2$eq). The other interventions cost between $50 and $100 per ton of CO$_2$eq. Indoor cook stove interventions affect large populations in China and South Asia as well as large parts of Africa. Consequently, pursuing indoor cook stove interventions is the most promising win–win strategy, followed by super-emitter repairs (although the latter may be limited in scope).

Before we delve into the question of how these costs compare with the benefits for the proposed interventions a few observations are in order. First, developing countries will benefit more immediately from health interventions, and so they will be more interested in health effectiveness measures, while it is in the interest of industrialized countries to intervene based on climate cost-effectiveness. Consequently, we show benefit-cost ratios (BCRs) for climate and health separately for each intervention in addition to the aggregate (health + climate) BCR.

B/C evaluation should ideally include a comprehensive assessment of social and environmental costs and benefits, and not be limited to climate and health alone. Therefore, there are other important benefits of proposed interventions that are not included in this analysis. For example, gasifier stoves may reduce the amount of time people spend gathering fuel (since they use less fuel overall), which in turn could influence the ability of women to engage in productive activities and allow more time for children (especially girls) to study.

Table 4.1.4 BCRs for the proposed black carbon reduction options

Black carbon reduction option	Health benefits DALYs ($) (BCR)	Climate benefit scenarios		Health + climate benefits	
		CO²eq ($) Stern (BCR)	CO²eq ($) Nordhaus (BCR)	Benefits ($) Stern BCR (range)	Benefits ($) Nordhaus BCR (range)
Indoor solid fuel use					
Biomass gasifier stove	760	880	100	1,600	870
	(13)	(14)	(1.7)	27 (16–54)	14 (9–28)
Coal to LPG stove switch	1,390	490	60	1,880	1,450
	(5)	(1.8)	(0.2)	7 (4–15)	6 (3–11)
Heavy Duty Vehicles					
Diesel to CNG fuel switch	30	5,130	600	5,160	630
	(0.03)	(0.9)	(0.1)	0.9 (0.5–2.6)	0.1 (0.06–0.3)
Diesel particulate traps	20	6,650	780	6,670	800
	(0.02)	(0.7)	(0.08)	0.7 (0.6–1.1)	0.1 (0.07–0.13)
Super-emitter control	40	17,270	2,030	17,320	2,070
	(0.01)	(5.8)	(0.7)	5.8 (3.5–17.3)	0.7 (0.4–2.1)

Notes: Benefits shown are per unit intervention (2005 dollars). The value of a DALY is taken to be $7,500, the average world GDP (PPP) *per capita* as per Smith and Haigler (2008). The two scenarios of $/CO$_2$eq are taken from the work of Nordhaus (2007) and Stern (2007). The costs of intervention shown in tables 4.1.2 and 4.1.3 are used to calculate BCRs.

Since the full social costs and benefits of the policy interventions cannot be easily accounted for, it is hard to pin down a comprehensive BCR for each intervention.

Table 4.1.4 shows the monetized benefits and corresponding BCRs for proposed interventions. While the conversion of DALYs to a monetary value is now relatively uncontroversial, putting a monetary value on the benefits from carbon reductions is not. In recent years this debate has crystallized around *The Stern Report* on the economics of climate change (Stern 2007) and responses to it (Nordhaus 2007, Weitzman 2007). It is beyond the scope of this Perspective paper to go into the details of the debate; we also eschew the question of who provides the "correct" estimate for the economic damages from carbon emissions.[7] For the purposes of this Perspective paper, it suffices to say that Stern and Nordhaus represent two strikingly different positions on economic damages from GHG emissions. Stern views climate change as a major threat to the global economy and estimates the damages at $310 per ton of carbon (i.e. $85 per tCO$_2$eq). Nordhaus estimates a far lower value of $35 per ton of carbon (around $10 per tCO$_2$eq). These differ-

ences are reflected in the benefits of interventions shown in table 4.1.4.

From table 4.1.4 it is clear that biomass gasifier stoves have BCRs much greater than 1 from a health perspective. It is also the only intervention that is justified from both the Stern and Nordhaus perspectives on climate damages. Interventions promoting the use of LPG fuel also have a large BCR, primarily because of the large health benefits of such a switch. Reducing emissions from super-emitting trucks and buses also may be justified from a BCR perspective, though that option does not pass muster when the lower (Nordhaus) value for carbon damages is used. The other options tend to have BCRs < 1 from combined health and climate perspectives, though switching to CNG might be justified from a climate perspective under specific assumptions – i.e. Stern's carbon damages and costs of implementation at the low end of the range. Interestingly, in all cases interventions are justified from either climate, or health, or both simultaneously. In no case do health and climate "add up" to justify an option that is not justified on either basis alone.

Policy Conclusions

One aspect of program effectiveness not easily captured by cost-based measures is the nature of

[7] Weitzman (2007) argues that Stern "gets it right for the wrong reasons," while Nordhaus strongly believes that Stern's position is incorrect.

institutions and governance that can facilitate the diffusion of these technological interventions. Institutional barriers in countries with resource constraints may be significant enough to cause a seemingly viable project to fail. The evidence from cook stove diffusion is instructive in this regard. A reassessment (Sinton *et al.* 2004) of the astonishing success of the Chinese biomass cook stove program in the 1980s (Smith *et al.* 1993) provides some important lessons. Sinton *et al.* (2004) found that China implemented successful programs that delivered improved biomass stoves to a majority of targeted households, and that those stoves continue to be used. Strong administrative, technical, and outreach competence, as well as the use of local resources and sustained national-level attention were critical to the success of these programs. The same cannot be said for coal stoves in China, which largely operate without flues even when they use cleaner-burning briquettes. Consequently, hundreds of millions of Chinese are currently exposed to very high levels of indoor air pollution from "improved" stoves.

The Chinese case of indoor coal-burning stoves highlights the difficulties faced by cook stove programs – they do not simply emerge from a market-based consumer demand. They need to be carefully thought through and supported by concerted institutional efforts. Other countries such as India have had much lower success in implementing stove programs despite their obvious and very large health benefits (Barnes *et al.* 1994). Reduction of black carbon emissions from HDDV also faces the challenge of weak institutions, coupled with rapid growth in the transport sector. Black carbon reduction provides an additional rationale – and potentially new funding – for putting programs in place, but the need for strong technical and administrative capacity and sound program design remains. These challenges will need to be overcome if black carbon reductions in the developing world and the concomitant health and climate benefits are to be realized.

Bibliography

Anderson, H.R., R. Atkinson, B. Chen, A. Cohen, D. Greenbaum, A.J. Hedley, W. Huang, J. Pande, C.A., Pope, and K. Smith, 2004: *Health Effects of Outdoor Air Pollution in Developing Countries of Asia: A Literature Review*, Health Effects Institute, Boston, MA

Andreae, M.O. and A. Gelencsér, 2006: Black carbon or brown carbon? The nature of light-absorbing carbonaceous aerosols, *Atmospheric Chemistry and Physics* **6**, 3131–48

Auffhammer, M., V. Ramanathan, and J.R. Vincent, 2006: Integrated model shows that atmospheric brown clouds and greenhouse gases have reduced rice harvests in India, *PNAS* **103**, 19668–72

Barnes, D.F., K. Openshaw, K.R. Smith, and R. Van Der Plas, 1994: *What Makes People Cook with Improved Biomass Stoves? A Comparative International Review of Stove Programs*, The World Bank, Washington, DC

Baron, R.E., W.D. Montgomery, and S.D. Tuladhar, 2009: Black carbon mitigation, chapter 4 in this volume

Bice, K., A. Eil, B. Habib, P. Heijmans, R. Kopp, J. Nogues, F. Norcross, M. Sweitzer-Hamilton, and A. Whitworth, 2009: Black carbon: a review and policy recommendations, Woodrow Wilson School of Public and International Affairs, Princeton University, Princeton, NJ

Bond, T.C., 2007: Can warming particles enter global climate discussions?, *Environmental Research Letters* **2**, 045030

Bond, T.C., D.G. Streets, K.F. Yarber, S.M. Nelson, J.H. Woo, and Z. Klimont, 2004: A technology-based global inventory of black and organic carbon emissions from combustion, *Journal of Geophysical Research – Atmospheres* **109**, D14203

Bond, T.C. and H.L. Sun, 2005: Can reducing black carbon emissions counteract global warming?, *Environmental Science and Technology* **39**, 5921–6

Ezzati, M., A.D. Lopez, A. Rodgers, S. Vander Hoorn, and C.J.L. Murray, 2002: Selected major risk factors and global and regional burden of disease. *Lancet* **360**, 1347–60

Ezzati, M., S. Vander Hoorn, A.D. Lopez, G. Danaei, A. Rodgers, C.D. Mathers, and C.J.L. Murray, 2006: Comparative quantification of mortality and burden of disease attributable to selected risk factors, in A.D. Lopez, C.D. Mathers, M. Ezzati, D.T. Jamison, and C.J.L. Murray (eds.), *Global Burden of Disease and Risk Factors*, Oxford University Press, Oxford, 241–68

Forster, P., V. Ramaswamy, P. Artaxo, T. Berntsen, R. Betts, D. Fahey, J. Haywood, J. Lean, D. Lowe, G. Myhre, J. Nganga, R. Prinn, G. Raga, M. Schulz, and R.V. Dorland, 2007: Changes in atmospheric constituents and in radiative forcing, in S. Solomon, D. Qin, M. Manning, Z. Chen, M. Marquis, K.B. Averyt, M. Tignor, and H.L. Miller (eds.), *Climate Change 2007: The Physical Science Basis – Contribution of Working Group I to the Fourth Assessment Report of the Intergovernmental Panel on Climate Change*, Cambridge University Press, Cambridge, 129–234

Grieshop, A.P., C.C.O. Reynolds, M. Kandlikar, and H. Dowlatabadi, 2009: A black-carbon mitigation wedge, *Nature Geoscience* **8**, 533–4

Hausker, K., 2004: Vehicle inspection and maintenance programs: international experience and best practice, office of Energy and Information Technology, US Agency for International Development, Washington, DC

Haywood, J.M. and V. Ramaswamy, 1998: Global sensitivity studies of the direct radiative forcing due to anthropogenic sulfate and black carbon aerosols, *Journal of Geophysical Research* **103**, 6043–58

Jacobson, M.Z., 2004: Climate response of fossil fuel and biofuel soot, accounting for soot's feedback to snow and sea ice albedo and emissivity, *Journal of Geophysical Research* **109**, D21201

2006: Effects of externally-through-internally-mixed soot inclusions within clouds and precipitation on global climate, *Journal of Physical Chemistry A* **110**, 6860–73

Koren, I., J.V. Martins, L.A. Remer, and H. Afargan, 2008: Smoke invigoration versus inhibition of clouds over the Amazon, *Science* **321**, 946–9

MacCarty, N., D. Ogle, D. Still, T. Bond, and C. Roden, 2008: A laboratory comparison of the global warming impact of five major types of biomass cooking stoves, *Energy for Sustainable Development* **12**, 56–65

Mehta, S. and C. Shahpar, 2004: The health benefits of interventions to reduce indoor air pollution from solid fuel use: a cost-effectiveness analysis, *Energy for Sustainable Development* **8**, 53–9

Molina, M. and L. Molina, 2004: Megacities and atmospheric pollution, *Journal of Air & Waste Management Association* **54**, 644–80

Nordhaus, W.D., 2007: A review of the Stern Review on the economics of climate change, *Journal of Economic Literature* **45**, 686–702

Pacala, S. and R. Socolow, 2004: Stabilization wedges: solving the climate problem for the next 50 years with current technologies, *Science* **305**, 968–72

Ramanathan, V. and G. Carmichael, 2008: Global and regional climate changes due to black carbon, *Nature Geoscience* **1**, 221–7

Ramanathan, V., M.V. Ramana, G. Roberts, D. Kim, C. Corrigan, C. Chung, and D. Winker, 2007: Warming trends in Asia amplified by brown cloud solar absorption, *Nature* **448**, 575–8

Reynolds, C.C.O. and M. Kandlikar, 2008: Climate impacts of air quality policy: switching to a natural gas-fueled public transportation system in New Delhi, *Environmental Science and Technology* **42**, 5860–5

Rodgers, A., P. Vaughan, T. Prentice, T.T. Edejer, D. Evans, and J. Lowe, 2002: *The World Health Report 2002: Reducing Risks, Promoting Healthy Life*, World Health Organization, Geneva

Shindell, D. and G. Faluvegi, 2009: Climate response to regional radiative forcing during the twentieth century, *Nature Geoscience* **2**, 294–300

Sinton, J.E., K.R. Smith, J.W. Peabody, L. Yaping, Z. Xiliang, R. Edwards, and G. Quan, 2004: An assessment of programs to promote improved household stoves in China, *Energy for Sustainable Development* **8**, 33–52

Smith, K.R., S. Gu, K. Huang, and D. Qiu, 1993: One hundred million improved cookstoves in China: how was it done?, *World Development*: **21**, 941–61

Smith, K.R. and E. Haigler, 2008: Co-benefits of climate mitigation and health protection in energy systems: scoping methods, *Annual Review of Public Health* **29**, 11–25

Stern, N., 2007: *The Economics of Climate Change: The Stern Review*, Cambridge University Press, Cambridge

Subramanian, R., E. Winijkul, T. Bond, W. Thiansathit, N.T. Kim Oanh, I. Paw-Armart, and K. Duleep, 2009: Climate-relevant properties of diesel particulate emissions: results from a piggyback study in Bangkok, Thailand, *Environmental Science and Technology* **43**, 4213–18

Weitzman, M.L., 2007: A review of the Stern review on the economics of climate change, *Journal of Economic Literature* **45**, 703–24

Wigley, T.M.L., 2006: A combined mitigation/geoengineering approach to climate stabilization, *Science* **314**, 452–4

Zhang, J.F., K.R. Smith, Y. Ma, S. Ye, F. Jiang, W. Qi, P. Liu, M.A.K. Khalil, R.A. Rasmussen, and S.A. Thorneloe, 2000: Greenhouse gases and other airborne pollutants from household stoves in China: a database for emission factors, *Atmospheric Environment* **34**, 4537–49

Zickfeld, K., A. Levermann, M.G. Morgan, T. Kuhlbrodt, S. Rahmstorf, and P. Keith, 2007: Expert judgements on the response of the Atlantic meridional overturning circulation to climate change, *Climatic Change* **82**, 235–65

Methane Mitigation

CLAUDIA KEMFERT AND WOLF-PETER SCHILL[*]

Introduction

Methane (CH_4) is a major anthropogenic greenhouse gas (GHG), second only to carbon dioxide (CO_2) in its impact on climate change. CH_4 has a high global warming potential that is twenty-five times as large as that of CO_2 on a 100-year time horizon according to the 2007 IPCC report (IPCC 2007a). Thus, CH_4 contributes significantly to anthropogenic radiative forcing, although it has a relatively short atmospheric perturbation lifetime of twelve years. CH_4 has a variety of sources that can be small, geographically dispersed, and not related to energy sectors.

In this chapter, we analyze CH_4 emission abatement options in five different sectors and identify economic mitigation potentials for different CO_2 prices. While mitigation potentials are generally large, there are substantial potentials at low marginal abatement costs (MACs). Drawing on different assumptions on the social costs of carbon (SCC), we calculate benefit-cost ratios (BCRs) for different sectors and mitigation levels.

We recommend an economically efficient global CH_4 mitigation portfolio for 2020 that includes the sectors of livestock and manure, rice management, solid waste, coal mine methane, and natural gas. Depending on SCC assumptions, this portfolio leads to global CH_4 mitigation levels of 1.5 or 1.9 $GtCO_2$-eq at overall costs of around $14 billion or $30 billion and BCRs of 1.4 and 3.0, respectively. We also develop an economically less efficient alternative portfolio that excludes cost-effective agricultural mitigation options. It leads to compara-

ble abatement levels, but has higher costs and lower BCRs.

If the global community wanted to spend an even larger amount of money – say, $250 billion – on CH_4 mitigation, much larger mitigation potentials could be realized, even such with very high marginal abatement costs. Nonetheless, this approach would be economically inefficient. If the global community wanted to spend such an amount, we recommend spreading the effort cost-effectively over different GHGs.

While CH_4 mitigation alone will not suffice to solve the climate problem, it is a vital part of a cost-effective climate policy. Due to the short atmospheric lifetime, CH_4 emission reductions have a rapid effect. CH_4 mitigation is indispensable for realizing ambitious emission scenarios like IPCC's "B1," which leads to a global temperature increase of less than 2°C by 2100. Policy makers should put more emphasis on CH_4 mitigation and aim for realizing low-cost CH_4 mitigation potentials by providing information to all relevant actors and by developing appropriate regulatory and market frameworks. We also recommend including CH_4 in market-based instruments, such as taxes or emission trading schemes (ETSs). The utilization of its energy value should be maximized.

Definition and Description of Climate Change

The latest Assessment Report of the IPCC (IPCC 2007a) states that warming of the global climate system is unequivocal. It reports that most of the observed increases in global average temperatures are very likely due to a rise in anthropogenic greenhouse gas (GHG) concentrations. Global GHG emissions due to human activities have grown since

* The authors would like to thank Kaspar Thede Anderskov, David Anthoff, Christian Bjornskov, Daniel Johansson, and Bjørn Lomborg for valuable comments. The authors are solely responsible for any remaining errors.

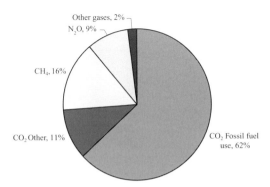

Figure 5.1 *Global anthropogenic GHG emissions, by gas (2005)*

Note: all emissions are in CO_2eq.
Source: IEA (2008).

preindustrial times, with an increase of around 70% between 1970 and 2004 (IPCC 2007a). As shown in figure 5.1 CO_2 is the most important anthropogenic GHG, followed by CH_4, nitrous oxide (N_2O), and other (fluorinated) gases.

CH_4 is second only to CO_2 in its impact on climate change. The radiative forcing of anthropogenic CH_4 contributes about 0.48 W/m^2 to total net anthropogenic radiative forcing of 1.6 W/m^2 (IPCC 2007b).[1] Including indirect CH_4 effects like enhancements of tropospheric ozone (O_3) or stratospheric water vapor further increases its total radiative impact.

If current emission trends persist, the global Earth surface temperature will increase substantially in the future. The IPCC reports that stabilizing atmospheric concentrations of carbon dioxide equivalents (CO_2-eq) at around 445–490 ppm would lead to a global average temperature increase above preindustrial levels of around 2.0–2.4°C. Stabilizing emissions at 855–1130 ppm CO_2-eq would lead to a temperature increase of around 4.9–6.1°C (IPCC 2007a).

According to the IPCC, further warming would induce many changes in the global climate system until 2100, such as changes in wind patterns, precipitation, weather extremes, and sea ice. A global temperature rise of more than 2°C compared to preindustrial levels might result in abrupt or irreversible changes. IPCC's "B1" emission scenario realizes this 2°C target. The IPCC has identified

five "reasons for concern", including risks to unique and threatened systems, risks of both more frequent and more violent extreme weather events, the distribution of impacts and vulnerabilities, aggregate impacts, and the risks of large-scale singularities (IPCC 2007a). In order to avoid such vulnerabilities and threats, it is necessary to reduce the global volume of GHG emissions significantly and stabilize global GHG concentrations at nearly today's level.

Extreme weather events are already causing enormous economic damages. However, estimates of future climate change damages and their economic consequences are highly uncertain (cf. Tol 2002a, 2002b; Nordhaus 2007; Stern 2007; Weizman 2007; OECD 2008). One reason for this is that the effects are subject to temporal and spatial disparities. For example, the benefits of climate protection policies pursued in Europe today may not necessarily also be felt in Europe. They could equally materialize in South East Asia, where exposed island nations might be spared a flood produced by a rising sea level. Moreover, as a result of the long atmospheric lifetime of several GHGs, many potential effects will emerge in the distant future.

While many publications on GHG mitigation have dealt with CO_2, we focus solely on different CH_4 emission mitigation solutions and assess their economic costs and benefits. We first give an overview of the characteristics of CH_4 and its emission sources. Subsequently, we describe several options for reducing CH_4 emissions. Economic mitigation potentials and MACs for specific solutions are listed, and we then estimate the economic costs and benefits of different options, drawing on different assumptions on the social costs of carbon emissions. Finally, we recommend a cost-effective portfolio of CH_4 mitigation options that could be implemented by 2020.

Most of the existing literature on CH_4 mitigation cost assessments focuses on time frames until about 2020 or 2030.[2] Accordingly, most costs described

---

[1] Total net anthropogenic forcing also contains some negative radiative forcing, for example caused by anthropogenic aerosols.

[2] One example for long-term cost assessment is provided by Lucas *et al.* (2007).

in this chapter are in this time range, while the benefits of lower global temperatures due to CH_4 mitigation will be visible over longer periods (cf. Hope 2005).

The Solution Category: CH_4 Mitigation

Background on CH_4 Emissions

Compared to CO_2, CH_4 is relatively short-lived. Its atmospheric perturbation lifetime is twelve years (IPCC 2007b). CH_4 is removed from the atmosphere mainly through a hydroxyl radical reaction process. As CH_4 is a much more short-lived GHG than CO_2, it has high reduction potentials and high impacts on radiative forcing within short time periods. On the other hand, CH_4 has a higher global warming potential (GWP) than CO_2, controlling for its shorter atmospheric lifetime. In the second IPCC Assessment Report of 1995 (IPCC 1995), CH_4 was estimated to trap heat twenty times more effectively than CO_2 on a 100-year time horizon. This value is also used for reporting under the United Nations Framework Convention on Climate Change (UNFCCC). According to the third IPCC Assessment Report of 2001 (IPCC 2001) the GWP of CH_4 is 23 relative to CO_2. The latest IPCC Assessment Report (IPCC 2007a) includes a GWP estimate of about 25 compared to CO_2 over a 100-year time horizon (USEPA 2006a; IPCC 2007b). The GWP for CH_4 calculated by the IPCC includes indirect effects from enhancements of tropospheric O_3 and stratospheric water vapor.[3]

CH_4 is generated when organic matter decays in anaerobic conditions. Natural CH_4 sources include wetlands, termites, oceans, and gas hydrates (cf. Milich 1999). Keppler *et al.* (2006) have suggested large-scale methanogenesis by plants in aerobic conditions. Given this newly detected emission source, it has been calculated that plants could account for up to 45% of global CH_4 emissions. However, Nisbet *et al.* (2009) refute Keppler *et al.* (2006) and conclude that there is no such biochemical pathway for aerobic CH_4 synthesis in plants,

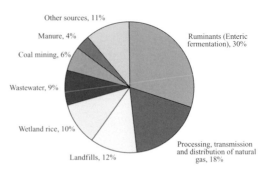

Figure 5.2 *Global anthropogenic CH_4 emissions, by source (2005)*

thereby rejecting the notion that plants may be a major source of global CH_4 production.

As shown in figure 5.2, major anthropogenic CH_4 sources in 2005 included enteric fermentation of ruminants (ca. 30% of anthropogenic CH_4 emissions), natural gas and oil systems (18%), landfills (12%), wetland (paddy) rice cultivation (10%), wastewater (9%), coal mining (6%) and livestock manure (4% according to USEPA 2006a, 2006b). That is, agriculture production (ruminant livestock, manures and rice grown under flooded conditions) currently accounts for about half of global anthropogenic CH_4 emissions. This is also confirmed by other sources (Povellato *et al.* 2007; Smith *et al.* 2009). However, the relative importance of anthropogenic CH_4 sources varies significantly between countries. For example, municipal solid waste landfills are the largest CH_4 source in the USA, while livestock dominates emissions in other countries (de la Chesnaye *et al.* 2001). The largest percentage of global coal mine CH_4 emissions comes from China (Yang 2009). China, the USA, India, Russia, the EU, and Brazil are the world's largest CH_4 emitters. Figure 5.3 shows emission trends for these countries/regions between 1970 and 2005.

CH_4 emissions and atmospheric concentrations have increased markedly since preindustrial times. Atmospheric concentrations of CH_4 increased from preindustrial values of about 715 ppb to about 1,774 ppb in 2005 and exceed by far the natural range over the last 650,000 years (IPCC 2007b). Bousquet *et al.* (2006) find that while anthropogenic CH_4 emissions were decreasing in the 1990s, they have been rising again since 1999.

---

[3] Regarding the assessment of benefits and costs of specific mitigation options, we do not draw on GWP, but directly use the values of CO_2-eq provided by the respective studies.

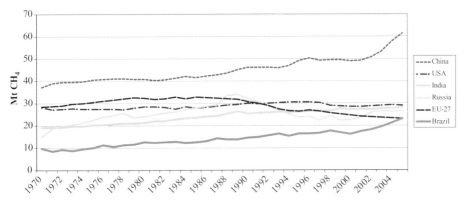

Figure 5.3 *CH₄ emissions, different regions (1970–2005)*

The latest increase in anthropogenic emissions has been masked by a coincident decrease in natural CH_4 emissions, mostly from wetlands. In general, Bousquet *et al.* (2006) find very large fluctuations in the growth rate of atmospheric CH_4 concentrations from one year to the next. The interannual variability seems to be dominated by wetland CH_4 emissions. USEPA (2006b) projects an increasing global anthropogenic emission trend until 2020.

Definition and Description of the Solution Category

The solution category "CH_4 mitigation" includes different measures for capturing CH_4 or for avoiding its release. In most cases, captured CH_4 will be oxidized to CO_2, which significantly reduces its climate impact. The oxidization energy may be utilized, which entails additional GHG mitigation if carbon-intensive fuels are substituted.

Anthropogenic CH_4 is emitted by various sources across different sectors and regions. Accordingly, mitigation potentials and cost vary widely. In this chapter, we focus on sectors that are characterized by both significant CH_4 emissions and substantial mitigation potentials. In order to identify these sectors, we survey the relevant literature. For example, Milich (1999) provides an early overview of CH_4 mitigation strategies. De la Chesnaye *et al.* (2001) survey US non-carbon GHG emission reductions strategies, including CH_4 mitigation. An IEA (2003) study builds global cost curves for industrial sources of several non-CO_2

GHGs. Povellato *et al.* (2007) review cost-effective GHG mitigation potentials in the European agro-forestry sector. Johnson *et al.* (2007) focus their review on agricultural GHG mitigation options for the USA. Smith *et al.* (2009) give another overview of GHG mitigation in the agricultural sector. They find that the largest CH_4 mitigation potentials are related to rice management and livestock, while potentials for manure management are lower. Delhotal *et al.* (2006) evaluate international CH_4 mitigation potentials and costs in the waste and energy sectors, including regional differentiations. Finally, USEPA (2006a) provides a very comprehensive analysis of mitigation options across sectors and world regions and a calculation of MAC curves.

The literature survey shows that the most important sectors for CH_4 mitigation include (1) livestock and manure management, (2) rice management, (3) solid waste management, (4) coal mining, and (5) processing, transmission, and distribution of natural gas.

Aside from these sectors, wastewater is also a significant global CH_4 source. We refrain from including wastewater in our analysis mainly due to a scarcity of mitigation costs data.[4] Moreover, while wastewater mitigation potentials might be significant, most technological options in

---

[4] IEA (2003) provides some short-term cost estimates for the wastewater CH_4 mitigation option of electricity generation from recovered CH_4. Lucas *et al.* (2007) estimate long-term costs for this category.

this sector are related to significant changes of wastewater management and infrastructure – e.g. the installation of sewerage systems in developing countries (cf. USEPA 2006a). Without such infrastructure measures, CH_4 mitigation potentials are low. In general there are other driving forces for installing wastewater collecting and treatment facilities, above all sanitary and hygienic ones (cf. Lucas et al. 2007). We have also excluded CH_4 emissions from the oil sector due to its comparatively low mitigation potential (cf. USEPA 2006a).

Our definition of "CH_4 mitigation" focuses on capturing CH_4 or avoiding its release. We exclude options for enhanced CH_4 removal from the atmosphere since existing technologies have very limited potentials due to very low concentrations of CH_4 in the atmosphere. For example, Johnson et al. (2007) and Smith et al. (2009) mention some examples for removing CH_4 from the atmosphere by specific agricultural practices. However, their effect is small compared to overall CH_4 fluxes. Another example is provided by Yoon et al. (2009) who analyze the feasibility of atmospheric CH_4 removal using methanotrophic biotrickling filters. They find that such measures are infeasible for removing atmospheric CH_4 since concentrations are far too low. Finally, we also refrain from exploring agriculture–climate interdependencies concerning GHG sources and sinks (cf. Povellato et al. 2007).

Description of Specific Solutions within the Solution Category

In the following section, we briefly describe five specific solutions for CH_4 mitigation. First, we point out strategies, technical definitions, and technical mitigation potentials. We then quickly discuss the feasibility of their application and mention interdependencies and side effects, if applicable. Specific mitigation costs and, accordingly, economic mitigation potentials are discussed afterwards.

Livestock and Manure Management

The most important livestock CH_4 mitigation strategies include improved feeding practices

(e.g. feeding concentrates), the use of specific agents or dietary additives (like antibiotics and antimethanogen that suppress methanogenesis), and long-term management changes and animal breeding. All these measures aim at improving feed conversion efficiency, increasing animal productivity, and decreasing specific CH_4 emissions (DeAngelo et al. 2006; USEPA 2006a; IPCC 2007d; Smith et al. 2009).

Manure mitigation includes both low-tech strategies like covering and cooling manure lagoons during storage and alternative techniques for manure dispersion and application (Weiske et al. 2006; USEPA 2006a; von Witzke and Noleppa 2007; IPCC 2007d). More advanced technologies include frequent manure removal from animal housing into covered storage using scraping systems (Weiske et al. 2006) as well as farm scale or centralized digesters for biogas generation and utilization (DeAngelo et al. 2006; USEPA 2006a). In small-scale farm digesters, biogas from local manure may be used for electricity and/or heat production. Larger, centralized digesters can also take in additional organic wastes. There are many different digester designs ranging from lowtech small-scale to high-tech large-scale models, for example polyethylene bag or covered lagoon digesters for cooking fuel, light flexible-bag digesters, and large-scale dome digesters (USEPA 2006a).

Mitigation potentials for livestock and manure are relatively high in some countries, such as Germany (cf. von Witzke and Noleppa 2007). In EU 15, the overall mitigation potential for milk production is around 3.5% of total EU 15 anthropogenic GHG emissions, of which a substantial share is related to CH_4 (Weiske et al. 2006). However, the applicability and technical efficiency of several measures varies by climate. For example, the technical mitigation potential of digesters is largest in warm climates (USEPA 2006a); nonetheless, the overall mitigation potential of digesters is limited (USEPA 2006a). In general, technical potentials for livestock and manure management are limited in many areas of the world due to feeding practices, widearea dispersion of livestock, and local farming techniques.

In manure management, complex interdependencies between CH_4 and N_2O exist, which might lead to tradeoffs. For example, while aerobic conditions during manure storage suppress CH_4 formation, they can promote N_2O formation (cf. USEPA 2006a; von Witzke and Noleppa 2007). When applying mitigation measures in livestock and manure management, it is important not to generate counter-effective emission increases of other GHGs. Some options related to livestock or manure management might potentially trigger increases in N_2O emissions in unfavorable circumstances. However, Smith *et al.* (2009) find that the measures cited above have no adverse N_2O impacts and thus a net emission mitigation effect.

Rice Management

This solution aims for reducing CH_4 generation from flooded rice paddies. A major mitigation strategy is improving water management through ways such as draining wetlands during rice seasons, avoiding water logging in off-seasons, and shallow flooding. Additional measures include upland rice cultivation and future cultivars with lower exudation rates (DeAngelo *et al.* 2006; USEPA 2006; IPCC 2007d; Smith *et al.* 2009).

Aside from CH_4, rice cultivation leads to emissions of other greenhouse gases like N_2O and soil CO_2 (USEPA 2006a; Wassmann and Pathak 2007). Such emissions may be mitigated by applying additives like phosphygypsum and nitrification inhibitors. In addition, the utilization of rice husks as fuel for heat and electricity generation can substitute carbon-intensive fossil fuels (Wassmann and Pathak 2007). However, these strategies are not further explored in this chapter.

In the case of rice management, some mitigation practices might lead to increases of N_2O emissions. However, according to Smith *et al.* (2009), there is general agreement and evidence for net mitigation effects of these measures. In addition, rice-related mitigation strategies might face social and institutional barriers as well as challenges regarding monitoring and enforcement. These issues are briefly discussed later on.

Solid Waste Management

The single most important specific solution in the category of solid municipal waste is preventing the release of landfill CH_4 into the atmosphere. Landfill CH_4 can be captured by installing a landfill cap and an active gas extraction system that uses vertical wells and optionally also horizontal collectors (Monni *et al.* 2006; IPCC 2007d). Captured CH_4 can be used directly as a gas or utilized for local heat and/or electricity generation. If carbon-intensive fuels are substituted, such measures have an additional GHG mitigation effect. If landfill CH_4 concentrations are low or if there is a lack of local energy demand, CH_4 can alternatively be oxidized to CO_2 by flaring (Gallaher *et al.* 2005; USEPA 2006a). Landfill CH_4 not captured may be oxidized by indigenous methanotrophic microorganisms in landfill cover soils. Moreover, "bioreactor landfill designs" allow enhanced CH_4 generation and capturing (IPCC 2007d).

There are also strategies that aim to reduce CH_4 generation in landfills by diverting organic matter from them. Such strategies include the application of anaerobic digestion or aerobic composting, mechanical biological treatment, waste incineration, as well as waste reduction, re-use and increased recycling activities (Monni *et al.* 2006; USEPA 2006a; IPCC 2007d). These strategies imply a structural change of waste management practices and the related infrastructure.

Solid waste CH_4 mitigation potentials vary substantially between countries. They are highest for China, followed by the USA and African nations (Delhotal *et al.* 2006). In general, CH_4 mitigation options of this category are highly dependent on the country-specific organization and structure of the waste management sector. Furthermore, the primary waste management objective is typically not GHG mitigation, but rather controlling environmental pollutants or mitigating health risks (Monni *et al.* 2006; IPCC 2007d).

Coal Mining

Depending on depth and geological conditions, coal seams can include significant amounts of CH_4. Since CH_4 is flammable in a concentration range from 5 to 16% in air, coal mine CH_4 is a safety hazard for mining operations. Thus, mine degasification by ventilation is a standard procedure in underground coal mining, resulting in substantial CH_4 releases to the atmosphere (USEPA 2006a).

While minor quantities of CH_4 are released in post-mining operations like processing and transportation, major emissions occur during mining operations. Accordingly, the most relevant mitigation strategies focus on mining operations. There are three major mitigation strategies (Gallaher et al. 2005; USEPA 2006a). First, degasification can be applied up to ten years before mining operations begin. This strategy aims for collecting and capturing CH_4 through vertical drills (at later stages, horizontal drills can also be used). Captured CH_4 may then be injected into pipelines or utilized for heat and/or electricity production. The second option is enhanced degasification, which follows the same principle, but includes advanced drilling and additional purification and enrichment of captured gas. The third major strategy is ventilation air CH_4 abatement. In contrast to degasification, this option is carried out during mining operations. It aims for oxidizing CH_4 in ventilation air which typically has much lower CH_4 concentrations than degasification air (mostly $<1\%$). Thus, catalytic CH_4 oxidation technologies are usually applied. The resulting oxidation heat may be used for space heating purposes (Gallaher et al. 2005; USEPA 2006a).

In 2000, coal mining accounted for 3.3% of global anthropogenic CH_4 emissions. China is the largest single emitter, followed by the USA, India, and Australia (USEPA 2006a). In 2004, China emitted about 190 million tons of CO_2-eq of coal mine CH_4, followed by the USA, with less than 60 million tons of CO_2-eq (Yang 2009). China also has by far the highest global coal mine mitigation potential (Delhotal et al. 2006). However, although several specific Chinese coal mine mitigation policies have been put into place, several country-specific barriers still remain, such as lack of suitable degasification technologies, shortage of micro-internal combustion-engine generators, and low amounts of capital investment from the private sector (Yang 2009).

It is important to note that safety concerns and not GHG mitigation is the driving force behind coal mine ventilation. However, safety concerns only give an incentive for mine operators to reduce CH_4 in the mines below flammable concentrations rather than fully mitigating its release into the atmosphere.

Processing, Transmission, and Distribution of Natural Gas

In the natural gas sector, CH_4 may be released during production, processing, transmission, storage, and distribution (Gallaher et al. 2005). Typical sources are leaks in natural gas pipelines, compressor stations, or venting of pipelines for maintenance reasons.

Mitigation strategies focus on the replacement of pipeline or compressor equipment, or on alternative management practices, such as increased maintenance and reduced venting (Delhotal 2006).

Selected measures include the use of gas turbines instead of reciprocating engines, the replacement of high-bleed pneumatic devices with low-bleed or compressed air systems, dry seals on centrifugal compressors, and catalytic converters (USEPA 2006a). The replacement of wet centrifugal compressor seal oil systems with dry seals and the installation of low-bleed pneumatic devices might be the most promising of options. Favorable management and operation practices include optimizing compressor shutdown, minimizing venting before pipeline maintenance, and periodic leak inspections (Lechtenbömer et al. 2007).

Of all sectors mentioned in this chapter, the natural gas sector might have the highest reduction potential in 2020. Most potential reductions are accumulated in a few world regions like Russia, the Middle East, Latin America, the USA, and the Commonwealth of Independent States (CIS) (Delhotal et al. 2006). Measurements along the world's largest gas-transmission system in Russia showed an overall CH_4 leakage of around 1.4%, which is comparable to US leakage rates (Lelieveld et al. 2005). Additional analyses showed that CH_4 emissions from the Russian natural gas long-distance network might be even smaller (approximately 0.6% of the natural gas delivered) (Lechtenbömer et al. 2007). It has been shown that with such low leakage rates, switching from coal or oil to natural gas as a fuel has positive overall GHG mitigation impacts even in the light of leakages (Lelieveld *et al.* 2005).

Importantly, the projected higher utilization of liquefied natural gas (LNG) could increase CH_4 emissions since liquefaction processes and LNG transportation provide new opportunities for CH_4 release.

Economic Evaluation of Specific Solutions

Methodology

Global economic mitigation potentials and marginal abatement costs

We identify overall global economic potentials for the different mitigation solutions outlined above and for different values of carbon between 0 and 200 US\$/tCO_2-eq.[5] Most studies refer to mitigation potentials in 2020, while some reach up to 2030. Where data is available, we also provide information on economic potentials and/or MACs of specific technologies within one sector. The method of research is an extensive literature survey of relevant bottom-up studies. Since economic abatement potentials vary significantly between some sources, we provide a range of different estimates that represents different strands of the literature.

MAC curves illustrate the potentials for reducing emissions at different cost levels. They are constructed by ordering different mitigation options from least to most expensive. Typically, MAC curves increase with an ascending slope. While emission abatement of the first units of CH_4 is often relatively cheap or even associated with negative costs, costs usually increase for additional abatement (cf. USEPA 2006a). There are static and dynamic MAC curves. For example, Stanford University's Energy Modeling Forum EMF21 used static MAC curves for a multi-gas mitigation modeling project. They were derived in cooperation with the US Environmental Protection Agency (EPA) from a global cost analysis of non-CO_2 GHGs, including CH_4. In contrast, Gallaher *et al.* (2005) have conducted a dynamic analysis of the costs and potentials of CH_4 mitigation strategies in

the solid waste, coal mining, and natural gas sectors. Incorporating firm-level data, their approach assumes technical change and decreasing costs, resulting in different MAC curves for 2010, 2020, and 2030.

The USEPA (2006a) report provides the most comprehensive calculation of global CH_4 MAC curves for different world regions and sectors. Using these MAC curves, technical and economic potentials of different mitigation strategies at different CO_2 prices are calculated. At breakeven CO_2 prices, the net present value (NPV) of a mitigation strategy is 0. For different CO_2 prices, related economic mitigation levels can be calculated. All numbers in the USEPA (2006a) report are provided in constant 2000 US\$. Typically, the report assumes a discount rate of 10% and a tax rate of 40%. This discount rate is also applied by the IEA (IEA 2003): it represents an industry perspective. From a social perspective, lower rates might be more appropriate, leading to even higher economic mitigation potentials. EPA provides more detailed technology-specific MAC curves with different discount and tax rates on their web site.[6]

Different approaches for B/C assessments

The most coherent way of estimating the costs and benefits of the CH_4 mitigation solutions discussed in this chapter would be the application of Computable General Equilibrium (CGE) models or integrated assessment models (IAMs). In the literature, a large number of such models has been applied for analyzing various mitigation policies, focusing on different GHGs and mitigation technologies.

A comprehensive modeling exercise that included CH_4 has been carried out in an international collaboration under the previously mentioned EMF21. The results are presented in the 2006 special issue of *The Energy Journal* entitled "Multi-Greenhouse Gas Mitigation and Climate Change." It includes various assessments of

---

[5] All numbers are in constant year 2000 US\$, if nothing else is provided.

[6] Technology-specific MAC curves for different discount rates are provided at www.epa.gov/methane/appendices. html. However, USEPA provides aggregate global MAC curves only for a tax rate of 40% and a discount rate of 10%. Thus, we stick to these numbers in the chapter.

economic and energy sector impacts of multi-gas mitigation strategies. Drawing on a range of different IAMs (for example, Aaheim *et al.* 2006; Jakeman and Fisher 2006; Kemfert *et al.* 2006; van Vuuren *et al.* 2006)[7], EMF21 includes but is not restricted to CH_4 mitigation measures. A general result is that including non-CO_2 GHGs like CH_4 and N_2O results in substantially lower mitigation cost compared to restricting GHG mitigation to CO_2. A more recent example for a CGE analysis of mitigation options in the agricultural and forestry sectors is provided by Golub *et al.* (2009). Using a global model that includes the opportunity costs of land use, the authors find that livestock and paddy rice CH_4 mitigation strategies are preferable agriculture-related GHG mitigation options.

However, such models have not been consistently applied to the specific CH_4 mitigation options discussed in this chapter. To our knowledge, there is no application of an IAM that explicitly analyses the costs and benefits of single CH_4 mitigation measures in the fields of livestock/manure, rice, solid waste, coal mining CH_4, and natural gas. Rather than assessing these mitigation measures separately, most models focus on integrated packages of different mitigation options. Moreover, in most cases a mixed mitigation strategy of CO_2 and a range of non-CO_2 GHGs, including CH_4, are applied.

Given this gap in the literature, we refrain from using IAM publications for estimating the benefits and costs of the specific CH_4 mitigation options discussed in this chapter. Rather, we estimate costs and benefits separately and then provide BCRs, as described in the following sections.

Estimating costs

Cost calculations are relatively straightforward if MAC data are available. Total mitigation costs up to a certain mitigation level equal the area under a MAC curve.

In the following, we calculate the total costs of applying specific mitigation solutions in two ways. One approach is multiplying technology-specific MACs and related mitigation potentials, where such data are available. Another approach is to look at the economic mitigation potentials at different CO_2 prices. Assuming carbon prices of 0 $/tCO_2$-eq, these price–quantity combinations can be interpreted as mitigation levels at different (marginal) mitigation costs. In steps of $15/tCO_2$-eq, we multiply these marginal costs with the related potentials and add the results up. This stepwise procedure is necessary due to a lack of information on the shape of the MAC curve between the intervals of 0, 15, 30, 45, and 60 $/tCO_2$-eq. Negative marginal costs are not considered, but regarded as costs of 0 $/tCO_2$-eq. This approach and the fact that MAC curves are usually convex leads to a systematic overestimation of costs.

Estimating benefits

Calculating the benefits of different mitigation measures is less straightforward than calculating costs. Different approaches might be chosen. For example, one might draw on model results and calculate the benefits of emission reductions by using shadow price estimates on CH_4.[8] For reasons of simplicity, traceability, and data availability, we focus on a different approach for estimating benefits. We look at the CO_2-equivalents of avoided CH_4 emissions and assign a value to these emission reductions with an estimate of the SCC.

While this procedure is very transparent, it involves a range of challenges. For example, choosing an appropriate SCC value is demanding. Depending on climate change projections, damage functions, and discount rates, SCC estimates in the literature vary significantly (Tol 2008). We use three different values in order to cover a range of different assumptions which we obtain from a literature survey (Tol 2008). Drawing only on a sample of peer-reviewed studies, we use the median, the mean, and the 90-percentile values calculated by Tol. The median SCC value is 48 $/tC, the mean 71 $/tC, and the 90-percentile is 170 $/tC. With a conversion factor of 3.667 tCO_2/tC, this translates to about 13.1, 19.4, and 46.4 $/tCO_2$, respectively.

[7] The full list of models applied in EMF21 includes AIM, AMIGA, COMBAT, EDGE, EPPA, FUND, GEMINI-E3, GRAPE, GTEM, IMAGE, IPAC, MERGE, MESSAGE, MiniCAM, PACE, POLES, SGM, and WIAGEM.

[8] For example, Nordhaus' DICE Model could be used, see www.econ.yale.edu/~nordhaus/homepage/dicemodels.htm.

Table 5.1 Livestock and manure: projected baseline emissions and economic mitigation potentials at different CO_2 prices

Source	Year	Baseline in MtCO$_2$-eq	Value of CO$_2$ in US$/tCO$_2$-eq						
			0	15	30	45	60	100	200
			Economic mitigation potentials in MtCO$_2$-eq						
DeAngelo *et al.* (2006)	2010	567	29						31
USEPA (2006a)	2020	2,867	83	126	158	175	192		
Smith *et al.* (2009)	2030	n/a						210	

Sources: USEPA (2006a); Smith *et al.* (2009); own calculations.
USEPA (2006a): overall livestock and manure; Smith *et al.* (2009): livestock.

Another challenge of this approach is the conversion of CH_4 to CO_2-eq, which depends on the time horizon, given the different atmospheric lifetime of CH_4 and CO_2 (cf. IPCC 2007b). We do not convert these values by ourselves, but rather take the CO_2-eq directly from the studies. However, the time horizons of CO_2-eq and the SCC values taken from Tol (2008) may differ. Finally, our approach might not consider important interdependencies, side effects and equilibrium issues that might be addressed in a more appropriate way with an IAM. Therefore, our B/C estimates should only be considered as first indications of the relative cost-effectiveness of different options.

Results: Global Economic Mitigation Potentials and Marginal Abatement Costs

Livestock and Manure Management

Estimations of costs and mitigation potentials in this category vary significantly between countries and world regions (cf. USEPA 2006a and Povellato *et al.* 2009). Table 5.1 provides an overview on different estimations of economic potentials at different CO_2 prices between 0 and 200 $/CO_2$-eq.

Table 5.1 indicates that a large share of the mitigation potential is in the low-cost range of less than 30 $/tCO_2$-eq. Measures with very high costs do not substantially increase mitigation potentials. The absolute numbers provided by DeAngelo (2006) for the shorter time frame until 2010 are much lower than the ones provided by USEPA (2006a) for 2020. However, since they also assume

Table 5.2 Livestock and manure: MACs of selected technologies

	Solution	US$/tCO$_2$eq
Livestock management	Feeding	60
	Additives	5
	Breeding	50
Manure management	Soil application	10
	Storage, biogas	200

Source: Smith *et al.* (2009).

lower baseline emissions, the relative shares are comparable.

There seem to be substantial mitigation potential at zero or even negative costs. In fact, MAC curves of some mitigation strategies become negative if the mitigation measures lead to increased efficiency in meat and milk production (cf. DeAngelo *et al.* 2006; Weiske *et al.* 2006). Smith *et al.* (2009) provide additional information on MACs of specific solutions that do not include negative values, as shown in table 5.2. Nonetheless, it can be seen that additives and improved soil application of manure are measures with particularly low costs.

Rice Management

As in the case of livestock and manure management, the feasibility and the costs of rice mitigation strategies depend on regional characteristics (Povellato *et al.* 2007). Table 5.3 provides an overview of mitigation potentials related to rice management.

Table 5.3 Rice: projected baseline emissions and economic mitigation potentials at different CO$_2$ prices

Source	Year	Baseline in MtCO$_2$-eq	Value of CO$_2$ in US$/tCO$_2$-eq						
			0	15	30	45	60	100	200
			Economic mitigation potentials in MtCO$_2$-eq						
DeAngelo *et al.* (2006)	2010	185	19						56
(USEPA 2006a)	2020	1,026	114	235	238	259	259		
Smith *et al.* (2009)	2030	n/a						230	

Sources: USEPA (2006a); Smith *et al.* (2009); own calculations.

Table 5.4 Solid waste: projected baseline emissions and economic mitigation potentials at different CO$_2$ prices

Source	Overall sector or specific measure	Year	Baseline in MtCO$_2$-eq	Value of CO$_2$ in US$/tCO$_2$-eq							
				0	15	30	45	50	60	100	200
				Economic mitigation potentials in MtCO$_2$-eq							
IEA (2003)	Overall	2020	1217	300	794	842	940	977	1,000	1,033	1,043
Delhotal *et al.* (2006)	Overall	2020	271								138
USEPA (2006a)	Overall	2020	817	97	332	405	464		717		
IPCC (2007e)	Overall	2020	910	109	373	455	519		801		
IPCC (2007e)	Overall	2030	1,500	300–500				375–1,000		400–1,000	
Monni *et al.* (2006)	Overall	2030	1,500	535					1,256	1,369	
	Anaerobic digestion	2030	n/a	0					94	124	
	Composting	2030	n/a	0					64	102	
	Mechanical biological treatment	2030	n/a	0					0	19	
	LFG recovery – energy	2030	n/a	411					162	65	
	LFG recovery – flaring	2030	n/a	0					0	0	
	Waste incineration with energy recovery	2030	n/a	124					936	1,059	

Sources: USEPA (2006a); Monni *et al.* (2006); IPCC (2007e drawing on Delhotal *et al.* 2006 and Monni *et al.* 2006); own calculations and interpolations. The studies take into account remaining CO$_2$ that results from CH$_4$ oxidation or waste incineration.

In the case of rice management, the largest share of mitigation potentials seems to be in the low-cost range of less than 15 $/tCO$_2$-eq. Potentials hardly increase with higher costs. Again, DeAngelo *et al.* (2006) assume much lower potentials than the other sources mentioned. Yet, since they also assume lower baseline emissions, the relative shares are comparable.

Solid Waste Management

Table 5.4 provides an overview of economic mitigation potentials in this category at different CO$_2$ prices. Since data availability in this category is high, it not only includes overall values, but also economic mitigation potentials for specific technologies.

Table 5.5 Solid waste: breakeven costs and mitigation potentials for selected technologies

Technology	Breakeven cost in US$/tCO$_2$-eq	Emission reduction in 2020 in MtCO$_2$-eq
LFG capture and heat production	−17	0.36
LFG capture and direct gas use (profitable at base price)	1	0.39
LFG capture and direct gas use (profitable above base price)	8	0.39
LFG capture and flaring	25	0.39
Anaerobic digestion (low-tech type)	36	0.16
LFG capture and electricity generation	73	0.39
Composting (average)	254	0.51
Increased oxidation	265	0.24
Mechanical biological treatment	363	0.16

Source: USEPA (2006a).

Table 5.6 Coal mining: projected baseline emissions and economic mitigation potentials at different CO$_2$ prices

Source	Year	Baseline in MtCO$_2$-eq	Value of CO$_2$ in US$/tCO$_2$-eq					
			0	15	30	45	60	200
			Economic mitigation potentials in MtCO$_2$-eq					
IEA (2003)	2020	648	140	418	418	418	418	418
Delhotal *et al.* (2006)	2020	161						129
USEPA (2006a)	2020	450	65	359	359	359	359	

Sources: USEPA (2006a); own calculations and interpolations.

The numbers vary between sources. Delhotal *et al.* (2006) seem to represent an outlier with much lower baseline emissions and lower economic potentials than other sources. However, there are some general findings. Baseline emissions will increase considerably until 2030. Monni *et al.* (2006) show that emission growth will be particularly strong in non-OECD countries. Overall, most of the potentials could be realized at costs of less than $50/tCO$_2$-eq. Several authors find substantial mitigation potentials at negative cost. This is mainly due to an assumed energy use of recovered landfill gas (LFG) or energy recovery from waste incineration. Gallaher *et al.* (2005) find very high relative mitigation potentials at zero cost until 2020 for US and Chinese emissions of 62% and 64%, respectively. As for specific technologies, LFG recovery and energy use has the largest potentials at low carbon prices, while waste incineration with energy recovery has very large potentials at higher carbon prices.

USEPA (2006a) provides additional information on marginal abatement costs in the form of breakeven costs and related mitigation potentials for some specific landfill CH$_4$ abatement measures for 2020. Table 5.5 provides an overview. Heat production and direct gas use have large mitigation potentials at low costs. In case of heat production, there are even negative costs.

Coal Mining

Table 5.6 shows economic mitigations at different carbon prices.

According to USEPA (2006a), the MAC curve is very steep to the right of a carbon price of $15/tCO$_2$-eq. That is, most mitigation measures are in the low-cost area. Spending additional money

Table 5.7 Coal mining: breakeven costs and emission reductions for selected technologies

Technology	Breakeven cost in US$/tCO$_2$-eq	Emission reduction in 2020 in MtCO$_2$-eq
Degasification and pipeline injection	−12	0.55
Catalytic oxidation (US technology)	14	0.94
Degasification and power production ("type C")	20	0.83

Source: USEPA (2006a).

Table 5.8 Natural gas: projected baseline emissions and economic mitigation potentials at different CO$_2$ prices

Source	Year	Baseline in MtCO$_2$-eq	Value of CO$_2$ in US$/tCO$_2$-eq					
			0	15	30	45	60	200
			Economic mitigation potentials in MtCO$_2$-eq					
IEA (2003)	2020	1,540	182	470	585	623	630	637
Delhotal *et al.* (2006)	2020	379						144
USEPA (2006a)	2020	1,696	173	428	564	651	913	

Sources: IEA (2003); Delhotal *et al.* (2006); USEPA (2006a); own calculations and interpolations.

does not result in increased mitigation. Gallaher *et al.* (2005) have similar findings when calculating the regional MAC curves of coal mining. They find that in the USA and China, large shares of overall reduction potentials can be achieved at zero cost. This is due to the energy value of captured coal mine CH$_4$. Delhotal *et al.* (2006) assume lower absolute mitigation potentials than USEPA (2006a). However, since they also assume lower baseline emissions, they find the same relative mitigation potential (80%) at costs of $200/tCO$_2$-eq as USEPA (2006a) for costs of $15–60/tCO$_2$-eq.

Additional information on breakeven prices and related potentials of some selected coal mining-related measures is provided by USEPA (2006a). Table 5.7 includes some selected technologies. While all listed options are relatively low-cost, they potentially lead to large emission reductions by 2020. Pipeline injection of captured coal mine CH$_4$ has negative abatement costs due to the revenues from selling the CH$_4$.

Processing, Transmission, and Distribution of Natural Gas

Table 5.8 provides some estimates on economic potentials at different carbon prices.

Delhotal *et al.* (2006) state that the natural gas sector offers many low-cost or no-regret options. However, compared to USEPA (2006a), they assume a much lower baseline and related lower potential mitigation potentials, even at high costs. On the other end of the spectrum, USEPA (2006a) estimates much larger mitigation potentials, with continuously increasing mitigation potentials at increasing costs. These numbers contrast with the analyses of Gallaher *et al.* (2005), which are slightly less optimistic than USEPA (2006a) in relative terms. For China, Russia and the USA, Gallaher *et al.* (2005) do not provide absolute numbers, but state that the MAC curves are relatively steep. They assume that for the three countries mentioned, most of the mitigation potential that is economic at $50/tCO$_2$-eq is also economic at zero cost.

USEPA (2006a) provide more detailed cost data for specific technologies in this category, as shown in table 5.9. There is a range of options with relatively low costs that lead to sizeable comparable emission reductions.

Lechtenbömer *et al.* (2007) have analyzed the Russian gas transportation system and provide some additional calculations. They find that in the Russian case more than 30% of CH$_4$ emissions (c. 15 MtCO$_2$-eq) could be mitigated at investment

Table 5.9 Natural gas: breakeven costs and emission reductions for selected natural gas mitigation technologies

Technology	Breakeven cost in US$/tCO$_2$-eq	Emission reduction in 2020 in MtCO$_2$-eq
Electronic monitoring at large surface facilities	1	0.33
Replace high-bleed pneumatic devices with low-bleed pneumatic devices	12	0.23
Enhanced inspection and maintenance in distribution	21	0.27
Dry seals on centrifugal compressors	37	0.20
Catalytic converter	77	0.20
Replace high-bleed pneumatic devices with compressed air systems	85	0.27
Gas turbines instead of reciprocating engines	113	0.27

Source: USEPA (2006a).

Table 5.10 Summary of absolute economic mitigation potentials at or below different CO$_2$ prices

Sector	Baseline 2020 in MtCO$_2$-eq	Value of CO$_2$ in US$/tCO$_2$-eq				
		0	15	30	45	60
		Absolute economic mitigation potentials in MtCO$_2$-eq				
Livestock management	2,867	83	126	158	175	192
Rice management	1,062	114	235	238	259	259
Solid waste management	817	97	332	405	464	717
Coal mine CH$_4$	450	65	359	359	359	359
Natural gas	1,696	173	428	564	651	913
Sum	**6,891**	**531**	**1,480**	**1,723**	**1,908**	**2,439**

Source: USEPA (2006a); own calculations.

costs below US$ 10/tCO$_2$-eq. Typical low-cost measures include operational practices like optimized compressor shutdown practices, minimized venting before maintenance, or cost-effective leak inspections.

Summary of Economic Mitigation Potentials

In the last section, we have provided economic CH$_4$ mitigation potentials in specific sectors at different carbon values. In the following, we provide a summary of these potentials over all sectors. For the summary, we focus on USEPA data, since USEPA (2006a) represents both the most detailed and the most consistent analysis of CH$_4$ mitigation costs and potentials. The data for absolute emission reductions at different carbon prices (i.e. different cost levels) for 2020 are summarized in table 5.10.

As before, the table provides the mitigation levels (in MtCO$_2$-eq or in percent) that economically break even at a given carbon price ("economic mitigation potentials"). The CO$_2$ prices can also be interpreted as MACs.

While baseline emissions are highest in the livestock sector, we find the largest absolute mitigation potentials in the categories of solid waste management and natural gas, in particular at high carbon prices. Interestingly, MAC curves for coal mine CH$_4$, rice management, and – to a lesser extent – livestock management are very steep at CO$_2$ prices of 15$/t. That is, spending additional money hardly increases mitigation levels. Figure 5.4 illustrates these findings.

As shown in table 5.11, the largest relative reduction potentials can be found in the categories of solid waste and coal mine CH$_4$, particularly in the

Table 5.11 Summary of relative economic mitigation potentials (%), at or below different CO_2 prices

Sector	Baseline 2020 in $MtCO_2$-eq	Value of CO_2 in US$/$tCO_2$-eq: 0	15	30	45	60
		Relative economic mitigation potentials in %				
Livestock management	2,867	3	4	6	6	7
Rice management	1,062	11	22	22	24	24
Solid waste management	817	12	41	50	57	88
Coal mine CH_4	450	15	80	80	80	80
Natural gas	1,696	10	25	33	38	54
Sum	**6,891**	**100**	**100**	**100**	**100**	**100**

Source: USEPA (2006a); own calculations.

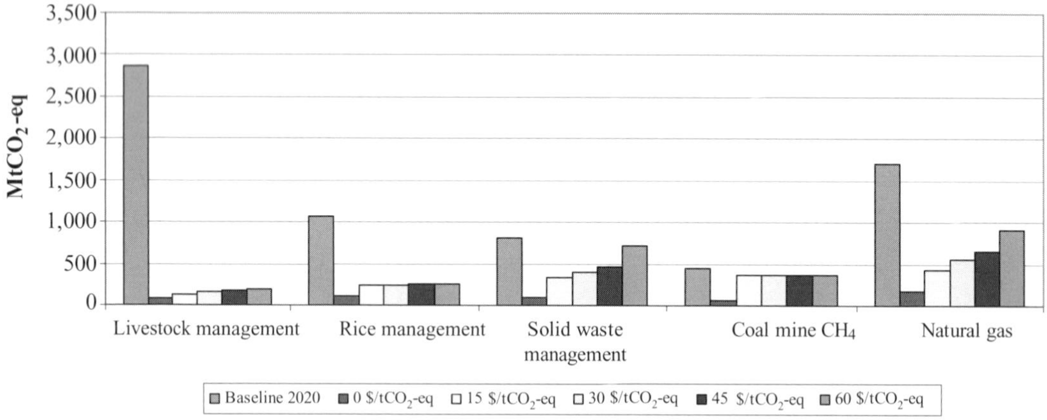

Figure 5.4 *Global CH_4 emissions: baseline and economic mitigation potentials (2020)*

case of high carbon prices. While natural gas also has substantial relative mitigation potentials, the values for livestock and rice management are much lower. Although these categories have high baseline emissions, the applicability of mitigation measures seems to be very restricted.

Results: BCRs

Table 5.12 provides an overview of BCRs for 2020, estimated according to the procedure outlined above. BCRs are shown for various levels of application of selected mitigation options – i.e. up to MACs of 15, 30, 45, and 60 $/$tCO_2$-eq. The table distinguishes between three SCC assumptions, as described above. We use SCC values of 13, 19, and 46 $/$tCO_2$-eq, which represent the median (13), mean (19), and 90-percentile (46), respectively, of Tol's literature survey of peer-reviewed studies (Tol 2008).

As expected, table 5.12 shows that BCRs decrease with increasing mitigation levels (i.e. increasing MACs). That is, BCRs are higher for the "first" mitigated CH_4 units in a sector that have low MACs. In contrast, BCRs for a given mitigation level increase with assumed SCC values, since higher SCC values represent larger benefits of avoided emissions. Accordingly, BCRs are particularly high under the assumption of high SCC.

Table 5.12 shows that BCRs are always greater than 1 if marginal abatement costs are smaller or

Table 5.12 BCRs for different solution categories, mitigation levels, and assumptions on SCC values

	Mitigation up to marginal abatement costs in US$/tCO$_2$-eq											
	15			30			45			60		
	SCC in US$/tCO$_2$-eq											
	13	19	46	13	19	46	13	19	46	13	19	46
Sector	B/C ratios											
Livestock management	2.6	3.8	9.1	1.3	1.9	4.6	1.0	1.4	3.4	0.7	1.1	2.6
Rice management	1.7	2.5	6.0	1.6	2.4	5.8	1.2	1.7	4.2	1.2	1.7	4.2
Solid waste management	1.2	1.8	4.4	0.9	1.4	3.3	0.7	1.1	2.6	0.4	0.6	1.4
Coal mining	1.1	1.6	3.8	1.1	1.6	3.8	1.1	1.6	3.8	1.1	1.6	3.8
Natural gas	1.5	2.2	5.2	0.9	1.4	3.3	0.7	1.1	2.6	0.4	0.6	1.5

Notes: SCC values of 13, 19, and 46$/tCO$_2$-eq represent the median (13), mean (19), and 90-percentile (46) of Tol's literature survey of peer-reviewed studies on SCC estimations (Tol 2008).
Sources: USEPA (2006a); own calculations.

roughly equal to the SCC. That is, the benefits of CH$_4$ mitigation outweigh the costs in these cases, which is an expected result. Nonetheless, BCRs can be significantly larger than 1.0 even in cases where MACs exceed SCC values. This is due to the fact that substantial mitigation potentials can be realized at zero cost in several sectors, which improves average BCRs.

In general, the livestock category has the highest BCRs for low mitigation levels, followed by rice management and natural gas. These categories also have large baseline emissions and substantial absolute economic mitigation potentials. For higher mitigation levels – i.e. up to MACs of 60 $/tCO$_2$-eq – rice management and coal mining have the highest BCRs. This is due to the fact that most of the reduction potentials in these sectors are in the low-cost range, i.e. moving towards higher MACs does not lead to additional mitigation and thus does not change the BCRs. Accordingly, BCRs should be used carefully. We recommend considering table 5.12 only in combination with table 5.10 and/or table 5.11.

The B/C values in table 5.12 refer to overall mitigation in the different sectors. It is complemented by table 5.13, which provides more detailed BCRs for selected mitigation technologies in the solid waste, coal mining, and natural gas sectors, where such data are available.

The technologies are listed in the order of increasing MACs. It is clear that the technologies with low MACs have high BCRs, and that BCRs increase with higher SCC. In general, only a few technologies within a category have BCRs greater than 1.0 for low social costs of carbon.

However, for the most cost-effective technologies "LFG capture and heat production" and "Degasification and pipeline injection," calculating BCRs is inappropriate since these technologies have negative marginal costs according to USEPA (2006a). These technologies should be the first to be implemented from a bottom-up point of view, since they involve only benefits and no costs.

Discussion

Economic Potentials and MAC Curves

In some cases, absolute mitigation potentials were calculated by multiplying relative potentials with projected baselines. This approach might be controversial. While most studies assume comparable relative mitigation potentials, the baselines vary considerably between the studies. The resulting absolute mitigation potentials (and the BCRs calculated from these potentials) are therefore sensitive to assumptions on future baseline emission scenarios.

Table 5.13 BCRs for selected technologies and different assumptions on SCC values

| | | SCC in US$/tCO$_2$-eq | | |
| | | 13 | 19 | 46 |
Sector	Technology	B/C ratios		
Solid waste management	LFG capture and heat production	*n/a – negative marginal costs*		
	LFG capture and direct gas use (profitable at base price)	14.5	21.5	51.5
	LFG capture and direct gas use (profitable above base price)	1.6	2.4	5.7
	LFG capture and flaring	0.5	0.8	1.9
	Anaerobic digestion (low-tech type)	0.4	0.5	1.3
	LFG capture and electricity generation	0.2	0.3	0.6
	Composting (average)	0.1	0.1	0.2
	Increased oxidation	0.0	0.1	0.2
	Mechanical biological treatment	0.0	0.1	0.1
Coal mining	Degasification and pipeline injection	*n/a – negative marginal costs*		
	Catalytic oxidation (US)	0.9	1.3	3.2
	Degasification and power production ("type C")	0.7	1.0	2.3
Natural gas	Electronic monitoring at large surface facilities	17.2	25.5	61.0
	Replace high-bleed pneumatic devices with low-bleed pneumatic devices	1.1	1.6	3.8
	Enhanced inspection and maintenance in distribution	0.6	0.9	2.2
	Dry seals on centrifugal compressors	0.4	0.5	1.3
	Catalytic converter	0.2	0.3	0.6
	Replace high-bleed pneumatic devices with compressed air systems	0.2	0.2	0.5
	Gas turbines instead of reciprocating engines	0.1	0.2	0.4

Sources: USEPA (2006a); own calculations.

Compared to other studies, data on abatement costs and economic mitigation potentials provided by USEPA (2006a) appear to be somewhat optimistic. However, to our knowledge USEPA (2006a) provides the most coherent and thorough analysis on global MACs of different CH$_4$ mitigation strategies. This data is calculated from an industry perspective with a 10% discount rate and a 40% tax rate. Lower discount rates would result in even higher mitigation potentials. In addition, our approach of calculating costs stepwise and treating negative abatement costs as zero costs systematically overestimates costs. Lastly, if we assume positive global carbon prices under future international climate agreements, mitigation costs for a given amount of CH$_4$ would be lower than calculated above. Considering these facts, our cost calculations (and the resulting BCRs in table 5.12) can be considered as conservative.

A weakness with the MAC curves provided by USEPA (2006a) is that they mainly represent technical or engineering costs and not economic costs. For example, the opportunity costs of some solutions might not be included, which may result in an underestimation of costs. However, combined with the cost-increasing factors discussed in the last section, we assume that our overall cost estimates are reasonable.

Agricultural Solutions

Livestock and rice management seem to be the most controversial sectors. The data provided in this chapter focus on technical feasibility and technical costs. Implementing mitigation strategies in these sectors might be infeasible due to geographic or social barriers (see also the next section). In contrast to landfills, waste management and natural gas systems, CH$_4$ sources in agricultural sectors can be very small and geographically widely dispersed. Accordingly, it will be challenging to regulate, monitor, and enforce CH_4 mitigation measures in these sectors.

While most of the livestock-related measures discussed in this chapter increase production efficiency, it has to be assured that productivity-related emission reductions are not counter-balanced by increasing overall production of meat and milk. Interactions with other GHGs also should not be neglected. While several possible interactions have been assessed (cf. Smith *et al.* 2009), more research is necessary on agriculture-related GHG interdependencies. For example, when applying large-scale CH_4 mitigation measures, it has to be ensured that there are no increases in emissions of other GHGs like N_2O. Another problem may be the costs of agriculture-related CH_4 mitigation measures, which may be prohibitive for farmers in developing countries. Solving this problem is a question of finding appropriate financing mechanisms like carbon trading. Policy makers should not put too much emphasis on agricultural CH_4 mitigation options for the reasons mentioned above, but they should consider them as a promising part of a broader CH_4 mitigation strategy.

Negative MACs and Implementation Barriers

By providing a summary of economic mitigation potentials in different sectors (shown in table 5.10), we do not suggest that all of these potentials will automatically be realized under the related carbon prices. There are several implementation barriers. This is most obvious in the case of negative MACs. The existence of such negative MACs is a well-known fact. For example, negative MACs are a frequent phenomenon in energy-related mitigation categories (compare IPCC 2007a and McKinsey 2007). Some mitigation potentials are not realized although it would be profitable – but why should there be "dollar bills left lying on the sidewalk"?

One possible answer to this question is that the mentioned bottom-up studies do not include all economic costs – for example, opportunity costs. Another answer is that there are several social and institutional implementation barriers – for example, a lack of knowledge and awareness. Moreover, the availability of crucial technologies might be lim-

ited, for example in the case of geographically dispersed, extensive livestock, but also in the case of coal mine CH_4 abatement in China (cf. Yang 2009). Financing might also be a problem if solutions are capital-intensive – e.g. in the waste management, coal mine CH_4, or natural gas categories. Finally, there might be institutional barriers in countries with weak institutional frameworks.

These barriers also have economic costs. Such costs are not included in B/C calculations because of high uncertainties and a lack of data. However, there is some evidence that negative MACs are not persistent in the long run. In the case of the Russian gas transportation system, repeated leakage measurements have indicated that CH_4 leakage rates tend to decrease over time, since related investments are profitable (cf. Lechtenbömer *et al.* 2007).

Overall, the implementation of CH_4 mitigation measures might be easier in such sectors where emission sources are geographically concentrated and a smaller number of owners and operators are involved. Solid waste management and coal mining CH_4 might be particularly promising in this respect.

BCRs

Our B/C estimations are sensitive to the calculation of costs and benefits as well as to the projection of emission baselines. Some controversial issues regarding cost calculations have been discussed previously, such as underestimations of costs that result in exaggerated BCRs and vice versa. As mentioned above, our approach of stepwise mitigation cost calculation and the fact that MAC curves are usually convex leads to a systematic overestimation of costs. It should also be noted that due to the procedure described above, the results implicitly assume carbon prices of 0 $/tCO_2$-eq. If future international agreements would lead to positive global carbon prices, the costs calculated in this chapter would decrease. For that reason, resulting B/C values might be conservative.

As for our calculation of benefits, the challenges of drawing on SCC estimates have already been discussed. Other concerns are the sensitivity to different discount rates and the timing of costs and benefits. Our approach is not very detailed regarding both of these issues. While our mitigation cost estimates implicitly assume relatively high discount rates – 10% in data provided by USEPA (2006a) – our discount rate in the benefit estimation is somewhat vague, since we use the median, mean, and 90-percentile values of SCC from a literature survey that includes many different studies with different assumptions on discount rates. According to Hope (2005), benefits are highly sensitive to the discount rate. For immediate CH_4 cutbacks, Hope (2005) estimates benefits of about 5$/t$CO_2$-eq using a pure time preference rate of 3%, but 18$/t$CO_2$-eq with a rate of only 1% (in 1990 US$ and using the IPCC conversion factor between CH_4 and CO_2 of 21). Hope (2005) also finds a regional disparity of CH_4 mitigation benefits. In his model, most benefits materialize outside the USA and the EU.

In general, mitigation options with BCRs below 1 should not be implemented. Due to the problems discussed above, we recommend dealing very cautiously with BCRs, particularly if they are calculated over long time horizons. Alternatively, mitigation policy decisions may directly be based on MACs. Options with low MACs should be preferred to such with high MACs in order to achieve cost-effectiveness. From an economic perspective each abatement option should be implemented up to such a level that marginal mitigation costs are equal over all mitigation solutions. It is important to note that the BCRs in table 5.12 include relatively large mitigation potentials that can be realized at zero costs in most categories. Large low-cost mitigation potentials can lead to BCRs larger than 1 even in such cases where MACs significantly exceed marginal benefits of mitigation – i.e. the avoided SCC.

Our estimations on BCRs certainly do not represent a comprehensive social benefit-cost analysis (CBA), but they may provide valuable indications of the relative cost-effectiveness of specific measures. More thorough research is necessary if the global community wanted to spend very large amounts of money on CH_4 mitigation. We recommend a more detailed and dedicated analysis of the benefits and costs of the solutions outlined in this chapter with appropriate IAMs in the near future.

Recommendations

If the international community wanted to spend a large amount of money – say, $250 billion – on CH_4 mitigation, how should it be done?

First, we recommend tackling "low-hanging fruit." Mitigation potentials at zero or even negative costs should be realized by removing institutional and social barriers. This includes educational efforts, making information and technology available in the right places, and developing appropriate legal frameworks. CH_4 mitigation needs to be taken seriously in the national and international climate policy debate.

Assuming restricted resources for mitigation monitoring and enforcement, it might be beneficial to focus on CH_4 emissions that come from relatively large and well-identified sources, for example landfills, coal mines, and natural gas systems. It might not be advisable to rely only on agricultural solutions given the challenges of monitoring and enforcing. Although the potentials for CH_4 mitigation in livestock, manure, and rice management are large, there are high uncertainties regarding implementation barriers and the short-term feasibility of some options. In addition, the effectiveness of several mitigation options in livestock has not yet been demonstrated on a large scale. More research is required on unintended side effects – for example, releases of other GHGs like N_2O. Nonetheless, it is clear that policy makers should not "put all their eggs in one basket" in order to diversify risks. We recommend spreading CH_4 mitigation efforts over several sectors instead of focusing on a single sector.

We recommend a global solution portfolio that covers all five sectors discussed in this report (Portfolio 1, tables 5.14 and 5.15). We also develop an alternative portfolio that leaves out the agricultural sectors for the reasons described earlier (Portfolio 2, tables 5.16 and 5.17). In both portfolios, MACs are equalized over all included sectors in order to ensure economic efficiency. We differentiate between two previously used extreme SCC values of 13 and 46 $/t$CO_2$-eq, which represent, respectively, the median and the 90-percentile SCC value in the literature survey of Tol (2008). In order to achieve economic efficiency, MACs should

Table 5.14 Portfolio 1: total abatement level, costs, and BCRs for SCC of 13 $/tCO$_2$-eq

| Sector | Total emission abatement | | Total abatement costs | | BCRs |
	in MtCO$_2$-eq	sector share (%)	in million $	sector share (%)	
Livestock management	126	9	645	5	2.6
Rice management	235	16	1,816	13	1.7
Solid waste management	332	22	3,536	25	1.2
Coal mine CH$_4$	359	24	4,403	31	1.1
Natural gas	428	29	3,831	27	1.5
Total	**1,480**	**100**	**14,231**	**100**	**1.4**

Note: This table shows global solution Portfolio 1 for 2020 with an SCC assumption of 13$/tCO$_2$-eq and related efficient abatement levels at MACs of 15$/tCO$_2$-eq over all sectors.
Sources: USEPA (2006a); own calculations.

Table 5.15 Portfolio 1: total abatement level, costs, and BCRs for SCC of 46 $/tCO$_2$-eq

| Sector | Total emission abatement | | Total abatement costs | | BCRs |
	in MtCO$_2$-eq	sector share (%)	in million $	sector share (%)	
Livestock management	175	9	2,365	8	3.4
Rice management	259	14	2,867	10	4.2
Solid waste management	464	24	8,381	28	2.6
Coal mine CH$_4$	359	19	4,403	15	3.8
Natural gas	651	34	11,833	40	2.6
Total	**1,908**	**100**	**29,850**	**100**	**3.0**

Note: This table shows global solution Portfolio 1 for 2020 with a SCC assumption of 46$/tCO$_2$-eq and related efficient abatement levels at MACs of 45$/tCO$_2$-eq over all sectors.
Sources: USEPA (2006a); own calculations.

Table 5.16 Portfolio 2: total abatement level, costs, and BCRs for SCC of 13 $/tCO$_2$-eq

| Sector | Total emission abatement | | Total abatement costs | | BCRs |
	in MtCO$_2$-eq	sector share (%)	in million $	sector share (%)	
Solid waste management	405	31	5,727	32	0.9
Coal mine CH$_4$	359	27	4,403	24	1.1
Natural gas	564	42	7,896	44	0.9
Total	**1,328**	**100**	**18,026**	**100**	**1.0**

Note: This table shows global solution Portfolio 2 for 2020 with a SCC assumption of 13$/tCO$_2$-eq and abatement levels at MACs of 30$/tCO$_2$-eq over all sectors, representing a more precautionary approach than Portfolio 1.
Sources: USEPA (2006a); own calculations.

Table 5.17 Portfolio 2: total abatement level, costs, and BCRs for SCC of 46 $/tCO$_2$-eq

Sector	Total emission abatement		Total abatement costs		BCRs
	in MtCO$_2$-eq	sector share (%)	in million $	sector share (%)	
Solid waste management	717	36	23,537	42	1.4
Coal mine CH$_4$	359	18	4,403	8	3.8
Natural gas	913	46	27,513	50	1.5
Total	**1,988**	**100**	**55,452**	**100**	**1.7**

Note: This table shows global solution Portfolio 2 for 2020 with a SCC assumption of 46$/tCO$_2$-eq and abatement levels at MACs of 60$/tCO$_2$-eq over all sectors, representing a more precautionary approach than Portfolio 1.
Sources: USEPA (2006a); own calculations.

equal the SCC. Thus, we choose efficient mitigation levels of 15 and 45 $/tCO$_2$-eq for Portfolio 1.[9] As for Portfolio 2, we give up this efficiency condition and mitigate up to such levels that the MACs exceed the assumed SCC values by about 15 $/tCO$_2$-eq. In doing so, the absence of agricultural mitigation options is roughly counterbalanced in terms of total abatement. This procedure also represents a security margin and thus a more precautionary approach towards SCC estimations and climate damages.

We recommend implementing the cost-effective Portfolio 1. However, policy makers may come to the conclusion that implementation barriers in the agricultural sectors are too high, or additional research may show that agricultural CH$_4$ mitigation is less feasible or more expensive than assumed today. In these cases, the more precautionary and less cost-effective Portfolio 2 could be implemented in order to achieve mitigation levels comparable to Portfolio 1.

Portfolio 1

Portfolio 1 includes all five sectors mentioned in this chapter. We choose mitigation levels such that MACs are equal over all categories and also roughly equal to the SCC emissions. We differentiate between two cases: a SCC assumption of 13 $/tCO$_2$-eq (and corresponding mitigation levels up to MACs of 15 $/tCO$_2$-eq) and a SCC assumption of 46 $/tCO$_2$-eq (and corresponding efficient

[9] These MACs do not exactly match the SCC values, but are the closest data points available.

MACs of 45 $/tCO$_2$-eq). Total abatement levels, total costs, and BCRs for 2020 for both cases are summarized in table 5.14 and table 5.15.

Drawing on our fairly conservative cost estimates as described above (stepwise calculation, no negative costs), it would be efficient to mitigate nearly 1.5 GtCO$_2$-eq at overall costs of around $14.2 billion at a SCC value of 13 $/tCO$_2$-eq. Assuming a SCC value of 46 $/tCO$_2$-eq, around 1.9 GtCO$_2$-eq could be efficiently mitigated at costs of about $29.9 billion. Most money should be spent in the sectors of solid waste management, coal mining, and natural gas in both cases. Overall BCRs are larger than 1.0 for all included sectors and both SCC assumptions. For the high SCC value, BCRs are much larger than in the case of the low SCC value, since benefits related to low-cost mitigation potentials increase with SCC.

Portfolio 2

Portfolio 2 disregards mitigation solutions in the livestock/manure and rice management sectors. It only includes waste management, coal mine CH$_4$, and natural gas. We once again distinguish between SCC values of 13 and 46 $/tCO$_2$-eq, but increase mitigation levels up to MACs of 30 and 60 $/tCO$_2$-eq, respectively. In doing so, we roughly compensate for the missing agricultural mitigation. As mentioned earlier, this procedure represents a more precautionary approach. The results for 2020 are summarized in table 5.16 and table 5.17.

We find comparable total abatement levels for Portfolios 1 and 2, but widely differing costs.

Portfolio 2 is less cost-effective due to the exclusion of low-cost agricultural solutions and due to the precautionary approach of abating up to MACs that exceed SCC by about 15 $/tCO$_2$-eq. This is also evident in the lower BCRs of Portfolio 2 compared to Portfolio 1.

Context

We want to put the total emission abatement levels that can be achieved with our portfolios into context. Precisely quantifying their effect on global temperatures is challenging due to timing and the importance of other factors like the development of the global population, the economy, and other GHG emission trends. Yet, we can put the numbers into the context of current emissions and future IPCC emission scenarios.

Global anthropogenic GHG emissions amounted to 44.7 GtCO$_2$-eq in 2000 and 49 GtCO$_2$-eq in 2004, with increasing trends (IPCC 2007a). Our portfolios lead to mitigation levels of around 1.3–2 GtCO$_2$-eq by 2020, which corresponds to about 3–4% of total global anthropogenic GHG emissions in 2000. These numbers indicate that CH$_4$ offers substantial cost-effective emission reduction opportunities.

IPCC has developed several emission scenarios. The most optimistic scenario (B1) assumes a convergent world with rapid economic change towards a service and information economy, large-scale adoption of efficient and clean technologies, and a global solution approach. In this scenario, global temperature increases less than 2°C by 2090–9 relative to 1980–99. In order to make B1 a reality, global emissions have to grow less than 10 GtCO$_2$-eq by 2030 compared to 2000 levels, and must peak around 2040. A portfolio of short-term CH$_4$ emission abatement measures, as outlined above, could play an important role in achieving such ambitious targets. While our portfolio alone will certainly not suffice to realize B1, it should be a cost-effective part of a larger mitigation strategy.

If the global community wanted to spend an even larger amount of money – say, $250 billion – on CH$_4$ mitigation, much larger CH$_4$ reductions than in Portfolio 1 or 2 could be realized. With such an amount of money it should be possible to realize virtually all CH$_4$ reduction potentials in the five sectors identified in this study, including the ones with very high MACs. However, it is clear that this approach would be inefficient. If the global community really wanted to spend such a large amount of money, we would recommend including other CH$_4$ mitigation options that have not been analyzed in this chapter – for example, in the wastewater sector. In order to ensure economic efficiency, we also recommend spreading such large amounts of money over a portfolio of different GHGs.

Summary and Conclusions

Several analyses have shown that including non-carbon GHG mitigation measures can decrease mitigation costs substantially compared to focusing exclusively on CO$_2$ (e.g. Kemfert *et al.* 2006 and other contributions of EMF21[10]). CH$_4$ emission abatement is a particularly promising supplement to CO$_2$ mitigation due to large global low-cost abatement potentials. CH$_4$ has the largest overall mitigation potentials among all non-CO$_2$ GHGs (USEPA 2006a). In addition, due to the short atmospheric lifetime of CH$_4$, the beneficial effects of mitigation will be more instantaneous than, for example, in the case of CO$_2$ mitigation.

In contrast to CO$_2$, some CH$_4$ emission sources are small, geographically dispersed, and not related to the energy sector. For example, there are several such CH$_4$ sources in the agricultural sector. CH$_4$ mitigation may therefore require different approaches regarding regulation, monitoring, and enforcement than CO$_2$. Another difference between CO$_2$ and CH$_4$ is that CH$_4$ is an energy carrier that has an energetic value, while CO$_2$ is mainly a waste product without a market value.

Several specific CH$_4$ mitigation solutions in different sectors have been identified in this report. The most important mitigation strategies regarding mitigation potentials and cost-effectiveness can be found in the sectors of livestock and manure management, rice management, solid waste management, coal mining, and natural gas.

[10] See de la Chesnaye and Weyant (2006).

Absolute economic CH_4 mitigation potentials in livestock management are limited relative to the large emissions of this sector, but most of the overall livestock-related potential reductions can be found in the low-cost range of up to c. 15 US$/$tCO_2$-eq. That is, spending more money does not provide much additional benefit. The same is true for rice management. However, geographical, social, and institutional barriers may impede the implementation of agriculture-related mitigation potentials. Solid waste management has higher absolute mitigation potentials than the agricultural sectors. MAC curves are flatter in this case, which means that more expensive measures (up to about 60 $/$tCO_2$-eq) still lead to substantial emission reductions. For coal mining CH_4 mitigation, absolute economic potentials are generally lower than for landfills. By far the largest part of mitigation measures related to coal mining is cheaper than 15 $/$tCO_2$-eq. Natural gas processing, transmission, and distribution are characterized by relatively high CH_4 emissions and large economic mitigation potentials over a broad cost range.

Total economic mitigation potentials identified in this report are subject to discussion. While USEPA (2006a) provides the most comprehensive and coherent data on economic potentials, the values are rather optimistic relative to other studies. The conversion of relative to absolute mitigation potentials is also highly sensitive to the baseline projection. In addition, there may be institutional barriers that prevent economic mitigation potentials from being realized.

Our rough estimation of costs and benefits of CH_4 mitigation in the categories livestock, rice, solid waste, coal mining, and natural gas shows that BCRs decrease with increasing mitigation levels since they involve higher marginal abatement costs. In contrast, BCRs increase with the assumed SCC since higher SCC values correspond to larger benefits of avoided emissions. BCRs can be significantly larger than 1 even in cases where MACs exceed SCC. This is due to the fact that substantial mitigation potentials can be realized at low or even zero cost in several sectors. Such low-cost potentials improve average BCRs.

We want to stress that our B/C values represent only rough estimates on the relative cost-effectiveness of different measures. We recommend that mitigation policies rather focus on MACs. From an economic point of view, the marginal cost of mitigating 1 ton of CO_2-eq should be equal over all different strategies and GHGs. Moreover, MACs should be equal to the SCC emissions. However, since estimations on SCC values per ton of CO_2-eq are challenging and highly uncertain, a more precautionary approach might be advisable where marginal abatement costs exceed SCC assumptions.

In this chapter, we have developed two solution portfolios for CH_4 mitigation. Portfolio 1 includes all the sectors discussed in this report. With two different assumptions on SCC, Portfolio 1 leads to economically efficient global CH_4 mitigation levels of 1.5 or 1.9 $GtCO_2$-eq by 2020 at costs of around $14 billion or $30 billion, and with overall BCRs of 1.4 and 3.0, respectively. Portfolio 2 not only disregards agricultural mitigation strategies, but also represents a more precautionary – and economically less efficient – approach by mitigating up to MACs that exceed assumed SCC by around 15 $/$tCO_2$-eq. Portfolio 2 leads to mitigation levels of 1.3 and 2.0 $GtCO_2$-eq by 2020, which are comparable to Portfolio 1. Yet, costs are much higher at around $18 billion or $55 billion, respectively, due to the inclusion of less cost-effective measures. BCRs are also lower, at 1.0 and 1.7, respectively. Comparing Portfolio 1 to Portfolio 2 provides a good illustration of the economic inefficiencies resulting from the exclusion of low-cost abatement options.

If the global community wanted to spend a large amount of money on mitigating GHG emissions, it should definitely include cost-effective CH_4 mitigation options, as described in this chapter. From a social perspective, there should be priority for such CH_4 mitigation solutions that involve large co-benefits – for example, increasing agricultural production, or health and security benefits, in coal mining and waste management. We recommend that policy makers focus on information and education of all the actors involved. CH_4 should urgently be included in market-based instruments, like taxes or emissions trading schemes, the utilization of its energy value should be maximized. There may be also a role for administrative rules and regulatory policies.

We want to conclude with some additional remarks. First, it should be noted that the comparison of CH_4 and to CO_2-eq remains challenging due to different time horizons. Many calculations in the literature are sensitive to this issue. Second, in order to fully assess the costs, benefits, and co-benefits of CH_4 mitigation strategies, more integrated modeling approaches should be applied. Next, while many CH_4 mitigation options are relatively low-cost, most of them do require positive carbon prices in order to break even economically. Accordingly, global carbon regulation, preferably in the form of carbon markets, is necessary for promoting these mitigation options. Institutional barriers impeding the implementation of some CH_4 mitigation options and the full realization of their technical potentials should be addressed by policy makers. In addition, new potential CH_4 emission sources should be avoided, as probably in the case of future undersea CH_4 clathrate mining. To conclude, several options mentioned above have long lead times – for example, coal mine degasification or waste management strategies. Thus, early action and clear policy signals are urgently required.

Bibliography

Aaheim, A. *et al.*, 2006: Costs savings of a flexible multi-gas climate policy, *The Energy Journal*, Multi-Greenhouse Gas Mitigation and Climate Policy, Special Issue **3**, 485–502

Bousquet, P. *et al.*, 2006: Contribution of anthropogenic and natural sources to atmospheric methane variability, *Nature* **443**, 439–43

DeAngelo, B.J. *et al.*, 2006: Methane and nitrous oxide mitigation in agriculture, *The Energy Journal*, Multi-Greenhouse Gas Mitigation and Climate Policy, Special Issue **3**, 89–108

de la Chesnaye, F.C. *et al.*, 2001: Cost-effective reductions of non-CO_2 greenhouse gases, *Energy Policy* **29**, 1325–31

de la Chesnaye, F.C. and J.P. Weyant, 2006: Multi-greenhouse gas mitigation and climate policy, *The Energy Journal*, Special Issue **3**, 1–520

Delhotal, K.C. *et al.*, 2006: Mitigation of methane and nitrous oxide emissions from waste, energy and industry, *The Energy Journal*, Special Issue **3**, 45–62

Gallaher, M. *et al.*, 2005: Region-specific marginal abatement costs for methane from coal, natural gas, and landfills through 2030, in E.S. Rubin, D.W. Keith, and C.F. Gilboy (eds.), *Greenhouse Gas Control Technologies*, vol. I, Oxford University Press, Oxford

Golub, A., T. Hertel, H.L. Lee *et al.*, 2009: The opportunity cost of land use and the global potential for greenhouse gas mitigation in agriculture and forestry, *Resource and Energy Economics* **31**(4), 299–319

Hope, C. 2005: The climate change benefits of reducing methane emissions, *Climatic Change* **86**, 21–39

European Commission, 2009: Joint Research Centre (JRC)/Netherlands Environmental Assessment Agency (PBL), Emission Database for Global Atmospheric Research (EDGAR), release version 4.0. http://edgar.jrc.ec.europa.eu

IEA, 2003: Building the cost curves for the industrial sources on non-CO_2-greenhouse gases, Report **PH4/25**, International Energy Agency Greenhouse Gas R&D Programme, October

2008: *CO_2 Emissions from Fuel Combustion* (2008 edn.), International Energy Agency, Paris

IPPC, 1995: Climate change 1995. The science of climate change, *Contribution of Working Group I to the Second Assessment Report of the Intergovernmental Panel on Climate Change*, J.T. Houghton, L.G. Meira Filho, B.A. Callander, N. Harris, A. Kattenberg, and K. Maskell (eds.), Cambridge University Press, Cambridge

2001: *Climate Change 2001: The Scientific Basis. Contribution of Working Group I to the Third Assessment Report of the Intergovernmental Panel on Climate Change*, Houghton, J.T., Y. Ding, D.J. Griggs, M. Noguer, P.J. van der Linden, X. Dai, K. Maskell, and C.A. Johnson (eds.), Cambridge University Press, Cambridge

2007a: *Climate Change 2007: Synthesis Report – Contribution of Working Groups I, II and III to the Fourth Assessment Report of the Intergovernmental Panel on Climate Change* (Core Writing Team, R.K. Pachauri and A. Reisinger (eds.), IPCC, Geneva

2007b: Technical summary, in *Climate Change 2007: The Physical Science Basis –*

Contribution of Working Group I to the Fourth Assessment Report of the Intergovernmental Panel on Climate Change, Solomon, S., D. Qin, M. Manning, Z. Chen, M. Marquis, K.B. Averyt, M. Tignor, and H.L. Miller (eds.), Cambridge University Press, Cambridge
2007c: Summary for policymakers, in *Climate Change 2007: Mitigation – Contribution of Working Group III to the Fourth Assessment Report of the Intergovernmental Panel on Climate Change*, B. Metz, O.R. Davidson, P.R. Bosch, R. Dave, and L.A. Meyer (eds), Cambridge University Press, Cambridge
2007d: Agriculture, in *Climate Change 2007: Mitigation – Contribution of Working Group III to the Fourth Assessment Report of the Intergovernmental Panel on Climate Change*, B. Metz, O.R. Davidson, P.R. Bosch, R. Dave, and L.A. Meyer (eds), Cambridge University Press, Cambridge
2007e: Waste management, in *Climate Change 2007: Mitigation – Contribution of Working Group III to the Fourth Assessment Report of the Intergovernmental Panel on Climate Change*, B. Metz, O.R. Davidson, P.R. Bosch, R. Dave, L.A. Meyer (eds), Cambridge University Press, Cambridge
Jakeman, G. and B.S. Fisher, 2006: Benefits of multi-gas mitigation: an application of the global trade and environment model (GTEM), *The Energy Journal*, Multi-Greenhouse Gas Mitigation and Climate Policy, Special Issue **3**, 323–42
Johnson, J.M.-F. *et al.*, 2007: Agricultural opportunities to mitigate greenhouse gas emissions, *Environmental Pollution* **150**, 107–24
Kemfert, C. *et al.*, 2006: Economic impact assessment of climate change – a multi-gas investigation with WIAGEM-GTAPEL-ICM, *The Energy Journal*, Multi-Greenhouse Gas Mitigation and Climate Policy, Special Issue **3**, 441–60
Keppler, F. *et al.*, 2006: Methane emissions from terrestrial plants under aerobic conditions, *Nature* **439**, 187–91
Lechtenbömer, S. *et al.*, 2007: Tapping the leakages: methane losses, mitigation options and policy issues for Russian long distance gas transmission pipelines, *International Journal of Greenhouse Gas Control* **1**, 387–95

Lelieveld, J. *et al.*, 2005: Low methane leakage from gas pipelines, *Nature*, **434**(7035), 841–2
Lucas, P.L. *et al.*, 2007: Long-term reduction potential of non-CO_2 greenhouse gases, *Environmental Science & Policy* **10**, 85–103
McKinsey & Co., 2007: Reducing US greenhouse gas emissions: how much at what cost? US Greenhouse Gas Abatement Mapping Initiative, Executive Report, Jon Creyts *et al.*, December, www.mckinsey.com/clientservice/ccsi/pdf/US_ghg_final_report.pdf
Methane to Markets, 2008: Global methane emissions and mitigation opportunities. Fact Sheet, www.methanetomarkets.org
Milich, L. 1999: The role of methane in global warming: where might mitigation strategies be focused?, *Global Environmental Change* **9**, 179–201
Monni, S., R.P. Patti, A. Letilä, I. Savolainen, and S. Syri, 2006: Global climate change mitigation scenarios for solid waste management, VTT Publications **603**, VTT Technical Research Center of Finland, www.vtt.fi/inf/pdf/publications2006/P603.pdf, accessed September 1, 2009
Nisbet, R.E.R. *et al.*, 2009: Emission of methane from plants, *Proceedings of the Royal Society B* **276**(1660), 1347–54
Nordhaus, W. 2007: *The Challenge of Global Warming: Economic Models and Environmental Policy*, MIT Press, Cambridge, MA
OECD, 2008: *Costs of Inaction on Key Environmental Challenges*, OECD, Paris
Povellato, A. *et al.*, 2007: Cost-effectiveness of greenhouse gases mitigation measures in the European agro-forestry sector: a literature survey, *Environmental Science & Policy* **10**, 474–90
Smith, P. *et al.*, 2009: Greenhouse gas mitigation in agriculture, *Philosophical Transactions of the Royal Society B* **363**, 789–813
Stern, N. 2007, *The Economics of Climate Change: The Stern Review*, Cambridge University Press, Cambridge
Tol, R.S.J. 2002a: Estimates of the damage costs of climate change, part 1: benchmark estimates, *Environmental and Resource Economics* **21**, 47–73
2002b: Estimates of the damage costs of climate change, part 2: dynamic estimates, *Environmental and Resource Economics* **21**, 135–60

2008: The social cost of carbon: trends, outliers and catastrophes, *Economics* **2**, 2008–25

USEPA, 2006a: *Global Mitigation of Non-CO₂ Greenhouse Gases*, US Environmental Protection Agency, Office of Atmospheric Programs **EPA 430-R-06-005**, Washington, DC

2006b: *Global Anthropogenic Non-CO₂ Greenhouse Gas Emissions: 1990–2020*, US Environmental Protection Agency, Office of Atmospheric Programs, Climate Change Division, Washington, DC

van Vuuren D. *et al.*, 2006: Long-term multi-gas scenarios to stabilise radiative forcing – exploring costs and benefits within an integrated assessment framework, *The Energy Journal*, Multi-Greenhouse Gas Mitigation and Climate Policy, Special Issue **3**, 201–34

Wassmann, R. and H. Pathak, 2007: Introducing greenhouse gas mitigation as a development objective in rice-based agriculture: II. Cost-benefit assessment for different technologies, regions and scales, *Agricultural Systems* **94**, 826–40

Weiske, A. *et al.*, 2006: Mitigation of greenhouse gas emissions in European conventional and organic dairy farming, *Agriculture, Ecosystems and Environment* **112**, 221–32

Weitzman, M. 2007: A review of *The Stern Review of the Economics of Climate Change*, *Journal of Economic Literature* **45**(3), 703–24

von Witzke, H. and S. Noleppa, 2007: *Methan und Lachgas: die vergessenen Klimagase*, Berlin, WWF

Yang, M., 2009: Climate change and energy policies, coal and coalmine methane in China, *Energy Policy* **37**, 2858–69

Yoon, S. *et al.*, 2009: Feasibility of atmospheric methane removal using methanotrophic biotrickling filters, *Applied Microbiology and Biotechnology* **83**, 949–56

Methane Mitigation

Alternative Perspective

DAVID ANTHOFF*

Introduction

Claudia Kemfert and Wolf-Peter Schill's chapter 5 provides a thorough overview of the details of methane (CH_4) emission mitigation options and relevant recent work on estimating costs of such emission mitigation. Their estimates of the benefits of CH_4 mitigation are less convincing: they use global warming potential (GWP) conversion rates to calculate equivalent emission reductions in terms of CO_2 and then use published estimates of the social cost of carbon (SCC) to arrive at monetized benefit estimates of CH_4 mitigation. Using GWP conversion rates is widely believed to be flawed in economic assessments of climate change (cf. Manne and Richels 2001) and the benefits estimated by Kemfert and Schill suffer from this weakness as well. Finally, the two solutions ("Portfolios") proposed are difficult to compare with any of the other solutions in the Copenhagen Consensus project, for two main reasons: first, the cost estimates are limited to just one year (2020) and do not seem to be discounted into net present value (NPV) terms. Second, the benefit-cost ratios (BCRs) calculated in the chapter are an inappropriate measure to rank solutions because costs (or, alternatively, benefits) are not held constant across solutions. This also prevents comparison of BCRs with solutions from other solution categories. Net benefit estimates (which would be the appropriate measure to rank solutions when neither costs nor benefits are fixed across solutions) are not provided in the chapter.

I provide alternative estimates of benefits and costs of CH_4 emission reductions in this Perspective paper. I use the integrated assessment model

FUND to calculate both benefits and costs of three different mitigation solutions for CH_4 emissions and investigate their relationship with CO_2 mitigation options.

In estimating the benefits of CH_4 emissions I do not rely on GWP conversion factors but rather employ a reduced form model of the CH_4 cycle and calculate changes in radiative forcing due to perturbations of the CH_4 stock in the atmosphere. This approach properly takes into account the very different atmospheric lifetime of CH_4 compared to other greenhouse gases (GHGs) and can, for example, account properly for the fact that CH_4 emission reductions that are limited to, say, the next ten years (as suggested by the Copenhagen Consensus guidelines) will have no effect on the climate in the long run.

Further, I follow the discounting guidelines of the Copenhagen Consensus project and discount all benefits and costs consistently at 6% and 3% per year. While this approach does not reflect best practice as found in the literature, in my opinion it does allow for a meaningful comparison of benefits and costs with estimates from other solution categories.

Finally, I follow the spending suggestion of the Copenhagen Consensus project of $250 billion per year for ten years in one of my solutions. The BCR of that solution can be compared in a meaningful way with BCRs from other solution categories where the same amount of money is spent. My other solutions do not follow this spending schedule: I solve for optimal mitigation paths without constraints in which decade money has to be spent and contrast this with solutions that conform to the Copenhagen Consensus spending schedule. I calculate net benefits for all solutions, making comparison across solutions feasible.

* I thank Richard Tol for his helpful comments on a draft of this Perspective paper. All remaining errors are my own.

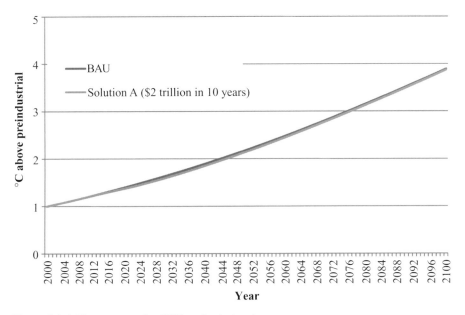

Figure 5.1.1 *Temperature for BAU and solution A*

Solutions

All benefits and costs are calculated with version 3.5 of the integrated assessment model FUND. (A full documentation of FUND can be found at www.fund-model.org; a brief description is contained in the appendix on p. 204. The CH_4 mitigation cost functions are described in Tol (2006). All dollar figures are in 1995 USD price levels. Numbers in tables are in billion USD.

In solution A, I try to roughly follow the spending schedule suggested by the Copenhagen Consensus project. The schedule is outlined as spending $250 billion per year for a time period of ten years. The NPV of that expenditure is roughly $2 trillion with a discount rate of 6%. I then search for a tax rate on CH_4 emissions in the FUND model at which the NPV of CH_4 mitigation costs equals $2 trillion.

Costs and benefits for this solution are presented in table 5.1.1. Costs are much higher than benefits for both discount rates. The reason for this is easily explained: in order to spend so much on CH_4 mitigation, CH_4 emissions would have to be eliminated almost completely during those ten years in the FUND model. Emission levels would be reduced to between 1% and 4% of today's emission levels.

Table 5.1.1 Benefits and costs of solution A

Solution A ($2 trillion in 10 years)		
	Discount rate	
Benefit and costs	Low (3%) ($)	High (6%) ($)
Benefit	1,179	365
Cost	2,126	2,081

The results for this solution are highly speculative: the cost function for CH_4 emission reduction is simply an extrapolation for such high emission reductions and overall a solution that would imply such radical emission reductions seems to be quite unrealistic.

There are two relevant results from this solution, nevertheless: first, spending the equivalent of $2 trillion just on CH_4 emission reductions in just the next ten years probably belongs in the realm of fiction, and certainly does not pass a benefit-cost (B/C) test. Second, even such a strong mitigation of CH_4 emissions in the short run has literally no effect on the long run temperature projection. Figure 5.1.1 plots temperature above preindustrial in °C as projected by FUND for a business-as-usual (BAU) scenario with no climate policy and the temperature projection for solution A. While there is a small reduction in temperatures right after the ten-year emission reduction of CH_4 in the next decade, there is no long-lasting effect of such a policy. This comes as no surprise: the atmospheric lifetime of CH_4 is much smaller than that of CO_2, and any effects of a policy that is restricted to just the next ten years will be gone by about mid-century. This is one of the key differences between a CH_4 and a CO_2 policy: the effects of CH_4 mitigations are limited to a much shorter time span (about ten years), while CO_2 policy has effects in the long run, due to the longer lifetime of CO_2 in the atmosphere.

Does this result imply that CH_4 mitigation is not a worthwhile option? In order to investigate this question I now look for proper optimal mitigation paths (solution C) that are not restricted by an arbitrary spending schedule that might in itself rule out the best solution for CH_4 mitigation. In doing so, I make a comparison with other solution categories in the Copenhagen Consensus difficult: if they restrict themselves to the spending schedule of the Copenhagen Consensus project as well, one would have to compare apples and oranges – namely, an optimal CH_4 mitigation solution presented in this Perspective paper to, e.g. a CO_2 mitigation solution that is restricted by the spending schedule of the Copenhagen Consensus project (and therefore almost certainly not a true optimal CO_2 mitigation solution). I therefore also compute an optimal CO_2 mitigation solution (solution B) in order to allow a proper comparison of net benefits of CH_4 vs. CO_2 mitigation options. I then finally compute the optimal joint mitigation path, in which both CO_2 and CH_4 emissions are regulated in an optimal fashion and compute net benefits for that solution (solution D).

The benefits and costs for those three solutions for two discount rates are presented in tables 5.1.2, 5.1.3, and 5.1.4. Table 5.1.5 presents net benefits for all solutions, the relevant measure to compare solutions that differ with respect to costs.

The first observation to make is that all solutions except for the one restricted by the Copenhagen Consensus spending schedule produce net benefits, and that in particular CH_4 emission reduction solutions can produce significant net benefits if not restricted to spending enormous amounts of money in just the next ten years.

Table 5.1.2 Benefits and costs of solution B

Solution B (optimal C tax)		
	Discount rate	
Benefit and costs	Low (3%) ($)	High (6%) ($)
Benefit	19,015	93
Cost	12,376	20

Table 5.1.3 Benefits and costs of solution C

Solution C (optimal CH_4 tax)		
	Discount rate	
Benefit and costs	Low (3%) ($)	High (6%) ($)
Benefit	8,818	295
Cost	3,718	134

Table 5.1.4 Benefits and costs of solution D

Solution D (optimal joint taxes)		
	Discount rate	
Benefit and costs	Low (3%) ($)	High (6%) ($)
Benefit	24,511	375
Cost	13,976	148

Table 5.1.5 Net benefits

	Discount rate	
Net benefits	Low (3%) ($)	High (6%) ($)
Solution A ($2 trillion in 10 years)	−947	−1,715
Solution B (optimal C tax)	6,639	73
Solution C (optimal CH_4 tax)	5,100	162
Solution D (optimal joint taxes)	10,534	227

The second observation is that net benefits for a joint solution that both mitigates CO_2 as well as CH_4 emissions at the same time is always ranked highest in terms of net benefits: neither a CO_2- or CH_4-only solution can achieve similar net benefits.

The third observation is that the ranking of a CO_2- or CH_4-only solution depends on the discount

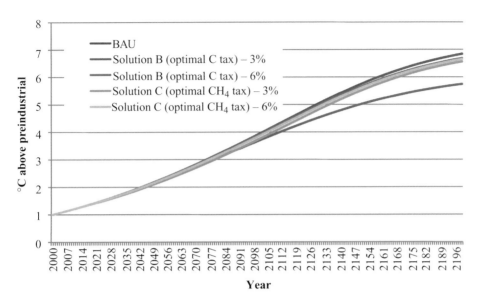

Figure 5.1.2 *Temperature*

rate. With a discount rate of 6% a CH_4-only solution yields higher net benefits than a CO_2 solution, while with a 3% discount rate the opposite holds. The explanation for this result lies in the interaction of the discount rate and the atmospheric lifetime of the two gases. CH_4 emissions stay in the atmosphere relatively briefly (for about ten years), while CO_2 has an atmospheric lifetime that is much longer. The benefit of reducing emissions of either gas is defined as the avoided damage from that emission reduction. The total damage caused by an emission at a specific point in time is the sum of damages caused by this emission in the future. With CH_4 emissions, those damages are all concentrated in a relatively short time frame after the emission, they will occur only in roughly the ten years after the initial emission of CH_4, given the short atmospheric lifetime of CH_4. The damages from a specific CO_2 emission are spread out over a much longer time frame, namely the atmospheric lifetime of CO_2, which amounts to many centuries. A change in discount rate therefore changes the damage estimates of CH_4 emissions much less than the damage estimates of CO_2 emissions. In fact, with a discount rate of 6%, the climate problem is simply not addressed in a significant way in either solution: for a CH_4-only reduction solution temperature is marginally changed along the BAU path (and this is highly

profitable) and for a CO_2-only reduction solution climate change is also not addressed in any comprehensive way, because most impacts from climate change are discounted away. I conclude from this that with a discount rate of 6%, climate change is simply not valued as an urgent problem in the first place. In such a situation, high net benefits can be gained by reducing CH_4 emissions, but those emission reductions occur at the margin and do not alter the general temperature trend. With a discount rate of 3%, a CH_4 emission reduction strategy can again be highly profitable, but again this will be gained by marginally changing the temperature along the BAU path.

I will now compare the three optimal solutions B–D with the solution A that conforms to the Copenhagen Consensus project. Table 5.1.6 shows the NPV of total expenditures on mitigation of either CH_4 or CO_2 in the time period 2010–19. The striking result here is that none of the optimal policies comes even close to spending as much as the solution that conforms to the Copenhagen Consensus spending suggestion in the first decade. At the same time total (i.e. not limited to 2010–19) expenditures of the optimal solutions B–D for a discount rate of 3% are much higher than what is spent in solution A. With a discount rate of 3%, this suggests that while expenditures of much more

Table 5.1.6 Costs (2010–19)

	Discount rate	
NPV cost in 2010–19	Low (3%) ($)	High (6%) ($)
Solution A ($2 trillion in 10 years)	2,126.39	2,080.54
Solution B (optimal C tax)	58.76	0.01
Solution C (optimal CH$_4$ tax)	267.04	21.24
Solution D (optimal joint taxes)	268.87	21.02

Table 5.1.7 BCRs for solutions A–D

	Discount rate	
B/C ratios	Low (3%)	High (6%)
Solution A ($2 trillion in 10 years)	0.6	0.2
Solution B (optimal C tax)	1.5	4.7
Solution C (optimal CH$_4$ tax)	2.4	2.2
Solution D (optimal joint taxes)	1.8	2.5

than $2 trillion in NPV terms are optimal, a large fraction of that should be spent after 2020. For a discount rate of 6% the total expenditure suggested by the Copenhagen Consensus project is overall too large. These results confirm an earlier suspicion: the spending suggestion of the Copenhagen Consensus itself is far away from an optimal response to climate change.

Limitations

In this section I will outline limitations both of the chapter and the results in this Perspective paper, and elaborate how I judge those to affect the applicability of the results for policy advice.

The first limitation concerns the published BCRs. The only solution that conforms to the spending schedule suggested by the Copenhagen Consensus project from both the chapter as well as this Perspective paper is my solution A. All other solutions spend vastly different sums on mitigation in the different solutions. This makes a ranking by BCRs arbitrary. The proper metric to rank solutions in this situation is net benefits, but those are not published for the chapter. Table 5.1.7 has BCRs for the solutions in this Perspective paper, and a comparison to the corresponding net benefits in table 5.1.5 shows clearly that a ranking by benefit-costs ratios would be misguided.

The second limitation is specific to the chapter, but again makes comparison with solutions from other categories almost impossible. The chapter only looks at mitigation costs in *one* year, namely 2020. Those costs do not seem to be discounted into NPV equivalents. Benefits are calculated by using the dubious GWP concept and thereby certainly misrepresent the specific dynamics of CH$_4$ stocks in the atmosphere. In summary, those two limitations make comparison with net benefits from other solution categories highly unconvincing.

The third limitation concerns the numerical results in this study. All estimates in this Perspective paper are calculated by using a deterministic version of the FUND model. In such a mode the model uses best-guess values for all input parameters and this study does not take into account any uncertainties surrounding climate change projections. This is clearly substandard: many previous studies have shown that taking proper account of uncertainty can have a significant effect on quantitative results from integrated assessment models (e.g. Stern 2007; Anthoff *et al.* 2009a, 2009b). The reason I have not incorporated uncertainty into this study is purely a technical one: such studies take considerable amount of computational time and could not be fitted into the tight time frame of the Copenhagen Consensus project. This limitation does mean that while most of the qualitative conclusions are sound, the precise quantitative magnitudes are not appropriate input into policy design. In particular, earlier studies suggest that incorporating uncertainty into the analysis would produce more aggressive emission mitigation paths for an optimal policy. The chapter also suffers from not accounting for uncertainty.

The final limitation concerns the discounting schemes employed in the Copenhagen Consensus project. A 6% constant consumption discount rate in particular seems highly inappropriate and not within the range of discount rates commonly employed in economic climate change analysis. In almost all integrated assessment models (IAMs), the standard approach to discounting is to specify a pure rate of time preference, and then calculate the interest rate as an endogenous variable as a function of the time preference rate, the *per capita* consumption rate, and the elasticity of marginal utility, using what is commonly referred to as the Ramsey equation (e.g. Nordhaus and Boyer 2000; Guo *et al.* 2006; Hope and Newbery 2007; Nordhaus 2008). There is an interesting and legitimate debate regarding how the pure time preference rate and the elasticity of marginal consumption should be chosen. Proponents of the so called "descriptive approach" usually calibrate their models to observed interest rates *today*. Those studies in general are those with the highest discount rates. But even these studies do *not* use consumption discount rates that are as high as 6% over the whole time horizon. Because *per capita* consumption growth rates fall in all those models in the second half of the century, the consumption interest rates employed by that approach in later years are smaller than in earlier periods. Using a 6% constant consumption discount rate seems clearly higher than the rate used in studies commonly thought to have high discount rates (e.g. Nordhaus 2008). The discounting schemes also ignore important research about discounting under uncertainty (cf. Weitzman 1998, 2001; Gollier 2004; Gollier and Zeckhauser 2005). All experience from climate change economics suggests that the choice of discount rate and scheme is one of the most important ones a modeler can make. Given that the Copenhagen Consensus project only considers two very basic discounting schemes, it already ignores at its outset what is probably the most relevant discussion in the economics of climate change. The quantitative results presented in this study follow the discounting guidelines of the Copenhagen Consensus project and are thereby, in my opinion, not reflecting state of the art as found in the literature.

Conclusion

This Perspective paper looked at the benefits and costs of CH_4 emission reductions. I looked at one solution that conformed to the guidelines given for the Copenhagen Consensus study and contrasted it with solutions that are optimal policy responses to the climate change problem. There are five main conclusions from this study.

First, a CH_4 emission reduction solution that follows the spending schedule suggested by the Copenhagen Consensus project does not pass a benefit-cost (B/C) test. I calculate BCRs well below 1 for such a solution, irrespective of the discount rate used in the assessment (the BCR is 0.6 for a discount rate of 3% and 0.2 for a discount rate of 6%). The specific quantitative results for this solution are extremely unreliable: in order to spend the enormous amounts of money suggested by the Copenhagen Consensus project in just ten years on CH_4 reductions, one would have to reduce CH_4 emissions to between 1% and 4% of today's emission levels, and any cost estimate for that range of emission reductions is highly speculative. The results for this solution say little to nothing about the desirability of CH_4 reductions from a policy point of view, they are mainly a consequence of the highly suboptimal nature of the spending schedule suggested by the Copenhagen Consensus project.

Second, the spending suggestion of the Copenhagen Consensus project of $250 billion per year for ten years excludes the solutions that create the greatest net benefit. In the optimal mitigation solutions that are not restricted to the spending schedule of the Copenhagen Consensus project, significantly less is spent in the years 2010–20 than in the solution following the $250 billion per year for ten years setup. At the same time, the total optimal expenditure on mitigation is much larger in NPV terms than the NPV of the high expenditures of the Copenhagen Consensus project in just a few years for a discount rate of 3%. The conclusion from this is simple: in NPV terms, one should spend a lot more than suggested by the Copenhagen Consensus project spending schedule, but that spending should not occur in some arbitrarily set time frame but rather should follow an optimal path over time, which is very different from the one suggested by the Copenhagen Consensus project. For a 6% discount rate, overall optimal spending in NPV terms is always lower than the NPV of the spending schedule of the Copenhagen Consensus project because, with such a high discount rate, climate change is not a problem almost by assumption.

Third, while an optimal CH_4 mitigation strategy is highly profitable at the margin, it does not significantly contribute to solving the climate problem. There are two lines of evidence on which I base this conclusion. The first is simply that with a discount rate of 3%, the net benefits of an optimal CH_4-only mitigation strategy are lower than the net benefits of a CO_2-only mitigation strategy. The second is a look at the temperature profile for the various solutions. The CH_4-only mitigation strategy does not alter the temperature trajectory in any significant way from that of BAU, while any solution that also includes CO_2 emission mitigation does. As discussed above, I ignore key uncertainties in my estimate of benefits of keeping temperatures below the BAU path in this study. Including such uncertainties would increase the net benefit estimates of a solution that changes the temperature trajectory in a significant way (which requires CO_2 mitigation) over a solution that reaps high net benefits at the margin, but would not alter the temperature profile (like a CH_4-only mitigation solution).

Fourth, CH_4 mitigation can add significant net benefits when combined with a CO_2 mitigation policy. The net benefits for a solution in which both CO_2 and CH_4 mitigation are chosen optimally are almost the sum of net benefits of doing either a CO_2-only or a CH_4-only mitigation solution, regardless of the discount rate chosen. This strongly suggests that an "either or" view which attempts to judge whether CO_2 or CH_4 emission mitigation is a better approach to climate change is misguided. The proper solution is a portfolio approach which combines various policy responses. This result also is in line with basic economic theory: if there are multiple significant externalities (like CH_4 emissions and CO_2 emissions), an optimal solution should internalize both externalities.

Fifth, the quantitative results in this Perspective paper, as well as in the chapter, suffer from strong limitations that make it difficult to compare them with other solution categories and make them of limited relevance for policy advice. The only solution that conforms to the spending schedule suggested by the Copenhagen Consensus project is solution A in this Perspective paper. All other solutions (in both the chapter and the Perspective paper) spend very different amounts of money on

mitigation in NPV terms. A ranking of solutions by BCRs would therefore be entirely arbitrary. The proper metric for ranking solutions that differ both in total costs as well as benefits is net benefits. The discounting schemes used for the Copenhagen Consensus project are not state of the art in climate change economics. Almost all IAMs today use an approach where the consumption discount rate is endogenously calculated using the Ramsey equation, thereby reflecting actual *per capita* consumption growth paths employed in the model. This is particularly relevant for the high discount rate of 6% in the Copenhagen Consensus project: in later (relevant) periods it is even higher than what is commonly assumed to be a high discounting scheme that calibrates interest rates to observed market rates. Finally, neither the chapter nor the Perspective paper factors uncertainty into the analysis. Previous studies have shown that including uncertainty significantly changes quantitative results (Anthoff *et al.* 2009a, 2009b). More recent work started a discussion whether highly unlikely but disastrous outcomes should drive rational climate change policy (Weitzman 2008). A thorough inclusion of uncertainty in a quantitative assessment would require significantly more time and resources than available for the Copenhagen Consensus project. While these limitations reduce the direct applicability of the quantitative results derived in this study for policy, the qualitative results would most likely hold in an analysis which included uncertainty. An optimal climate change policy consists of a portfolio of mitigation measures; the allocation of costs over time should not follow an arbitrary rule but rather an optimal time path; and CH_4 mitigation by itself cannot make a significant impact on climate change overall, but adds significant net benefits when combined with a CO_2 mitigation strategy.

Appendix: the FUND Model

FUND (the Climate Framework for Uncertainty, Negotiation, and Distribution) is an IAM linking projections of populations, economic activity, and emissions to a simple carbon cycle and climate model, and to a model predicting and monetizing welfare impacts. Climate change welfare impacts are monetarized in $1995 and are modelled over sixteen regions. Modeled welfare impacts include agriculture, forestry, sea level rise, cardiovascular and respiratory disorders influenced by cold and heat stress, malaria, dengue fever, schistosomiasis, diarrhea, energy consumption, water resources, and unmanaged ecosystems (Link and Tol 2004). (The source code, data, and a technical description of the model can be found at www.fund-model.org.)

Essentially, FUND consists of a set of exogenous scenarios and endogenous perturbations. The model distinguishes sixteen major regions of the world – the USA, Canada, Western Europe, Japan and South Korea, Australia and New Zealand, Central and Eastern Europe, the former Soviet Union (FSU), the Middle East, Central America, South America, South Asia, Southeast Asia, China, North Africa, sub-Saharan Africa, and Small Island States. Version 3.5, used in this Perspective paper, runs from 1950 to 3000 in time steps of one year. The primary reason for starting in 1950 is to initialize the climate change impact module. In FUND, the welfare impacts of climate change are assumed to depend in part on the impacts during the previous year, reflecting the process of adjustment to climate change. Because the initial values to be used for the year 1950 cannot be approximated very well, both physical impacts and monetized welfare impacts of climate change tend to be misrepresented in the first few decades of the model runs. The twenty-second and twenty-third centuries are included to provide a proper long-term perspective. The remaining centuries are included to avoid end point problems for low discount rates, they have only a very minor impact on overall results.

The period 1950–90 is used for the calibration of the model, which is based on the IMAGE 100-year database (Batjes and Goldewijk 1994). The period 1990–2000 is based on observations (http://earthtrends.wri.org). The 2000–10 period is interpolated from the immediate past. The climate scenarios for the period 2010–2100 are based on the EMF14 Standardized Scenario, which lies somewhere in between IS92a and IS92f (Leggett *et al.* 1992). The period 2100–3000 is extrapolated.

The scenarios are defined by varied rates of population growth, economic growth, autonomous energy efficiency improvements, and decarbonization of energy use (autonomous carbon efficiency improvements), as well as by emissions of CH_2 from land-use change, CH_4 emissions, and N_2O emissions.

Emission reduction of CH_2, CH_4, and N_2O is specified as in Tol (2006). Simple cost curves are used for the economic impact of abatement, with limited scope for endogenous technological progress and interregional spillovers (Tol 2005).

The scenarios of economic growth are perturbed by the effects of climatic change. Climate-induced migration between the regions of the world causes the population sizes to change. Immigrants are assumed to assimilate immediately and completely with the respective host population.

The tangible welfare impacts are deadweight losses to the economy. Consumption and investment are reduced without changing the savings rate. As a result, climate change reduces long-term economic growth, although consumption is particularly affected in the short term. Economic growth is also reduced by CO_2 abatement measures. The energy intensity of the economy and the carbon intensity of the energy supply autonomously decrease over time. This process can be accelerated by abatement policies.

The endogenous parts of FUND consist of the atmospheric concentrations of CO_2, CH_4, and N_2O, the global mean temperature, the effect of CO_2 emission reductions on the economy and on emissions, and the effect of the damages on the economy caused by climate change. CH_4 and N_2O are taken up in the atmosphere, and then geometrically depleted. The atmospheric concentration of CO_2, measured in ppmbv, is represented by the five-box model of Kemfert and Schill (2009). Its parameters are taken from the chapter.

The radiative forcing of CO_2, CH_4, N_2O, and sulfur aerosols is determined based on chapter 5. The global mean temperature, T, is governed by a geometric build-up to its equilibrium (determined by the radiative forcing, RF), with a half-life of fifty years. In the base case, the global mean temperature rises in equilibrium by 2.5°C for a doubling of CO_2 equivalents. Regional temperature is derived by multiplying the global mean temperature by a fixed factor, which corresponds to the spatial climate change pattern averaged over fourteen GCMs (Mendelsohn *et al.* 2000). The global mean sea level is also geometric, with its equilibrium level determined by the temperature and a half-life of fifty years. Both temperature and sea level are calibrated to correspond to the best-guess temperature and sea level for the IS92a scenario of the chapter.

The climate welfare impact module, based on Tol (2002a, 2002b) includes the following categories: agriculture, forestry, hurricanes, sea level rise, cardiovascular and respiratory disorders related to cold and heat stress, malaria, dengue fever, schistosomiasis, diarrhea, energy consumption, water resources, and unmanaged eco-systems. Climate change-related damages are triggered by either the rate of temperature change (benchmarked at 0.04°C/year) or the level of temperature change (benchmarked at 1.0°C). Damages from the rate of temperature change slowly fade, reflecting adaptation (cf. Tol 2002b).

In the model, individuals can die prematurely due to temperature stress or vector-borne diseases, or they can migrate because of sea level rise. Like all welfare impacts of climate change, these effects are monetized. The value of a statistical life (VSL) is set to be 200 times the annual *per capita* income. The resulting VSL lies in the middle of the observed range of values in the literature (cf. Cline 1992). The value of emigration is set to be three times the *per capita* income (Tol 1995, 1996), the value of immigration is 40% of the *per capita* income in the host region (Cline 1992). Losses of dryland and wetlands due to sea level rise are modeled explicitly. The monetary value of a loss of 1 km^2 of dryland was on average $4 million in OECD countries in 1990 (cf. Fankhauser 1994). Dryland value is assumed to be proportional to GDP km^2. Wetland losses are valued at $2 million per km^2 on average in the OECD in 1990 (cf. Fankhauser 1994). The wetland value is assumed to have a logistic relation to *per capita* income. Coastal protection is based on a cost-benefit analysis (CBA), including the value of additional wetland lost due to the construction of dikes and subsequent coastal squeeze.

Other welfare impact categories, such as agriculture, forestry, hurricanes, energy, water, and ecosystems, are directly expressed in monetary values without an intermediate layer of impacts measured in their "natural" units (cf. Tol 2002a). Modeled effects of climate change on energy consumption, agriculture, and cardiovascular and respiratory diseases explicitly recognize that there is a climatic optimum, which is determined by a variety of factors, including plant physiology and the behavior of farmers. Impacts are positive or negative depending on whether the actual climate conditions are moving closer to or away from that optimum climate: impacts are larger if the initial climate conditions are further away from the optimum climate. The optimum climate is of importance with regard to the potential impacts: the actual impacts lag behind the potential impacts, depending on the speed of adaptation. The impacts of not being fully adapted to new climate conditions are always negative (cf. Tol 2002b).

The welfare impacts of climate change on coastal zones, forestry, hurricanes, unmanaged eco-systems, water resources, diarrhea, malaria, dengue fever, and schistosomiasis are modeled as simple power functions. Impacts are either negative or positive, and they do not change sign (cf. Tol 2002a).

Vulnerability to climate change changes with population growth, economic growth, and technological progress. Some systems are expected to become more vulnerable, such as water resources (with population growth) and heat-related disorders (with urbanization), or more valuable, such as ecosystems and health (with higher *per capita* incomes). Other systems are projected to become less vulnerable, such as energy consumption (with technological progress), agriculture (with economic growth) and vector- and water-borne diseases (with improved health care) (cf. Tol 2002b).

Bibliography

Anthoff, D. *et al.*, 2009a: Discounting for climate change, *Economics: The Open-Access, Open-Assessment E-Journal* **3**, 2009–24

2009b: Risk aversion, time preference, and the social cost of carbon, *Environmental Research Letters* **4**(2)

Batjes, J.J. and C.G.M. Goldewijk, 1994: *The IMAGE 2 Hundred Year (1890–1990) Database of the Global Environment (HYDE)*, RIVM, Bilthoven, 410100082

Cline, W.R., 1992: *The Economics of Global Warming*, Institute for International Economics, Washington, DC

Fankhauser, S., 1994: Protection vs. retreat – the economic costs of sea level rise, *Environment and Planning A* **27**, 299–319

Gollier, C., 2004: Maximizing the expected net future value as an alternative strategy to gamma discounting, *Finance Research Letters* **1**, 85–9

Gollier, C. and R. Zeckhauser, 2005: Aggregation of heterogeneous time preferences, *Journal of Political Economy* **113**(4), 878–96

Guo, J. *et al.*, 2006: Discounting and the social cost of carbon: a closer look at uncertainty, *Environmental Science & Policy* **9**(3), 205–16

Hammitt, J.K. *et al.*, 1992: A sequential-decision strategy for abating climate change, *Nature* **357**, 315–18

Hope, C. and D. Newbery 2007: Calculating the social cost of carbon, in *Delivering a Low Carbon Electricity System: Technologies, Economics and Policy*, M. Grubb, T. Jamasb and M.G. Pollitt (eds.), Cambridge University Press, Cambridge

Kattenberg, A. *et al.*, 1996: Climate models – projections of future climate, in *Climate Change 1995: The Science of Climate Change – Contribution of Working Group I to the Second Assessment Report of the Intergovernmental Panel on Climate Change*, J.T. Houghton, L.G. Meiro Filho, B.A. Callander *et al.* (eds.), Cambridge University Press, Cambridge, 285–357

Kemfert, C. and W.-P. Schill, 2009: Methane mitigation, chapter 5 in this volume

Leggett, J. *et al.*, 1992: Emissions scenarios for the IPCC: an update in *Climate Change 1992 – The Supplementary Report to the IPCC Scientific Assessment*, J.T. Houghton, B.A. Callander and S.K. Varney (eds.), Cambridge University Press, Cambridge, 71–95

Link, P.M. and R.S.J. Tol, 2004: Possible economic impacts of a shutdown of the thermohaline circulation: an application of FUND, *Portuguese Economic Journal* **3**(2), 99–114

Maier-Reimer, E. and K. Hasselmann, 1987: Transport and storage of carbon dioxide in the ocean: an inorganic ocean circulation carbon cycle model, *Climate Dynamics* **2**, 63–90

Manne, A.S. and R.G. Richels, 2001: An alternative approach to establishing trade-offs among greenhouse gases, *Nature* **410**, 675–77

Mendelsohn, R.O. *et al.*, 2000: Comparing impacts across climate models, *Integrated Assessment* **1**, 37–48

Nordhaus, W., 2008: *A Question of Balance: Weighing the Options on Global Warming Policies*, Yale University Press, New Haven, CT

Nordhaus, W. and J. Boyer, 2000: *Warming the World: Economics Models of Global Warming*, MIT Press, Cambridge, MA

Shine, K.P. *et al.*, 1990: Radiative forcing of climate, in *climate change – The IPCC Scientific Assessment*, J.T. Houghton, G.J. Jenkins and J.J. Ephraums (eds.), Cambridge University Press, Cambridge, 41–68

Stern, N., 2007: *The Economics of Climate Change: The Stern Review*, Cambridge University Press, Cambridge

Tol, R.S.J., 1995: The damage costs of climate change – towards more comprehensive calculations, *Environmental and Resource Economics* **5**, 353–74

1996: The damage costs of climate change: towards a dynamic representation, *Ecological Economics* **19**, 67–90

2002a: Estimates of the damage costs of climate change. part 1: benchmark estimates, *Environmental and Resource Economics* **21**(2), 47–73

2002b: Estimates of the damage costs of climate change. part 2: dynamic estimates, *Environmental and Resource Economics* **21**(2), 135–60

2005: An emission intensity protocol for climate change: an application of FUND, *Climate Policy* **4**, 269–87

2006: Multi-gas emission reduction for climate change policy: an application of FUND, *The Energy Journal: Multi-Greenhouse Gas Mitigation and Climate Policy* (Special Issue **3**), 235–50

Weitzman, M.L., 1998: Why the far-distant future should be discounted at its lowest possible rate, *Journal of Environmental Economics and Management* **36**, 201–8

2001: Gamma discounting, *American Economic Review* **91**(1), 260–71

2008: On modeling and interpreting the economics of catastrophic climate change, *Review of Economics and Statistics* **91**(1), 1–19

Methane Mitigation

Alternative Perspective

DANIEL J.A. JOHANSSON AND FREDRIK HEDENUS*

Introduction

Chapter 5 on Methane Mitigation by Kemfert and Schill (Kemfert and Schill 2009) presents an up-to-date and very comprehensive overview of estimates of methane (CH_4) abatement costs. We want to focus in this Perspective paper on four issues which we believe are important and that have not been dealt with in any considerably length in chapter 5:

- The effect of CH_4 and carbon dioxide (CO_2) emissions on the global average surface temperature
- The shadow prices and marginal social costs of CH_4 and CO_2
- The impact of CH_4 on the global tropospheric ozone (O_3) level and the economic benefits related to tropospheric O_3 of CH_4 abatement
- (to present one additional abatement solution) Output taxes on beef meat as an option to reduce CH_4 (and other greenhouse gases, GHGs).

For all calculations in the Perspective paper we focus on 2020. This is within the time horizon considered in this Copenhagen Consensus Project.

However, we start to present our perspective on climate change and on the issue of cost-benefit (C/B) vs. cost-effectiveness analysis when evaluating abatement options.

The Climate Challenge

Climate change is a reality, with a warming of the Earth as a result. Current trends in the most relevant

climate indicators are in line with what can be expected from climate models – e.g. the global heat content of oceans has increased significantly in recent decades, the global average sea level is rising, and the global average surface temperature has an increasing long-term trend. If nothing is done to reduce the emissions of GHGs the global average surface temperature is estimated by the Intergovernmental Panel on Climate Change (IPCC) to increase between about 2 and 7°C above the preindustrial level by 2100 and with continuing warming thereafter (IPCC 2007; Solomon *et al.* 2009). Temperature scenarios at the mid- and high range of the interval would imply a global average temperature level and a rate of change in the temperature not witnessed for millions of years (Jansen *et al.* 2007).

Even for less severe scenarios, serious negative impacts on ecosystems and society can be expected to occur – see, for example, Warren (2006), Parry *et al.* (2007), and Smith *et al.* (2009). Even though there may be some positive economic impacts for small changes in the climate, these will mainly occur in developed countries – while developing countries – which in general are more dependent on agricultural activities, already have a warmer climate, and have less resources for adapting to changes in it – are expected to suffer rapidly from small changes in the climate. The overall picture concerning the negative impacts of climate change is more serious today than it was about a decade ago (see Smith *et al.* 2009 where this is illustrated in an updated version of the "reasons for concern" diagram).

Benefit Calculations and Targets

Due to the complexity of climate change and the deep structural uncertainty in the science, the

* The authors would like to thank Christian Azar and Olof Johansson-Stenman for valuable comments. The Göteborg Energi Research Foundation and the Swedish Energy Agency are acknowledged for complementary funding. The authors are solely responsible for any remaining errors.

expected benefits of emission reductions can be found to be very large and strongly dependent on arbitrarily set upper bounds on the climate sensitivity or on the damages caused by an increase in the global average surface temperature (Dasgupta 2008; Weitzman 2009). Similar arguments concerning the application of a cost-benefit analysis (CBA) for climate change have been around in the literature for more than a decade (e.g. Azar 1998), although it has not been shown as rigorously before in Weitzman (2009). Given this, it is not controversial, even for economists working in the field – see, for example, Dasgupta (2008) and Tol (2009) – if one argues for more stringent climate policies than can be justified by a formal B/C calculation. How large such an uncertainty premium that warrants a more stringent climate policy should be is, however, not easily quantified.

Moreover, besides the problems with structural uncertainties, existing benefit functions are likely to underestimate the benefits of emission reductions. Very few of these benefit estimates includes the cost of large-scale surprises or non-market costs, and no existing benefit function tries to capture socially contingent damages (Warren et al. 2006; Watkiss and Downing 2008). In the few cases where the costs of large-scale surprises are considered, the probabilities of these are likely to be underestimated (see Kriegler et al. 2009).

An alternative approach to CBA for approaching the challenge of climate change is to base the reasoning on the almost globally adopted United Nations Framework Convention on Climate Change (UNFCCC). The overarching aim of the UNFCCC is "stabilisation of greenhouse gas (GHG) concentrations in the atmosphere at a level that would prevent dangerous anthropogenic interference with the climate system." Even though this rather vague political aim is hard to transform into more clear-cut formulations on targets, there is a growing support for a long-term stabilization of the global average surface temperature at or about 2 K above the preindustrial level. This target is supported by a large group of scientists (see, for example, Richardson et al. 2009 and Allan et al. 2007), and has recently been given support by the G8 and MEF countries.[1] Currently, countries that

contain a majority of the world's population have now expressed support for a 2 K target.

Given the widespread support for the 2 K target and the well-known problems with benefit functions discussed above, we will complement the benefit calculations in chapter 5 with a cost-effectiveness approach. By this, we mean that we calculate shadow prices on CH_4 and CO_2 in an optimizing integrated assessment model (IAM) where the 2 K target is implemented as a constraint and use these shadow prices as "benefits" for calculating benefit-cost ratios (BCRs). We believe that this is a more policy-relevant approach than the use of social cost estimates based on damage functions.

Methane vs. Carbon Dioxide

As discussed in the chapter, it is important to abate a portfolio of GHGs and not only CH_4 or CO_2. Clearly it would be a waste of money if not as many sources of emissions of GHGs as possible were targeted for abatement strategies. This multi-gas approach to climate change is not new, cost-effectiveness was the main reason why a "basket" approach was adopted in the Kyoto Protocol.

When allocating resources to mitigate the adverse effects of climate change it is crucial to understand the dynamics (both climate and economic) of different mitigation options. Even though the chapter discusses this briefly we believe it deserves more attention and devote two sections to discuss temperature dynamics and social costs/shadow prices. In this section we focus on the temperature response of an emissions pulse of a short-lived GHG being CH_4 and of the most important and long-lived GHG being CO_2 and in the next section we turn to the economic side of the question.

As discussed in the chapter, the atmospheric perturbation lifetime of CH_4 is about twelve years, while the perturbation life-time for CO_2 can not be accurately described by a single time constant.

[1] Canada, France, Germany, Italy, Japan, Russia, the UK, the USA, China, India, Brazil, Korea, Mexico, South Africa, Australia, and the remaining countries in the EU.

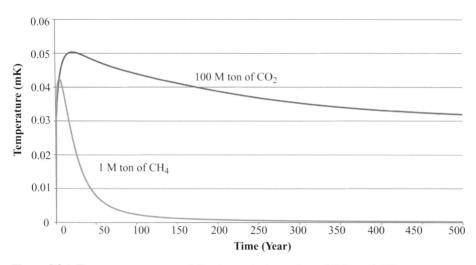

Figure 5.2.1 *Temperature response following emission pulses of CO_2 and CH_4*

Note: The CO_2 emissions pulse is 100 times larger than the CH_4 emissions pulse.

Rather, a multitude of different time constants are needed in order to reflect the different time scales at which CO_2 equilibrates between atmosphere, oceans, biomass, soil, sediments, and rocks (Archer *et al.* 2009).

So as to compare and illustrate the effect on the temperature of emitting CO_2 and CH_4 we calculate the temperature response of a 100 M ton CO_2 emissions pulse and of a 1 M ton CH_4 emissions pulse by using an upwelling-diffusion energy balance model where the climate sensitivity is set to 3 K. The reason for assuming unequally large pulses is that CH_4 is a considerably stronger GHG than CO_2 and that we want to fit the two curves on the same diagram. Note that the indirect effects on radiative forcing induced by CH_4 emissions are taken into account. As the chapter notes, CH_4 contributes to an increased level of tropospheric O_3 and to stratospheric water vapor and these enhance the direct forcing strength of CH_4 by some 30–40% (Forster *et al.* 2007).[2]

The result of this calculation is summarised in figure 5.2.1 and in the following list:

- CH_4 is a considerably stronger GHG than CO_2. For short time horizons (less than ten years) the effect on the temperature is about 100 times as strong for equally sized emission pulses.
- An emission pulse of CH_4 has an effect on the global average surface temperature far longer in time than the atmospheric perturbation lifetime of CH_4. This is due to inertia in the climate system.
- Even though the temperature response of a CH_4 pulse lingers on for more than a century, the effect on the temperature decays considerably faster than for an emissions pulse of CO_2. Emissions of CO_2 have in principle an irreversible effect on the global average surface temperature, while CH_4 has not (see also Solomon *et al.* 2009 and Matthews and Caldeira 2008).
- For equally sized emissions pulses of CH_4 and CO_2 the effect of the CO_2 pulse on global average surface temperature would surpass that of CH_4 after about 400 years in this model.

Social Costs and Shadow Prices

Chapter 5 uses the social cost of carbon estimates from Tol (2008) together with the CO_2 equivalent abatement potential for CH_4 as reported in the abatement cost studies (primarily USEPA 2006) when calculating the BCRs. Although the chapter is not explicit that it uses Global Warming Potential (GWP) calculated over a time horizon of 100 years,

[2] The total radiative forcing contribution from CH_4 emissions are 0.6–0.7 W/m^2 – i.e. close to half of that of CO_2, if the indirect effects are taken into account.

it implicitly does so since that is the approach taken in the abatement cost studies. In USEPA (2006) the conversion factor for 1 ton of CH_4 to 1 ton of CO_2eq is 21 – i.e. the climate impact of 1 ton of CH_4 is said to be equal to 21 tons of CO_2. As noted in the chapter, this combination of the social cost of carbon (SCC) and GWP is inconsistent.

In this section we will calculate the social cost of CH_4 (and CO_2) using a B/C approach and the shadow price of CH_4 (and CO_2) assuming a globally adopted 2 K target and analyze how these numbers depend on the discount rate and the climate sensitivity.[3] This is done to get consistent estimates on the social cost and shadow price of CH_4 (and CO_2), to illustrate the very large uncertainty in such estimates, and to show the strong dependence on the discount rate.

We will not perform any new calculations concerning GWP and its physically based alternative metrics; we refer to Fuglestvedt *et al.* (2003) and Forster *et al.* (2007) for such discussions.

We use an updated version of the globally aggregated climate–economy model MiMiC when estimating the social costs and shadow prices. The model is presented in detail in Johansson *et al.* (2006, 2008), see also the appendix (p. 218) for a brief presentation. We estimate the social costs and shadow prices given three different climate sensitivities,[4] 2 K (a low value), 3 K (a best estimate), 4.5 K (a high value). For simplicity (and lack of time) we do not separate out discounting due to economic growth, elasticity of marginal utility of consumption, and pure rate of time preference, instead we presuppose three different discount rates – 1%, 3%, and 5% per year. The low rate is in line with the rates used by, for example, Stern (2007) and the high rate is in line with the rates used by, for example, Nordhaus (2008). In the recommendations from the Copenhagen Consensus Center a discount rate of 3% is suggested. We adopt this as the main case in this Perspective paper.

When estimating the social cost of CH_4 and CO_2 we adopt the quadratic damage function used in Nordhaus (2008). Our baseline scenarios for gross world production and emissions of GHGs are from IIASA A2r (IIASA 2009). Economic growth is exogenous in MiMiC. As discussed above, existing damage functions (including the one used here) are likely to underestimate the damage of climate

Table 5.2.1 The cost-benefit case

Discount rate (%)	Climate sensitivity (K)		
	2 K (US$/t CH$_4$)	3 K (US$/t CH$_4$)	4.5 K (US$/t CH$_4$)
1	600 (21)	1,000 (22)	1,700 (22)
3	320 (42)	520 (42)	780 (42)
5	210 (58)	320 (58)	470 (58)

Note: The social cost of CH_4 in 2020 (US$/per ton CH_4). The ratio of the social cost of CH_4 to that of CO_2 is shown within the brackets. The social cost of CO_2 is obtained by dividing the social cost of CH_4 by the ratio given within the brackets.

change. The objective of the MiMiC model is to minimize the net present value (NPV) of the sum of the climate damages and the abatement costs for the three most important well-mixed GHGs – CO_2, CH_4, and N_2O. Consequently, the emissions of these gases are endogenously determined in the model.

When estimating the shadow price for CH_4 and CO_2 we run the MiMiC model with the 2 K target as a constraint and minimize the NPV of the cost of abatement. In this case, the damages of the temperature increase are not considered.

Both the B/C approach and the cost-effectiveness approach suffer from the large uncertainties concerning the cost of abatement. Technical improvements leading to declining abatement costs are exogenous in the model.

Social Costs

The social cost of CH_4 and the ratio of the social cost of CH_4 to the social cost of CO_2 depends strongly on the climate sensitivity and the discount rate, see table 5.2.1. The ratio is clearly declining

---

[3] In a technical sense both the social cost and the shadow price are shadow prices; we, however, refer to "social cost" when discussing results from the cost-benefit cases and "shadow price" when discussing results from the cost-effectiveness cases.

[4] The climate sensitivity is explained by the IPCC (2007) as follows: "The equilibrium climate sensitivity is a measure of the climate system response to sustained radiative forcing. It is not a projection but is defined as the global average surface warming following a doubling of CO_2 concentrations. It is likely to be in the range 2°C to 4.5°C with a best estimate of about 3°C, and is very unlikely to be less than 1.5°C."

with a declining discount rate. This comes as no surprise since emissions of CO_2 have a significantly longer-lasting effect on the temperature than emissions of CH_4. Hence, given the shorter lifetime of CH_4 (and the shorter corresponding effect on the temperature) the social cost of CH_4 is less sensitive to the discount rate as compared to the social cost of CO_2. The ratio of the social costs of CH_4 to CO_2 can be interpreted as an alternative conversion factor to the GWP (see Reilly and Richards 1993 and Kandlikar 1995). Hence, if the GWP value is 21 (as used in the Kyoto Protocol), or 25 (as in the latest IPCC assessment, IPCC 2007), the GWP approach undervalues the relative importance of reducing CH_4 as compared to CO_2 unless the discount rate is low. As can be expected, the social cost of CH_4 and CO_2 increases strongly with climate sensitivity.

The two extremes for the social cost of CH_4 in our calculation are 210 US$/ton CH_4 at the lower end and 1,700 US$/ton CH_4 at the higher end (while the corresponding numbers for CO_2 are 3.7 US$/ton CO_2 and 78 US$/ton CO_2). These numbers can be compared to the numbers (implicitly) assumed in the chapter, which are 275 US$/t CH_4, 407 US$/t CH_4, and 974 US$/t CH_4[5] (while their numbers for CO_2, taken from Tol 2008, are 13.1 US$/t CO_2, 19.4 US$/t CO_2, and 46.4 US$/t CO_2, respectively). Hence, even though the chapter uses an inconsistent approach, its assumptions on the benefit side seem to be roughly in line with our results on the social costs of CH_4, but without including the upper level of our estimate.

Shadow Prices

Given the widespread political support for a global temperature target of 2 K above the preindustrial level, we believe that it is more policy-relevant to focus on shadow prices obtained from models where such a target is taken into account. Taking such an approach alters the relative importance of reducing CH_4 as compared to CO_2. Hence, the

[5] The values for the social cost of CO_2 that chapter 5 uses are converted to estimates of the social cost of CH_4 by using a GWP equal to 21. As noted above, this conversion is somewhat inappropriate, but is the methodology (implicitly) used by chapter 5.

Table 5.2.2 The cost-effectiveness case

Discount rate (%)	Climate sensitivity (K)		
	2 K (US$/t CH_4)	3 K (US$/t CH_4)	4.5 K (US$/t CH_4)
1	260 (3.3)	550 (3.7)	3,700 (5.0)
3	120 (5.5)	330 (5.7)	980 (6.3)
5	65 (7.9)	250 (9.2)	740 (9.8)

Note: The shadow price of CH_4 in 2020. The ratio of the shadow price of CH_4 to that of CO_2 is shown within the brackets. The shadow price of CO_2 is obtained by dividing the shadow price of CH_4 by the ratio given within the brackets.

ratio of the shadow price of CH_4 to CO_2 in 2020 is considerably lower than for the ratio of social costs discussed in the previous subsection (compare table 5.2.1 and 5.2.2, see also Manne and Richels 2001). The reason is that the temperature response prior to the date that the constraint (i.e. the 2 K targets) starts to bite does not influence the shadow price of an emission. Given the relatively short lifetime of the temperature response of CH_4 reductions, and that the target will be met beyond the middle of this century, the shadow price of CH_4 will be relatively low compared to what is found in the CBA or to its GWP value calculated over 100 years. The case is different for CO_2, since it has an almost irreversible effect on temperature. Hence, given a cost-effectiveness approach (with a 2 K target) the use of GWP overvalues the importance of reducing CH_4 in 2020 and, correspondingly, relatively more economic resources should be devoted to reduce long-lived GHGs such as CO_2; see also van Vuuren et al. (2006).

A cost-effectiveness approach does in general imply lower shadow prices on CH_4 as compared to the social costs obtained from the B/C approach. It is only in the case where a high climate sensitivity is assumed that the shadow price is higher in the cost-effectiveness case than in the B/C case. The situation is different for CO_2, where the shadow prices are higher than the social costs for all separate cases.

As noted above, our recommendation concerning the main case is to use a discount rate of 3%, a climate sensitivity of 3 K, and a cost-effectiveness approach with a 2 K target for the global average surface temperature. This implies a shadow price of CH_4 emissions equal to 330 US\$/t CH_4 and a shadow price of CO_2 emissions equal to 57 US\$/t CO_2 in 2020.[6]

Finally, if economic efficiency is a primary aim, GWP should not be used to assess and compare the benefits of CH_4 abatement with other abatement options, such as CO_2 abatement. GWP calculated over 100 years will overvalue the importance of reducing short-lived GHGs as compared to long-lived gases if the aim is to stabilize the global average surface temperature at 2 K. Given the use of a B/C approach GWP will in general undervalue the importance of reducing CH_4 as compared to long-lived GHGs such as CO_2.[7] Setting aside efficiency, there is political support for the GWP value calculated over a 100-year time period; this approach is adopted within the Kyoto Protocol and it would most likely be politically difficult to change the metric. Also, estimates on the costs of using the GWP approach instead of an optimal approach of valuing different GHG emissions are seemed to be rather small, less than about 5–10% (Johansson *et al.* 2006).

Methane and Tropospheric Ozone

CH_4 is an important precursor to the increased background level of tropospheric O_3. Tropospheric O_3[8] carries a lot of other impacts besides being a GHG. It has serious negative impacts on human health, ecosystems, and forest and agricultural productivity.[9] It is only recently that the abatement of CH_4 has been considered as an option to reduce tropospheric O_3 levels. Historically, tropospheric O_3 has been approached as a local and/or regional atmospheric environmental problem and the policies in place to reduce the tropospheric O_3 load have focused on a precursor important on such a spatial scale (West and Fiore, 2005; The Royal Society, 2008). CH_4, on the other hand, is globally well mixed due to its relatively long atmospheric lifetime and therefore affects the level of tropospheric O_3 all over the globe, although not uniformly.

We will touch upon two aspects concerning tropospheric O_3:

- Tropospheric O_3 has a negative impact on the biospheric carbon stock.
- Tropospheric O_3 carries a range of health and economic problems. The economic impacts of these have been quantified in the academic literature and we briefly summarize what they imply for the social cost of CH_4.

Tropospheric Ozone and the Biospheric Carbon Stock

Tropospheric O_3 is well known to have important impacts on plant physiology (Stitch *et al.* 2007; The Royal Society 2008). Recent estimates have pointed to the fact that tropospheric O_3 has a strong negative impact on the carbon stock in biomass and soil. Stitch *et al.* (2007) suggest that by 2100 the radiative forcing caused by elevated atmospheric CO_2 levels which are caused by a decrease in the CO_2 sink induced by tropospheric O_3 may be higher than the direct global average radiative forcing of tropospheric O_3. The direct radiative forcing of tropospheric O_3 is estimated to be about 0.5–0.7 W/m^2 by 2100. However, it is hard to judge, given existing integrated assessment models, how large this effect is on the social cost and/or shadow price of CH_4, but it would certainly raise the price.

Non-Climate Co-Benefits of Methane Mitigation

The non-climate-related economic benefits of reducing tropospheric O_3 through CH_4 abatement have been assessed in West and Fiore (2005) and West *et al.* (2006). The health impacts of tropospheric O_3 are primarily associated with acute and

---

[6] These values will increase over time beyond 2020.
[7] Note that we have assumed full global cooperation in both the cost-benefit cases and the cost-effectiveness case. The general result concerning the relative valuation of CH_4 to CO_2 per ton emission should not change considerably if partial cooperation were assumed in the modeling.
[8] Note that tropospheric O_3 is not a primary pollutant but created through reactions by precursors, such as CH_4, NO_x, carbon monoxide, and volatile organic compounds (VOCs).
[9] Stratospheric O_3 is important for capturing UV radiation, tropospheric O_3 is not.

chronic effects on the respiratory system and daily premature mortality, while its impact on agricultural and forestry production is that it reduces yields (West and Fiore 2005; West *et al.* 2006; The Royal Society 2008).

We base our calculations of the benefits of reducing the O_3 level on West and Fiore (2005) and West *et al.* (2006, 2007). However, we update their calculations so that numbers consistent (concerning the discount rate) with the social costs/shadow prices presented above can be presented.

According to West and Fiore (2005) the non-mortality benefits of reducing tropospheric O_3 are close to linear in concentration and can be divided into the following categories:

- Agricultural benefits = US$ 2.8 billion /yr/ppb O_3.
- Forestry benefits = US$ 1.7 billion /yr/ppb O_3.
- Human health (non-mortality) = US$ 3 billion /yr/ppb O_3.

West *et al.* (2006, 2007) estimate the global mortality effects of changes in the global O_3 concentration in 2030. From these papers we estimate a simple relationship between premature mortalities and the global average tropospheric O_3 concentration. Further, we scale this relationship with the assumed global population scenario. We estimate from Shindell *et al.* (2005), West and Fiore (2005), and Fiore *et al.* (2008) that 1 ppb change in atmospheric CH_4 gives on average over the globe a change of 0.004 ppb O_3. As noted at the beginning of this section, the effect of changing the atmospheric concentration of CH_4 has a global impact on the O_3 concentration. However, the local impact on the O_3 level depends on chemical and metrological conditions and is not uniform over the globe. Also, 1 M ton of atmospheric CH_4 corresponds to 0.3646 ppb CH_4 (Tanaka 2008).

Table 5.2.3 **The social cost of CH_4 through its effect on non-climate impacts of tropospheric O_3**

Discount rate (%)	VSL = 1 million (US$/t CH_4)	VSL = 3 million (US$/t CH_4)
1	580	1,500
3	470	1,200
5	390	930

In order to estimate the social cost of CH_4 through its effect on the non-climate impact of tropospheric O_3 we have to assign a Value of a Statistical Life (VSL). We take the assumption in West *et al.* (2006) and set global average VSL to US$1 million. As an alternative we include a rather high global average VSL of US$ 3 million. We scale the mortalities per ppb tropospheric O_3 to the global population. The population is assumed to be 7.8 billion by 2020 and thereafter to grow at 1% per year.[10]

From these calculations we estimate, in very round numbers, that the anthropogenic emissions of CH_4 are annually accountable for 100,000 premature mortalities. As seen in table 5.2.3, the non-climate economic benefits of abating CH_4 through its effect on tropospheric O_3 is comparable in size to the social cost estimates in table 5.2.1 and the shadow prices in table 5.2.2. The non-climate benefits of CH_4 reduction are strongly dependent on the VSL assumption, but not very strongly dependent on the discount rate, due to the relative short lifetime of CH_4.[11] To get the numbers in table 5.2.3 directly comparable to benefit numbers in the chapter, they should be divided by methane's (old) GWP value of 21.

If these non-climate benefits of CH_4 abatement were taken into account in the BCRs presented in the chapter, these ratios would roughly double. However, even if we believe that it is important to recognize these benefits when suggesting climate-related measures we hold the position that they should be of second-order importance since they are not climate benefits. Besides, the literature on this topic is sparse and the numbers uncertain.

Finally, the calculation presented in this section was done given immensely large simplifications of the atmospheric chemistry and O_3 mortality relationship. However, the calculations produce

---

[10] The population projection is set to roughly equal the scenario in the IIASA A2r scenario.

[11] Note that there is a difference in the time dynamics of the impacts on O_3 and temperature following changes in CH_4 emissions. In the latter case the effect is dependent on the inertia of the climate system while in the former case it is not. Hence, the effect on the O_3 level decays with methane's perturbation lifetime of twelve years.

meaningful results since they are based on results from advanced models (Fiore and West 2005; West *et al.* 2006, 2007). There are also a range of additional uncertainties that we have not assessed. In particular, there are large uncertainties for the O_3– mortality relationships, which are not very well studied in epidemiological studies outside the USA. Thus, the uncertainty is larger than what is presented in table 5.2.3, perhaps in the order of +/– 100% in each cell. Our benefit estimates are higher than those in West *et al.* (2006). The reason for this is mainly because they only considered the benefits of CH_4 abatement over a rather limited period of time.

Climate Tax on Beef Meat

As discussed in the chapter, some abatement options may have considerable implementation barriers. For example, there are cheap options to reduce CH_4 from ruminants, and measures to reduce emissions from rice fields, but how should a policy be constructed so that these abatement options are realized efficiently? Actors are in general small in scale, geographically scattered, and the emissions hard to monitor. Also, for livestock management the low-cost abatement potential is small – e.g. the abatement potential below a marginal cost of 60 US$/t CO_2eq is less than 10% of the CH_4 emissions from that subsector.

Beef production does not only cause large CH_4 emissions but also indirectly large emissions of N_2O, and some CO_2, leaving aside induced deforestation. In total the GHG emission per eatable unit of energy of beef is around eight times higher than for poultry, and fifty times higher than for beans when emissions are converted to CO_2eq using GWPs calculated over 100 years. Due to these large differences in emissions between different foodstuffs, a changed diet, containing less beef, could decrease GHG emissions considerably (Carlsson-Kanyama and González 2009; Stehfest *et al.* 2009; Wirsenius *et al.* 2009). Using a nutritious and healthy diet as the norm, it is obvious that there is a considerable substitutability between different sorts of food from a nutritional perspective. Substitutability is still significant when considering

the prevailing preferences for meaty texture, since several different meat types are available, as well as vegetable-based meat substitutes.

These aspects point towards the conclusion that output-based policies may be a realistic alternative, at least in the developed countries where food security is less of a problem. See, for example, Schmutzler and Goulder (1997) and Sterner (2003) for discussions on when output taxes may be the suitable policy of choice to curb emissions. Changing the diet of the people is a difficult and controversial issue. However, output taxes on gasoline have changed people's driving patterns as well as the energy efficiency of vehicles. Similar effects could be achieved by introducing a greenhouse-weighted consumption tax on beef. In this section we will analyze the BCR of a tax on beef in OECD countries as a policy to reduce beef consumption and thereby CH_4 and other GHG emissions.

To give a tentative back-of-the-envelope estimate of the cost of the tax we calculate the deadweight loss of the tax under two cases and interpret those as the cost of the policy.[12] The two cases are:

1 The ruminant market in the OECD is assumed to be a closed economy and we only account for own-price effects.
2 The ruminant market in the OECD is assumed to be a small open economy. In this case we also only account for own-price effects.

Given the size of the OECD, the former case is probably a better approximation than the latter.

A demand price elasticity of –1.3 is assumed, based on Allais and Nichele (2007) and Burton and Young (1992) and a supply price elasticity of 1 based on Banse *et al.* (2005). Both the demand and supply elasticities are based on data for the EU. The supply elasticity is only of importance in the closed-economy case since the producer price is unaffected in the small open-economy case. The OECD average retail price of beef meat products is estimated to be US$12 per kg and a simple linear

[12] This should be seen as a very rough and first estimate; to get better results an agriculture sector model where existing subsidies and policies are taken into account should be used. Given the time frame for this project there was no time to do such a calculation.

Table 5.2.4 Reductions in beef meat consumption due to a beef tax in the OECD countries, and the GHG mitigation expressed in GWP calculated over 100 years

Tax on ruminant meat	Open small economy		Closed economy		B/C ratios		
(US$/kg beef)	Reduction (kt meat)	Reduction (M ton CO_2-eq)	Reduction (kt meat)	Reduction (M ton CO_2-eq)	Kemfert & Schill	Cost-effective	Cost-eff and O_3
0.5	1,455	36	657	16	4.7	3.3	4.7
1	2,783	70	1,302	33	2.4	1.7	2.4
2	5,114	128	2,555	64	1.2	0.9	1.2
3	7,091	177	3,759	94	0.8	0.6	0.8

Note: The BCRs are presented for chapter 5's high-benefit case, assuming a carbon price of 46US$/$CO_2$-eq cost-effective case with a climate sensitivity of 3 K and a discount rate of 3%, and finally a cost-effective case with the O_3 co-benefit included, assuming a VSL of US$ 1 million.

extrapolation is used to project the baseline beef consumption in the OECD countries to 28,160 kton carcass weight in 2020 (FAO 2009).

The life-cycle GHG emissions from reduced beef consumption are estimated to be about 25 kg CO_2-eq/kg beef in carcass weight (Williams *et al.* 2006). Combining with data from Cederberg *et al.* (2009), we estimate the emission of CH_4 to 0.7 kg/kg beef and the N_2O emissions to 0.02 kg/kg beef and 3 kg of CO_2/kg beef. Since several GHGs are involved, their relative weight is of crucial importance for the BCRs. For that reason, we study three cases, one that corresponds to the chapter's case with a CO_2 price of 46 US$/t CO_2 and using GWP calculated over 100 years as the relative weights for the different gases, a second where shadow prices of emissions are based on a cost-effective approach with a climate sensitivity of 3 K and a discount rate of 3%[13] (see table 5.2.2), and a third where the non-climate co-benefits of tropospheric O_3 assuming a VSP of US$ 1 million are also included (see table 5.2.3).

Our results show that the abatement level is about twice as large in the small open-economy case as in the closed-economy case for a given tax level. However, the BCRs differ very little between the two cases, less than 0.1. For that reason only the BCR for the closed-economy case is presented in table 5.2.4. We can also see that a low tax on beef has a fairly high BCR.

Reduced beef consumption has additional benefits to those discussed above. Most importantly, land required for the global agricultural system would be reduced if beef consumption were reduced. Cattle ranching is a major driver of tropical deforestation, and reduced consumption of beef in the OECD countries would alleviate some of the pressure on the tropical forests as land prices would probably drop. This aspect suggests that the BCR would be higher for a beef tax than that we have estimated here. Furthermore, decreased land demand and reduced land prices will increase the cost-effective potential of using bioenergy as a carbon mitigation option in the energy system (Wirsenius *et al.* 2009).

Recommendations and Conclusions

As noted in the introduction we find ourselves in agreement with the abatement estimates presented in the chapter. Instead of discussing these abatement measures in detail, we have mainly discussed aspects related to the benefit estimates of CH_4 abatement. As we wrote above there are serious problems with benefit estimates concerning climate measures. For that reason, we recommend the use of shadow prices estimated from models with specific climate targets. We therefore suggest that the Copenhagen Consensus should use shadow prices estimated from IAMs where the widely supported 2 K target is taken into account. We further suggest that the shadow prices are estimated assuming a climate sensitivity of 3 K and a discount rate of 3%.

[13] In this case, the relative value of N_2O to CO_2 is 300.

By using these mid-range estimates for our calculation, we end up with shadow prices that slightly exceed the high SCC presented in the chapter of 46 US$/t CO_2. To support shadow prices of around 15 US$/ton CO_2, also presented in the chapter, we either have to assume a less stringent climate target or assume a low climate sensitivity and a rather high discount rate of 5%.

Even though we support the assumption of a carbon price of about 50 US$/t CO_2, the cost-effective approach prescribes that the relative valuation of CH_4 is considerable lower than estimated using GWP calculated over 100 years. Instead of valuing CH_4 as 21 times as high as CO_2 per ton emission as, the chapter implicitly does, we suggest that CH_4 should only be valued about 6 times as high (see table 5.2.2).

When considering that CH_4 is an important precursor to the global level of tropospheric O_3, the relative value of CH_4 emissions should, however, increase. By assuming that a global average VSL is US$ 1 million, the O_3-related benefit of CH_4 mitigation more than doubles the cost-effective valuation of CH_4 (see table 5.2.3). However, the literature concerning the non-climate tropospheric O_3 benefits of CH_4 abatement is sparse. The numbers should therefore be seen as rather preliminary.

Thus, taking a cost-effectiveness approach (assuming a 2 K target, a climate sensitivity of 3 K, and a discount rate of 3%) and valuing the non-climate benefit of tropospheric O_3, the B/C numbers presented in the chapter with an SCC of 46 US$/t CO_2 should roughly be in line with what we suggest. If the Expert Panel appointed by the Copenhagen Consensus Center prefers another approach, the relative weight of CH_4 should be adjusted accordingly, based on tables 5.2.1, 5.2.2, and 5.2.3. Also, we find little support for using either of the chapter's two lowest (implicit assumptions) on the social cost/shadow price of CH_4 if methane's impact on tropospheric O_3 is considered.

As discussed in the chapter, many sources of CH_4 are non-point emission sources. This makes it harder to regulate and control CH_4 emissions from these sources than, for example, the pricing of CO_2 emissions from fossil fuels. Still, the chapter recommends using their Portfolio 1, which includes several mitigation options in the agricultural sector, unless policy makers find the implementation barriers too large. The chapter also argues that the mitigation efforts of CH_4 should be spread over several sectors to diversify risk. We claim that there are reasons to diversify risk for climate mitigation as a whole, but not for measures targeting only CH_4. Furthermore, we argue that there are three reasons why policy makers should not rely to any large extent on technical mitigation options in the agricultural sector for now. First, as the chapter also points out, the engineering cost estimates may be in reality higher for several reasons – e.g. transaction costs and intangible costs are not taken into account and may be large. Secondly, we are not convinced that there are not significant indirect emissions of GHGs for some mitigation options. For instance, adding fat to cattle feed to reduce the CH_4 emissions from enteric fermentation could lead to large indirect emissions. Oil crops often cause quite large N_2O emissions – and, perhaps even more important – palm oil is a major driver of deforestation in Malaysia and Indonesia, thus causing large CO_2 emissions. Finally, there are yet no convincing policy instruments suggested in the literature that would induce CH_4 abatement measures in the agricultural sector. As the emissions can hardly be taxed directly or included in permit trading schemes, due to their high monitoring costs, it is hard to provide reliable incentives to farmers to adopt these measures.

For these reasons we suggest that one should focus primarily on CH_4 emissions from solid waste management, coal mine CH_4, and natural gas, thus aiming at large-scale sources which are easy to monitor. The possibility for successful implementation is much larger. If emissions in the agricultural sector are to be targeted we suggest a tax of around 1 US$/kg beef (carcass weight) to affect the diets of people in the OECD countries. This policy would be, if it gained political acceptance, fairly easy to implement.

Just as the chapter points out that its estimates are very rough we would like to do the same concerning our estimates. We have tried to give crude numbers on how to adjust the chapter's BCRs to be consistent with a cost-effectiveness approach, and have added the benefit CH_4 abatement has on tropospheric O_3. If the Expert Panel appointed by the Copenhagen Consensus Center prefers a B/C approach to climate change we have provided numbers so that BCRs can be calculated. In addition, we have also provided rough and tentative numbers on the BCR of a beef tax. All these calculations are inherently uncertain due to both parametric and structural uncertainties and simplifications. Still, we think that our estimates complement the data provided in the chapter and also give guidance for the Copenhagen Consensus on some crucial aspects in order to make a consistent assessment of different mitigation efforts.

Appendix: The MiMiC Model

The Multi-gas Mitigation Climate (MiMiC) model is a globally aggregated optimizing IAM. What is used here is an updated version of the MiMiC model presented and used in Johansson *et al.* (2006, 2008). The main differences between the model used here and the versions in Johansson *et al.* (2006, 2008) are that the energy balance model has been improved (by the use of an upwelling-diffusion energy balance model), the carbon sink has been recalibrated, climate feedbacks on the carbon cycle are taken into account, and updated data are used to initialize and fit the model to historical global average radiative forcing and surface temperature levels.

The model runs between the years 1880 and 2200, with yearly time-steps over the period 1880–2004 and with five-year time-steps over the period 2005–2200. The period 1880–2004 is used to calibrate the forcing strength of aerosols and initialize the carbon-cycle model and the energy balance model.

CO_2 concentrations are modeled by a linear pulse representation of the Bern carbon-cycle model based on Joos *et al.* (1996). CH_4 and N_2O concentrations are modeled using the global mean mass-balance equations (Prather *et al.* 2001), taking the feedback effect CH_4 has on its own atmospheric lifetime into account. The equations for radiative forcing are the expressions given in the IPCC Third Assessment Report (Ramaswamy *et al.* 2001). We also include the indirect effect of CH_4 concentrations on tropospheric O_3 and stratospheric water vapor concentrations (Wigley *et al.* 2002). The

relationship between aerosols emission and their direct and indirect radiative forcing is assumed to be linear.

The energy balance model used to calculate the temperature response from changes in the radiative forcing is based on a linear upwelling-diffusion energy balance model with polar overturning. The model is calibrated to emulate the global average surface response of Atmosphere/Ocean General Circulation Models (AOGCMs) (see Johansson 2009 for more details).

Abatement costs are modeled with the aid of abatement cost functions. The abatement costs of CO_2 abatement are based on the EPPA model and the GET model, while the abatement cost of reducing CH_4 and N_2O is primarily based on USEPA (2006) and EMF21.

Baseline scenarios for the period 2010–2100 for CO_2, CH_4, and N_2O and for GWP are taken from the IIASA A2r scenario which is an updated version of the SRES A2 scenario. After 2100, these scenarios are extrapolated. Abatement of emissions is only allowed from 2015 onwards. CO_2 emissions from land-use change follow the A2r scenario and abatement of these emissions is not considered.

The radiative forcing for halocarbons and aerosols are assumed to exogenously decline over time. For halocarbons the radiative forcing declines at 1% per year. This decline rate corresponds to the inverse of the atmospheric lifetime of the CFC with the highest forcing – i.e. CFC-12. For aerosols the radiative impact is constant at the 2000 level up until 2015 and then declines at 2% per annum.

In the case when the model is run as a cost-benefit model the damage function from Nordhaus (2008) is used. The climate sensitivity and discount rate is varied in this Perspective paper in order to show the great importance of these two parameters.

Bibliography

Allais, O. and V. Nichele, 2007: Capturing structural changes in french meat and fish demand over the period 1991–2002, *European Review of Agricultural Economics* **34**(4), 517–38

Allan, R. *et al.*, 2007: *Bali Climate Declaration by Scientists*, Climate Change Research Centre, University of New South Wales (UNSW),

Sydney, www.ccrc.unsw.edu.au/news/2007/Bali.html

Archer, D., M. Eby, V. Brovkin, A. Ridgwell, L. Cao, U. Mikolajewicz, K. Caldeira, K. Matsumoto, G. Munhoven, A. Montenegro, and K. Tokos, 2009: Atmospheric lifetime of fossil fuel carbon dioxide, *Annual Review of Earth and Planetary Sciences* **37**, 117–34

Azar, C., 1998: Are optimal CO_2 emissions really optimal? – Four critical issues for economists in the greenhouse, *Environmental and Resource Economics* **11**(3–4), 301–15

Banse, M., H. Grethe, and S. Nolte, 2005: *Documentation of ESIM Model Structure, Base Data and Parameters*, in GAMS, User Handbook, Berlin and Göttingen

Burton, M. and T. Young, 1992: The structure of changing tastes for meat and fish in Great Britain, *European Review of Agricultural Economics* **19**(2), 165–80

Carlsson-Kanyama, A. and A.D. González, 2009: Potential contributions of food consumption patterns to climate change, *American Journal of Clinical Nutrition* **89**(5), 1704–9

Cederberg, C., U. Sonesson, M. Henriksson, V. Sund, and J. Davis, 2009: *Greenhouse Gas Emissions from Production of Meat, Milk and Eggs in Sweden 1990 and 2005*, SIK Report **793**, Swedish Institute for Food and Biotechnology (SIK), Gothenburg

Dasgupta P., 2008: Discounting climate change, *Journal of Risk and Uncertainty* **37**, 141–69

FAO, 2009: FAO statistical database, http://faostat.fao.org/default.aspx

Fiore A.M., J.J. West, L.W. Horowitz, V. Naik, and M.D. Schwarzkopf, 2008: Characterizing the tropospheric ozone response to methane emission control and the benefits to climate and air quality, *Journal of Geophysical Research* **113**(D8), D08307

Forster, P., V. Ramaswamy, P. Artaxo, T. Berntsen, R. Betts, D.W. Fahey, J. Haywood, J. Lean, D.C. Lowe, G. Myhre, J. Nganga, R. Prinn, G. Raga, M. Schulz, and R. Van Dorland, 2007: Changes in atmospheric constituents and in radiative forcing, *Climate Change 2007: The Physical Science Basis – Contribution of Working Group I to the Fourth Assessment Report of the Intergovernmental Panel on Climate Change*, S. Solomon, D. Qin, M. Manning, Z. Chen, M. Marquis, K.B. Averyt,

M. Tignor, and H.L. Miller (eds.), Cambridge University Press, Cambridge

Fuglestvedt, J.S., T.K. Berntsen, O. Godal, R. Sausen, K.P. Shine, and T. Skodvin, 2003: Metrics of climate change: assessing radiative forcing and emission indices, *Climatic Change* **58**, 267–331

IIASA, 2009: GGI Scenario Database, www.iiasa.ac.at/Research/GGI/DB/

IPCC, 2007: Summary for Policymakers, *Climate Change 2007: The Physical Science Basis – Contribution of Working Group I to the Fourth Assessment Report of the Intergovernmental Panel on Climate Change*, S. Solomon, D. Qin, M. Manning, Z. Chen, M. Marquis, K.B. Averyt, M. Tignor, and H.L. Miller (eds.), Cambridge University Press, Cambridge

Jansen, E., J. Overpeck, K.R. Briffa, J.-C. Duplessy, F. Joos, V. Masson-Delmotte, D. Olago, B. Otto-Bliesner, W.R. Peltier, S. Rahmstorf, R. Ramesh, D. Raynaud, D. Rind, O. Solomina, R. Villalba, and D. Zhang, 2007: Palaeoclimate, *Climate Change 2007: The Physical Science Basis – Contribution of Working Group I to the Fourth Assessment Report of the Intergovernmental Panel on Climate Change*, S. Solomon, D. Qin, M. Manning, Z. Chen, M. Marquis, K.B. Averyt, M. Tignor, and H.L. Miller (eds.), Cambridge University Press, Cambridge

Johansson, D.J.A., 2009: Temperature stabilization, ocean heat uptake and radiative forcing profiles, submitted to *Climate Change*

Johansson, D.J.A., U.M. Persson, and C. Azar, 2006: The cost of using global warming potentials: analysing the trade off between CO_2, CH_4, and N_2O, *Climatic Change* **77**, 291–309

2008: Uncertainty and learning: implications for the trade-off between short-lived and long-lived greenhouse gases, *Climatic Change* **88**, 293–308

Joos, F., M. Bruno, R. Fink, U. Siegenthaler, T.F. Stocker, C. Le Quéré, and J.L. Sarmiento, 1996: An efficient and accurate representation of complex oceanic and biospheric models of anthropogenic carbon uptake, *Tellus B* **48**(3), 397–417

Kandlikar, M., 1996: Indices for comparing greenhouse gas emissions: Integrating science and economics, *Energy Economics* **18**, 265–81

Kemfert, C. and W.-P. Schill 2009: Methane mitigation, chapter 5 in this volume

Kriegler, E., J.W. Hall, H. Held, R. Dawson, and H.J. Schellnhuber, 2009: Imprecise probability assessment of tipping points in the climate system, *Proceedings of the National Academy of Sciences* **106**(13), 5041–6

Manne, A.S. and R.G. Richels, 2001: An alternative approach to establishing trade-offs among greenhouse gases, *Nature* **410**(6829), 675–77

Matthews, H.D. and K. Caldeira, 2008: Stabilizing climate requires near-zero emissions, *Geophysical Research Letters* **35**, L04705

Nordhaus, W.D., 2008: *A Question of Balance: Economic Modeling of Global Warming*, Yale University Press, New Haven, CT

Parry M.L., O.F. Canziani, J.P. Palutikof, P.J. Van Der Linden, and C.E. Hanson (eds), 2007: *Climate Change 2007: Impacts, Adaptation and Vulnerability – Contribution of Working Group II to the Fourth Assessment Report of the Intergovernmental Panel on Climate Change*, Cambridge University Press, Cambridge

Prather, M., D.H. Ehhalt, F. Dentener, R. Derwent, E. Dlugokencky, E. Holland, I. Isaksen, J. Katima, V. Kirchhoff, P. Matson, P. Midgley, and M. Wang, 2001: Atmospheric chemistry and greenhouse gases, *Climate Change 2001: The Scientific Basis – Contribution of Working Group I to the Third Assessment Report of the Intergovernmental Panel on Climate Change*, J.T. Houghton, Y. Ding, D.J. Griggs, M. Noguer, P.J. van der Linden, and D. Xiaosu (eds.), Cambridge University Press, Cambridge

Ramaswamy, V., O. Boucher, J. Haigh, D. Hauglustaine, J. Haywood, G. Myhre, T. Nakajima, G.Y. Shi, and S. Solomon, 2001: Radiative forcing of climate change, *Climate Change 2001: The Scientific Basis – Contribution of Working Group I to the Third Assessment Report of the Intergovernmental Panel on Climate Change*, J.T. Houghton, Y. Ding, D.J. Griggs, M. Noguer, P.J. van der Linden, and D. Xiaosu (eds.), Cambridge University Press, Cambridge

Reilly, J.M. and K.R. Richards, 1993: Climate change damage and the trace gas index issue, *Environmental and Resource Economics* **3**, 41–61

Richardson, K., W. Steffen, H.J. Schellnhuber, J. Alcamo, T. Barker, D. Kammen, R. Leemans, D. Liverman, M. Munasinghe, B.

Osman-Elasha, N. Stern, and O. Waever, 2009: *Synthesis Report: Climate Change – Global Risks, Challenges and Decisions*, University of Copenhagen, Copenhagen, www.climatecongress.ku.dk

Schmutzler, A. and L.H. Goulder, 1997: The choice between emission taxes and output taxes under imperfect monitoring, *Journal of Environmental Economics and Management* **32**(1), 51–64

Shindell, D.T., G. Faluvegi, W. Bell, and G.A. Schmidt, 2005: An emissions-based view of climate forcing by methane and tropospheric ozone, *Geophysical Research Letters* **32**(4), L048031

Stehfest, E., L. Bouwman, D.P. van Vuuren, M.G.J. den Elzen, B. Eickhout, and P. Kabat, 2009: Climate benefits of changing diet, *Climatic Change* **95**(1–2), 83–102

Smith, J.B., S.H. Schneider, M. Oppenheimer, G.W. Yohe, W. Hare, M.D. Mastrandrea, A. Patwardhan, I. Burton, J. Corfee-Morlot, C.H.D. Magadza, H.M. Füssel, A.B. Pittock, A. Rahman, A. Suarez, and J.P. van Ypersele, 2009: Assessing dangerous climate change through an update of the Intergovernmental Panel on Climate Change (IPCC) "reasons for concern," *Proceedings of National Academy Science* **106**, 4133–7

Solomon, S., G.K. Plattner, R. Knutti, and P. Friedlingstein, 2009: Irreversible climate change due to carbon dioxide emissions, *Proceedings of National Academy of Sciences* **106**(6), 1704–9

Stern, N.H., 2007: *The Economics of Climate Change: The Stern Review*, Cambridge University Press, Cambridge

Sterner, T., 2003: *Policy Instruments for Environmental and Natural Resource Management*, RFF Press; Washington, DC, USA

Stitch, S., P.M. Cox, W.J. Collins, and C. Huntingford, 2007: Indirect radiative forcing of climate change through effects on the land-carbon sink, *Nature* **448**, 791–4

Tanaka, K., 2008: *Inverse Estimation for the Simple Earth System Model ACC2 and Applications*, PhD thesis, International Max Planck Research School on Earth System Modelling, Hamburg

The Royal Society, 2008: *Ground-Level Ozone in the 21st Century: Future Trends, Impacts and Policy Implications*, Science Policy Report **15/08**, royalsociety.org

Tol, R.S.J., 2008: The social cost of carbon: trends, outliers, and catastrophes, *Economics* **2**, 2005–25

2009: The economic effects of climate change, *Journal of Economic Perspectives* **23**(2), 29–51

USEPA, 2006: *Global Mitigation of Non-CO$_2$ Greenhouse Gases*, US Environmental Protection Agency, Office of Atmospheric Programs (**6207J**), Washington, DC

van Vuuren, D.P., J. Weyant, and F. de la Chesnaye, 2006: Multi-gas scenarios to stabilize radiative forcing, *Energy Economics* **28**, 102–20

Warren, R., 2006: Impacts of global climate change at different annual mean global temperature increases, in H.J. Schellnhuber, W. Cramer, N. Nakicenovic, T. Wigley, and G. Yohe (eds.), *Avoiding Dangerous Climate Change*, Cambridge University Press, Cambridge

Warren, R., C. Hope, M. Mastrandrea, P.S.J. Tol, N. Adger, and I. Lorenzoni, 2006: *Spotlighting Impact Functions in Integrated Assessment: Research Report Prepared for the Stern Review on the Economics of Climate Change*, Tyndall Centre for Climate Change Research Working Paper **91**

Watkiss, P. and T. Downing 2008: The social cost of carbon: valuation estimates and their use in UK policy, *Integrated Assessment* **8**(1), 85–105

Weitzman, M.L., 2009: On Modeling and interpreting the economics of catastrophic climate change, *The Review of Economics and Statistics* **91**(1), 1–19

West, J.J. and A.M. Fiore, 2005: Management of tropospheric ozone by reducing methane, *Enviornmental Science & Technology* **39**(13), 4685–91

West, J.J., A.M. Fiore, L.W. Horowitz, and D.L. Mauzerall, 2006: Global health benefits of mitigating ozone pollution with methane emissions control, *Proceedings of National Academy of Sciences* **103**(11), 3988–93

West, J.J, S. Szopa, and D.A. Hauglustaine, 2007: Human mortality effects of future concentrations of tropospheric ozone, *Comptes Rendus Geosciences* **339**, 775–83

Williams, A.G., E. Audsley, and D.L. Sandars, 2006: *Determining the Environmental Burdens and Resource Use in the Production of Agricultural and Horticultural Commodities*, Silsoe Research Institute, Cranfield University, Bedford

Wigley, T.M.L., S.J. Smith, and M.J. Prather, 2002: Radiative forcing due to reactive gas emissions, *Journal of Climate* **15**, 2690–6

Wirsenius, S., F. Hedeus, and K. Mohlin, 2009: Greenhouse gas taxes on animal food products: rationale, tax scheme and climate mitigation effects, submitted to *Climatic Change*

Market- and Policy-Driven Adaptation

FRANCESCO BOSELLO, CARLO CARRARO, AND ENRICA DE CIAN*

Introduction

Adaptation to climate change has become a strategic negotiation issue only recently, although the UNFCCC has referred to it in Article 2 and Article 4. The difficulty of implementing national and international mitigation policies and the increasing awareness of climate inertia eventually put adaptation under the spotlight of science and policy. The EU released a Green Paper on Adaptation (European Commission 2007b) and many EU countries have prepared and started to implement national adaptation plans. The Bali action plan adopted at the December 2007 conference has identified the need for enhanced adaptation action by the Parties of the Convention, and adaptation is among the five key building blocks for a strengthened response to climate change. COP 13 has established the Adaptation Fund Board with the role of managing the Adaptation Fund, established at COP 7. COP 14, held in Poznán in 2008, also made some progress on a number of important issues concerning adaptation.

The ultimate question that interests policy makers is how to reduce the climate change vulnerability of socioeconomic systems in the most cost-effective way. This objective can be achieved with both mitigation and adaptation. It also requires, on the one hand, a thorough knowledge of the size

and the regional distribution of damages and, on the other hand a precise assessment of the cost-effectiveness of alternative policies.

Given its local- and project-specific nature, a cost-benefit analysis (CBA) of adaptation strategies has been treated from a micro-perspective. Although this approach can inform us about the economic performance of specific projects, it lacks a broader perspective on the interactions with other economic activities. Adaptation is only one of the possible responses to global warming, within a range of possible options. In order to maximize the benefit from a portfolio of alternatives, a joint analysis can be more informative.

If an extended literature has investigated the different dimensions of mitigation strategies, much less can be found on adaptation. Even less attention has been paid to the interactions between adaptation and mitigation. At the same time, the interest in defining their strategic complementarity or trade-off in a macroeconomic cost-benefit (C/B) context has constantly risen (EEA 2007; Parry *et al.* 2007; Stern 2007; Parry 2009).

The increasing emphasis on the role of adaptation as an aggregated climate change strategy raises a set of still unanswered questions concerning the design of an optimal mix of mitigation and adaptation measures, the benefit-cost ratio (BCR) of different adaptation/mitigation options, and their regional characterization. New relevant insights need to be provided on the optimal resource allocation between mitigation and adaptation, on the optimal timing of mitigation and adaptation measures, and on their marginal contribution to reducing vulnerability to climate change. However, as recently observed (Parry 2009), a framework that explicitly models the connections between mitigation, climate change impacts, and adaptation is still missing.

* AD-WITCH, the model used in this study, has been developed by FEEM in cooperation with the OECD. The authors gratefully acknowledge their financial support. They are also grateful to Shardul Agrawala, Rob Dellink, Kelly de Bruin, and Richard Tol for helpful comments. Nonetheless, the views expressed in this chapter are the authors' sole responsibility. Finally, the contribution of all colleagues who worked to the development of the original WITCH model – in particular, Valentina Bosetti, Emanuele Massetti, and Massimo Tavoni – is gratefully acknowledged.

Against this background, the present research adopts a macro-angle. The chapter addresses these and other issues using the AD-WITCH model, an integrated assessment model (IAM) that has been developed for the joint analysis of adaptation and mitigation. Compared to the few existing studies in the field, the proposed modeling framework provides a more detailed characterization of the adaptation process, which is disaggregated into three components: anticipatory, reactive, and innovative adaptation. In addition, it provides updated quantitative support for the calibration of adaptation costs and benefits at the regional level. Therefore, in this chapter, we will be able to:

- Analyze adaptation to climate change both in isolation and jointly with mitigation strategies
- Provide a comparative CBA of both adaptation and mitigation
- Assess the marginal contribution to the BCR of different adaptation modes
- Emphasize region-specific characteristics.

We will start with a CBA of macro-, policy-driven responses to climate change, namely adaptation, mitigation, and joint adaptation and mitigation. By narrowing down the focus on policy-driven adaptation, we will then compute the benefit-cost ratios (BCRs) of three macro-adaptation strategies (reactive, anticipatory or proactive, and knowledge adaptation). We will also assess how market-driven adaptation reduces the vulnerability of economic systems to climate change. Finally, we will re-compute the BCRs for different policy-driven adaptation strategies net of market-driven, autonomous adaptation to climate change.

AD-WITCH, the model used to carry out most of the analysis, is an optimal growth IAM endowed with an adaptation module to compute the costs and benefits of policy-driven mitigation and adaptation strategies. Given the game-theoretic and regional structure of AD-WITCH (see appendix 1, p. 260), both first-best and second-best climate policies can be computed. In this chapter, we focus on a first-best world in which all externalities are internalized. The social planner implements the optimal levels of adaptation and mitigation – that is, the levels that equalize marginal costs and benefits.

This chapter emphasizes that adaptation can also be driven by changes in relative prices, which lead to what can be defined as market-driven adaptation. Market-driven adaptation may affect the size and the regional distribution of climate change damages. As a consequence, policy-driven adaptation should be planned on the basis of climate change damages net of market adjustments.

To account for both market-driven and policy-driven adaptation, two different modeling tools have been used. The ICES model, which is a highly disaggregated computable general equilibrium (CGE) model, has been used to identify the effects of market–driven adaptation. ICES and AD-WITCH have then been integrated to provide a full assessment of both market- and policy-driven adaptation. More precisely, the effects of market-driven adaptation on regional climate damages have been estimated using the ICES model. These estimates have been used to modify climate change damage functions in the WITCH model to compute climate damages net of market-driven adaptation.

The final part of this chapter describes specific adaptation proposals. These are consistent with the analysis carried out in the first part of the chapter, and build upon existing estimates of the costs and benefits of specific adaptation strategies.

Background Concepts

In this chapter, climate change is defined as a set of alterations in the average weather caused by global warming, which is due to emissions of greenhouse gases (GHGs). Climate change affects not only average surface temperature but also involves other physical modifications, such as changes in precipitation, intensity and frequency of storms, and the occurrence of droughts and floods.

Average temperature is already $0.7°C$ above the preindustrial level and further warming might be substantial if no immediate global action is undertaken. Even if all radiative forcing agents were held constant at the 2000 level, a further warming would be observed due to the inertia of oceans. According to the main IPCC scenarios,[1] world-average

[1] The SRES scenarios A2, B1, and B2 are from www.iiasa. ac.at/Research/GGI/DB/.

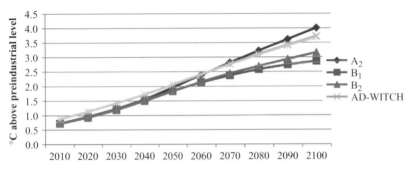

Figure 6.1 *Temperature estimates of the IPCC SRES (IIASA), the WITCH model and the AD-WITCH baseline scenario*

temperature is likely to increase in the business-as-usual (BAU) scenarios as shown in figure 6.1, which also shows our projections. Projected global temperature increases above preindustrial levels range between 2.8 and 4°C.

Anthropogenic climate change, accelerating the natural trend, will induce a series of impacts on natural and social ecosystems with potentially both negative and positive consequences on human well-being. As highlighted in the IPCC Fourth Assessment Report (Parry *et al.* 2007), already a moderate warming has produced negative consequences: increasing number of people exposed to water stresses, extinction of species and ecosystems, decrease in cereal productivity at low latitudes, land loss due to sea level rise in coastal areas, increases in mortality and morbidity associated with change in the incidence of vector-borne diseases or to increased frequency and intensity of heatwaves, infrastructural disruption and mortality increase due to more frequent and intense extreme weather event occurrence.

A first classification of climate change impacts distinguishes between market and non-market impacts. Market effects can be valued using prices and observed changes in demand and supply, whereas non-market effects have no observable prices and therefore require other methods such as valuations based on willingness-to-pay (WTP).

The recent literature points to the large potential damages from climate change, especially in developing countries and in non-market sectors (Parry *et al.* 2007; Stern 2007). In particular, important non-market impacts are those on health,

although current estimates are largely incomplete and most assessments have looked at specific diseases (vector-borne diseases, cardiovascular and respiratory diseases). Moreover, indirect economic implications may be relevant. Nonetheless, for the USA only, Hanemann (2008) estimates large impacts on health, reporting a loss of 1990 $US10 billion per year against the $US 2 billion reported in Nordhaus and Boyer (2000).

Climate change can lead to a significant rise in sea level and catastrophic events, with implications for migrations and the stock of capital. Insurance companies are an important source of information regarding estimates of capital losses due to climate change impacts. The United Nations UNFCCC (UNFCCC 2007) reports a cost of protecting infrastructure from climate change in North America between 1990 $US4 and 64 billion already in 2030, when temperature increase is likely to be far below 2.5°C.

The Munich Re insurance company has developed a database which catalogues great natural catastrophes that have had severe impacts on the economic system. Such a database underestimates damages from climate because only large events are included. Yet estimated losses are in the order of 0.5% of current world GDP, and damages are increasing at a rate of 6% a year in real terms. Using this information and adjusting for the underreporting of other minor impacts, UNFCCC (2007) extrapolated a cost between 1 and 1.5% of world GDP in 2030, which corresponds to 1990 $US850–1350 billion. Nordhaus and Boyer (2000) reported similar figures for total impacts, and for a

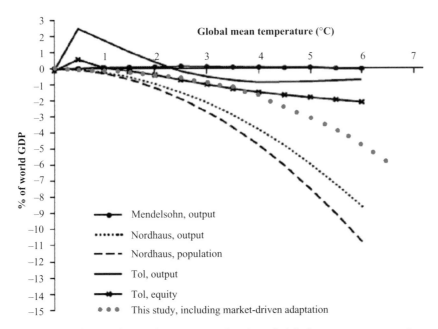

Figure 6.2 *Climate change damages as a function of global mean temperature increase (above preindustrial levels)*

temperature increase of 2.5°C, which is likely to occur at least several decades after 2030.

For a temperature increase above 2.5°C, the majority of IAMs currently used to evaluate the full cost of climate change forecast net losses from climatic changes ranging roughly from a tiny percentage to 2% of world GDP (figure 6.2).

Climate change is not uniform over the world, though; moreover, impacts are diverse and highly differentiated by region. Regions themselves differ in their intrinsic adaptive capacity. These dimensions – i.e. exposure, sensitivity, and autonomous adaptive capacity – determine a highly differentiated regional vulnerability to climate change. Accordingly, the global picture can provide only a very partial and potentially misleading insight into the true economic cost of climate change. Aggregation can indeed conceal vulnerability and climate change costs "hot spots", as depicted in table 6.1. As a general rule, developing countries would be more affected than their developed country counterparts.

Notwithstanding the differences in results, – driven by different model specifications, modeling approaches, and underlying assumptions – the

inspection of table 6.1 highlights two robust messages:

• Even an almost null aggregate loss potentially experienced by the world as a whole, and associated with a moderate climatic change, entails high costs for some regions. It is even more so in the case of moderate-to-high aggregate economic losses.
• There is a clear equity-adverse effect from the distribution of climate change impacts: higher costs are experienced by developing regions which are already facing serious challenges to their social and economic development; moreover, within a country or region, climate change adverse effects hit more severely the weaker social groups which are both more exposed and less able to adapt.

What is true at the world level applies at the regional level as well. Even a net gain for a region compounds both positive and negative effects. Some of these negative effects can be particularly concerning also for a developed region. Think, for instance, of an increase in mortality due to more frequent

Table 6.1 Regional climate change impacts as a percentage of GDP corresponding to a temperature increase of 2.5°C (negative figures are gains)

	ICES model (Bosello et al. 2009)[a]	AD-WITCH model (Bosello et al. 2009)[b]	Fankhauser (Fankhauser and Tol 1996)	Tol (Fankhauser and Tol 1996)	Nordhaus and Boyer (2000)	Mendelsohn et al. (2000)	Pearce et al. (1996)
USA	0.2	0.4	1.3	1.5	0.4	−0.3	1
WEURO	−1.3	1.6	1.4	1.6	2.8	n.a.	1.4
EEURO	0.8	0.5	n.a.	0	0.7	n.a.	−0.3
KOSAU	0.9	0.8	n.a.	0	−0.4	n.a.	1.4
CAJANZ	−0.8	0.5	n.a.	3.8	0.5	0.1	1.4
TE	0.9	0.8	0.4	−0.4	−0.7	−11	0.7
MENA	0.2	2.9	n.a.	5.5	1.9	n.a.	4.1
SSA	2.0	5.1	n.a.	6.9	3.9	n.a.	8.7
SASIA	3.0	5.5	n.a.	0	4.9	2	n.a.
CHINA	1.7	0.5	2.9	−0.1	0.2	−1.8	5
EASIA	2.3	4.2	n.a.	5.3	1.8	n.a.	8.6
LACA	1.8	2.3	n.a.	3.1	2.4	1.4	4.3

Notes:
[a] This study includes market-driven adaptation.
[b] This study includes only policy-driven adaptation.
Source: Our adaptation from the quoted studies.

and intense heatwaves, hitting an aged population; loss of coastal areas due to sea level rise; increase in hydrogeological risk due to an increase in frequency and intensity of extreme weather events. Table 6.2 summarizes the damage estimates for a 2.5°C increase in global temperature above its 1900 level, both for the whole economy (Total column) and broken down by sector, as estimated in Nordhaus and Boyer (2000).

Among rich countries, Europe is estimated to suffer most from climate change because of the assumption of high vulnerability to catastrophic events. Among developing regions, Africa and India face larger climate impacts due to effects on health and catastrophic events, respectively. Impacts on agriculture vary a lot with the climatic conditions of the regions and become positive for cold or mild regions (e.g. Russia, China). A similar pattern can be identified for impacts on energy use, with cold regions being more positively affected (Russia).

Current climate change policies – under discussion within the EU (European Commission 2005, 2007a) – aim at setting a prudential 2°C threshold to temperature increase above the prein-

dustrial level within the century. The aim is to limit the impacts of climate change and the likelihood of massive and irreversible disruptions of the global eco-system (European Commission 2007a: 2). Thus, even assuming a successful accomplishment, the world will be in any case exposed to a certain degree of climate change and to its negative consequences for the century to come. Moreover, the stated target is considered particularly ambitious: it requires aggressive mitigation actions from developed regions, coupled with an extended international participation involving a still-to-reach explicit commitment to binding emission reduction from major polluters among developing countries. Accordingly, it is very likely that the world will face a higher temperature increase and more damaging consequences than those expected from a 2°C warming.

In the light of this, as stressed by the EU White Paper on Adaptation (European Commission 2009), mitigation needs to be necessarily coupled with adaptation actions. These, be they anticipatory or reactive, represent the only viable option to cope with the unavoidable climate change impacts that mitigation cannot eliminate.

Table 6.2   Climate change impacts in different world regions under a 2.5°C increase in global temperature above preindustrial levels

Region	Total	Agriculture	Other vulnerable market	Coastal	Health	Non-market time use	Catastrophic	Settlements
USA	0.45	0.06	0	0.11	0.02	−0.28	0.44	0.1
China	0.22	−0.37	0.13	0.07	0.09	−0.26	0.52	0.05
Japan	0.5	−0.46	0	0.56	0.02	−0.31	0.45	0.25
EU	2.83	0.49	0	0.6	0.02	−0.43	1.91	0.25
Russia	−0.65	−0.69	−0.37	0.09	0.02	−0.75	0.99	0.05
India	4.93	1.08	0.4	0.09	0.69	0.3	2.27	0.1
Other high-income	−0.39	−0.95	−0.31	0.16	0.02	−0.35	0.94	0.1
High-income OPEC	1.95	0	0.91	0.06	0.23	0.24	0.46	0.05
Eastern Europe	0.71	0.46	0	0.01	0.02	−0.36	0.47	0.1
Middle-income	2.44	1.13	0.41	0.04	0.32	−0.04	0.47	0.1
Lower middle-income	1.81	0.04	0.29	0.09	0.32	−0.04	1.01	0.1
Africa	3.91	0.05	0.09	0.02	3	0.25	0.39	0.1
Low-income Global	2.64	0.04	0.46	0.09	0.66	0.2	1.09	0.1
Output-weighted	1.5	0.13	0.05	0.32	0.1	−0.29	0.17	1.02
Population weighted	2.19	0.17	0.23	0.12	0.56	−0.03	0.1	1.05

Source: Nordhaus and Boyer (2000).

Defining Adaptation: A Multi-dimensional Concept

Adaptation to climate change received a wide set of definitions from the scientific and the policy environment (among the first group see, e.g., Burton 1992; Smit 1993; Smithers and Smit 1997; Smit *et al.* 2000; among the second group see, e.g., EEA 2005; Lim and Spanger-Siegfred 2005; UNFCCC 2006). The large number of not always coincident definitions already highlights a specific problem concerning adaptation: it is a process that can take the most diverse form depending on where and when it occurs, and on who is adapting to what.

Indeed, probably the most comprehensive, known, and widely accepted definition of adaptation is the one provided by the IPCC Third Assessment Report of 2001, which states that adaptation is any "adjustment in ecological, social, or economic systems in response to actual or expected climatic stimuli, and their effects or impacts. This term refers to changes in processes, practices or structures to moderate or offset potential damages or to take advantages of opportunities associated with changes in climate" (McCarthy *et al.* 2001), which is general enough to encompass the widest spectrum of options.

Adaptation can be identified along three dimensions:

- the *subject* of adaptation (who or what adapts)
- the *object* of adaptation (what they adapt to)
- the *way* in which adaptation takes place (how they adapt).

This last dimension includes what resources are used, when and how they are used, and with what results (Wheaton and Maciver 1999).

The subject of adaptation: who or what adapts. Adaptation materializes in changes in ecological, social, and/or economic systems. These changes can be the result of natural responses, and in this case they usually involve organisms or species, or of socioeconomic or institutional reactions, in which case they are undertaken by individual or collective actors, private or public agents.

Table 6.3 Adaptation: possible criteria for classification

Concept or attribute	
Purposefulness	Autonomous → Planned
Timing	Anticipatory → Reactive, Responsive
Temporal scope	Short term → Long term
Spatial scope	Localized → Widespread
Function/Effects	Retreat – Accommodate – Protect – Prevent
Form	Structural – Legal – Institutional
Valuation of performance	Effectiveness – Efficiency – Equity – Feasibility

Source: Our adaptation from Smit *et al.* (1999).

The object of adaptation: what they adapt to. In the case of climate change, adaptive responses can be induced either by changes in average conditions or by changes in the variability of extreme events. While in the first case the change is slow and usually falls within the coping range of systems, in the second case changes are abrupt and outside this coping range (Smit and Pilifosova 2001).

How adaptation occurs: modes, resources and results. The existing literature (see, e.g., Klein and Tol 1997; Fankhauser *et al.* 1999; Smit *et al.* 1999; McCarthy *et al.* 2001) proposes several criteria that can be used to identify the different adaptation processes. Table 6.3 offers a tentative summary of this classification based upon spatial and temporal aspects, forms, and evaluation of performances.

This chapter focuses on a different way of classifying adaptation to climate change, by distinguishing between autonomous or market-driven and planned or policy-driven adaptation. Inside policy-driven adaptation, we will distinguish between anticipatory or proactive and responsive or reactive adaptation.

The IPCC Third Assessment Report defines autonomous adaptation as "adaptation that does not

constitute a conscious response to climatic stimuli but is triggered by ecological changes in natural systems and by market or welfare changes in human systems" and planned adaptation as: "adaptation that is the result of a deliberate policy decision based on an awareness that conditions have changed or are about to change and that action is required to return to, maintain, or achieve a desired state" (McCarthy *et al.* 2001).

This apparently clear distinction may originate some confusion when adaptation involves socioeconomic agents. Indeed, climate change may induce market- or welfare effect-triggering reactions in private agents without the necessity of a planned strategy designed by a public agency, but just as a response to scarcity signals provided by changes in relative prices. A typical example of this is the effect of climate change on crops' productivity. This has both physical effects (changing yields) and economic effects (changing agricultural goods' prices) that can induce farmers to some adaptation (for example, changes in cultivation type or timing). This form of private socioeconomic adaptation, even though responding to a plan and originated by (rational) economic decisions, is considered autonomous or market-driven (see, e.g., Smit 1993; Leary 1999). On the contrary, the term "planned adaptation" is reserved for public interventions by governments or agencies.[2]

Another important distinction is the one based on the timing of adaptation actions which distinguishes between anticipatory or proactive adaptation and reactive or responsive adaptation. They are defined by the IPCC Third Assessment Report (McCarthy *et al.* 2001) as "adaptation that takes place before and after impacts of climate change are observed," respectively. There can be circumstances when an anticipatory intervention is less costly and more effective than a reactive action (a typical example is that of flood or coastal protection), and this is particularly relevant for planned adaptation. Reactive adaptation is a major characteristic of unmanaged natural system and of autonomous adaptation reactions of social and economic systems.

The temporal scope defines long-term and short-term adaptation. This distinction can also be referred to as tactical as opposed to strategic, or

[2] The IPCC (McCarthy *et al.* 2001) also provides the definition of private adaptation: "adaptation that is initiated and implemented by individuals, households or private companies. Private adaptation is usually in the actors' rational self interest" and of public adaptation: "adaptation that is initiated and implemented by governments at all levels. Public adaptation is usually directed at collective needs."

as instantaneous vs. cumulative. In the natural hazard field it is adjustment vs. adaptation (Smit *et al.* 2000).

For the sake of completeness, let us mention other classifications of adaptation. Based on spatial scope, adaptation can be localized or widespread, even though it is noted that adaptation has an intrinsic local nature (Füssel and Klein 2006). Several attributes can also characterize the effects of adaptation. According to Smit (1993) they can be: accommodate, retreat, protect, prevent, tolerate, etc. Based on the form adaptations can take they can be distinguished according to whether they are primarily technological, behavioral, financial, institutional, or informational.

Finally the performance of adaptation processes can be evaluated according to the generic principles of policy appraisal: cost-efficiency,[3] cost-effectiveness, administrative feasibility, and equity. As noted by Adger *et al.* (2005), in such appraisal effectiveness has to be considered *lato sensu*. Indeed, it is important to account for spatial and temporal spillovers of adaptation measures. Basically, a locally effective adaptation policy may negatively affect neighboring regions, and a temporary successful adaptation policy may weaken vulnerability in the longer term; both constitute examples of maladaptation. By the same token efficiency, effectiveness, and equity are not absolute, but context-specific, varying between countries, sectors within countries, and actors engaged in adaptation processes.

Mitigation and Adaptation as a Single Integrated Policy Process

Mitigation and adaptation are both viable strategies to combat damages due to climate change. However they tackle the problem from completely different angles.

Mitigation and adaptation work at different spatial and time scales. Mitigation is global and long-term while adaptation is local and short-term (Klein *et al.* 2003; Ingham *et al.* 2005a; Tol 2005; Wilbanks 2005; Füssel and Klein 2006). This has several important implications.

First, mitigation can be considered as a permanent solution to anthropogenic climate change. Indeed, once abated, 1 ton of, say, carbon dioxide (CO_2), cannot produce damage any longer (unless its removal is temporary as in the case of the carbon capture and sequestration provided by forests or agricultural land). In contrast, adaptation is more temporary as it typically addresses *current* or *expected* damages. It may require adjustments, if climate change damage varies or if it is substantially different from what was originally expected.

Secondly, the effects of mitigation and adaptation occur at different times (Wilbanks 2005; Klein *et al.* 2003; Füssel and Klein 2006). Mitigation is constrained by long-term climatic inertia, adaptation by a shorter term, social–economic inertia. In other words, emission reductions today will translate into a lower temperature increase and ultimately lower damage only in the (far) future, whereas adaptation measures, once implemented, are immediately effective in reducing the damage.[4] This differentiation is particularly relevant from the perspective of policy makers. The stronger reason for the low appeal of mitigation policies is probably due to their certain and present costs and future and uncertain benefits.[5] This issue is less problematic for adaptation. Moreover the different intertemporal characteristics tend to expose mitigation more than adaptation to subjective assumptions in policy decision making, like the choice of discount rates. It can be expected that a lower discount rate, putting

---

[3] The concept of cost-efficiency implies that resources are used in the best possible way. Cost-effectiveness means that resources to reach a given target – that can be suboptimal – are used in the best possible way. The practical implementation of both concepts requires that actions respond to some kind of B/C criterion.

[4] It has to be stressed that economic inertias can be long as well – e.g. implementing coastal protection interventions can take many years (or even decades) – and that adaptation may not be immediately effective as is the case for anticipatory adaptation.

[5] Füssel and Klein (2006) note that monitoring mitigation effectiveness is easier than monitoring adaptation. They refer to the fact that it is easier to measure emission reduction than quantify the avoided climate change damage due to adaptation. They do not refer to the quantification of the avoided future damage due to emission reduction.

more weight on future damages, can increase the appeal of mitigation with respect to adaptation.

Thirdly, mitigation provides a global good, whereas adaptation is a local response to anthropogenic climate change. The benefits induced by a ton of carbon abated are experienced irrespectively of where this ton has been abated. Put differently, adaptation entails measures implemented locally whose benefits advantage primarily the local communities targeted. The global public good nature of emissions reduction creates the well-known incentive to free ride. This is one of the biggest problems in reaching a large and sustainable international mitigation agreement (Carraro and Siniscalco 1998; Bosetti et al. 2009). Again this should be less of a problem in the case of adaptation policies.

It is worth mentioning that mitigation involves decision making at the highest level, such as by national governments. Mitigation is implemented at the country level (Tol 2005) and it concerns large, highly concentrated sectors – for example, energy and energy-intensive industries (Klein et al. 2003). Adaptation needs to be implemented at an atomistic level, involving a much larger number of stakeholders. Thus, at least in principle, the design of an international policy effort could be easier and the related coordination and transaction costs lower.

In the absence of international coordination, substantial unilateral mitigation actions are unlikely to occur. Here the concern is two-fold. On the one hand, the environmental effectiveness of unilateral action is likely to be small. On the other hand, the national goods and services of the abating country can lose competitiveness in international markets if their prices incorporate the cost of the tighter emission standards. This is not necessarily true with adaptation. Its smaller scale and the excludability of its benefits can make unilateral effort a viable choice.

The different regional effectiveness of mitigation and adaptation is also relevant in light of the spatial uncertainty of climate change damages (Lecoq and Shalizi 2007). Not knowing exactly where and with what intensity negative climatic impacts are going to hit, policy decision should be biased toward mitigation which is globally effective. On the contrary, adaptation should be used to deal with reasonably well understood local phenomena.

Finally, there is an equity dimension. Abatement intrinsically endorses the polluter-pays principle. Each one abates her own emissions (directly or indirectly if flexibility is allowed).[6] This is not necessarily the case with adaptation. It can well alleviate damages which are not directly provoked by the affected community. This is particularly important for international, especially North–South, climate negotiations. Indeed adaptation is particularly needed in developing countries which are either more exposed or vulnerable (higher sensitivity, lower capacity to adapt) to climate change (Watson et al. 1995; McCarthy et al. 2001; Parry et al. 2007), while historically they have contributed relatively less to the problem. Adaptation in developing countries thus calls objectively for strong international support.

Following a widely accepted efficiency principle according to which a wider portfolio of options should be preferred to a narrower one, the integration of mitigation and adaptation should increase the cost-effectiveness of a policy aimed at facing climate change (Ingham et al. 2005a; Kane and Yohe 2000; Parry et al. 2001). This is particularly true in light of the overall uncertainty that still surrounds our understanding of climatic, environmental, and social–economic processes, which ultimately determines the uncertainty in the assessment of the costs and benefits of climate change policy. In an uncertain framework, a precautionary policy would avoid the extremes of both total inaction and of drastic immediate mitigation. The optimal strategy would be a combination of mitigation and adaptation measures (Kane and Shogren 2000; McKibbin and Wilcoxen 2004). In other words, the decision maker needs to place herself somewhere inside the decision space represented by the triangle of figure 6.3. Vertexes are possible, but unlikely.

How should mitigation and adaptation be combined? This intuitively depends on their degree of substitutability or complementarity. Kane and Shogren (2000) analyze this issue in the context of the economic theory of endogenous risk.

---

[6] Again this is not necessarily so in the case of sequestration activities.

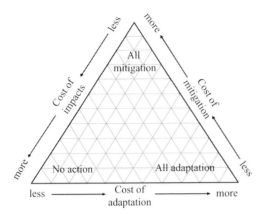

Figure 6.3 *Mitigation and adaptation impacts: a schematic "decision space"*

They demonstrate that when both mitigation and adaptation reduce the risk of adverse effects of climate change, they are used by agents until expected marginal benefits and costs are equated across strategies. Corner solutions (adaptation- or mitigation-only outcomes) are also discussed as theoretical possibilities. They could occur if, for instance, an international mitigation agreement failed to be signed, making agents aware of the practical ineffectiveness of (unilateral) mitigation action or if, conversely, the climate regime is so strict as to eliminate the necessity to adapt to any climate change damage. The analysis of agents' response to increased climate change risk is more complex. It depends on two effects: a direct effect of risk on the marginal productivity of a strategy and

an indirect effect of risk which is determined by risk impacts on the other strategy and by the relationship between the two. The indirect effect amplifies (dampens) the direct effect if the marginal productivity of one strategy increases (decreases) and the two strategies are complements (substitutes) or if marginal productivity decreases (increases) and the strategies are substitutes (complements). Kane and Shogren (2000) suggest that the actual relationship between mitigation and adaptation strategies is an empirical matter.

Figure 6.4 provides a neat representation of the tradeoff between mitigation and adaptation, taking into account the potential effects of technical change. The role of technical change as a key element in reducing abatement costs and therefore in encouraging a cheaper abatement effort has long been studied in the climate-economy literature (e.g. Bosetti *et al.* 2009). However, such analyses have neglected the potential interactions that may arise in the presence of adaptation responses. Technical change, as conceived by most IAMs featuring endogenous technical change, would reduce marginal abatement cost from MC to MC' (see figure 6.4). In the absence of any adaptation effort, abatement would increase to a''. However, adaptation affects the optimal level of mitigation and thus of abatement, because it increases the damage that can be tolerated, thus reducing the marginal benefit from abatement. Should adaptation shift the marginal benefit curve downward (from MB to MB'), then final abatement could be even lower than

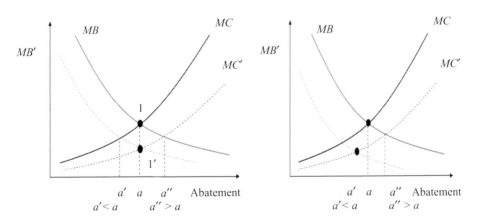

Figure 6.4 *Technical change and optimal abatement in the presence of mitigation and adaptation*

the initial level a (see the right-hand side panel of figure 6.4 where the final equilibrium a' is smaller than a).

It is crucial to assess the exact nature of the relationship between mitigation and adaptation. However, the literature on this topic, either that focusing on the general characteristics of mitigation and adaptation or that proposing specific case studies, does not seem to converge on a consistent characterization of the tradeoff between the two.

According to Klein *et al.* (2003) complementarity can be invoked as important synergies can be created between the two strategies when measures that control GHG concentration also reduce adverse effects of climate change, or vice versa. In addition, there is the possibility that many adaptation measures implemented specifically in developing countries may also promote the sustainability of their development (see, e.g., Dang *et al.* 2003; Huq *et al.* 2003).

Parry *et al.* (2001) highlights that mitigation delaying climate change impacts can buy more time to reduce vulnerability through adaptation (the converse is more controversial, see Klein *et al.* 2007). Symmetrically, adaptation can raise the thresholds which need to be avoided by mitigation (Yohe and Strzepek 2007). Consequently there is an intuitive appeal to exploit and foster synergies by integrating mitigation and adaptation.

An excessive emphasis on synergies can present some risks as well (Dang *et al.* 2003; Klein *et al.* 2003, 2007; Tol 2005). Adaptation measures could pose institutional or coordination difficulties, especially at the international level, and these may be transmitted to the implementation of mitigation measures if the two are conceived as tightly linked. Synergetic interventions can be less cost-effective than separate mitigation, adaptation, and especially (sustainable) development interventions.

There are finally tradeoffs between mitigation and adaptation (Tol 2005; de Bruin *et al.* 2007; Bosello 2008). Resources are scarce. If some of them are used for mitigation, fewer are available for adaptation, and vice versa. This point is clarified by Ingham *et al.* (2005a, 2005b), who demonstrate that mitigation and adaptation are substitutes

in economic terms, implying that if the cost of mitigation falls, agents' optimal response would be to increase mitigation and decrease adaptation.

It is worth noting that substitutability is not in contradiction with the fact that mitigation and adaptation should be both used in climate change policies. Substitutability justifies an integrated approach because either mitigation or adaptation alone cannot optimally deal with climate change (Watson *et al.* 1995; Pielke 1998). The point is that an increase in climate-related damage costs would increase both mitigation and adaptation efforts, which is exactly the typical income effect with normal goods. Finally, as noted by Tol (2005), if adaptation is successful, a lower need to mitigate could be perceived.

Turning to more case-specific examples, Klein *et al.* (2007) discuss many circumstances in which mitigation and adaptation can complement (facilitate) or substitute (conflict with) each other. In general, each time adaptation implies an increased energy use from fossil sources, emissions will increase and mitigation becomes more costly. This is the case, for instance, of adaptation to changing hydrological regimes and water scarcity. This form of adaptation takes place through increasing re-use of wastewater and the associated treatment, deep-well pumping, and especially large-scale desalination. These adaptation measures increase energy use in the water sector, leading to increased emissions and mitigation costs (Boutkan and Stikker 2004, quoted by Klein *et al.* 2007). Another example is the case of indoor cooling, which is proposed as a typical adaptation in a warming world (Smith and Tirpak 1989, quoted by Klein *et al.* 2007).

However, there are also adaptation practices that decrease energy use and thus facilitate mitigation. For instance, the new design principles for commercial and residential buildings could simultaneously reduce vulnerability to extreme weather events and energy needs for heating and/or cooling. Carbon sequestration in agricultural soils also highlights a positive link from mitigation to adaptation. It creates an economic commodity for farmers (sequestered carbon) and it makes the land more valuable, by improving soil and water conservation. In this way, it enhances both the economic and environmental components of adaptive capacity (Butt and McCarl 2004; Klein *et al.* 2007).

There are, finally, ambiguous cases. For instance, avoided forest degradation implies in most cases an increased adaptive capacity of ecosystems through biodiversity preservation and climate benefits. However, if incentives to sequester carbon by afforestation and reforestation spur an overplantation of fast-growing alien species, biodiversity can be harmed (Caparrós and Jacquemont 2003, quoted by Klein *et al.* 2007) and the natural system can become less adaptable.

These examples demonstrate the intricate interrelationships between mitigation and adaptation, and also the links with other environmental concerns such as water resources and biodiversity, with profound policy implications.

Adaptation Strategies and Macro-, Policy-Driven Integrated Measures

Given the multi-faceted features of adaptation, and the difficulty of comparing the very different adaptation actions or even the same adaptation strategy in different locations, the choice of this chapter is to aggregate adaptation responses into three main categories: anticipatory adaptation, reactive adaptation, and adaptation research and development (R&D).

Anticipatory adaptation implies building a stock of defensive capital that must be ready when the damage materializes. It is subject to economic inertia: investment in defensive capital translates into protection capital after some years. Hence, it needs to be undertaken before the damage occurs. By contrast, reactive adaptation is immediately effective and can be put in place when the damage effectively materializes.

Reactive adaptation is represented by all those actions that need to be undertaken every period in response to those climate change damages that can not be or were not accommodated by anticipatory adaptation. They usually need to be constantly adjusted to changes in climatic conditions. Examples of these actions are energy expenditures for air conditioning or farmers' yearly changes in seasonal crops' mix.

Investing in R&D and knowledge can be seen as a peculiar form of anticipatory adaptation. Innovation activity in adaptation, or simply knowledge adaptation, is represented by all those R&D activities and investments that make adaptation responses more effective. These are especially important in sectors such as agriculture and health, where the discovery of new crops and vaccines is crucial to reduce vulnerability to climate change (Barrett 2008).[7]

These three groups of adaptation measures will be contrasted one with the other and with mitigation in a CBA in both a non-cooperative and cooperative (first-best) setting. The analysis will be conducted with the AD-WITCH model (see appendix 1 for more information). AD-WITCH is a climate-economic, dynamic-optimization, IAM that can be solved under two alternative game-theoretic scenarios:

- In a non-cooperative scenario, each of the twelve regions in which the world is disaggregated maximizes its own private welfare (defined as the present value (PV) of the logarithm of *per capita* consumption), taking other regions' choices as given. This yields a Nash equilibrium, which is also chosen as the baseline. In this context, externalities are not internalized.
- In a cooperative scenario, a social planner maximizes global welfare and takes into account the full social cost of climate change. In this scenario, the first-best cooperative outcome in which all externalities are internalized can be achieved.

The climate change damage function used by the AD-WITCH model includes a reduced form relationship between temperature and gross world product (GWP) which follows closely Nordhaus and Boyer (2000), in both the functional form and the parameter values. The resulting patterns of regional damages are thus in line with that depicted in tables 6.1 and 6.2. Higher losses are estimated in developing countries: in South Asia (including India) and sub-Saharan Africa, especially because

---

[7] To test the generality of results, appendix 3 proposes an alternative specification in which R&D contributes to building adaptive capacity that improves the effectiveness of all adaptation actions, be they proactive or reactive.

of higher damages in agriculture, from vector-borne diseases, and because of catastrophic climate impacts.

Damage estimates in agriculture, coastal settlements, and catastrophic climate impacts are significant in western Europe, resulting in higher damages than in other developed regions. In China, eastern EU countries, non-EU eastern European countries (including Russia), and Japan–Korea, climate change up to 2.5°C would bring small benefits, essentially because of a reduction in energy demand for heating purposes (non-EU eastern European countries including Russia) or positive effects on agricultural productivity (China).

Nonetheless recent evidence – an important contribution on this is *The Stern Review* (Stern 2007), but also UNFCCC (2007) and the IPCC Fourth Assessment Report (Parry *et al.* 2007) – suggests that climate change damages may probably be higher than the values proposed in the RICE model by Nordhaus and Boyer (2000). Probably, the most important reason is that RICE, as well as AD-WITCH and many other IAMs), only partially captures non-market impacts, which are confined to the recreational value of leisure. Important climate-related impacts on biodiversity and eco-system losses or on cultural heritage are not part of the damage assessment.

Secondly the model abstracts from very rapid warming and large-scale changes of the climate system (system surprises). As a consequence, AD-WITCH yields climate-related impacts that, on average, are smaller than those described in studies like *The Stern Report* (Stern 2007) or 2007 UNFCCC report (UNFCCC 2007), which do consider the possibility of abrupt climate changes.

Thirdly, the time horizon considered in this chapter also plays a role. The longer it is, the larger the observed damages from climate change, as temperature is projected to keep an increasing trend. Like

most IAMs, AD-WITCH considers the dynamics of economic and climatic variables up to 2100, while, for instance, *The Stern Report* reaches 2200.

Finally, the AD-WITCH model is partly based on out-of-date evidence, as many regional estimates contained in Nordhaus and Boyer (2000) are extrapolations from studies that have been carried out for one or two regions, typically the USA.

In order to account for new evidence on climate-related damages and economic impacts, the CBA of adaptation has been performed under two different specifications of the damage functions. The standard one, based on the assessments contained in Nordhaus and Boyer (2000), and a new one, characterized by a much higher damage from climate change, about twice the standard figure. This new specification of the damage function yields values of damages larger than those contained in UNFCCC (2007) and close to those in Stern (2007).

As suggested by Stern (2007), we have also assessed the BCRs of adaptation under two possible values of the pure rate of time preference. The standard one, again based on Nordhaus and Boyer (2000), is equal to 3% declining. The new one is much lower and equal to 0.1%, as in Stern (2007). Still the AD-WITCH model does not perform a risk assessment on threshold effects or on discontinuous low-probability, high-damage impacts, which go beyond the scope of this chapter.[8]

Summing up, four cases will be considered when analyzing the costs and benefits of mitigation, adaptation, and of different types of adaptation:

1 **LDAM_HDR**: low damage–high discount rate. This is the baseline scenario with a discount rate set initially at 3% and then declining over time, as in WITCH, DICE, and RICE (see Nordhaus and Boyer, 2000).
2 **LDAM_LDR**: low damage–low discount rate. The damage is the same as in the baseline; the discount rate is 0.1% and then declining, as in Stern (2007).
3 **HDAM_LDR**: high damage–low discount rate. The damage is about twice the damage in the baseline; the discount rate is 0.1% and then declining, as in Stern (2007).
4 **HDAM_HDR**: high damage–high discount rate. The damage is about twice the damage in

---

[8] However, it is likely that the general conclusions of the present study would not change. What can change is the relative weight of adaptation and mitigation in the optimal policy mix. As adaptation to catastrophic events can only be partial, and given that the probability of their occurrence can be lowered only by reducing temperature increase, mitigation could become more appealing than adaptation when the occurrence of catastrophic events is accounted for.

(a)

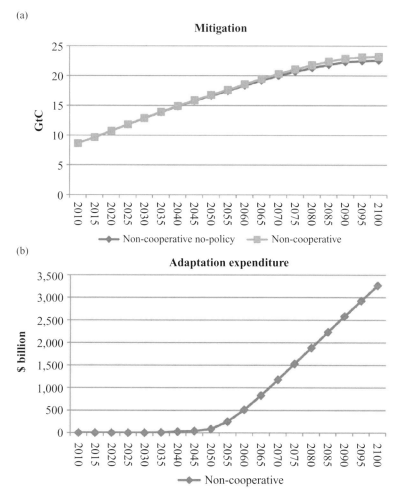

Figure 6.5 *Equilibrium (a) mitigation and (b) adaptation in the non-cooperative scenario*

the baseline; the discount rate is 3% and then declining over time, as in WITCH, DICE, and RICE.

Optimal Integrated Climate Change Strategy in a Non-Cooperative Scenario

The main strategic difference between mitigation and adaptation responses to global warming can be summarized as follows. Mitigation provides a public good that can be enjoyed globally, while adaptation provides private or club goods. Mitigation is thus affected by the well-known free-riding curse, while this is much less of an issue for adaptation.

In the absence of climate change international cooperation, climate change policies at the regional level are chosen to equalize marginal private benefits and marginal private costs, without internalizing negative externalities imposed globally. Because of the free-riding incentive, little mitigation effort is thus undertaken.

In practice, in a non-cooperative scenario, when both adaptation and mitigation are chosen optimally, equilibrium abatement (mitigation) is so low that emissions almost coincide with the no-policy case (figure 6.5a). Optimal (non-cooperative) adaptation reduces climate change damages and therefore provides an incentive to increase emissions compared to the no-policy case (non-cooperative

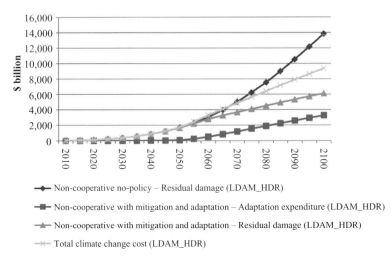

Non-cooperative no-policy – Residual damage (LDAM_HDR)

Non-cooperative with mitigation and adaptation – Adaptation expenditure (LDAM_HDR)

Non-cooperative with mitigation and adaptation – Residual damage (LDAM_HDR)

Total climate change cost (LDAM_HDR)

Figure 6.6 *Residual damage in the non-cooperative scenario*

no-policy scenario). By contrast, the full appropriability of benefits from adaptation induces regional planners to implement adaptation measures even in the non-cooperative equilibrium. Expenditures for adaptation reach US$ 3.2 trillion or, 0.8% of world GDP in 2100 (figure 6.5b). Cumulated over the century and discounted at the 3% discount rate, they total about US$ 9 trillion, 77% of which taking place in developing countries, and the remaining part in developed countries.

Figure 6.6 shows total climate change damage (residual damages + adaptation expenditure) in the absence of any policy. It amounts to an annual average of US$ 584 billion already in 2035, and increases exponentially over time. Adaptation reduces substantively residual damages (see, again, figure 6.6), up to 55% in 2100. Adaptation starts slowly in the first two decades. Consistently with the AD-WITCH damage function, damages from climate change are indeed low in the first two decades. Hence, adaptation, typically addressing current and near-term damages, is only marginally needed. This applies also to anticipatory adapta-

tion. Economic inertia in the model is about five years. As a consequence, adaptation investments do not need to start too far in advance. When considering higher damages and higher preferences for the future (the high-damage and low-discount rate case), adaptation starts earlier – already in 2020 US$ 60 billion are allocated to the reduction of damage. Hence, total damage reduction increases – it amounts to more than 70% in 2100 (see figure 6.7).

The BCRs of adaptation, measured as the discounted sum of avoided damages over the discounted sum of total adaptation expenditures, are reported in table 6.4. On a sufficiently long-term perspective, they are larger than 1. Had we chosen a longer time period they could have been even higher, as in the model benefits increase more than costs, due to the stronger convexity of the damage function with respect to the adaptation cost function.[9]

Figure 6.4 also shows that adaptation BCRs increase more when climate damage increases than when the discount rate decreases. When damages become more relevant all along the simulation period, and not only at its later stages, adaptation becomes relatively more useful.

Summing up: the theoretical insight[10] that, in a non-cooperative setting, adaptation is the main climate policy tool, is confirmed by our results. Mitigation is negligible in the non-cooperative

---

[9] This result is driven by our model assumptions, which are anchored on calibration data.

[10] There is an extensive literature on international environmental agreements showing that the non-cooperative abatement level is negligible at the equilibrium. Therefore, adaptation remains the only option to reduce climate damages.

Table 6.4 BCRs of adaptation in four scenarios (non-cooperative scenario with mitigation and adaptation)

US$ 2005 trillion 3% discounting 2010–2105	LDAM_HDR	HDAM_HDR	LDAM_LDR	HDAM_LDR
Benefits	16	62	227	695
Costs	10	25	134	270
BCR	1.67	2.41	1.69	2.57

Notes: Benefits are measured as total discounted avoided damages compared to the non-cooperative no-policy case. Costs are measured as total discounted expenditures on adaptation.

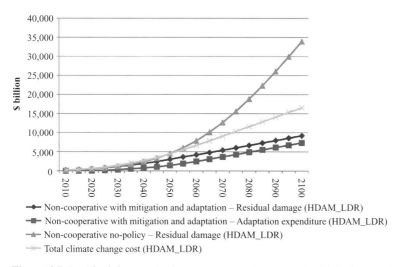

Figure 6.7 *Residual damage in the non-cooperative scenario: high damage low discount rate*

equilibrium. As a consequence, adaptation investments are high and increasing over time. Most importantly, the BCR is larger than 1. Higher emissions in the presence of adaptation, and the relatively higher sensitivity of adaptation to the level of climate damages, already highlight the potential strategic complementarity between mitigation and adaptation. This issue will be addressed more deeply in the following sections.

Optimal Integrated Climate Change Strategy in a Cooperative Scenario

In a cooperative scenario, all externalities originated by emissions are internalized. Accordingly, emission abatement (mitigation) is considerably higher than in the non-cooperative scenario (figure 6.8a). Adaptation is still undertaken, but

slightly less than in the non-cooperative case (figure 6.8b). Higher cooperative mitigation efforts reduce the need to adapt with respect to the non-cooperative scenario. This result is robust to different discount factors and damage levels (see figure 6.9a and 6.9b).

As expected, abatement is further increased when the discount rate decreases or the damage from climate change increases. Adaptation is reduced accordingly. This effect is not proportional to emission reduction, though. The discounting effect, which tends to favor mitigation by increasing the weight of future damages, is partly offset by the damage effect, which increases future and present damages and calls for both mitigation and adaptation.

The tradeoff between optimal mitigation and adaptation emerges also when analyzing cooperative mitigation with and without adaptation. As

(a)

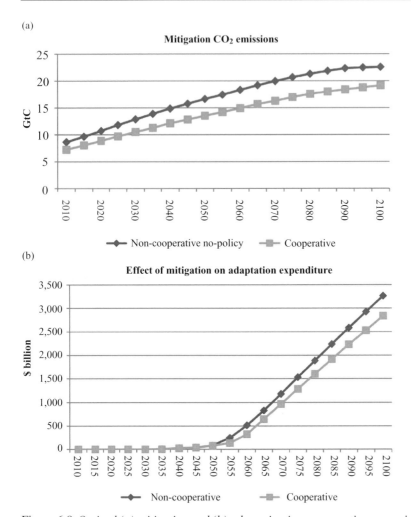

(b)

Figure 6.8 *Optimal (a) mitigation and (b) adaptation in a cooperative scenario*

shown by figure 6.10, adaptation reduces the need to mitigate – i.e. cooperative emissions in the presence of adaptation are higher. Nonetheless, even in the presence of adaptation, which can potentially reduce climate change damage by 50%, mitigation remains an important and far from negligible component of the optimal response to climate change.

After 2050, on a five-year average, optimal emission reduction is approximately 17% compared to the no-policy case. This stresses again the strategic complementarity between mitigation and adaptation. Both reduce climate-related damages. Therefore their integration can increase total welfare (proxied by cumulated discounted consumption),

as shown by figure 6.11. Notice also that cumulated consumption decreases less by giving up adaptation than mitigation. Indeed, investments in (proactive) adaptation crowd out consumption. This effect is amplified by the discounting process in earlier periods.

Further information on the relation between mitigation and adaptation is provided by table 6.5. In 2100, mitigation cuts the potential climatic damage by roughly US$ 3 trillion, whereas adaptation by nearly US$ 8 trillion. Interestingly, the two strategies, when jointly chosen, reduce climate change damages by US$ 8.2 trillion, which is less than the sum of what the two strategies could accomplish

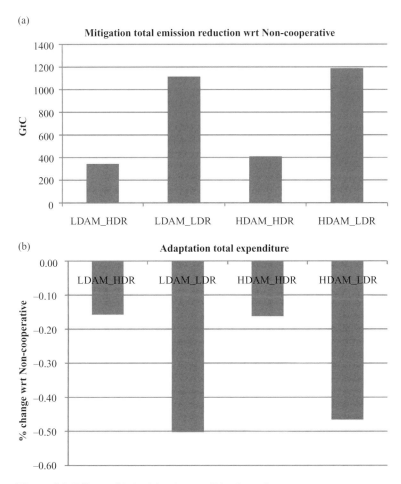

Figure 6.9 *Effects of (a) mitigation on (b) adaptation*

if adopted separately. Mitigation and adaptation remain indeed competing strategies: on the benefit side, because adaptation reduces the marginal benefit of mitigation, and on the cost side, because both compete for scarce resources. Accordingly, when they are used jointly, there is a lower incentive to use each of the two.

Table 6.6 highlights another important difference between adaptation and mitigation: their timing. Mitigation starts well in advance with respect to adaptation. Abatement is substantial when adaptation expenditure is still low. Mitigation needs to be implemented earlier than adaptation. It works through carbon-cycle inertia. Accordingly action needs to start soon to grasp some benefits in

the future. By contrast, adaptation measures work through the much shorter economic inertia, and can thus be implemented when relevant damages occur, which is from the third decade of the century.

Table 6.7 disaggregates the effectiveness of mitigation and adaptation when they are chosen optimally. It shows clearly that mitigation is preferred when the discount rate is low, whereas adaptation prevails when damages are high.

Table 6.8 shows the BCR of adaptation in the non-cooperative and in the cooperative scenarios. The BCR of adaptation improves when it is optimally complemented by mitigation.[11] This

---

[11] This happens also to mitigation, not shown.

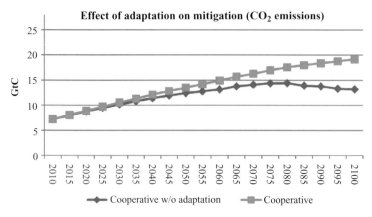

Figure 6.10 *CO_2 emissions*

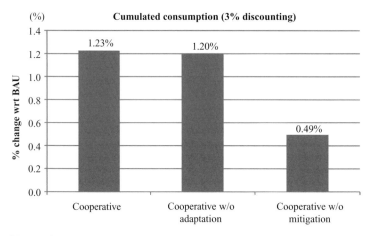

Figure 6.11 *Global welfare*

is another way of expressing the rule that two instruments are better than one instrument in the first best – i.e. (net) welfare can be enhanced by increasing the degrees of freedom of the policy maker. When combined, both adaptation and mitigation can better be used than in isolation – i.e. with a higher BCR.

The sensitivity analysis reported in table 6.9 highlights the fact that adaptation becomes more profitable when climate-related damages increase. Indeed, compared to mitigation which reduces mainly future damages, adaptation is more rapidly effective in contrasting future and present damages. Accordingly, in a high-damage world (but without climate catastrophes), adaptation becomes the preferred strategy and this is reflected in an

increasing BCR. When the discount rate declines, the opposite occurs: future damages become more relevant; mitigation is thus preferred; the BCR of adaptation declines accordingly. As shown in table 6.7, with low discounting, a larger share of damage reduction is achieved with mitigation. Similar results hold also when adaptation and mitigation are implemented jointly.

Summing up, mitigation and adaptation are *strategic complements*. Therefore, they should be integrated in a welfare maximizing climate policy. It is worth stressing again that the possibility to mitigate (adapt) reduces, but does not eliminate, the need to adapt (mitigate). The optimal climate policy mix comprises both mitigation and adaptation measures. The BCR of a policy mix where

Table 6.5 Strategic complementarity between mitigation and adaptation (2035–2100)

	Damage reduction in the cooperative case wrt baseline (2005 US$ trillion)				
	Mitigation-only	Adaptation-only	Sum	Adaptation and Mitigation	Interaction effect
2035	0.04	0.00	0.05	0.04	0.00
2050	0.20	0.10	0.30	0.23	−0.06
2075	0.99	2.24	3.23	2.43	−0.80
2100	3.05	7.92	10.97	8.23	−2.74

Table 6.6 Timing of mitigation and adaptation in the cooperative scenario (2035–2100)

	2035	2050	2100
Adaptation (total protection costs – billion US$ 2005)	2	78	2838
Mitigation (emission % change wrt BAU)	−18.8	−18.7	−15.1

Table 6.7 Damage reduction due to different strategies (2050–2100), %

	Adaptation and Mitigation	Mitigation	Adaptation
LDAM_LDR			
2050	34	31	3
2075	56	39	17
2100	72	45	27
LDAM_HDR			
2050	14	11	3
2075	39	11	28
2100	59	9	50
HDAM_LDR			
2050	49	32	17
2075	72	43	29
2100	82	47	35
HDAM_HDR			
2050	33	12	21
2075	61	10	51
2100	74	8	66

adaptation and mitigation are optimally integrated is larger than that in which mitigation and adaptation are implemented alone.

Unraveling the Optimal Adaptation Strategy Mix

The analysis performed so far does not disaggregate the role of different adaptation strategies. This is the aim of this section. Let us consider first the relationship between proactive (anticipatory) and reactive adaptation. As shown by figure 6.12a, 6.12b, and table 6.10, the non-cooperative and the cooperative scenarios highlight the same qualitative behavior: not surprisingly, anticipatory adaptation is undertaken in advance with respect to reactive adaptation.

Consequently, until 2085 the bulk of adaptation expenditure is devoted to anticipatory measures; reactive adaptation becomes the major budget item afterwards. This is the optimal response to climate damage dynamics. When it is sufficiently low, it is worth preparing to face future damages. When eventually it becomes high and increasing, a larger amount of resources needs to be invested in reactive interventions, coping with what cannot be accommodated *ex ante*.

Notice that investments in adaptation R&D show a behavior similar to anticipatory adaptation, but the scale of dedicated resources is much smaller. This result depends on the calibration data: we relied on quantitative estimates provided by UNFCCC (2007) for the aggregate amount of money that could be spent on R&D in agriculture, which is estimated to be around US$ 7 billion in 2060, a very tiny amount compared to world GDP.[12]

The results shown in figure 6.12a, 6.12b, and table 6.10 are based on the full availability of resources and political consensus to implement the optimal policy mix. What happens when first-best options are not available? In other words, what kind of adaptation strategy should a decision maker

[12] UNFCCC (2007) provides estimates for 2030. We scale this number up proportionally to the temperature gap between 2030 and our reference, 2.5°C, which is our calibration point.

Table 6.8 BCR of adaptation and of joint mitigation and adaptation

US$ 2005 trillion 3% discounting 2010–2105	BCR adaptation		BCR joint mitigation and adaptation
	Non-cooperative	Cooperative	Cooperative
Benefits	16	14	19
Costs	10	8	10
BCR	**1.67**	**1.73**	**1.93**

Notes: Benefits are measured as discounted avoided damages compared to the non-cooperative no-policy case. Adaptation costs are measured as discounted expenditures on adaptation Mitigation costs are measured as additional investments in carbon-free technologies and energy efficiency compared to the non-cooperative no-policy case

Table 6.9 Sensitivity analysis: BCR of adaptation and of joint mitigation and adaptation in the cooperative scenario

US$ 2005 trillion 3% discounting 2010–2105	LDAM_HDR	HDAM_HDR	LDAM_LDR	HDAM_LDR
Adaptation				
Benefits	14	55	99	337
Costs	8	21	65	144
BCR	1.73	2.63	1.52	2.33
Joint mitigation and adaptation				
Benefits	19	67	294	811
Costs	10	24	266	347
BCR	1.93	2.82	1.10	2.34

Table 6.10 Expenditure composition of the adaptation mix

	Non-cooperative setting (%)	Cooperative setting (%)
2035		
Reactive adaptation	0.2	0.6
Anticipatory adaptation	99.6	99.1
Knowledge adaptation	0.2	0.2
2050		
Reactive adaptation	19.5	17.2
Anticipatory adaptation	80.3	82.6
Knowledge adaptation	0.2	0.2
2100		
Reactive adaptation	56.8	55.8
Anticipatory adaptation	42.7	43.8
Knowledge adaptation	0.5	0.5

Table 6.11 BCR of the adaptation strategy mix in the cooperative scenario

US$ 2005 billion 3% discounting 2010–2105	Option excluded from the optimal mix		
	Reactive adaptation	Anticipatory adaptation	Knowledge adaptation
Benefits	789	7.4	13657
Costs	771	5.7	7938
BCR	**1.02**	**1.30**	**1.72**

prefer were she forced to make a choice between different adaptation measures because of resource scarcity? The answer to this question is summarized by table 6.11. It reports the BCR when either one of the three options is forgone.

If just only one adaptation strategy were to be chosen, reactive adaptation should be privileged.

Indeed, the non-implementation of reactive adaptation would induce a worsening of the BCR of the whole climate change strategy by 41% (and by 45% in welfare terms). By contrast, the impossibility of using anticipatory adaptation would decrease the BCR by 24% (33% in welfare terms).

R&D adaptation appears to be the less crucial adaptation option, but this depends on the way it is modeled. R&D adaptation improves the productivity of reactive adaptation. Hence, its elimination does not excessively impair reactive adaptation itself. Appendix 3 (p. 270) illustrates an alternative

(a)

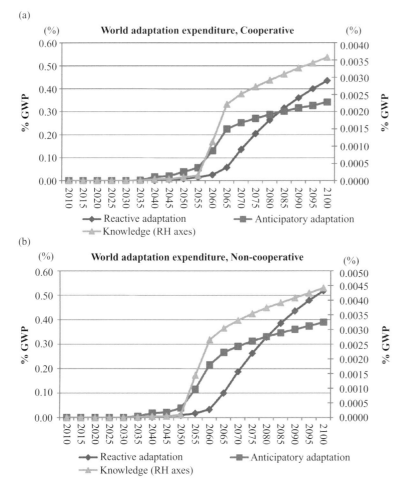

Figure 6.12 *Scale and timing of adaptation investments, (a) cooperative, (b) non-cooperative*

formulation in which R&D augments the productivity of both proactive and reactive adaptation and in which adaptation R&D investments are therefore much larger. Nonetheless, all other conclusions are robust to changes in the model specification as described in appendix 3.

Regional Analysis

In order to provide insights on regional specificities, this section disaggregates the above results between developed and developing countries. Even this broad disaggregation is sufficient to highlight substantial differences.

Figure 6.13a, 6.13b, and table 6.12 stress the higher vulnerability and the higher need to adapt of developing countries. Not surprisingly, non-OECD countries spend a higher share of their GDP on adaptation than OECD countries. This is driven by their higher damages – by the end of the century, and also in absolute terms, optimal adaptation expenditure is nearly five times higher in non-OECD than in OECD countries – and by their lower GDP.

It is also worth noting the different composition and timing of the optimal adaptation mix between the two regions. Non-OECD countries rely mainly on reactive measures, which in 2100 contribute 65% of total adaptation expenditure,

(a)

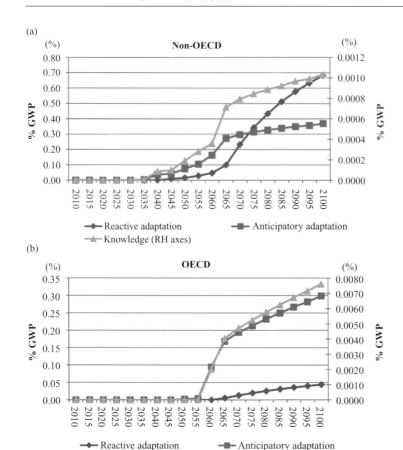

Figure 6.13 *Adaptation expenditures in (a) non-OECD and (b) OECD countries*

whereas OECD countries focus on anticipatory measures, which constitute 85% of their total expenditure on adaptation. As for the timing, adaptation in non-OECD is undertaken much earlier than in OECD regions.

The different composition of adaptation responses depends upon two facts[13]: first, the regional characteristics of climate vulnerability. In OECD countries, the higher share of climate change damages originates from loss of infrastructures and coastal areas, whose protection requires a form of adaptation that is largely anticipatory. In non-OECD countries, a higher share of damages originates from agriculture, health, and the

[13] More on the calibration procedure can be found in appendix 1 and in another annex available upon request.

energy sectors (space heating and cooling). These damages can be accommodated more effectively through reactive measures.

Secondly, OECD countries are richer. Thus, they can give up their present consumption relatively more easily to invest in adaptation measures that will become productive in the future. By contrast, non-OECD countries are compelled by resource scarcity to act in an emergency.

Only the expenditure on adaptation R&D is higher in OECD countries than in non-OECD ones. Data on R&D and innovation aimed at improving the effectiveness of adaptation are very scarce. Starting from UNFCCC (2007), we decided to distribute adaptation R&D to different regions on the basis of current expenditure on total R&D, which is concentrated in OECD countries. This explains

Table 6.12 Mitigation and adaptation in OECD and non-OECD regions in the cooperative scenario

	OECD	NON-OECD
2035		
Reactive adaptation (US$ billion)	0	0
Anticipatory adaptation (US$ billion)	0	2
Knowledge adaptation (US$ billion)	0	0
Total adaptation expenditure (US$ billion)	0	2
Mitigation (emission reduction %)	−24%	−15%
2050		
Reactive adaptation (US$ billion)	0	13
Anticipatory adaptation (US$ billion)	2	62
Knowledge adaptation (US$ billion)	0	0
Total adaptation expenditure (US$ billion)	2	76
Mitigation (emission reduction %)	−24%	−16%
2100		
Reactive adaptation (US$ billion)	62	1520
Anticipatory adaptation (US$ billion)	421	821
Knowledge adaptation (US$ billion)	11	2
Total adaptation expenditure (US$ billion)	494	2344
Mitigation (emission reduction %)	−18%	−14%

why adaptation R&D investments in developing countries in 2100 is roughly one-tenth and one-fifth of that of developed regions – as a share of their GDP and in absolute terms, respectively.

Table 6.13 and 6.14 show the BCR of adaptation, and of adaptation and mitigation jointly. In non-OECD countries, the combination of the two strategies always shows a higher BCR than adaptation alone (table 6.14). By contrast, in OECD regions (table 6.13) this remains true only with a high discount rate. With lower discounting, mitigation increases its weight in the policy mix. The additional effort undertaken by OECD countries, which is the group of countries investing

more on low-carbon technologies, benefits mostly non-OECD regions. In other words, in a cooperative setting OECD countries are called to abate partly on behalf of non-OECD countries. For example, consider the low-damage, low-discount case (LDAM_LDR). The global benefits of joint adaptation and mitigation amount to US$ 294 trillion (see table 6.9): 75% of these benefits occur in non-OECD countries, for a total benefit of US$ 226 trillion, whereas OECD countries receive the remaining 25% (US$ 68 trillion), though they bear slightly higher costs.

Again, what happens if first best-options are not fully available? If just one adaptation strategy were to be chosen, anticipatory adaptation should be privileged by OECD countries, whereas non-OECD countries should prioritize expenditure on reactive adaptation (see table 6.15).

Indeed, the elimination of anticipatory adaptation from the adaptation option "basket" of OECD countries induces a worsening of the BCR of the whole climate change strategy equal to 72%. The impossibility of using reactive adaptation in non-OECD countries reduces the overall BCR by 48% (table 6.15).

The difference between developing and developed regions is notable. Forgoing reactive adaptation is much more damaging for developing than for developed countries, consistently with what is observed about the regional structure of damages and adaptation expenditure, whereas the opposite holds for anticipatory adaptation. Again, R&D adaptation appears to be the option one can give up less regretfully.

These results, although driven by our model specification and calibration, contain some preliminary policy implications:

- OECD countries invest heavily in anticipatory adaptation measures. This depends on their damage structure. Planned anticipatory adaptation is particularly suited to cope with sea level rise, but also with the hydrogeological risks induced by more frequent and intense extreme events, which are a major source of negative impacts in the developed economies. Thus, it is more convenient to act *ex ante* rather than *ex post* in OECD countries.

Table 6.13 Sensitivity analysis: BCR of adaptation and of joint mitigation and adaptation in the cooperative scenario – OECD regions

US$ 2005 trillion 3% discounting 2010–2105	LDAM_HDR	HDAM_HDR	LDAM_LDR	HDAM_LDR
Adaptation				
Benefits	2.2	16	14	93
Costs	1.5	5.9	12	39
BCR	**1.45**	**2.64**	**1.12**	**2.38**
Joint adaptation and mitigation				
Benefits	4.2	21	68	238
Costs	1.8	6.6	146	164
BCR	**2.23**	**3.17**	**0.46**	**1.45**

Table 6.14 Sensitivity analysis: BCR of adaptation and of joint mitigation and adaptation in the cooperative scenario – non-OECD regions

US$ 2005 trillion 3% discounting 2010–2105	LDAM_HDR	HDAM_HDR	LDAM_LDR	HDAM_LDR
Adaptation				
Benefits	11	40	86	243
Costs	6	15	53	105
BCR	**1.79**	**2.63**	**1.61**	**2.31**
Joint adaptation and mitigation				
Benefits	15	46	226	573
Costs	6.9	16	128	183
BCR	**2.11**	**2.85**	**1.77**	**3.13**

Table 6.15 Marginal contribution of specific policy-driven strategies

	World (%)	OECD (%)	Non-OECD (%)
Reactive adaptation	−41	−29	−48
Anticipatory adaptation	−24	−72	−24
Knowledge adaptation	−0.36	−2	−0.1

- In non-OECD countries, climate change adaptation needs are presently relatively low, but will rise dramatically after the mid-century, as long as climate change damages increase. In 2050, they could amount to US$ 78 billion, in 2065 they will be above US$ 500 billion, to peak at more than US$ 2 trillion by the end of the century. It is sufficient to recall that in 2007 total overseas development aid (ODA) was slightly above US$ 100 billion to understand by how much climate change can stress adaptive capacity in the developing world. Non-OECD countries are unlikely to have the resources to meet their adaptation needs, which will call for international aid and cooperation on adaptation to climate change.

- At the equilibrium, non-OECD countries place little effort on adaptation R&D and rely primarily on reactive adaptation. This outcome, however, depends on the particular structure of non-OECD economic systems. Being poor, other forms of adaptation expenditures, more rapidly effective, mainly of the reactive type, are to be preferred. This suggests that richer countries can also help developing countries by supporting their adaptation R&D (e.g. by technology transfer or TT; see also chapter 8 in this volume) and their adaptation planning.

Comparison with the Existing Modeling Literature

The modeling literature that analyzes the optimal investments in adaptation, their time profile, and the tradeoff between adaptation and mitigation is thin and still mainly in the grey area. To our knowledge, it is confined to Hope *et al.* (1993, 2007), de Bruin *et al.* (2007, 2009), and Bosello (2008).

In the PAGE model (Hope *et al.* 1993; Hope and Newbery 2007) adaptive policies operate in three ways: they increase the slope of the tolerable temperature profile, they increase its plateau, and finally they can decrease the adverse impact of climate change when the temperature eventually exceeds the tolerable threshold. However, adaptation is exogenously imposed and costs and benefits are given: the default adaptation strategy has a cost in the EU of US$ 3, 12, and 25 billion a year (minimum, mode, and maximum, respectively) to achieve an increase of $1°C$ of temperature tolerability and of an additional US$ 0.4, 1.6, and 3.2 billion a year to achieve a 1% reduction in climate change impacts. At the world level, this implies, at a discount rate of 3% declining, a cost of nearly US$ 3 trillion to achieve a damage reduction of roughly US$ 35 trillion within the period 2000–2200. Impact reduction ranges from 90% in OECD countries to 50% elsewhere.

With the given assumptions, the PAGE model could easily justify aggressive adaptation policies (see, e.g., Hope *et al.* 1993), implicitly decreasing the appeal of mitigation. Due to the huge uncertainty about the cost and effectiveness of adaptation, rather than questioning the credibility of these assumptions it is worth emphasizing that in the PAGE model adaptation is exogenous. It is not determined by the model, but decided at the outset. Accordingly, mitigation and adaptation can not be really compared in an optimizing framework.

De Bruin *et al.* (2007) enrich the Nordhaus (1994) DICE model with explicit cost and benefit functions of a world adaptation strategy. They model adaptation as a flow variable: it needs to be adjusted period by period but also, once adopted in one period, it does not affect damages in the next period. De Bruin *et al.* (2007) show that adaptation and mitigation are strategic complements: optimal policy consists of a mix of adaptation measures and investments in mitigation. This result holds also in the short term, even though mitigation will only decrease damages in later periods. Adaptation is the main climate change cost-reducer until 2100, whereas mitigation prevails afterwards. In addition,

it is shown that benefits of adaptation are higher than those of mitigation until 2130.

The authors highlight the tradeoff between the two strategies: the introduction of mitigation decreases the need to adapt, and vice versa. However, the second effect is notably stronger than the first. Indeed, mitigation lowers only slightly climate-related damages, especially in the short to medium term. Therefore, it does little to decrease the need to adapt, particularly during the first decades.

Sensitivity over the discount rate highlights the fact that mitigation becomes relatively more preferable as the discount rate becomes lower. Intuitively, mitigation reduces long-term climatic damages: thus, it becomes the preferred policy instrument as these damages become more relevant.

All these results are consolidated in de Bruin *et al.* (2009), which repeats the analysis with an updated calibration of adaptation costs and benefits and also proposes regional results. This shows that in terms of utility for a low level of damages adapting-only is preferable than mitigating-only. However, the relationship is reversed when climate damages increase.

Bosello (2008) compares adaptation and mitigation using the FEEM–RICE model (Buonanno *et al.* 2000), a modified version of Nordhaus' 1996 RICE model in which technical progress is endogenous. Unlike de Bruin *et al.* (2007, 2009), adaptation is modeled as a stock of defensive capital that is accrued over time by a periodical protection investment. First, it is shown that mitigation should be optimally implemented in early periods whereas adaptation should be postponed to later stages. Accordingly, and this is the first key qualitative difference with de Bruin *et al.* (2007), the main damage-reducer is mitigation and not adaptation, at least in the first decades. Mitigation has to be anticipated because of its delayed effects driven by environmental inertia; adaptation can be postponed partly because it is more rapidly effective, but mainly because it is not worth reducing consumption by investing in adaptation when damage is low. Adaptation becomes cost-efficient only when climate-related damage is sufficiently high.

The second important difference with respect to de Bruin *et al.* (2007) is that when climate damage becomes large, albeit both adaptation and mitigation increase, the share of total damage reduction due to adaptation increases. In Bosello (2008) adaptation does not vanish after one period, as in de Bruin *et al.* (2007). Therefore, it is more cost-effective to cope with incremental damages than in de Bruin *et al.* (2007).

In all these papers, adaptation emerges as a powerful strategy to cope with climate change damage. However, irrespectively of its effectiveness, mitigation is always undertaken. Mitigation and adaptation are again strategic complements. They are also economic substitutes: more investments in mitigation reduce the equilibrium expenditure in adaptation, and vice versa. However, mitigation is more responsive to adaptation than vice versa. Finally, an increased (decreased) intertemporal preference for the future (a lower (higher) discount rate) shifts the policy emphasis into mitigation (adaptation).

Assessing the Role of Market-Driven Adaptation

The analysis conducted so far has abstracted from any role potentially played by market-driven adaptation. In other words, either the economic impact assessment or the design of the optimal mix between mitigation and adaptation strategies are based on damage functions not accounting for behavioral changes induced by market or welfare changes in human systems.

Modeling and then quantifying market-driven adaptation is extremely challenging. In economic terms, this means representing supply and demand reactions to scarcity signals conveyed by prices and

triggered by climate-related impacts. Even assuming a satisfactory knowledge of these impacts, this requires us to assess substitution elasticities in consumers' preferences and transformation elasticities in production functions for all goods and services. This then needs to be coupled with a realistic picture of intersectoral and international trade flows. Some seminal studies in this field exist, which try to capture the autonomous reactions of demand and supply to climate-induced changes in relative prices and/or in the availability of resources. Most studies use applied or CGE models (see, for example, Darwin and Tol 2001; Deke *et al.* 2001; Bosello *et al.* 2006).

Initially, CGE models were developed mainly to analyze international trade policies and, partially, public sector economic issues (e.g. fiscal policies). Soon, because of their great flexibility, they became a common tool for economists to investigate the consequences of the most diverse economic perturbations, including those provoked by climate change. Indeed, notwithstanding its complexity, as long as climate-related physical impacts can be translated into a change in productivity, production, or demand for the different inputs and outputs of the model, their GDP implications can be determined by a CGE model.[14]

The structure of an integrated climate impact assessment exercise within a general equilibrium framework is presented in figure 6.14. Economics is not independent from other disciplines, in particular it comes into play only after climatic changes have been translated into physical consequences (impacts) and then into changes of activities relevant for human welfare.

Using a CGE approach for the economic evaluation of climate impacts implies an explicit modeling of sectors and of trade in production factors, goods, and services. Changes in relative prices induce sectoral adjustments and changes in trade flows, thus triggering autonomous adaptation all over the world economic system.

Studies in this field, however, share one or both of the following shortcomings: they analyze climate change impacts in a static framework; and they consider only one or a very limited number of impacts. A static approach fails to capture important cumulative effects – think, for example, of a loss of

[14] In principle, the CGE model also offers the possibility of measuring welfare changes captured by changes in indicators other than GDP, like the Hicksian equivalent variation or consumers' surplus from a pre- to a post-perturbation state. However, great care should be placed on their interpretation. Here it is sufficient to mention that CGE models only partially capture changes in stock values (like property), and that they usually miss non-market aspects, to understand the important limitation of these assessments. Nevertheless a CGE approach has the merit to depict explicitly resource relocation, a crucial aspect of which is international trade, which is not captured by traditional direct-costing methodologies.

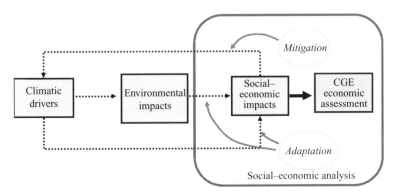

Figure 6.14 *The structure of an integrated impact assessment exercise*

Table 6.16 Impacts analyzed with the ICES model

Supply- side impacts
- Impact on labor quantity (change in mortality – health effect of climate change)
- Impacts on labor productivity (change in morbidity – health effect of climate change)
- Impacts on land quantity (land loss due to sea level rise)
- Impacts on land productivity (yield changes due to temperature and CO_2 concentration changes)

Demand-side impacts
- Impacts on energy demand (change in households' energy consumption patterns for heating and cooling)
- Impacts on recreational services demand (change in tourism flows induced by climate change)
- Impacts on health care expenditure

Table 6.17 Regional and sectoral disaggregation of the ICES model

Regional disaggregation of the ICES model (this chapter)	
USA	USA
Med_Europe	Mediterranean Europe
North_Europe	Northern Europe
East_Europe	Eastern Europe
FSU	Former Soviet Union (FSU)
KOSAU	Korea, South Africa, Australia
CAJANZ	Canada, Japan, New Zealand
NAF	North Africa
MDE	Middle East
SSA	sub-Saharan Africa
SASIA	India and South Asia
CHINA	China
EASIA	East Asia
LACA	Latin and Central America

Sectoral disaggregation of the ICES model (this chapter)	
Rice	Gas
Wheat	Oil products
Other cereal crops	Electricity
Vegetable fruits	Water
Animals	Energy-intensive industries
Forestry	Other industries
Fishing	Market services
Coal	Non-market services
Oil	

productive capital that needs to be compensated by an increased investment rate; it is thus severely limited especially for analyzing long-term climate impacts. As to the second issue, albeit some market-driven adaptation mechanisms can be described even in a single-impact case, interactions among impacts and the full potential of market-driven adaptation are neglected by focusing on only one or few impacts.

A research effort conducted at FEEM tackled these two limitations. ICES, a recursive-dynamic CGE model, has been developed and then used as an investigation tool to analyze the higher-order costs of an extended set of climate-related impacts (see table 6.16) considered one at a time, but also jointly. The study is still in a preliminary phase (many relevant impacts have still to be included, and the methodological approach can be improved by a more realistic representation of many features of market functioning), however, it can already offer

an interesting glimpse of the possible role played by market-driven adaptation.

In this chapter, the ICES model replicates the same geographical disaggregation of the WITCH and AD-WITCH models. The only difference is that WITCH WEURO (western Europe) is now divided into Mediterranean and Northern Europe, while MENA (Middle East and North Africa) is split into Middle East and North Africa. Seventeen production sectors are considered in our analysis (see table 6.17).

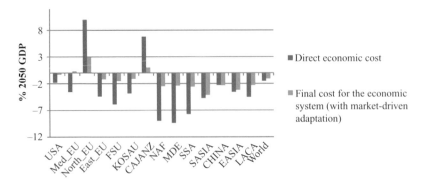

Figure 6.15 *Direct vs. final climate change costs as percentage of regional GDP (2050)*

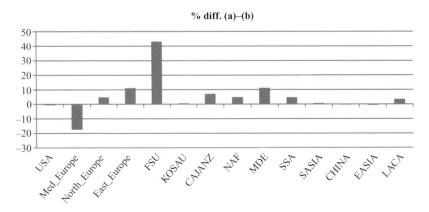

Figure 6.16 *Role of impact interaction: percentage difference between GDP costs of all climatic impacts implemented jointly and the sum of GDP costs associated with each impact implemented individually*

The model, running from 2001 to 2050, has been calibrated to replicate regional GDP growth paths consistent with the A2 IPCC scenario, and has then been used to assess climate change economic impacts for 1.2 and 3.1°C increase in 2050 with respect to 2000, which is the likely temperature range associated with that scenario. The difference between these values and initial direct costs provides an indication of the possible role of autonomous adaptation. This information is then allowed to calibrate world and macro-regional climate change damage functions by explicitly considering market-driven adaptation.

Our main results can be summarized as follows (see appendix 2, p. 267 for a more detailed presentation).

Socioeconomic systems share a great potential to adapt to climate change. Figure 6.15 shows the difference between the direct cost of climate change impacts (all jointly considered) and the final impact on regional GDP after sectoral and international adjustments have taken place. Resource reallocation smooths initial direct costs (in some cases, turns them into gains). Nevertheless, it is worth highlighting that in some regions (SASIA, EASIA, and CHINA) final costs are very close to direct costs, and that in China they are *higher*. This means that some market adjustment mechanisms – primarily international capital flows and terms of trade effects – can exacerbate initial impacts.

Interactions among impacts are also relevant (see figure 6.16). In general, costs of impacts together

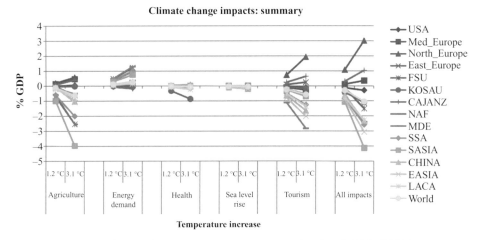

Figure 6.17 *Final climate change impact as percentage of regional GDP (2050)*

are higher than the sum of the cost associated with each single impact. This also provides an important justification for performing a joint impact analysis instead of collecting the results provided by a set of single-impact studies.

Finally, as clearly shown by figure 6.17, climate change impacts at the world level induce costs, even when market-driven adaptation is accounted for. Impacts and adaptive capacity are highly differentiated, though – i.e. a relatively small loss at the world level may hide large regional losses. In particular, developing countries remain the most vulnerable to climate change, particularly because of adverse impacts on the agricultural sector and food production.

Let us underline that the above results have been computed only for a subset of potential adverse effects of climate change (possible consequences of increased intensity and frequency of extreme weather events and of biodiversity losses, for instance, are not included). Irreversibilities or abrupt climate and catastrophic changes to which adaptation can be only limited are also neglected. Then, the model assumes costless adjustments and no frictions. Finally, the world is currently on an emission path leading to higher temperature increases than the ones consistent with the A2 scenario. Hence, for these four reasons, our analysis is likely to yield a lower bound of climate change costs. It can be considered as at the same time both optimistic and cautious.

Nonetheless, the main conclusion can be phrased as follows:

> Despite its impact smoothing potential, market-driven adaptation cannot be the solution to the climate change problem. The distributional and scale implications of climate-related economic impacts need to be addressed by adequate policy-driven mitigation and adaptation strategies.

Our study of market-driven adaptation enabled us to recompute the damage functions for the different regions modeled in WITCH. We have been able to compute the residual damage after market-driven adaptation has displayed its effects and a new equilibrium has been reached in the economic systems. Figure 6.18 reports our new estimates of world and regional climate damage functions. These new damage functions can be used to recompute the BCRs of different policy-driven adaptation and mitigation strategies.

Re-Examining Policy-Driven Adaptation: Effects of Including Market Adjustments

In this section, previous results obtained with the AD-WITCH model are re-examined by accounting for the contribution of market-driven adaptation. To do so, first the AD-WITCH climate damage function has been recalibrated in order to replicate regional damage patterns estimated by the ICES

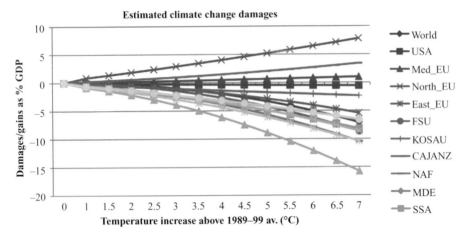

Figure 6.18 *Economic cost of climate change including market-driven adaptation*

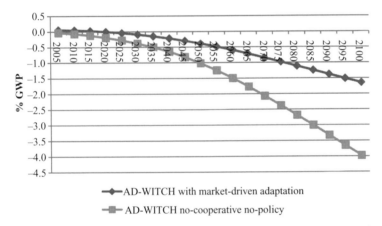

Figure 6.19 *Climate change damage with and without market-driven adaptation*

model. Then, optimal mitigation and adaptation strategies have been recomputed.

The first clear insight is that market-driven adaptation has a strong damage-smoothing potential at the global level (see figure 6.19). This result hides some important distributional changes. Market-driven adaptation reranks winners and losers. In particular (see figure 6.18, 6.20a, 6.20b), the main OECD countries are likely to gain from climate change, while all non-OECD countries still lose (even though less than with previous estimates of climate damages). It also hides the fact that a positive effect can be the sum of positive and negative impacts. Accordingly the need to adapt can persist

even in the presence of a net gain from climate change.

The policy implications are relevant. Non-OECD countries still face positive damages, but smaller than in the absence of market-driven adaptation, thus leading also to lower adaptation spending in these countries.

Accordingly, optimal mitigation and policy-driven adaptation expenditures are smaller (see figure 6.21). In particular, by the end of the century, adaptation expenditure is half of what it would have been in the absence of market-driven adaptation, even though adaptation expenditure reaches the remarkable amount of US$ 1.5 trillion in any

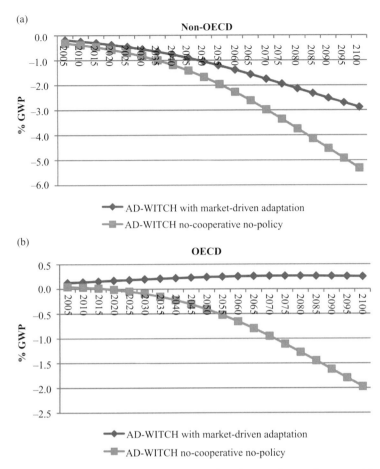

Figure 6.20 *(a) Non-OECD and (b) OECD climate change damage with and without market-driven adaptation*

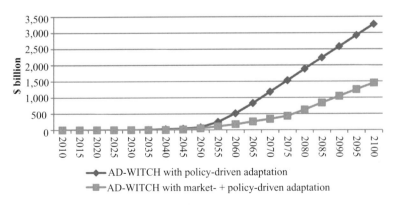

Figure 6.21 *Total protection expenditure*

Table 6.18 BCR of policy-driven adaptation in the presence of market-driven adaptation

US$ 2005 billion 3% discounting 2010–2105	World	OECD	Non-OECD
With market-driven adaptation			
Benefits	5,282	202	5,079
Costs	3,123	164	2,959
BCR	**1.69**	**1.24**	**1.72**
Without market-driven adaptation			
Benefits	14	2,250	11,535
Costs	8	1,550	6,434
BCR	**1.73**	**1.45**	**1.79**

case. Almost all this expenditure is concentrated in developing countries.

As a consequence, BCRs are slightly lower than in the absence of market-driven adaptation, both regionally and globally. The upper part of table 6.18 shows global and regional BCRs of adaptation, in comparison with those obtained without accounting for market-driven adaptation (the lower part of table 6.18). The largest difference can be seen in the OECD regions, where aggregate regional damages have turned positive (overall they have a benefit, see figure 6.20a, 6.20b). Only a few OECD regions still face negative damages, and therefore find it optimal to spend resources on adaptation. BCRs are also lower in developing regions (non-OECD), reflecting the fact that market-driven adaptation can reduce overall climate change impacts.

Specific Adaptation Strategies: Insights from the Existing Literature

There are two main policy implications that emerge from the analysis carried out in the previous sections of this chapter. First of all, the optimal response to climate change entails both mitigation and adaptation measures. Second, the adaptation mix consists of different strategies, and such a mix is region-specific. In OECD countries most resources are devoted to anticipatory adaptation, whereas non-OECD countries spend more in reactive adaptation.

As for specific adaptation measures, priority should be given to those measures offering no-regret opportunities – i.e. benefits higher than costs

irrespective of the adaptation (damage-reducing) potential. Some of these measures are already well identified – e.g. better insulation of old buildings, improved insulation standards for new buildings, and more efficient air conditioning systems (McKinsey 2009).

These measures offer three advantages: they improve adaptation of urban areas to warmer climates, they create important energy savings opportunities which on their own can motivate their adoption, and they finally entail carbon emission reductions. Indeed, they are primarily considered mitigation strategies. It would thus be wise to use scarce resources to foster the adoption of these measures first.

The composition of the optimal adaptation mix is related essentially to regional and sectoral vulnerability, as different types of climate change impacts call for specific interventions. Moreover, whereas some adjustments can take place autonomously through markets, other responses require interventions by policy makers.

In developed countries, the higher share of climate change damages seems to be related to extreme and catastrophic events. Damages from sea level rise also pose a risk for high-income countries. Accordingly, resources can be conveniently used to improve the extreme-climate resilience of infrastructures – from settlements to transportation routes – but also to mainstream climate change adaptation into long-term spatial/landscape planning to reduce from scratch the probability of experiencing extreme losses from hydrogeological risk with respect to which, by definition, adaptation can only be partial. A network of accurate and efficient early warning systems seems to provide a particularly high BCR. These forms of adaptation can be classified as "anticipatory," as they can be put in place before the occurrence of the damage.

The World Bank (World Bank 2006) quantifies the costs of adapting vulnerable infrastructures to the impacts of climate change as a 5–20% increase in investments in 2030, which is reported to amount to US$ 10–100 billion. According to the Association of British Insurers in the UK, accounting for climate change in flood management policies, and including developments in floodplains and increasing investments in flood defences, could limit the rising costs of flood damage to a possible four-fold increase (to US$ 9.7 billion) rather than 10–20-fold by the 2080s. If all properties in south Florida met the stronger building code requirements of some counties, property damages from another Hurricane Andrew (taking the same track in 2002 as it did in 1992) would drop by nearly 45% (ABI 2005).

These adaptation responses include better flood protection, stronger land-use planning, and catchmentwide flood storage schemes. A specific study on costs of flooding for the new developments in East London showed that proactive steps to prepare for climate change could reduce annual flooding costs by 80–90%, saving almost US$ 1 billion.

The major forms of adaptation to sea level rise are protect, accommodate, or retreat. Nicholls and Klein (2003) noted that the benefits of adaptation to sea level rise far outweigh the costs, though it is not clear up to what sea level rise human beings can adapt. Total costs including investment costs (beach nourishment and sea dikes) and losses (inundation and flooding) are estimated to be US$ 21–22 billion in 2030 (UNFCCC 2007). Building a sea dike coast is the most expensive option (US$ 8 billion). However, costs in isolation are not very informative and what is to be considered is the BCR.

According to Nicholls and Klein (2003), the costs of coastal protection are justified in most European countries. The avoided damage without protection, at least in the case of the Netherlands, Germany, and Poland, would amount to 69%, 30%, and 24% of GDP, respectively. These benefits largely offset the costs even in the case of the highest protection costs. Although average estimates report costs below 1% of GDP (McCarthy et al. 2001; Nicholls and Klein 2003); Bosello et al. 2006) found much higher costs, about 14% of GDP, which still remain low relative to the potential benefits. Smith and Lazo (2001) report BCRs[15] for the protection of the entire coastlines of Poland and Uruguay, the Estonian cities of Tallin and Pärnu, and the Zhujian Delta in China. They are in the range of 2.6 to around 20 for a sea level rise of 0.3–1 m.

In developing countries, in addition to catastrophic events, high losses, and thus adaptation needs, are associated with adverse impacts on agricultural activity and, particularly in sub-Saharan Africa (SSA), on health. Assessing the cost benefits of health care policies is always difficult, but these are associated with relatively low BCRs as well.

Many studies describe the possible adaptation strategies that can be implemented by health sectors in developed and developing countries (see, e.g., Kirch et al. 2005; Bettina and Ebi 2006). Nevertheless, very few researchers try a quantitative cost assessment of these measures. The problem here is two-fold: first, there is a general lack of information concerning the potential costs of some interventions. Secondly, it is very difficult conceptually and practically to disentangle the costs of adaptation to changes in health status induced by climate change from those related to changes in health status per se.

Agrawala and Fankhauser (2008) report just one study (see Ebi 2008), providing direct adaptation costs for the treatment of additional number of cases of diarrheal diseases, malnutrition, and malaria related to climate change. The additional cost for the world as a whole ranges between US$ 4 and 12.6 billion by 2030. In 2000, the additional mortality attributable to climate change was estimated to be 154,000 deaths (0.3%), with a burden of 5.5 million (0.4%) DALYs.[16] According to the World Health Organization (WHO), in developing countries the most sensitive diseases to climate change are malnutrition, diarrheal disease, and malaria. Assuming GHG stabilization at 750 ppm CO_2 by 2200, Ebi (2008) estimates an increase in incidence of

---

[15] They represent the ratio between the monetized avoided damage and the cost of the intervention.

[16] The WHO define a disability adjusted life year (DALY) as: "a measure of overall disease burden. One DALY can be thought of as one lost year of 'healthy,' life. DALYs for a disease or health condition are calculated as the sum of the years of life lost (YLL) due to premature mortality in the population and the years lost due to disability (YLD) for incident cases of the health condition. The YLL basically correspond to the number of deaths multiplied by the standard life expectancy at the age at which death occurs. To estimate YLD for a particular cause in a particular time period, the number of incident cases in that period is multiplied by the average duration of the disease and a weight factor that reflects the severity of the disease on a scale from 0 (perfect health) to 1 (dead)" (www.who.int/healthinfo/global_burden_disease/metrics_daly/en/index.html).

Table 6.19 Most cost-effective strategies against diarrheal disease

Strategies	Cost-effectiveness ($US for DALY averted)
Breastfeeding promotion	527–2,001
Measles immunization	257–4,565
Oral rehydratation therapy	132–2,570
Water and sanitation in rural areas	1,974

Source: Jamison *et al.* (2006).

Table 6.20 Most cost-effective strategies against malaria in SSA

Strategies	Cost-effectiveness ($US for DALY averted)
Preventive treatment in pregnancy with newer drugs	2–11
Insecticide-treated bed nets	5–17
Residual household spraying	9–24
Preventive treatment in pregnancy with sulfa drugs	13–24

Source: Jamison *et al.* (2006).

diarrheal disease, malnutrition, and malaria due climate change in 2030, respectively, of 3%, 10%, and 5%. Almost all the malnutrition and malaria cases would be in developing countries, with 1–5% of the diarrheal disease affecting developed countries (UNFCC 2007).

According to Ebi (2008)'s analysis, the adaptation response corresponds to an increase of both preventive (anticipatory adaptation) and therapy costs (reactive adaptation). In the 750 ppm scenario, the projected climate change-driven expenditure in 2030 would be US$ 2–7 billion for diarrheal disease, US$ 81–108 million for malnutrition, and US$ 2–5.5 billion for malaria.

Tables 6.19–6.21 rank alternative adaptation strategies in the health sector according to the cost-effectiveness criterion. It is worth noting that several strategies are considered even though not strictly related to the health sector. This is because, despite their lower cost-effectiveness, they may also have advantages in the health sector. For example, in the case of diarrheal disease, within

Table 6.21 Most cost-effective strategies against malnutrition

Strategies	Cost-effectiveness ($US for DALY averted)
Breastfeeding support programs	3–11
Growth monitoring and counseling	8–11
Capsule distribution	6–12
Sugar fortification	33–35

Source: Jamison *et al.* (2006).

the improvement of water and sanitation facilities there exist interventions such as the installation of hand pumps, corresponding to US$ 94 per DALY averted, and the provision and promotion of basic sanitation facilities, corresponding to US$ 270 per DALY averted, that are cost-effective (Jamison *et al.* 2006). Therefore, these may be considered no-regret options, also increasing development and health benefits at the society level in the absence of climate change.

Agriculture is another sector particularly vulnerable in developing countries. In the literature on adaptation, what is largely missing is the quantification of the costs of adaptation in agriculture (EEA 2007; Agrawala and Fankhauser 2008). This is mostly due to the fact that a large part of agricultural adaptation practices are implemented at the farm level and are decided autonomously by the farmers without the direct intervention of public agencies suggesting long-term planning or investment activities.

Typical examples of these practices are seasonal adjustments in the crop mix or timing, which in the literature are assumed to entail very low if not zero costs. Probably the most significant cost component of climate change adaptation in agriculture is related to the improvement of irrigation, or water conservation systems. According to the OECD ENV-Linkage model, which simulates projections of the International Energy Agency World Energy Outlook (IEA WEO) scenario, the additional expenditure on adaptation to adverse impacts of climate change will be about US$ 7 billion in 2030; the highest share (about US$ 5.8 billion) is estimated to be needed to purchase

Table 6.22 PV of benefits and costs and IRR under three ENSO frequency scenarios ($ million)

ENSO event probabilities	Accuracy of information	PV of benefits ($)	PV of costs ($)	NPV of project ($)	IRR (%)
19-year period	Perfect	479.9	51.5	428.4	227.5
	70%	87.5	51.5	36.0	22.9
51-year project	Perfect	486.7	51.5	435.2	233.6
	70%	106.4	51.5	55.0	30.4
Climate change included	Perfect	637.2	51.5	587.5	441
ENSO frequency	70%	255.8	51.5	204.3	90

Note: The values reported here are converted from pesos to dollars using the 2001 conversion rate of approximately 9 Pesos to the $.

new capital – for example to improve irrigation systems and adopt more efficient agricultural practices (UNFCCC 2007). As regarding the effectiveness of adaptation, Kirshen and Anderson (2006) reported broad ranges, depending on the type of measure adopted. Callaway *et al.* (2006), analyzing management adaptation costs for the Berg River in South Africa, emphasized the role of water management system efficiency, which can increase the benefits of improved water storage capacity by 40%.

A case study on Mexican agriculture suggests high BCRs for proactive adaptation measures in the agricultural sector (Adams *et al.* 2000). This study assessed the effectiveness of establishing accurate early warning systems, capable of detecting climate disturbances sufficiently in advance. Adams *et al.* (2000) found that the benefits of an ENSO early warning system for Mexico were approximately US$ 10 million annually, measured in terms of the saved cost for the agricultural sector that can plan crop timing and mix in advance. Table 6.22 summarizes the PV of benefits and costs under different assumptions of information accuracy. Benefits, under different assumptions of information accuracy, far outweigh costs, leading to an internal rate of return (IRR) of at least 30%. BCRs are even higher for a better level of accuracy.

The National Adaptation Programmes of Action (NAPA) Project database contains a list of ranked priority adaptation activities and projects in thirty-nine least-developed countries (LDCs). Projects on agriculture and food security have the highest priority for one-third of LDCs, the main adaptation activities in this sector are the introduction of drought-prone-tolerant or rainfall-resilient crops.

Another important area of intervention is R&D in both agriculture and health. Innovation is needed to develop climate-ready crops (heat-tolerant, drought-escaping, water-proof crops) and to advance tropical medicine. This type of adaptation strategy requires some kind of North–South cooperation, because those who need these interventions lack the financial and technical resources to implement them. UNFCCC (2007) reported an additional expenditure on agriculture-related R&D of about US$ 3 billion out of the US$ 14 billion required to cope with climate change in agriculture in 2030. The case of innovation exemplifies how market-driven adaptation can accommodate damages only partly, and how policy-driven adaptation is needed to complement other forms of adjustments.

Conclusions and Policy Implications

Currently debated mitigation targets, such as keeping global warming below 2°C, as endorsed during the L'Aquila G8 summit (July 8–10 2009), are particularly ambitious and require aggressive and immediate mitigation actions in both developed and developing regions. Given the reluctance of some large emitters to subscribe to binding commitments, the world will likely be facing a temperature increase above the proposed 2°C ceiling. Even in the case in which the 2°C target is met, a series of negative consequences for social and economic systems are likely to be observed in both the near and in the far future.

Therefore, it is important to analyze how to deal with the damages induced by climate change:

- Is market-driven adaptation sufficient to control climate-related damages?
- Is it worth investing in short-term ambitious mitigation policies?
- Or should we postpone action by focusing more on policy-driven adaptation?
- Is there an optimal level of adaptation and mitigation?
- Will the focus on adaptation crowd out investments in mitigation?

This chapter addresses these issues using an IAM framework. Let us summarize in this final section the main conclusions. First, markets cannot deal with all climate damages. Even under the optimistic assumptions of this chapter, market-driven adaptation can attenuate the total damage from climate change, but not eliminate it. Globally, direct impacts of climate change in 2050 amount to a loss of about 1.55% of GWP. Market-driven adaptation reduces this loss to 1.1% of GWP. In addition, important distributional impacts remain. Therefore, policy intervention, in the form of either mitigation or adaptation, or both, is necessary.

Second, under a social optimum perspective (global cooperation to internalize the social cost of climate change), the optimal strategy to deal with climate change entails the adoption of both adaptation and mitigation measures. Mitigation is always needed to avoid irreversible and potentially unmanageable consequences, whereas adaptation is necessary to address unavoidable climate change damages. The optimal mix of strategies has been shown to be welfare-improving. At the global level, their joint implementation increases the BCR of each of them.

Third, there is a tradeoff between mitigation and adaptation. The use of mitigation (adaptation) decreases the need to adapt (mitigate). In addition, resources are scarce. If some resources are used for mitigation (adaptation), fewer are available for adaptation (mitigation). Nonetheless, in the optimal policy mix, the possibility to abate never eliminates the need to adapt, and vice versa.

Fourth, in terms of timing, mitigation needs to be carried out earlier, because of its delayed

effects driven by environmental inertia, while adaptation can be postponed until damages are effectively higher. Were damages considerable in earlier period, adaptation would also be carried out earlier.

Fifth, both higher damages and lower discount rates foster mitigation and adaptation efforts. However, in the first case, adaptation expenditures increase more than mitigation ones, while in the second case mitigation becomes relatively more important. The intuition goes as follows. If present and future damages increase uniformly, adaptation, which deals effectively with both, is to be preferred. If future damages increase relatively more (because of lower discounting), mitigation, which is more effective in the distant future, is to be preferred.

Sixth, OECD countries should invest heavily in anticipatory adaptation measures. This depends on their damage structure. Planned anticipatory adaptation is particularly suited to cope with sea level rise, but also with the hydrogeological risks induced by more frequent and intense extreme events, which are a major source of negative impacts in the developed economies. Thus, in OECD countries it would be more convenient to act *ex ante* rather than *ex post*.

In non-OECD countries, climate change adaptation needs are presently relatively low, but will rise dramatically after the mid-century, as long as climate change damages increase. In 2050, they will amount to US$ 78 billion, in 2065 they will be above US$ 500 billion, to peak at more than US$ 2 trillion by the end of the century. Non-OECD countries are unlikely to have the resources to meet their adaptation needs, which will call for international aid and cooperation on adaptation and adaptation planning. In light of the current development deficit of developing countries, these resources are to be considered additional to the development aid required to fill this gap. They can also offer an additional opportunity to foster development itself when they take the form of educational programs, easier access to bank credit for dedicated projects, etc.

Non-OECD countries place little effort on adaptation R&D and rely primarily on reactive adaptation. This outcome, however, depends on the particular structure of non-OECD economic system. Being poor, other forms of adaptation expenditure,

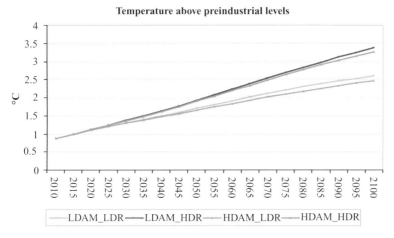

Figure 6.22 *Temperature change in the four scenarios*

more rapidly effective and mainly of the reactive type, are to be preferred. This suggests that richer countries can also help developing countries by supporting their adaptation R&D (e.g. by technology transfer or TT). The success of this policy is crucially dependent on the design of the TT program that must take into account developing country absorptive capacity (see chapter 8 in this volume).

As shown by our sensitivity analysis, these results are robust to different model specifications and parameterizations.

There is a final important issue to be emphasized. We have shown that both mitigation and adaptation belong to the optimal policy mix to deal with climate change, even though with different timing (mitigation comes first) and different distribution across world regions (more mitigation in developed countries, more adaptation in developing countries). In this optimal policy mix, the balance between adaptation and mitigation depends on the discount rate and the level of damages. This is clearly shown by table 6.23. With low discounting, a larger share of damage reduction is achieved with mitigation. With high damage, a larger share of damage reduction is achieved with adaptation.

What are the environmental implications of the optimal policy mix? Given that adaptation partly replaces mitigation, thus enabling countries to grow more, but also to emit more, the optimal tempera-

Table 6.23   Share of damage reduction in the optimal policy mix

	Total damage reduction (Undiscounted cumulative sum 2010–2100) (%)	Adaptation (%)	Mitigation (%)
LDAM_HDR	44	77	23
HDAM_LDR	73	41	59
LDAM_LDR	60	33	67
HDAM_HDR	62	85	15

Source: Our elaboration.

ture target is higher than 2°C and lies between 2.5 and 3°C, as shown in figure 6.22 (let us recall that we do not include catastrophic damages and tipping points in the model). The economic cost of achieving this target is very limited, because mitigation exploits low-cost options in developed countries and low MACs in developing countries, whereas adaptation takes place far in the future and therefore at low discounted costs.

Residual damages are nonetheless low (between 1 and 2% of GWP), because of the role of adaptation in offsetting them.

Therefore, the optimal strategy seems to be the one in which mitigation is undertaken (and starts immediately) to offset the most dangerous damages from climate change – i.e. to the level that future damages can be dealt with through

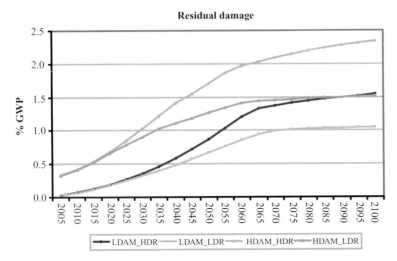

Figure 6.23 *Residual damages from climate change in the four scenarios*

adaptation. Then, adaptation, if well prepared in advance, will protect our socioeconomic systems, from climate change. The mitigation target could be slightly larger than 2°C and compensated by a commitment to invest in adaptation.

To conclude, it is worth stressing again the important qualifications of our findings. Firstly, the damage function used is highly stylized. Aggregating different damage categories hides the existence of "hot spots" for costs at the sectoral level, with the risk of underestimating adaptation needs. Secondly, it only partially considers non-market impacts and does not incorporate the very recent damage estimates highlighting higher figures than those shaping the WITCH model. Thirdly, the study refers to a smooth world. It neither considers irreversibility and tipping points nor analyzes extreme temperature scenarios. Ours is also a perfect information world in which uncertainty does not play a role. These two issues introduce another downward bias to mitigation needs and anticipatory strategies of adaptation, which are also mainly driven by precautionary motives.

Appendix 1: The AD-WITCH Model

The WITCH model developed by the climate change group at FEEM (Bosetti *et al.* 2006, 2007) is an energy–economy–climate model designed to deal explicitly with the main features of climate change. It is a regional model in which the non-cooperative nature of international relationships is explicitly accounted for. It is a truly intertemporal optimization model, with a long-term horizon covering the whole century until 2100. The regional and intertemporal dimensions of the model make it possible to differentiate climate policies across regions and over time. Finally, the model includes a wide range of energy technology options, with different assumptions on their future development, which is also related to the level of innovation effort undertaken by countries.

The core structure of the model is described at length in the technical report (Bosetti *et al.* 2007). The focus of this appendix is on the new elements of the latest version used in this chapter, and in particular on the adaptation module of WITCH.

Overall Model Structure

WITCH is a dynamic optimal growth general equilibrium model with a detailed (bottom-up) representation of the energy sector, thus belonging to a new class of hybrid (both top-down and bottom-up) models. It is a global model, divided into twelve macro-regions.

The world economy is disaggregated into twelve macro-regions: USA (United States),

WEURO (Western Europe), EEURO (Eastern Europe), KOSAU (Korea, South Africa, Australia), CAJANZ (Canada, Japan, New Zealand), TE (Transition Economies), MENA (Middle East and North Africa), SSA (sub-Saharan Africa), SASIA (South Asia), CHINA (China and Taiwan), EASIA (South East Asia), and LACA (Latin America, Mexico and Caribbean). This grouping has been determined by economic, geographic, resource endowment, and energy market similarities.

The model proposes a bottom-up characterization of the energy sector. Seven different energy-generating technologies are modeled: coal, oil, gas, wind and solar, nuclear, electricity, and biofuels. Their penetration rate is driven also by endogenous country- and sector-specific innovation. The model distinguishes between dedicated R&D investments for enhancing energy efficiency from investment aimed at facilitating the competitiveness of innovative low-carbon technologies in both the electric and non-electric sectors (backstops). R&D processes are subject to stand on shoulders' as well on neighbors' effects. Specifically, international spillovers of knowledge are accounted for to mimic the flow of ideas and knowledge across countries. Finally, experience processes via learning-by-doing (LBD) are accounted for in the development of niche technologies such as renewable energy (wind and solar) and the backstops. Through the optimization process regions choose the optimal dynamic path of different investments, namely in physical capital, in R&D, in energy technologies, and in the consumption of fossil fuels.

We updated the model base year to 2005, and use the most recent estimates of population growth. The annual estimates and projections produced by the UN Population Division are used for the first fifty years.[17] For the period 2050–2100, the updated data is not available, and less recent long-term projections, also produced by the UN Population Division[18] are adopted instead. The differences in the two datasets are smoothed by extrapolating population levels at five-year periods for 2050–2100, using average 2050–2100 growth rates. Similar techniques are used to project population trends beyond 2100.

The GDP data for the new base year are from the *World Bank Development Indicators 2007*, and

are reported in 2005 US$. We maintain the use of market exchange rates (MER). World GDP in 2005 equals to US$ 44.2 trillion. Although GDP dynamics is partly endogenously determined in the WITCH model, it is possible to calibrate growth of different countries by adjusting the growth rate of total factor productivity (TFP), the main engine of macroeconomic growth.

The prices of fossil fuels and exhaustible resources have been revised, following the dynamics of market prices between 2002 and 2005. Base-year prices have been calibrated following Enerdata, IEA WEO2007, and EIA AEO2008.

Climate Module and GHG Emissions

We continue to use the MAGICC 3-box layer climate model[19] as described in Nordhaus and Boyer (2000). CO_2 concentrations in the atmosphere have been updated to 2005 at roughly 385 ppm and temperature increase above preindustrial at 0.76°C, in accordance with the IPCC Fourth Assessment Report (Parry *et al.* 2007). Other parameters governing the climate equations have been adjusted following Nordhaus (2007).[20] We have replaced the exogenous non-CO_2 radiative forcing in equations with specific representation of other GHGs and sulphates. The damage function of climate change on economic activity is left unchanged.

In this version of WITCH we maintain the same initial stoichiometric coefficients as in previous versions. However, in order to differentiate the higher emission content of non-conventional oils as opposed to conventional ones, we link the carbon emission coefficient for oil to its availability. Specifically, the stoichiometric coefficient for oil increases with the cumulative oil consumed so that it increases by 25% when 2000 billion barrels are reached. An upper bound of 50% is assumed. The 2000 figure is calibrated on IEA 2005[21]

---

[17] Data are available from http://unstats.un.org/unsd/cdb/cdb_simple_data_extract.asp?strSearch=&srID=13660&from=simple.
[18] UN (2004).
[19] Wigley (1994).
[20] http://nordhaus.econ.yale.edu/DICE2007.htm.
[21] IEA (2005).

estimates for conventional oil resource availability. The 25% increase is chosen given that estimates[22] range between 14% and 39%.

Non-CO_2 GHGs are important contributors to global warming, and might offer economically attractive ways of mitigating it.[23] Previous versions of WITCH only consider explicitly industrial CO_2 emissions, while other GHGs, together with aerosols, enter the model in an exogenous and aggregated manner, as a single radiative forcing component.

In this version of WITCH, we take a step forward and specify non-CO_2 gases, modeling explicitly emissions of CH_4, N_2O, SLF (short-lived fluorinated gases, i.e. HFCs with lifetimes under 100 years) and LLF (long-lived fluorinated gases, i.e. HFCs with long lifetime, PFCs, and SF_6). We also distinguish SO_2 aerosols, which have a cooling effect on temperature.

Since most of these gases are determined by agricultural practices, we rely on estimates for reference emissions and a top-down approach for mitigation supply curves. For the baseline projections of non-CO_2 GHGs, we use EPA regional estimates.[24] The regional estimates and projections are available until 2020 only: beyond that date, we use growth rates for each gas as specified in the IIASA–MESSAGE-B2 scenario,[25] that has underlying assumptions similar to the WITCH ones. SO_2 emissions are taken from MERGE v.5[26] and MESSAGE B2: given the very large uncertainty associated with aerosols, they are translated directly into the temperature effect (cooling), so that we only report the radiative forcing deriving from GHGs. In any case, sulfates are expected to be gradually phased out over the next decades, so

that eventually the two radiative forcing measures will converge to similar values.

The equations translating non-CO_2 emissions into radiative forcing are taken from MERGE v.5. The GWP methodology is employed, and figures for GWP as well as base-year stock of the various GHGs are taken from the IPCC Fourth Assessment Report, Working Group I (Solomon *et al.* 2007). The simplified equation translating CO_2 concentrations into radiative forcing has been modified from WITCH06 and is now in line with the IPCC (Solomon *et al.* 2007).

We introduce end-of-pipe-type of abatement possibilities via MACs for non-CO_2 GHG mitigation. We use the MAC provided by EPA for the EMF21 project,[27] aggregated for the WITCH regions. MACs are available for eleven cost categories ranging from 10 to 200 US$/tC. We have ruled out zero or negative cost abatement options. MACs are static projections for 2010 and 2020, and for many regions they show very low upper values, such that even at maximum abatement emissions would keep growing over time. We thus introduce exogenous technological improvements: for the highest cost category only (the 200 US$/tC) we assume a technical progress factor that reaches 2 in 2050 and the upper bound of 3 in 2075.

We do, however, set an upper bound to the amount of emissions which can be abated, assuming that no more than 90% of each gas emission can be mitigated. Such a framework enables us to keep non-CO_2 GHG emissions somewhat stable in a stringent mitigation scenario (530-CO_2-eq) in the first half of the century, and subsequently decline gradually. This path is similar to what is found in the CCSP report,[28] as well as in the MESSAGE stabilization scenarios. Nonetheless, the very little evidence on technology improvements potential in non-CO_2 GHG sectors indicates that sensitivity analysis should be performed to verify the impact on policy costs.

Technological Innovation

WITCH is enhanced by the inclusion of two backstop technologies that necessitate dedicated innovation investments to become economically

---

[22] Farrell and Brandt (2005).

[23] See *The Energy Journal*, Special Issue (2006) (EMF-21), and the IPCC *Fourth Assessment Report – Working Group III* (Metz *et al.* 2007).

[24] EPA Report **430-R-06–003**, June 2006, www.epa.gov/climatechange/economics/mitigation.html.

[25] Available at http://www.iiasa.ac.at/web-apps/ggi/GgiDb/dsd?Action=htmlpage&page=regions.

[26] www.stanford.edu/group/MERGE/m5ccsp.html.

[27] www.stanford.edu/group/EMF/projects/projectemf21.htm.

[28] www.climatescience.gov/Library/sap/sap2–1/finalreport/default.htm.

competitive, even in a scenario with a climate policy. We follow the most recent characterization in the technology and climate change literature, modeling the costs of the backstop technologies with a two-factor learning curve in which their price declines with both investments in dedicated R&D and with technology diffusion. This improved formulation is meant to overcome the main criticism of the single-factor experience curves[29] by providing a more structural – R&D investment-led – approach to the penetration of new technologies, and thus to ultimately better inform policy makers on the innovation needs in the energy sector.

More specifically, we model the investment cost in a backstop technology as being influenced by a learning-by-researching (LBR) process (the main driving force before adoption) and by an LBD process (the main driving force after adoption), the so-called two-factor learning curve formulation.[30]

We set the initial prices of the backstop technologies at roughly ten times the 2005 price of commercial equivalents (16,000 US$/kW for electric, and 550 US$/bbl for non-electric). The cumulative deployment of the technology is initiated at 1000twh and 1000EJ, respectively, for electric and non-electric, an arbitrarily low value.[31] The backstop technologies are assumed to be renewable in the sense that the fuel cost component is negligible; for power generation, it is assumed to operate at load factors comparable with those of baseload power generation.

Backstops linearly substitute nuclear power in the electric sector and oil in the non-electric one. We assume that once the backstop technologies become competitive thanks to dedicated R&D investment and pilot deployments, their uptake will not be immediate and complete, but rather there will be a transition/adjustment period. The upper limit on penetration is set equivalent to 5% of the total consumption in the previous period by technologies other than the backstop, plus the electricity produced by the backstop itself.

Adaptation

Our goal with the AD-WITCH model is first to disaggregate the different components of climate

change costs, separating adaptation costs from residual damage; and, secondly, to attribute adaptation costs and benefits to different adaptation strategies. In the AD-WITCH model these have been clustered in three large categories.

Proactive or anticipatory adaptation is represented by all those actions taken in anticipation of the materialization of the expected damage, aiming at reducing its severity once manifested. Typical examples of these activities are coastal protection, or infrastructure and settlement climate-proving measures. They need some anticipatory planning and (if well designed) are effective in the medium and long-term.

Reactive adaptation is represented by all those actions that need to be undertaken every period in response to those climate change damages that cannot be or were not accommodated by anticipatory adaptation. They usually need to be constantly adjusted to changes in climatic conditions. Examples of these actions are energy expenditures for air conditioning or farmers' yearly changes in seasonal crops' mix.

Innovation activity is activity in adaptation or simply knowledge adaptation, represented by all those R&D activities making adaptation responses more effective. These are especially important in some sectors such as agriculture and health, where the discovery of new crops and vaccines are keys to reducing vulnerability to climate change.

The adaptation "basket," which exhibits decreasing marginal productivity, reduces the negative impact from climate change on gross output, reducing the climate change damage coefficient in the WITCH damage function. It is composed of the different adaptation activities which are modeled as a sequence of constant elasticity of substitution (CES) nested functions (see figure 6A1.1).

In the first CES nest, total adaptation is a combination of proactive and reactive adaptation. Proactive adaptation is modeled as a stock variable: some defensive capital accumulates over time because of adaptation-specific investment activity. As defensive capital does not disappear, investment

---

[29] Nemet (2006).
[30] Kouvaritakis *et al.* (2000).
[31] Kypreos (2007).

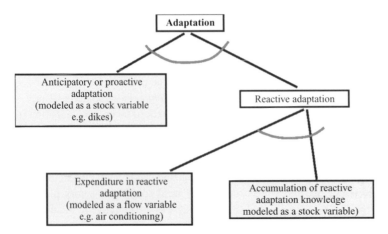

Figure 6A1.1 *Adaptation tree in the AD-WITCH model*

is needed to cope with incremental climate change damage. Proactive adaptation is also subjected to economic inertia: an initial investment in adaptation takes five years to accrue to the defensive stock and thus to become effectively damage-reducing.

Services from reactive adaptation are described by a second CES nest compounding reactive adaptation expenditures *stricto sensu* and improvements in adaptation knowledge. Expenditure on reactive adaptation is modeled as a flow variable: each simulation period, some expenditure is needed to cope with climate change damages irrespective of the expenditure in the previous period. Accumulation of adaptation knowledge is modeled as a stock accrued by a periodical adaptation-specific investment in R&D, representing an endogenous progress in reactive adaptation technologies.[32]

Then the cost of each of the adaptation activities considered (i.e. investment in proactive adaptation, investment in adaptation knowledge, and expenditure in reactive adaptation) are included in the national accounting identity. Investment in proactive adaptation, in adaptation knowledge, and in reactive adaptation expenditure are three additional control variables with which the AD-WITCH regional decision makers are endowed, which compete with alternative uses of regional income in

<hr/>

[32] In fact, adaptation R&D could also improve the effectiveness of proactive adaptation. However, we consider mostly R&D activities in the health care sector, which in the model is related to the treatment of climate-related diseases and in agriculture, which are both reactive.

Table 6A1.1 Different adaptation strategies

Proactive adaptation activities → Modeled as "stock" variable
Coastal protection activities
Settlements, other infrastructures (excluding water), and eco-system protection activities
Water supply (agriculture and other) protection activities
Setting-up of early warning systems
Reactive adaptation activities → Modeled as "flow" variable
Agricultural adaptation practices
Treatment of climate-related diseases
Space heating and cooling expenditure
Innovation in adaptation constituting → Modeled as "stock" variable
Research activities for the development of climate-resilient crops
Research activities in the health sector

the maximization of welfare. These alternative uses are: consumption, investments in physical capital, investments in different energy technologies, and investments in energy efficiency R&D.

Calibration of AD-WITCH

As in the DICE/RICE model, the WITCH climate change damage function includes the cost of both adaptation and residual damages from climate

Table 6A1.2 Adaptation costs in response to a doubling of CO_2 concentration in absolute values and as percentage of GDP

	Water in agriculture (irrigation) (billion $)	Water in other vulnerable markets (billion $)	Early-warning systems (million $)	Coastal protection (billion $)	Settlements (billion $)	Cooling expenditure (billion $)	Disease treatment costs (billion $)	Adaptation R&D (billion $)	Total (billion $)	Total (% of GDP)	AD-WITCH (% of GDP)
USA	5.0	2.1	5.0	3.6	31.3	1.1	2.9	2.92	49.0	0.12	0.15
WEURO	7.8	3.3	5.0	5.0	63.3	−0.7	2.4	2.44	83.6	0.21	0.38
EEURO	12.3	5.3	5.0	0.3	2.4	−0.1	0.0	0.03	20.3	0.54	0.17
KOSAU	0.1	0.1	5.0	1.8	3.7	1.9	0.3	0.29	8.1	0.29	0.27
CAJANZ	2.7	1.1	5.0	2.9	23.1	3.0	1.7	1.66	36.1	0.21	0.22
TE	16.9	7.2	5.0	1.7	2.0	0.1	0.1	0.06	28.1	0.40	0.26
MENA	79.1	33.9	5.0	1.2	3.2	2.1	0.1	0.14	119.8	1.48	1.01
SSA	16.1	6.9	5.0	2.7	4.0	0.5	0.0	0.01	30.2	0.78	0.96
SASIA	28.4	12.2	5.0	1.3	12.8	1.1	0.0	0.04	55.9	0.54	0.66
CHINA	12.5	5.4	5.0	1.3	9.7	0.3	0.2	0.16	29.4	0.22	0.08
EASIA	31.2	13.4	5.0	4.3	6.0	4.7	0.0	0.04	59.6	0.84	0.65
LACA	7.2	3.1	5.0	7.7	15.0	5.7	0.1	0.07	38.9	0.19	0.52

Note: Extrapolation from the literature and calibrated values with the AD-WITCH model.

Table 6A1.3 Effectiveness of adaptation (1 = 100% damage reduction) against doubling of CO$_2$ concentration

	Agriculture	Other vulnerable markets	Catastrophic events	Coastal systems	Settlements	Non-market time use	Health	Weighted total ([a])	AD-WITCH
USA	0.48	0.80	0.100	0.75	0.40	0.90	0.90	0.25	0.23
WEURO	0.43	0.80	0.100	0.54	0.40	0.80	0.90	0.20	0.26
EEURO	0.43	0.80	0.100	0.63	0.40	0.80	0.60	0.34	0.35
KOSAU	0.27	0.80	0.100	0.62	0.40	0.80	0.81	0.24	0.25
CAJANZ	0.38	0.80	0.100	0.37	0.40	0.90	0.69	0.25	0.25
TE	0.38	0.80	0.100	0.37	0.40	0.80	0.70	0.20	0.16
MENA	0.33	0.40	0.100	0.55	0.40	0.63	0.60	0.38	0.52
SSA	0.23	0.40	0.001	0.30	0.40	0.30	0.20	0.21	0.14
SASIA	0.33	0.40	0.001	0.47	0.40	0.50	0.35	0.19	0.08
CHINA	0.33	0.40	0.100	0.76	0.40	0.70	0.40	0.22	0.14
EASIA	0.33	0.40	0.010	0.25	0.40	0.43	0.40	0.19	0.11
LACA	0.38	0.40	0.001	0.46	0.40	0.70	0.90	0.38	0.31

Notes: Extrapolation from the literature and calibrated values with the AD-WITCH model.
[a] Reduction in each category of damage is weighted by the % contribution of that damage type to total damage. Weighted damages are then summed.

change. As a consequence, calibrating adaptation in the AD-WITCH model requires the separation of those two components, which requires implementing an adaptation function explicating the costs and benefits of the different forms of adaptation. The adaptation function is then to be parameterized so as to replicate the damage of the original WITCH model. A detailed description of the calibration process is reported in an appendix available upon request. Here it is worth mentioning three major points.

First, we gathered new information on climate change damages consistent with the existence of adaptation costs and tried to calibrate AD-WITCH on these new values, and not on the original values of the WITCH model.

Secondly, due to the optimizing behavior of the AD-WITCH model, when a region gains from climate change it is impossible to replicate in that region any adaptive behavior and positive adaptation costs. Accordingly, when our data estimate gains from climate change we rather referred to the Nordhaus and Boyer (2000) results if they reported costs. If both sources reported gains (as in the case of TE and KOSAU) we calibrated a damage with

the AD-WITCH model originating adaptation costs consistent with the observations.

Thirdly, the calibrated total climate change costs are reasonably similar to the reference values; however, correspondence is far from perfect. The main explanation is that consistency needs to be guaranteed between three interconnected items: adaptation costs, total damage, and protection levels. Adaptation costs and damages move together – thus, for instance, it is not possible to lower WEURO adaptation costs to bring them closer to their reference value (see table 6A1.2) without decreasing total damage, which is already lower than the reference.

Table 6A1.1 summarizes the different adaptation activities for which data were available; table 6A1.2 reports the costs of each of these strategies as they emerged from the available literature and the values calibrated for the AD-WITCH model; table 6A1.3 summarizes estimated and calibrated protection levels; table 6A1.4 introduces total damages proposed by Nordhaus and Boyer (2000), by the original WITCH model, those newly estimated by this study, and the calibration results from the AD-WITCH model.

Table 6A1.4 Total climate change costs (residual damages and adaptation cost) for a doubling of CO_2 concentration

	Nordhaus and Boyer (2000)	WITCH model	This chapter	AD-WITCH model
USA	0.45	0.41	0.37	0.44
WEURO	2.84	2.79	2.25	1.58
EEURO	0.70	−0.34	0.82	0.55
KOSAU	−0.39	0.12	−0.05	0.82
CAJANZ	0.51	0.12	0.01	0.52
TE	−0.66	−0.34	−0.01	0.80
MENA	1.95	1.78	2.41	2.93
SSA	3.90	4.17	4.19	5.09
SASIA	4.93	4.17	4.76	5.51
CHINA	0.23	0.22	0.22	0.50
EASIA	1.81	2.16	1.93	4.17
LACA	2.43	2.16	2.13	2.31

Appendix 2: Estimating Market-Driven Adaptation with the ICES Model

Through a meta-analysis and extrapolations from the existing impact literature, the set of direct impacts reported in table 6A2.1 has been computed for the regions of the ICES model.

It is first evident that, except for the case of land losses to sea level rise, the impacts are not all necessarily negative. For instance, labor productivity decreases in some regions (at the lower latitude) where the decrease in cold-related mortality/morbidity cannot compensate the increase in heat-related mortality/morbidity, but increases in others (typically at the medium-to-high latitudes), where the opposite happens. The same applies to crop productivity: in hotter regions it decreases (note that the loss of the aggregate KOSAU is mainly due to agricultural losses in Australia) whereas in the cooler regions it tends to increase, as for cereal crops in Northern Europe. Climatic stimuli are indeed regionally differentiated and affect populations or crops with different sensitivity.

Secondly, impacts concern both the supply and the demand side of the economic system. In the first case they can be unambiguously defined as

positive or negative: a decrease in labor productivity due to adverse health impacts is a certain initial loss for the economic system. In the second case, when agents' preferences change, assigning a positive or negative label to an impact is more difficult. For instance when, due to warmer climates, oil and gas demand for heating purposes decreases, this cannot be considered straightforwardly a cost or a gain before the redistributional effects have been analyzed.

This said, the larger supply-side impacts in percentage terms concern agricultural markets, whereas labor productivity and land losses to sea level rise are much smaller. Among demand shifts, the larger relate to household energy consumption: electricity demand for space cooling could increase up to 50% in hot regions depending on the climate scenario; it decreases in the cooler regions like Northern Europe and in CAJANZ, this last being dominated by the Canada effect. Natural gas and oil demand for heating purposes declines everywhere. Also highly relevant are demand changes for market services, driven by redistribution of tourism flows, accompanied by income inflows (outflows) in those regions where climatic attractiveness increases (decreases). The larger beneficiaries are cooler regions, Northern Europe, and CAJANZ (this last again dominated by the Canada effect), whereas China, East Asia, and the Middle East experience a loss.

When all these impacts are used as an input to the CGE model, figure 6.17 is obtained.

Final effects are dominated by impacts on crop productivity and on the tourism industry. It can be surprising that sea level rise and health impacts appear so negligible. This depends on two facts:

(a) *The initially low estimates of the impacts themselves.* In the case of sea level rise, only land losses are part of the assessment and capital losses or people displacement are not considered. In the case of health, both heat- and cold-related diseases are considered, thus the increase in the first is partly counterbalanced by the decrease in the second.

(b) *The nature of the analysis.* Here what is shown is the reduced (or increased) ability of economic systems to produce goods and services because

Table 6A2.1 Climate change impacts (% change, 2000–50)

	Health						Land productivity					
	Labor product.		Public exp.		Private exp.		Wheat		Rice		Cereal crops	
	1.2°C	3.2°C	1.2°C	3.2°C	1.2°C	3.2°C	1.2°C	3.2°C	1.2°C	3.2°C	1.2°C	3.2°C
USA	−0.06	−0.18	−0.15	−0.28	−0.02	−0.03	−5.66	−18.89	−6.19	−20.37	−8.18	−25.15
Med_Europe	0.01	0.01	−0.10	−0.18	0.00	−0.01	−1.14	−8.33	−4.62	−18.94	−2.00	−11.84
North_Europe	0.06	0.16	−0.35	−0.88	−0.01	−0.03	1.50	−7.74	−5.90	−26.01	50.00	107.82
East_Europe	0.09	0.23	−0.47	−1.17	−0.01	−0.02	−1.13	−10.50	−2.64	−13.57	−4.60	−18.35
FSU	0.11	0.28	−0.41	−1.03	−0.01	−0.03	−6.12	−21.92	−7.47	−24.64	−9.73	−30.10
KOSAU	−0.43	−1.14	0.57	1.62	0.04	0.11	−7.78	−17.00	−2.90	−7.41	−3.11	−7.38
CAJANZ	0.09	0.22	0.03	0.24	0.00	0.00	−0.74	−12.33	−1.87	−14.31	−2.24	−15.17
NAF	−0.28	−0.69	2.02	4.41	0.10	0.23	−12.81	−42.14	−10.78	−41.00	−12.62	−45.97
MDE	−0.22	−0.34	1.34	1.81	0.10	0.14	−8.40	−32.40	−11.73	−38.52	−13.60	−43.12
SSA	−0.31	−0.84	0.47	1.34	0.07	0.19	−9.89	−15.02	−7.17	−7.42	−8.81	−10.59
SASIA	−0.11	−0.30	0.28	0.76	0.06	0.17	−2.96	−13.37	−4.89	−17.39	−6.61	−21.43
CHINA	0.14	0.37	0.65	1.80	0.06	0.17	0.93	2.69	0.50	1.79	−1.42	−2.37
EASIA	−0.11	−0.32	1.05	2.96	0.06	0.17	2.45	9.82	0.34	5.04	−1.15	1.93
LACA	−0.14	−0.39	0.68	1.98	0.07	0.19	−6.69	−68.10	−6.61	−55.65	−8.25	−76.37

| | Sea lev. rise | | Tourism | | | | Households' energy demand | | | | | |
| | Land Losses | | Market Serv. Demand | | Income Flows | | Natural Gas | | Oil Products | | Electricity | |
	1.2°C	3.2°C	1.2°C	3.2°C	1.2°C	3.2°C	1.2°C	3.2°C	1.2°C	3.2°C	1.2°C	3.2°C
USA	−0.026	−0.055	−0.68	−1.76	−0.17	−0.43	−13.67	−35.31	−18.52	−47.84	0.76	1.96
Med.Europe	−0.007	−0.015	−1.86	−4.81	−0.40	−1.02	−12.68	−32.76	−15.84	−40.91	0.76	1.96
North.Europe	−0.020	−0.041	7.54	19.47	1.78	4.61	−13.75	−35.51	−15.52	−40.09	−2.20	−5.68
East.Europe	−0.022	−0.046	−2.46	−6.36	−0.33	−0.86	−12.93	−33.41	−17.39	−44.92	0.76	1.97
FSU	−0.007	−0.015	0.00	−0.01	0.00	0.00	−13.02	−33.65	−17.39	−44.92	0.75	1.94
KOSAU	−0.005	−0.011	−1.31	−3.39	−0.32	−0.82	nss	nss	−13.03	−33.66	12.31	31.81
CAJANZ	−0.004	−0.009	5.54	14.30	1.40	3.61	−5.05	−13.04	−12.63	−32.63	−4.80	−12.40
NAF	−0.017	−0.036	−2.52	−6.52	−0.24	−0.63	−8.60	−22.22	−13.25	−34.22	5.95	15.37
MDE	−0.004	−0.007	−4.67	−12.06	−0.91	−2.34	−13.12	−33.89	−17.39	−44.92	0.74	1.92
SSA	−0.066	−0.139	−4.43	−11.45	−0.37	−0.96	nss	nss	−6.51	−16.83	16.35	42.23
SASIA	−0.204	−0.427	−1.21	−3.12	−0.10	−0.25	nss	nss	nss	nss	20.38	52.65
CHINA	−0.045	−0.094	−4.99	−12.89	−0.33	−0.85	nss	nss	nss	nss	20.38	52.65
EASIA	−0.316	−0.662	−4.69	−12.10	−0.53	−1.38	nss	nss	nss	nss	20.38	52.66
LACA	−0.025	−0.052	−2.68	−6.91	−0.56	−1.45	nss	nss	nss	nss	21.37	55.20

Note:
nss: Non-statistically significant.

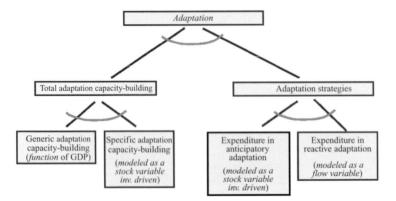

Figure 6A3.1 *Adaptation tree in the AD-WITCH model: an alternative specification*

of climate change. This is what GDP, typically a flow variable, measures. Thus, say, a land loss is not evaluated in terms of loss of property value, which can be very high, but in terms of the lower capacity of the economic system hit by that land loss to produce (agricultural) goods. Given the possibility to substitute at least partially a scarcer input with one more abundant, effects on GDP are usually smaller.[33]

Final effects also present Northern Europe, CAJANZ, and Mediterranean Europe as winners from climate change. In Northern Europe, all impacts except sea level rise bring gains. In CAJANZ, huge positive impacts on tourism demand can explain its gain. More interesting is to note the case of Mediterranean Europe, which benefits from climate change even though, except for a slight gain in labor productivity, all impacts are negative. Indeed if measured in terms of direct costs, climate change entails a net loss higher than the 3% of GDP (see figure 6.15) for the region. However two mechanisms turn this into a small gain. First, an improvement in terms of trade. This is driven by the decrease in energy prices due to the global contraction of GDP and thus of world energy demand, and by the increased agricultural goods prices induced by their reduced supply. This benefits particularly a net energy importer and food exporter like Mediterranean Europe. Secondly, foreign capital inflows. In the model, these are driven

by the expected rate of return to capital. Mediterranean Europe is one region attracting capital as its rental prices are decreasing, but less than in other regions. These resources spur investment and growth. These two second-order effects are stronger than the direct effect.

It is worth stressing that this kind of analysis cannot be performed with models like RICE (or WITCH) which lack some economic detail (the most important being sectoral and international trade) and where damages are summarized by reduced form equations. While these *assume* a given relation between damage and temperature, and the damage usually includes property losses, our exercise *estimates* the relation quantifying the change in the capacity of an economic system hit by a joint set of impacts to produce goods and services.

As a final remark: the analysis performed does not include the effect of catastrophic losses; we decided to omit them due to the uncertainty of those estimates. They are extremely relevant in other studies, though – e.g. in Nordhaus and Boyer (2000) they constitute from 10%–90% of total regional damages (see table 6.2). This means that slightly different assumptions on catastrophic outcomes may change results considerably.

Appendix 3: An Alternative Formulation of Adaptation

Two critical aspects of our exercise relate to the choice to model adaptation knowledge as an

---

[33] This is, for instance, why today catastrophic events, entailing huge property losses, translate in no or only very little effects on GDP.

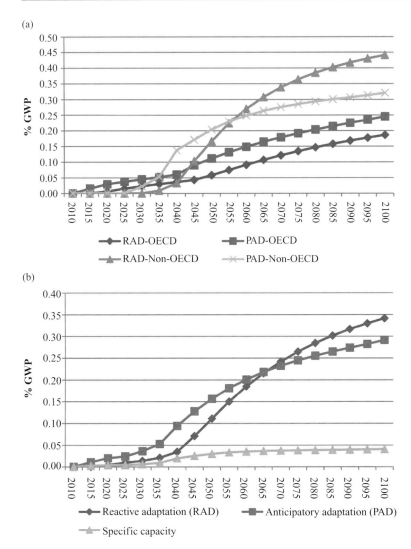

Figure 6A3.2 *(a) and (b) Adaptation expenditure*

efficiency improver of reactive adaptation only, and on the assumption of very low damage until 2040. The first assumption is driven by data evidence as investment in adaptation knowledge basically takes place in the agriculture and health sectors where reactive adaptation is preponderant, the second is an assumption embedded in Nordhaus' damage function. The main consequences are that investment in adaptation knowledge remains very small, that it is performed mainly by developed countries, and that adaptation (either proactive or reactive) starts only after 2040.

To test the robustness of our result we propose here a different specification and calibrate the damage in order to have some climate change impacts already at the beginning of the century. Adaptation strategies are now clustered in four large categories, as depicted in figure 6A3.1. A first decision is whether to spend resources on activities (adaptation strategies) or capacity-building. Both groups contain some further categorization into other subinvestments or activities. Total capacity consists of two components: generic capacity, which is not necessarily related to adaptation, and

Table 6A3.1 Benefits and costs of adaptation without mitigation (non-cooperative)

US$ 2005 billion 3% discounting 2010–2105	World	OECD	Non-OECD
Benefits	29444	8641	20802
Costs	11237	3548	7690
BCR	2.62	2.44	2.71

specific capacity, which instead includes capacity specific to adaptation. Adaptation activities include reactive and proactive adaptation measures, as in the main specification considered in the text.

Using this new specification, we have recomputed the optimal adaptation–mitigation mix in the non-cooperative scenario. All the qualitative results found with the old specification hold: mitigation is close to 0; the optimal adaptation mix is composed by reactive, proactive, and specific capacity (figure 6A3.2b). Anticipatory adaptation is undertaken in advance, because of its stock nature, whereas reactive adaptation becomes more important when the damage is sufficiently large. In the long run, anticipatory adaptation stabilizes, whereas reactive adaptation keeps increasing.

The regional differentiation of the adaptation "basket" is also robust to the new specification. Non-OECD countries spend more on adaptation than OECD regions. In the second half of the century, reactive adaptation becomes the main adaptation form in non-OECD countries, whereas in OECD countries anticipatory measures are always the dominant strategy. Once more, the explanation lies in the different climate vulnerability (figure 6A3.2a).

What changes is the path of adaptation. It starts immediately and is smoother. To conclude, table 6A3.1 reports the BCRs of all adaptation strategies jointly in the non-cooperative scenario. They show the same ranking as the previous analysis.

Therefore, even under a different structural specification of the model – i.e. even when testing the sensitivity to a different model functional form – our results are largely confirmed and seem to be robust to changes in the specification of the adaptation module.

Bibliography

Association of British Insurers, 2005: *Financial Risk of Climate Change*, Summary Report

Adams, R.M., L.L. Houston, and B.A. McCarl, 2000: The benefits to Mexican agriculture of an El Niño-southern oscillation (ENSO) early warning system, *Agricultural and Forest Meteorology* **115**, 183–94

Adger, N.W., N.W. Arnell, and E.L. Thompkins, 2005: Successful adaptation to climate change across scales, *Global Environmental Change* **15**, 77–86

Agrawala, S. and S. Fankhauser (eds.), 2008: *Economic Aspects of Adaptation to Climate Change. Costs, Benefits and Policy Instrument*, OECD, Paris

Barrett, S., 2008: A portfolio system of climate treaties, Discussion Paper **08–13**, The Harvard Project on International Climate Change Agreements

Bettina, M. and K.L. Ebi (eds.), 2006: *Climate Change and Adaptation Strategies for Human Health*, Steinkopff Verlag, Darmstadt

Bosello, F., 2008: Adaptation, mitigation and green R&D to combat global climate change. Insights from an empirical integrated assessment exercise, CMCC Research Paper **20**

Bosello, F., E. De Cian, F. Eboli, and R. Parrado, 2009: Macro economic assessment of climate change impacts: a regional and sectoral prespective, in *Impacts of Climate Change and Biodiversity Effects*, Final Report of the CLIBO project, European Bank, University Research Sponsorship Programme

Bosello, F., R. Roson, and R.S.J. Tol, 2006: Economy wide estimates of the implication of climate change: human health, *Ecological Economics* **58**, 579–91

Bosetti, V., C. Carraro, E. De Cian, R. Duval, E. Massetti, and M. Tavoni, 2009: The incentives to participate in and the stability of international climate coalitions: a game-theoretic approach using the WITCH model, OECD Working Paper **702**, OECD, Paris

Bosetti, V., C. Carraro, R. Duval, A. Sgobbi, and M. Tavoni, 2009: The role of R&D and technology diffusion in climate change mitigation: new perspectives using the WITCH model, OECD Working Paper **664**, OECD, Paris

Bosetti, V., C. Carraro, M. Galeotti, E. Massetti, and M. Tavoni, 2006: WITCH: a world induced technical change hybrid model, *The Energy Journal, Special Issue on Hybrid Modeling of Energy-Environment Policies: Reconciling Bottom-up and Top-down*, 13–38

Bosetti, V., E. Massetti, and M. Tavoni, 2007: The WITCH model. Structure, baseline, solutions, FEEM Working Paper **10–2007**

Boutkan, E. and A. Stikker, 2004: Enhanced water resource base for sustainable integrated water resource management, *Natural Resources Forum* **28**, 150–4

Buonanno, P., C. Carraro, E. Castelnuovo, and M. Galeotti, 2000: Efficiency and equity of emission trading with endogenous environmental technical change, in C. Carraro (ed.), *Efficiency and Equity of Climate Change Policy*, Kluwer Academic Publishers, Dordrecht

Burton, I., 1992: *Adapt and Thrive*, Unpublished manuscript, Canadian Climate Centre, Downsview

Butt, T.A. and B.A. McCarl, 2004: Farm and forest carbon sequestration: can producers employ it to make some money?, *Choices* **19**, 3–11

Caparrós, A. and F. Jacquemont, 2003: Conflicts between biodiversity and carbon offset programs: economic and legal implications, *Ecological Economics* **46**, 143–57

Callaway, J.M., D.B. Louw, J.C. Nkomo, E.M. Hellmuth, and D.A. Sparks, 2006: The Berg River Dynamic Spatial Equilibrium Model: a new tool for assessing the benefits and costs of alternatives for coping with water demand growth, climate variability, and climate change in the Western Cape, AIACC Working Paper **31**, The AIACC Project Office, International START Secretariat, Washington, DC, 41, www. aiaccproject.org

Carraro, C. and D. Siniscalco, 1998: International environmental agreements: incentives and political economy, *European Economic Review* **42**, 561–72

Dang, H.H., A. Michaelowa, and D.D. Tuan, 2003: Synergy of adaptation and mitigation strategies in the context of sustainable development: the case of Vietnam, *Climate Policy* **3**, S81–S96

Darwin, R. and R.S.J. Tol, 2001: Estimates of the economic effects of sea level rise, *Environmental and Resource Economics* **19**, 113–29

de Bruin, K.C., R.B. Dellink, and S. Agrawala, 2009: Economic aspects of adaptation to climate change: integrated assessment modelling of adaptation costs and benefits, OECD Environment Working Paper **6**, OECD, Paris

de Bruin, K.C., R.B. Dellink, and R.S.J. Tol, 2007: AD-DICE: an implementation of adaptation in the DICE model, FEEM Working Paper **51.2007**

Deke, O., K.G. Hooss, C. Kasten, G. Klepper, and K. Springer, 2001: Economic impact of climate change: simulations with a regionalized climate-economy model, Kiel Working Paper **1065**,

Ebi, K.L., 2008: Adaptation costs for climate change cases of diarrhoeal disease, malnutrition and malaria in 2030, *Globalization and Health 2008* **4**(9)

EIA, 2008: *Annual Energy Outlook. Energy Information Administration*, EIA, Washington, DC

European Commission, 2005: *Winning the Battle Against Global Climate Change*, COM/2005/ 0035, Brussels
2007a: *Limiting Global Climate Change to 2 degrees Celsius – The Way Ahead for 2020 and Beyond*, COM2007/0002, Brussels
2007b: *Green Paper from the Commission to the Council, the European Parliament, the European Economic and Social Committee and the Committee of the Regions*, COM2007/0354, Brussels
2009: *White paper – adapting to climate change: towards a European framework for action*, COM/2009/0147, Brussels

European Environment Agency (EEA), 2005: Vulnerability and adaptation to climate change in Europe, EEA Technical Report **7/2005**
2007: Climate change: the cost of inaction and the cost of adaptation, EEA Technical Report **13/2007**

Fankhauser, S. and R.S.J. Tol, 1996: The social costs of climate change, the IPCC Assessment Report and beyond, *Mitigation and Adaptation Strategies for Global change*, **1**(4), 386–403

Fankhauser, S., J.B. Smith, and R.S.J. Tol, 1999: Weathering climate change: some simple rules to guide adaptation decisions, *Ecological Economics* **30**, 67–78

Farrell, A.E. and A.R. Brandt, 2005: Risk of the oil transition, *Environmental Research Letters* **1**(1),

Füssel, H.M. and R.J.T. Klein, 2006: Climate change vulnerability assessments: an evolution of conceptual thinking, *Climatic Change* **75**, 301–29

Hanemann, W.M., 2008: What is the cost of climate change?, CUDARE Working Paper **1027**, University of California, Berkeley, CA

Hope, C., 2003: The marginal impacts of CO_2, CH_4 and SF_6 emissions, Judge Institute of Management Research Paper **2003/10**, University of Cambridge, Judge Institute of Management, Cambridge

Hope, C., J. Anderson, and P. Wenman, 1993: Policy analysis of the greenhouse effect: an application of the PAGE model, *Energy Policy* **21**, 327–38

Hope, C. and D. Newbery, 2007: Calculating the social cost of carbon, Electricity Policy Research Group Working Papers **EPRG 07/20**, University of Cambridge, Cambridge

Huq, S., A. Rahman, M. Konate, Y. Sokona, and H. Reid, 2003: Mainstreaming adaptation is climate change in least developed countries (LDCs), International Institute for Environment and Development, London

IEA, 2005: *Resources to Reserves – Oil & Gas Technologies for the Energy Markets of the Future*, IEA, Paris

2007: *World Energy Outlook 2007*, OECD/IEA, Paris

Ingham, A., J. Ma, and A.M. Ulph, 2005a: Can adaptation and mitigation be complements?, Working Paper **79**, Tyndall Centre for Climate Change Research, University of East Anglia

2005b: How do the costs of adaptation affect optimal mitigation when there is uncertainty, irreversibility and learning?, Working Paper **74**, Tyndall Centre for Climate Change Research, University of East Anglia

Jamison, D.T., J.G. Breman, A.R. Measham, G. Alleyne, M. Claeson, D.B. Evans, P. Jha, A. Mills, and P. Musgrove (eds.), 2006: *Disease Control Priorities in Developing Countries*, World Bank, Washington, DC

Kane, S. and J. Shogren, 2000: Linking adaptation and mitigation in climate change policy, *Climatic Change* **45**, 75–102

Kane, S. and G. Yohe, 2000: Societal adaptation to climate variability and change: an introduction, *Climatic Change* **45**, 1–4

Kirch, W., B. Menne, and R. Bertollini (eds.), 2005: *Extreme Weather Events and Public Health Responses*, Springer Verlag, Berlin

Kirshen, P., 2007: *Adaptation Options and Coast of Water Supply*, Tuft University, Cambridge, MA

Kirshen, P., M. Ruth and W. Anderson, 2006: Climate's long-term impacts on urban infrastructures and services: the case of metro Boston, in M. Ruth (ed.), *Regional Climate Change and Variability: Local Impacts and Responses*, Edward Elgar, Cheltenham

Klein, R.J.T., S. Huq, F. Denton, T.E. Downing, R.G. Richels, J.B. Robinson, and F.L. Toth, 2007: Inter-relationships between adaptation and mitigation. Climate change 2007: Impacts, Adaptation and Vulnerability, *Contribution of Working Group II to the Fourth Assessment Report of the Intergovernmental Panel on Climate Change*, in M.L. Parry, O.F. Canziani, J.P. Palutikof, P.J. van der Linden, and C.E. Hanson (eds.), Cambridge University Press, Cambridge

Klein, R.J.T., E.L. Schipper, and S. Dessai, 2003: Integrating mitigation and adaptation into climate and development policy: three research questions, Working Paper **40**, Tyndall Centre for Climate Change Research, University of East Anglia

Klein, R.J.T. and R.S.J. Tol, 1997: Adaptation to climate change: options and technologies – an overview paper, Technical Paper **FCCC/TP/1997/3**, UNFCC Secretariat, Bonn, Germany

Kouvaritakis, N., A. Soria, and S. Isoard, 2000: Endogenous learning in world post-Kyoto scenarios: application of the POLES Model under adaptive expectations, *International Journal of Global E* **14**(1–4), 228–48

Kurukulasuriya P. and R. Mendelsohn, 2008: How will climate change shift agro-ecological zones and impact African agriculture? Policy Research Working Paper **WPS4717**, The World Bank, Washington, DC

Kypreos, S., 2007: A MERGE model with endogenous technical change and the cost of carbon stabilization, *Energy Policy* **35**, 5327–36

Leary, N.A., 1999: A framework for benefit-cost analysis of adaptation to climate change and climate variability, *Mitigation and Adaptation Strategies for Global Change* **4**, 307–18

Lecoq, F. and Z. Shalizi, 2007: Balancing expenditures on mitigation and adaptation to climate change: an exploration of issues relevant for developing countries, World Bank Policy Research Working Paper **4299**, World Bank, Washington, DC

Lim, B. and E. Spanger-Siegfred (eds.), 2005: *Adaptation Policy Framework for Climate Change: Developing Policies Strategies and*

Measures, Cambridge University Press, Cambridge

McCarthy, J.J, O.F. Canziani, N.A. Leary, D.J. Dokken, and K.S. White (eds.), 2001: *Climate Change 2001: Impacts, Adaptation and Vulnerability – Contribution of Working Group II to the Third Assessment Report of the Intergovernmental Panel on Climate Change*, Cambridge University Press, Cambridge

McKibbin, W.J. and P.J. Wilcoxen, 2004: Climate policy and uncertainty: the roles of adaptation versus mitigation, *Brookings Discussion papers in International Economics* **161**

McKinsey & Co., 2009: *Pathways to a Low-Carbon Economy. Version 2 of the Global Greenhouse Gas Abatement Curve*, McKinsey & Co

Mendelsohn, P.O., W.N. Morrison, M.E. Schlesinger, and N.G. Adronova, 2000: Country-specific impacts of climate change, *Climatic Change* **45**(3–4), 553–69

Metz, B., O.R. Davidson, P.R. Bosch, R. Dave, and L.A. Meyer (eds.), 2007: *Climate Change 2007: Mitigation of Climate Change – Contribution of Working Group III to the Fourth Assessment Report of the Intergovernmental Panel on Climate Change*, Cambridge University Press, Cambridge

Nemet, G.F., 2006: Beyond the learning curve: factors influencing cost reductions in photovoltaics, *Energy Policy* **34**, 3218–32

Nicholls, R.J. and R.J.T. Klein, 2003: Climate change and coastal management on Europe's Coast, EVA Working Paper **3**

Nordhaus, W.D., 1994: *Managing the Global Commons: The Economics of the Greenhouse Effect*, MIT Press, Cambridge, MA
2007: http://nordhaus.econ.yale.edu/DICE2007.htm

Nordhaus, W.D. and J.G. Boyer, 2000: *Warming The World: The Economics of The Greenhouse Effect*, MIT Press, Cambridge, MA

Palutikof, P.J. van Der Linden, and C.E. Hanson (eds.), *Climate Change 2007: Impacts, Adaptation and Vulnerability – Contribution of Working Group II to the Fourth Assessment Report of the Intergovernmental Panel on Climate Change*, Cambridge University Press, Cambridge, 745–77

Parry, M., 2009: Closing the loop between mitigation, impacts and adaptation, *Climatic Change* **96**, 23–7

Parry, M., N. Arnell, T. McMichael, R. Nicholls, P. Martens, S. Kovats, M. Livermore, C. Rosenzweig, A. Iglesias, and G. Fischer, 2001: Millions at risk: defining critical climate change threats and targets, *Global Environmental Change* **11**, 181–3

Parry, M., O. Canziani, J. Palutikof, P. van der Linden, and C. Hanson (eds.), 2007: *Climate Change 2007: Impacts, Adaptation and Vulnerability – Contribution of Working Group II to the Fourth Assessment Report on Climate Change*, Cambridge University Press, Cambridge

Pearce, D.W., W.R. Cline, A.N. Achanta, S. Fankhauser, R.K. Pachauri, R.S.J. Tol, and P. Vellinga, 1996: The social costs of climate change: greenhouse damage and the benefits of control, in Bruce and E.F. Haites (eds.), climate change 1995: Economic and Social Dimensions, *Contribution of Group III to the Second Assessment Report of the Intergovernmental Panel on Climate Change*, Cambridge University, Cambridge

Pielke, R.A., 1998: Rethinking the role of adaptation in climate policies, *Global Environmental Change* **8**, 159–70

Smit, B. (ed.), 1993: *Adaptation to Climatic Variability and Change*, Environment Canada, Guelph

Smit, B., I. Burton, R.J.T. Klein, and R. Street, 1999: The science of adaptation: a framework for assessment, *Mitigation and Adaption Strategies for Global Change* **4**(3–4), 199–213

Smit, B., I. Burton, J.T. Klein, and J. Wandel, 2000: An anatomy of adaptation to climate change and variability, *Climatic Change* **45**, 223–51

Smit, B. and O. Pilifosova, 2001: Adaptation to climate change in the context of sustainable development and equity, in *Climate Change 2001 – Contribution of Working Group II to the Third Assessment Report of the IPCC* (eds.), Cambridge University Press, Cambridge, 877–912

Smith, J.B. and J.K. Lazo, 2001: A summary of climate change impact assessments from the US Country Studies Programme, *Climatic Change* **50**, 1–29

Smith, J.B. and D.A. Tirpak (eds.), 1989: *The Potential Effects of Global Climate Change on the United States. Executive Summary*, US Environmental Protection Agency. Washington, DC

Smithers, J. and B. Smit, 1997: Human adaptation to climatic variability and change, *Global Environmental Change* **7**, 129–46

Solomon, S., D. Qin, M. Manning, Z. Chen, M. Marquis, K.B. Averyt, M. Tignor, and H.L. Miller (eds.), 2007: *Climate Change 2007: The Physical Science Basis – Contribution of Working Group I to the Fourth Assessment Report of the Intergovernmental Panel on Climate Change*. Cambridge University Press, Cambridge

Stern, N., 2007: *The Economics of Climate Change: The Stern Review*, Cambridge University Press, Cambridge

Tol, R.S.J., 2005: Emission abatement versus development as strategies to reduce vulnerability to climate change: an application of FUND, *Environment and Development Economics* **10**, 615–29

UNFCCC, 2006: Secretariat website, quoted by E. Lerina and D.A. Tirpak, in Adaptation to climate change: key terms, OECD, Paris

UN, 2004: *World Population to 2300*, Report **ST/ESA/SER.A/236**, Department of Economic and Social Affairs, Population Division, New York

UNFCCC, 2007: *Investments and Financial Flows to Address Climate Change*, Climate Change Secretariat, Bonn

Watson, R.T., M.C. Zinyowera, Richard H. Moss, and D.J. Dokken (eds.), 1995: *Climate Change, 1995: Impacts, Adaptations, and Mitigation of Climate Change: Scientific-Technical Analyses – Contribution of Working Group II to the Second Assessment Report of the Intergovernmental Panel on Climate Change*, Cambridge University Press, Cambridge

Wheaton, E.E. and D.C. Maciver, 1999: A framework and key questions for adapting to climate variability and change, *Mitigation and Adaptation Strategies for Global Change* **4**, 215–25

Wigley, T.M.L., 1994: MAGICC (Model for the Assessment of Greenhouse-gas Induced Climate Change): User's Guide and Scientific Reference Manual. National Center for Atmospheric Research, Boulder, Colorado

Wilbanks, T.J., 2005: Issues in developing a capacity for integrated analysis of mitigation and adaptation, *Environmental Science & Policy* **8**, 541–7

World Bank, 2006: *Clean Energy and Development: Towards an Investment Framework*, World Bank, Washington, DC

Yohe, G. and K. Strzepek, 2007: Adaptation and mitigation as complementary tools for reducing the risk of climate impacts, *Mitigation and Adaptation Strategies for Global Change* **12**, 727–39

Market- and Policy-Driven Adaptation

Alternative Perspective

SAMUEL FANKHAUSER

Introduction

The policy debate on climate change distinguishes two generic response options. The first (and more prominent) option is mitigation. Mitigation addresses the *causes* of climate change by reducing the emission of harmful greenhouse gases (GHGs). The second response is adaptation. Adaptation deals with the *consequences* of climate change and seeks to reduce the vulnerability of human and natural systems to a shift in climate regime.

This Perspective paper sets out the case for adaptation, complementing and building on chapter 6 by Bosello, Carraro, and De Cian (Bosello *et al*. 2010). Both chapter 6 and this Perspective paper aim to answer the same question: What is the role of adaptation in the international policy response to climate change? Bosello *et al*. 2010 approach the question from a modeling point of view, using an integrated assessment model (IAM) that explicitly includes both adaptation and mitigation. This Perspective paper seeks to extract answers from the wider literature, rather than through bespoke modeling work.[1]

The Perspective paper is structured as a set of six theses that I believe are central to the adaptation debate and can help to frame the question at hand, and deals with each of them in turn:

1. A minimum level of adaptation is now unavoidable
2. Adaptation and mitigation are complements, but making the tradeoff is hard
3. Adaptation can have massive net benefits
4. Adaptation goes hand in hand with development
5. The timing and sequencing of adaptation action matters
6. Uncertainty matters.

A Minimum Level of Adaptation is Now Unavoidable

The need to adapt to climatic conditions has been a feature of human life since the beginning of time. It is an ongoing challenge that affects the way we live, how we design our infrastructure, and how we produce our goods and services. Adaptation is not a new activity introduced as a consequence of climate change. What climate change forces us to do is to readjust our economies and our behavior to reflect the new climate realities. Adaptation to climate change is a challenge not because the concept is new but because the scale and speed of the adjustments required is unprecedented and because the exact nature of the anticipated changes remains highly uncertain.

Yet much of that change is already in the pipeline. Global mean temperatures today are already about three-quarters of a degree warmer than in preindustrial times, and even if carbon emissions completely ceased today the warming trend would continue for many decades. In other words, the mitigation measures currently discussed will determine the climate (and adaptation needs) towards the end of the century. The adaptation needs over the next couple of decades are already pretty much set.

[1] This Perspective paper draws heavily on Fankhauser *et al*. (1999), Agrawala and Fankhauser (2008), and Fankhauser (2009).

Even over the longer term it looks pretty certain that the world will have to adapt to climate change of at best 2°C. There are few realistic policy scenarios that entail equilibrium warming of less than that. Both a 2°C world and the temperature changes already committed to will require considerable adaptation.

Short-term adaptation needs (up to 2015–30) have been costed at anywhere between US$4 billion and over US$100 billion a year. These numbers are crude and at best indicative. At the low end, they almost certainly underestimate true adaptation needs. The high end is more realistic, but sometimes also includes "social adaptation" activities that could arguably be part of baseline economic development (Fankhauser 2009).

Mitigation and Adaptation are Complements, but Making the Tradeoff is Hard

While the short-term need for adaptation is pretty much predetermined, there is policy flexibility in the longer term. At least conceptually, policy makers may choose between different combinations of adaptation and mitigation. From an economic point of view the policy choice is an intertemporal optimization problem. An imaginary global social planner seeks to minimize the costs of climate change through a judicious mix of mitigation policies and adaptation action.

For example, the social planner may decide to limit the overall temperature increase to 2°C (mitigation) and invest in items like flood protection, coastal defense, and drought-tolerant cultivars to limit the negative impacts of 2°C warming (adaptation). There would be some residual damages – for example, the loss of certain coastlines and lower agricultural yields because this cannot be avoided at reasonable cost. If the social planner chooses correctly, the combined costs of mitigation, adaptation, and residual damage are kept as low as possible.

Chapter 6 is firmly in this vein. It is the basic approach most economists would apply to the problem, although it is well recognized that more complex frameworks should also consider

reasons for concern other than net costs, such as the unfair distribution of impacts, the risk of "tipping points," excessive climate variability, and the threat to unique natural systems (see Smith *et al.* 2001, 2009).

IAMs that include both adaptation and mitigation policies are still fairly novel, and they provide new and interesting insights. However, they are too stylized and not yet robust enough to allow firm policy conclusions. Very little is known, for example, about the shape of the climate change damage function. Similarly, most adaptation estimates are point estimates. We do not know how adaptation costs vary as a function of temperature rise, and to what extent there are limits to adaptation.

Moreover, policy decisions about adaptation and mitigation are often not made by the same people. Mitigation decisions are reached globally in international negotiations, backed up subsequently through national legislation. Adaptation decisions are made, more often than not, at the local level (e.g. by municipal governments) and by private agents (households and firms), perhaps incentivized by national policy. These people are "climate takers" in the sense that global emissions are outside their control. Their own GHG output has no noticeable impact on total emissions.

In practice, therefore, no explicit choice, or trade-off, will be made between the optimal levels of mitigation and adaptation.

Adaptation can have Massive Net Benefits

Much of what we know about the costs and benefits of adaptation comes from case studies of particular sectors or countries. A survey carried out by the OECD found that our knowledge about adaptation at the sector level is growing, but information is unevenly distributed (Agrawala and Fankhauser 2008). Although our knowledge is increasing all the time, outside coastal zones and agriculture our knowledge base is still limited.

Nevertheless, the available evidence shows that adaptation is very powerful for dealing with moderate amounts of warming at least. For example:

- In agriculture there is broad evidence that low-cost adaptation measures like changes in planting dates, cultivars, fertilizer use, and management practices can often reduce the impact on crop yields by more than half, relative to the no-adaptation case (see figure 6.1.1).
- Coastal protection is one of the few sectors where adaptation costs (usually sea walls and beach nourishment) and adaptation benefits (avoided land loss, flooding) are routinely compared. The resulting benefit-cost ratios (BCRs) are not always reported, but one study, on coastal protection in the EU, suggests BCRs of 1.1–2.6 by 2020, rising to 4.3–6.5 by 2080 (Commission of the European Communities 2007).
- In the health sector, it has been estimated that preventing some 133 million climate-related deaths from malaria, malnutrition and, diarrhea would cost around $3.8–4.4 billion, or less than $33 per life saved (UNFCCC 2007).

Since the focus of many of these studies is on low-cost adaptation, high BCRs are not unexpected. The question is how the return on adaptation changes as we move up the adaptation cost curve and start to implement more expensive measures. A study by McKinsey and Swiss Re in eight countries confirms that BCRs will eventually drop below 1 (McKinsey 2009). There is a limit to cost-effective adaptation, however, the study also found that in the eight cases considered most of the expected impacts may be avoided through cost-effective adaptation.

Two caveats are in order. First, cost-effectiveness, while a key consideration, is not the only concern in the allocation of adaptation funding. The equitable distribution of funds is equally important. In particular, developed countries have an obligation, acknowledged in the UN Framework Convention on Climate Change (UNFCCC), to support adaptation in developed countries that are particularly vulnerable to climate change. Providing sufficient adaptation funding to developing countries is a key concern that goes well beyond cost-effectiveness considerations.

Second, practically all the available evidence on adaptation effectiveness concerns adaptation to a "moderate" amount of climate change of perhaps 2–3°C. Very little is known about the effectiveness of adaptation to the more severe levels of change that will occur if global GHG emissions are not curtailed. It would therefore be dangerous to rely on adaptation as a large-scale substitute for mitigation.

Adaptation goes Hand in Hand with Development

Since adaptation to the prevailing climate is nothing new, it is often difficult in practice to delineate where "normal" socioeconomic development ends and adaptation to anthropogenic climate change begins. Socioeconomic trends over the coming decades – population growth, economic expansion, the deployment of new technologies – will both shape and be shaped by our vulnerability to climate conditions.

This is particularly the case for developing countries, where there is a well-documented adaptation deficit – that is, insufficient adaptation to the current climate. Poor people and poor countries are less well prepared to deal with current climate variability than rich people and rich countries. There is evidence that development indicators such as *per capita* income, literacy, and institutional capacity are associated with lower vulnerability to climate events (see, for example, Noy 2009). This has led authors like Schelling (1992) to conclude that good development is one of the best forms of adaptation.

More broadly, we can think of adaptation as a "pyramid of needs," where certain development conditions have to be fulfilled before it makes sense to move to the next response level. McGray *et al.* (2007) distinguish four levels in the development–adaptation continuum:

- Policies to reduce vulnerability to stress more broadly (whether climate-related or not), including core human development objectives like education, health, sanitation, and poverty eradication.
- Creation of "response capacity," such as resource management practices, planning systems, and effective public institutions.

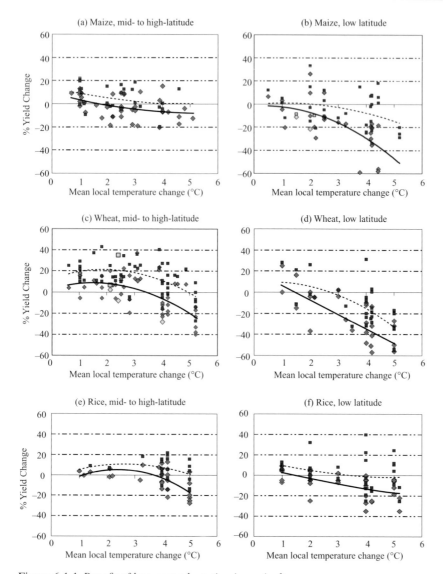

Figure 6.1.1 *Benefit of low-cost adaptation in agriculture:*

(a) Maize, mid- to high-latitude;
(b) Maize, low-latitude;
(c) Wheat, mid- to high-latitude;
(d) Wheat, low-latitude;
(e) Rice, mid- to high-latitude;
(f) Rice, low-latitude.

Note: The **bold** line shows yield change without adaptation; the dashed line shows yield change including basic adaptation measures. Lines are derived from sixty-nine published studies.
Source: Easterling *et al.* (2007).

- The management of current climate risks, including flood and drought prevention, disaster preparedness, and risk management.
- Policies specifically addressing anthropogenic climate change, such as accelerated sea level rise and an increased incidence of extreme weather events.

Although only the last of these sets of activities is "adaptation to climate change," strictly defined, effective strategies (and spending decisions) to reduce climate vulnerability have to address the entire pyramid and recognize synergies between the different levels.

The Timing and Sequencing of Adaptation Action Matters

While some impacts can already be felt, climate change is essentially a long-term problem. The worst effects are not expected to materialize for a couple of decades. This makes the timing and the sequencing of response measures an important part of adaptation decisions.

In deciding the optimal timing for adaptation, decision makers will compare the net present value (NPV) of adaptation now with the NPV of adaptation at a later stage. The two present values (PVs) consist of adaptation costs (incurred either now or later) plus a stream of climate costs (say, the costs of flooding), which is reduced once adaptation takes place. Comparing the two PVs, there are three cost components that will determine adaptation timing:

- *The difference in adaptation costs over time.* The effect of discounting would normally favor delayed action, but there is also a class of adaptations where proactive action (e.g. during the design phase of a project) is cheaper than costly retrofits at a later point. Long-term development plans – for example, the development of a coastal zone – and long-lived infrastructure investments – such as water and sanitation systems, bridges, and ports – fall into this category. For such investments it makes sense to already incorporate climate change considerations today. This was the view taken, for example, by the Canadian authorities when they built

sea level rise into the design of the Confederation Bridge that links Prince Edward Island with New Brunswick (Smith *et al.* 1998).
- *The short-term benefits of adaptation.* Early adaptation will be justified if it has immediate benefits that later action would forgo. The prime example is measures that address current climate variability as well as future change. Similarly, many of the more developmental measures in the adaptation pyramid (see above) have immediate development benefits and are a precondition for effective adaptation later on.
- *Long-term irreversibilities or cumulative effects.* Early adaptation is justified if it can lock-in lasting long-term benefits. For example, failure to protect ecosystems from current-day stress may leave them in too weakened a state to cope with future climate change.

These points suggest a preference, in the short term, for adaptations that have immediate benefits, are long-lived, and prevent costly retrofits or even irreversible loss. These conditions are met by most measures to close existing adaptation gaps.

Uncertainty Matters

Timing decisions, in fact all adaptation decisions, are complicated by uncertainty about the exact nature of climate change impacts, especially at the local level (for example, in terms of precipitation and storminess). This makes it difficult to fine-tune adaptation measures proactively.

Uncertainty will favor measures with strong near-term benefits, which are easier to ascertain, and win–win measures that are justifiable independently of the climate outcome. Measures to close existing adaptation gaps clearly fall into this category.

Others have argued that given the prevailing uncertainties, the best way to account for potential climate change in current investment decisions is to increase the *flexibility* of systems – that is, allowing them to adjust to a range of climate outcomes – and/or their *robustness* – that is, designing them to function under a wide range of climatic conditions and to withstand more severe climatic shocks (Fankhauser *et al.* 1999; Hallegatte 2009).

The call for increased flexibility and robustness applies to physical, natural, and social systems. In the case of physical capital, the capacity of water storage systems may be increased in anticipation of possible future droughts and sewage systems may be enlarged to deal with heavy downpours. In the case of natural capital, measures to protect the environment may increase the ability of species to adapt to a changing climate. Institutionally, creating regulatory frameworks that encourage individual adaptability would help to increase the flexibility and robustness of economic systems. It has been argued, for example, that opening agricultural markets to competition and trade would help to dampen the negative shock of a bad harvest in individual regions.

Conclusion

This Perspective paper sets out the case for adaptation as a core aspect of the global policy response to climate change. The case for adaptation is made through a set of six propositions.

The Perspective paper argues that some adaptation is unavoidable. There are no realistic mitigation policies that restrict warming to a level that does not require substantial adaptation. Moreover, the adaptation needs over the coming decades are already set. They are predetermined by the amount of warming that is already in the pipeline.

In the longer term there is a choice between adaptation and mitigation. The two policy options are complements. The Perspective paper shows that adaptation is an important part of the policy mix. The net benefits of basic adaptations – such as coastal defence and adjustments in agricultural practices – are often substantial. However, we know very little about the effectiveness of adaptation under more severe climate scenarios, which makes a strategy that relies too heavily on adaptation (at the expense of mitigation) rather risky.

Moreover, cost-effectiveness should not be the only criterion in making adaptation decisions. In the international negotiations, adaptation is often linked to questions of fairness and compensation.

In practice, proactive adaptation is also made difficult by uncertainty about the exact nature of the expected change. A key area where proactive adap-tation has strong and unequivocal benefits independent of climate change outcomes is action to close prevailing "adaptation gaps" – that is, measures that simultaneously address development and adaptation needs. In developing countries, adaptation and development have to go hand in hand. Or in the words of Stern (2009), adaptation is development in a hostile climate.

Bibliography

Agrawala, S. and S. Fankhauser, 2008: *Economic Aspects of Adaptation to Climate Change. Costs, Benefits and Policy Instruments*, OECD, Paris

Bosello, F., C. Carraro, and E. De Cian, 2010: *Market- and Policy-Driven Adaptation*, chapter 6 in this volume

Commission of the European Communities, 2007: *Commission Staff Document. Accompanying Document to the Communication of the Commission to the Council, the European Parliament, the European Economic and Social Committee and the Committee on the Regions on Limiting Global Climate Change to 2 degrees Celsius. Impact Assessment* **SEC (2007) 8**, http://ec.europa.eu/environment/climat/pdf/ia_sec_8.pdf

Easterling, W.E. *et al.*, 2007: Food, fibre and forest products, in *IPCC, Climate Change 2007: Impacts, Adaptation and Vulnerability – Contribution of Working Group II to the Fourth Assessment Report of the Intergovernmental Panel on Climate Change*. Cambridge University Press, Cambridge

Fankhauser, S., 2009: The costs of adaptation, in *Wiley Interdisciplinary Review Climate Change* **1**(1), 23–30

Fankhauser, S., J.B. Smith, and R.S.J. Tol, 1999: Weathering climate change. Some simple rules to guide adaptation investments, *Ecological Economics* **30**(1), 67–78

Hallegatte, S., 2009: Strategies to adapt to an uncertain climate change, *Global Environmental Change – Human and Policy Dimensions* **19**(2), 240–7

McGray, H., A. Hamill, R. Bradley, E.L. Schipper, and J.-O. Parry, 2007: *Weathering the Storm. Options for Framing Adaptation and Development*, World Resources Institute, Washington DC

McKinsey & Company, 2009: *Shaping Climate-Resilient Development. A Framework for Decision-Making*, Report of the Economics of Adaptation Working Group, McKinsey & Company and Swiss Re

Noy, I., 2009: The macro-economic consequences of disaster; *Journal of Development Economics* **88**, 221–31

Schelling, T., 1992: Some economics of global warming, *American Economic Review* **82**(1), 1–14

Smith, J.B. *et al.*, 1998: *Proactive Adaptation to Climate Change. Three Case Studies on Infrastructure Investments*, Working Paper **D-98/03**, Institute for Environmental Studies, Free University, Amsterdam

2001: Lines of evidence for vulnerability of climate change: a synthesis, IPCC, *Climate Change: Impacts, Adaptation and Vulnerability – Contribution of Working Group II to the Third Assessment Report of the IPCC*, Cambridge: CUP

2009: Assessing dangerous climate change through an update of the Intergovernmental Panel on Climate Change (IPCC) "reasons for concern", *Proceedings of the National Academy of Science of the United States of America* **106**(11), 4133–7

Stern, N., 2009: *Blueprint for a Safer Planet*, The Bodley Head, London

UNFCCC, 2007: *Investment and Financial Flows to Address Climate Change*, UNFCCC, Bonn

6.2 Market- and Policy-Driven Adaptation

Alternative Perspective

FRANK JOTZO[*]

Introduction

Climate change is highly likely to have substantial impacts on natural and human systems in decades and centuries to come, and has the potential to severely disrupt economic activities. It could force the relocation of large numbers of people between and within countries, and change the location, extent, and nature of economic activities including agriculture. In some cases, adaptation to changed climatic conditions could be feasible simply through changed practices, not necessarily incurring significant economic costs. In others, it will require expansion and remodeling of service systems such as public health and necessitate the early retirement of existing and construction of new infrastructure including in housing, transport, water supply, energy, and so forth.

These facts drive a demand for qualitative and quantitative economic analysis on an optimum degree of climate change mitigation (that is, reducing greenhouse gas, GHG, emissions) and adaptation to the effects of climate change, the optimal timing of such actions, and the optimum distribution of adaptation action between countries and sectors. This Perspective paper discusses what is possible for economic modeling in this field and what is not, with specific reference to chapter 6 by Bosello, Carraro, and De Cian.

Adaptation to climate change impacts will be necessary, will occur both through individual and

* The author thanks the authors of chapter 6 and the members of the Expert Panel for discussion and comments. Thanks also go to presenters and participants at the session "Economics of Climate Change Adaptation" that I convened at the Copenhagen Climate Change Congress in March 2009. Sabit Otor provided valuable research assistance.

policy-driven actions, and may require substantial economic resources; but the economic analysis – and, in particular, aggregate economic modeling, of climate change impacts and adaptation – is in many respects a long way from being directly useful for the formulation of policy decisions. I argue that to make such analysis relevant for policy decisions, it must incorporate three factors that define the economics of climate change. The first is uncertainty – in particular, the risk of abrupt climate change, which is a major reason for urgency in addressing the problem. The second is improved calibration of economic climate change impacts, and the inclusion of non-market impacts. The third is equity and differential climate impacts at a fine scale, which will define adaptation actions in practice. Analyses that leave out these factors tend to yield results biased against mitigation, and their more detailed quantitative results are of very limited use as a guide to policy.

Economics of Adaptation and its Analysis: Some Basics

Climate change is already being observed, further change is already loaded into the system, and there will be the persistence of a GHG-emitting system even if the world were set on a path of strong climate change mitigation soon. It appears almost certain that adverse climate change impacts will occur, and will continue to occur for a long period of time. Hence, the role of mitigation is to reduce the extent of future damages and to limit the risk of dangerous climate change. Adaptation to the impacts that are already unavoidable will be necessary alongside mitigation.

Adaptation is going to occur. This is in contrast to mitigation, which suffers severe problems of coordination and free-riding, so that the extent of global mitigation might forever remain below the social optimum. Adaptation will typically consist of localized or regional actions, changing economic systems (and more broadly human systems) to deal with observed or anticipated environmental changes. Taking adaptive action is in the direct self-interest of those individuals and communities affected by the changes. While externalities will exist within communities or nations for some adaptation action, a much greater share of benefits will be captured by those groups taking action than for mitigation, and the lag between investment and return will generally be much shorter than with mitigation.

A number of factors will inhibit efficient adaptive action – among them, credit constraints, imperfect information or wrong perceptions about climate change impacts and risks, the use of socially non-optimal discount rates, and shortcomings in public policy making. As a consequence, optimal abatement responses may not be achieved, but a large extent of useful adaptation will nevertheless take place.

Both market-driven and policy-driven adaptation are needed, as pointed out by chapter 6, and it is a fair expectation that both will occur. Individuals will react to climate change impacts as they do to any other changes in their physical, social, and economic environment. And policy action to facilitate adaptation will for the most part be of a similar type as government reactions to other societal needs.

It is also clear that benefit-cost ratios (BCRs) will in aggregate be positive for adaptation, at least in expectation terms (that is, the anticipated benefits and costs when adaptation is decided on). As a principle, this follows directly from the nature of adaptation as specific actions in response to specific changes and risks at a local level. Adaptation options whose costs exceed their anticipated benefits will, on the whole, not be implemented, but options with positive ratios will. Hence BCRs somewhere above 1, as posited by chapter 6, are unsurprising.

Much harder and less tractable questions revolve around the optimal mix between adaptation and mitigation, and the more "classic" topic in the economic literature, the optimal amount of mitigation *per se*. While it is clear that some degree of mitigation action is economically beneficial, debate is lively (and unlikely to ever be resolved) over what that optimal amount is, exemplified by the conflicting conclusions reached by Stern (2006) or Garnaut (2008) on the one hand, and Nordhaus (2007) or Tol (2009) on the other (for more on these, see below). But, as I will argue, the data and tools available to economic analysis at this point in time are insufficient for a reliable empirical analysis of the adaptation–mitigation tradeoff – for related arguments on the more general theme of a cost-benefit analysis (CBA) of climate change and mitigation see, for example, Spash (2007) and Ackerman *et al.* (2009).

An important point often overlooked is that adaptation and mitigation are in important respects not substitutes. They are treated as substitutes in economic analyses and models that revolve around single aggregate welfare measures. But, in reality, mitigation and adaptation will in important respects serve different objectives. A key objective of mitigation is precaution, reducing the risk of irreversible impacts. For many climate change damages, there is no adaptation. This is particularly true for the natural environment.

Finally, the question of how, when, and where to adapt cannot be answered with confidence, because of pervasive and persistent uncertainty about future climate change and its impacts. Aggregate analyses are at a particular disadvantage as they do not have fine-grained information about whether particular adaptation actions are possible and beneficial in a particular setting. Hence, predictions about the extent, type, and timing of adaptation based on aggregate economic models cannot be more than illustrative scenarios.

Quantitative Modeling of the Economics of Climate Change

The likely costs of climate change adaptation are beginning to be estimated in detailed sector-by-sector studies (as an example, see Ciscar *et al.*

2009 and, for an overview, Parry *et al.* 2009). Only some of the cost studies include an explicit assessment of the benefits from adaptive action, and thus make it possible to assess the BCR of adaptation. Where they do, the result is typically that the costs of adapting are far smaller than the economic losses that would be incurred without adaptation. The focus in such studies is typically on options for adaptive action that will pay large dividends (think of expanded water storage and improved fire prevention in areas that become drier with climate change). One could also think of adaptive investments that are economically wasteful (an example might be sea walls to shield existing infrastructure from sea level rise when it would be cheaper to rebuild at a higher elevation), but for obvious reasons these are typically not included in studies of adaptation options.

At the other end of the spectrum of quantitative economic analysis, aggregated models of the economy overall, in particular computable general equilibrium (CGE) models, are beginning to be used for the analysis of climate change impacts as well as the analysis of adaptive responses. Where both mitigation and impacts/adaptation are modeled together, these models are referred to as "integrated assessment models" (IAMs). Chapter 6 is a specific example of the application of such a model.

Such modeling, in principle, has decisive advantages over micro-level, partial equilibrium modeling: it gives an integrated representation of benefits and costs adaptation over many different sectors and countries; a representation of economic flow-through effects, such as changes in relative prices, trade, production, and consumption patterns that may result from climate change impacts, and mitigation and adaptation actions; and it can be used in simultaneous analysis of mitigation and adaptation. But IAM modeling also brings great abstractions, generalizations, and reliance on assumptions about parameters that drive the aggregate results but are difficult or impossible to estimate or determine.

Some Results in Chapter 6

Chapter 6 makes an important contribution in showing that the optimal policy mix for the world entails both mitigation and adaptation, that an increase in mitigation action reduces the optimal level of adaptation action, that both market- and policy-driven adaptation is needed, and that the degree of climate change damages as well as time preferences affect the extent of optimal adaptation and mitigation, as well as the optimal mix.

Regarding the time dimensions of mitigation and adaptation, chapter 6 concludes that mitigation action needs to come first, and little adaptation action is needed until the middle of the century, when climate impacts are assumed to begin. However, it stands to reason that many adaptation actions would need to take place ahead of time, to manage the risk of future climate change. This relates in particular to long-lived infrastructure, including transport. A current real-world example is desalination plants, which in Australia and elsewhere are now being planned, and in some cases built, to come on line if and when drought and water shortages become worse.

The findings of chapter 6 on the BCRs of different scenarios of adaptation and mitigation show, first, the overwhelming role played by the discount rate. Under the "low-discount rate" scenario, both the benefits and the costs of adaptation, and in particular of joint mitigation and adaptation, are greatly higher than for high discount rates. This is of course a familiar result, especially in the wake of The Stern Review (Stern 2007), and is a core difficulty with CBA in climate change (Quiggin 2008). Along with the overall magnitude of benefits and costs, the absolute difference between them increases greatly under a lower discount rate. The B/C *ratio*, however, is lower with low discount rates. Hence, considering only the BCR could lead to the fundamentally wrong impression that greater concern for the future *reduces* the desirability of climate change adaptation and mitigation compared to other investments.

Using higher damage functions greatly increases both the absolute size of benefits and costs, and the BCRs. This is intuitive: if climate change is more of a problem then the payoff from addressing it is greater. However, the stark differences in BCRs between the "low" and "high" damage scenarios show that to a great extent these ratios are driven by the assumptions about climate change damages. As discussed below, leaving out the risk of extreme or catastrophic climate outcomes biases the damage estimates downward, perhaps severely. Leaving out non-market values and equity impacts will generally bias the results in the same direction.

A fundamental point to note in assessing the BCRs and other quantitative results is the damage cost estimates and functions, which go back to Nordhaus and Boyer (2000), and which assume only relatively small impacts from climate change on economic activity and welfare, with any economic damages swamped by increases in economic growth over time. In chapter 6 only modest GDP impacts are shown even at temperature increases around 4°C, which is now commonly regarded as carrying a significant risk of large-scale, highly disruptive, and possibly catastrophic climate change (Schellnhuber 2009).

Remarkably, when market-driven adaptation is considered in chapter 6, OECD countries as a group *benefit* from climate change (and presumably net benefits are even greater when taking into account government-driven adaptation). This result could be seen to imply that OECD countries' interest, as a group, is in increasing global emissions, not reducing them, and that only the developing world has an interest in mitigation – which is in obvious conflict with actual climate policy.

Limits of Economic Modeling of Climate Change

These results derive from the climate change damage functions used in the model, and the fact that the risk of abrupt or catastrophic climate change is not considered. Much of the relevant economic modeling literature incorporates a similar assumption that climate change damages are small relative to economic growth over time, and ignores the risk of catastrophic change. The conclusion that follows from such assumptions is that the globally optimal amount of mitigation is rather small – a conclusion that is at odds with the dominant view in the natural sciences, and with the precautionary considerations that are clearly an important motivator for climate policy in the real world.

Chapter 2 by Tol in this volume is an example of a modeling analysis that finds small amounts of mitigation to be optimal, as a direct consequence of assumptions about climate change damages. Tol assumes that the social cost of carbon (SCC) is only \$2/ton of carbon (equivalent to \$0.5/tCO$_2$), far less than mainstream views of the SCC, and an extreme outlier in the literature.[1] By comparison, the marginal cost of emissions already in place under the EU emissions trading system (ETS) has been in the range of \$15–30/tCO$_2$. From such an assumption inevitably follows the conclusion that the costs of mitigation exceed the benefits for anything more than very small efforts.

Any modeling analysis is defined and limited by the choice of features of reality that are represented and ignored, and the calibration of parameters, for which empirical evidence is often scarce. I argue that the aggregated modeling tools at the disposal to the economics community, and including those applied in chapter 6, are not nearly sophisticated enough to yield quantitative answers that are useful to policy makers. They may be able to give important qualitative indications – such as about the complementarity of mitigation and adaptation – but the quantitative results are under a heavy cloud of doubt even for broad aggregate results, and are generally of no use as a guide to policy at a disaggregated level.

Below, I discuss three aspects that would need to be included in any quantitative economic modeling of climate change in order for the quantitative results to usefully speak to policy. The first is uncertainty, in particular the risk of abrupt climate change, which is a major reason for urgency in addressing climate change but difficult to capture in economic models. The second is improved calibration of economic climate change impacts, and the inclusion of non-market impacts, which motivate much of public concern about climate change and for which adaptation options are typically

---

[1] For example Tol's own survey (Tol 2005) showed the mean of twenty-eight studies assumed a SCC of \$97/tC, and the subsample of studies published in peer-reviewed journals \$43/tC, denoted in 1995 US\$ (and thus higher in current value terms). An SCC of \$2/tC or below is found only at the extreme end of the range of assumptions in some of the studies.

much narrower than for market impacts. The third is equity and differential climate impacts on the fine scale, which will define adaptation actions in practice, but cannot be represented in aggregate models.

Uncertainty

Most modeling of whole economies – in particular, that using CGE models – takes places in a deterministic framework. CGE models consist of a set of parameters that describe observed economic data and relationships (such as inputs to production processes and trade flows), and fixed assumptions for behavioral responses (such as responses to changes in prices). In typical applications, including chapter 6, the model is then subjected to "shocks" in the form of sets of changes in exogenous variables. In modeling of climate change, a set of assumptions about the impacts of such change is imposed – for example, through changes in the productivity of certain sectors of the economy; and a price (tax) on emissions is imposed which results in shifts in production and consumption away from emissions-intensive processes, goods, and services. The myriad effects and interactions in the model can then be presented in an aggregate measure such as GDP or consumption. It is generally thought that responses in an economy to changes in relative prices – for example, through changes in taxation or tariffs – can be modeled in this way with at least some degree of confidence.

Any extension to the modeling of climate change impacts, however, brings hugely more complex issues into play. The nature and extent of future physical climate change impacts is unknown. Climate change science increasingly indicates that there may be strong feedback mechanisms in the system, making the correlation between GHG emissions, temperature increase, and physical impacts highly non-linear (Richardson *et al.* 2009). In other words, there is a wide probability distribution for the possible climate impacts (and their economic effects or damages) of any given level of emissions or global temperature increase. Consequently, modeling that deterministically maps emissions to climate change damages lacks the crucial dimension

of uncertainty about what the actual effect might be, and in particular risk of very strong damages.

The risk of extreme climate change is in fact the main reason why the mainstream of climate change scientists urge fast and strong action to rein in emissions, and the key reason why a range of governments are pursuing urgent global mitigation action. A central objective of climate change mitigation, already evident in the 1992 UN Framework Convention on Climate Change (UNFCCC), is to reduce the risk of extreme climate change in an expression of societal risk aversion.

It has been shown that under assumptions about the probability distribution of climate change damages that appear plausible given current knowledge, the (low) probability of catastrophic climate change alone could be the single overwhelming factor in an economic analysis of climate change, and for considerations relevant to economic decision making about mitigation. In the words of Weitzman (2009), the problem is characterized by "deep structural uncertainty in the science coupled with an economic inability to evaluate meaningfully the catastrophic losses from disastrous temperature changes." Thus, avoiding the risk of very large-scale economic damage dominates the effect even of the choice of discount rate, traditionally seen as the main variable driving the optimal level of mitigation.

Similar arguments, though likely to a lesser extent, also apply to the modeling of adaptation. Abrupt climate change could necessitate very different adaptation responses, and at a different timescale, requiring a greater extent of anticipatory adaptation to achieve greater readiness for possible climate change impacts. Furthermore, it must be questioned whether current assumptions about behavioral parameters built into economic models are an accurate guide to what may happen in the future, particularly under scenarios of significant change in the structure of economies.

The upshot for economic modeling of climate change, its economic effects, and policy responses is that, at a minimum, stochastic modeling of climate impacts is needed, rather than using only the median of the presumed probability distribution, as is so often done. In the first instance, this would involve the modeling of a large number of different scenarios of climate change impacts, ranging from very small to catastrophic changes according to an assumed probability distribution. Such stochastic modeling was undertaken, for example, by Stern (2007), then conflated into an aggregate measure of expected economic impacts from climate change. Such a stochastic approach does not overcome structural uncertainty and the inability to economically evaluate catastrophe, but at least it can give a sense of the range of possible outcomes.

Economic Impacts and Valuation of Climate Change Impacts

A second set of fundamental issues for economic modeling of climate change and adaptation options relates to the likely economic effect of environmental change, especially if and where such change is large in scale; and the inclusion and valuation of non-market impacts. Most current modeling exercises, chapter 6 included, rely on highly aggregate climate change damage functions that may underestimate feedback effects within economies, and do not represent non-market impacts such as the loss of species or natural icons.

CGE models typically assume a strong degree of substitutability in both production and consumption structures, and aggregate welfare measures such as GDP and consumption are driven much more by assumed underlying growth in productivity than by changes in productivity because of a shift in structure away from the optimum. Physical factors, such as the need to produce and consume a certain amount of food per person, are often inadequately represented, or not at all. Similarly, and using a related example, possible feedback effects such as escalating food prices during times of shortage are generally not well represented. Hence, even large-scale physical impacts from climate change tend to be translated into only small changes in welfare, especially when compared to the assumed increase over time.

A striking result from chapter 6 is that adverse impacts on tourism are the approximately equal largest category of economic damages, alongside agriculture. By contrast, the impacts from sea level rise and health are almost insignificant. Total net climate change damages are less than half a percentage point of GDP at 2050, compared to GDP typically expected to more than treble over that time-span. This is in a scenario of a 3°C increase in mean temperatures, which is now generally regarded to herald unacceptable risks from climate change for humanity.

These damage estimates originate in the damage functions taken from other studies, in interaction with the data and assumptions in the models. While it is impossible to confirm or refute any particular pattern of climate change damages, this particular result provokes doubts over the damages functions used. Alternative specifications need to be explored that accord with notions that impacts on coastal infrastructure, health and agriculture, and so forth would be so serious that they would likely far outweigh economic impacts on the tourism industry.

A well-understood, yet extremely difficult to address shortcoming of standard economic modeling of climate change is the omission of non-market impacts, including amenity value to people and the existence value of natural and cultural icons. These aspects are difficult to quantify, and leaving them out is in the mainstream modeling tradition. Nevertheless, an analysis that speaks to actual policy decisions on climate change cannot afford to set aside non-market impacts. In an illustration from Australia, it appears that the possible or indeed likely loss of the Great Barrier Reef, the world's largest coral reef, is a major factor in public concern about climate change. While it will be impossible to reliably quantify the amenity and existence value of such natural icons, they must figure in the overall evaluation of mitigation and adaptation strategies.

Adaptation options will typically be more restricted for issues revolving around non-market values than for market impacts. The coral reef example is obvious in that there are no apparent adaptation options. The situation may be similar if somewhat different for issues such as the survival of species, where assisted relocation may be an option in some instances. The inclusion of non-market values in the analysis thus shows greater importance for mitigation, rather than adaptation.

Equity and Scale

A third set of issues critical for the modeling relates to the distribution of the impacts of climate change, and the costs and benefits of mitigation and adaptation.

Mainstream economic modeling exercises aggregate welfare measures across countries, and implicitly within countries, and derive optima over the globally aggregated result. The implicit assumption is that an extra dollar of income provides the same utility to each person in the world. Given the stark differences in income and living standards, this is self-evidently untrue, and subnational equity aspects also have important implications for welfare analysis around climate change (Baer 2009). The point is generally recognized in the broader climate policy debate, where there is heavy emphasis – at least in the rhetoric of international negotiations and domestic politics – on shielding the poorest countries and people from climate change damages.

One way to deal with this in a modeling context is to give equity weighting to welfare results. In a multi-country model, this would result in a different global optimum, namely one that gives greater emphasis to the best outcome in poor countries. On the basis of the numbers reported in chapter 6, this would probably mean a greater optimum amount of both mitigation and adaptation, and a changed mix between the two.

A final issue to note here relates to the scale of the modeling. The sectoral and regional detail in the economic models used for climate change analysis is much coarser than the likely pattern of damages and benefits from climate change and adaptation. For example, a net loss within agriculture in one country could in fact consist of gains in some regions and for some types of agriculture, offset by larger losses in other areas. Similarly, there would be pertinent and highly cost-effective adaptation options in some activities and regions, whereas none might exist elsewhere. The design and implementation of policy must and will take the fine scale into account. Data from much coarser aggregate economic modeling will be of limited value in guiding such policy.

Conclusion

Integrated assessment modeling, such as in chapter 6 can provide powerful qualitative insights – for example, about the need for both mitigation and adaptation and the interactions between the two, or the need for both individual and policy-driven adaptation. However, the more detailed quantitative results from such studies are subject to such strong limitations as to be virtually irrelevant as a guide to policy.

The Copenhagen Consensus exercise places heavy emphasis on BCRs. These ratios come about as a result of highly contestable assumptions about climate change impacts, economic damage functions, and societal valuations and preferences, with interactions between them shaped by assumptions about behavioral relationships in economies decades in the future. Consequently, the estimated BCRs are highly unreliable as a guide for policy.

This Perspective paper has argued that three important features are needed in economic models of climate change in order for them to be useful representations of reality: representation of uncertainty about impacts, in particular the risk of abrupt climate change; fuller representation of economic impacts from climate change and inclusion of non-market impacts; and modeling of equity dimensions. Where these features are absent, it tends to result in quantitative results that are biased against mitigation as an option to address climate change, and in favor of other alternatives including adaptation. A stark example is the analysis by Tol in chapter 2, which assumes an extremely low SCC, and by virtue of that assumption concludes that only very small mitigation efforts would be cost-effective. Insofar as the recommendations from the convenors of the Copenhagen Consensus – in particular, the low ranking for mitigation – are based on such modeling, there must be strong doubts over the validity of their conclusions.

For adaptation, the type of quantitative analysis that will be most useful for policy makers will not be aggregate estimates of economic benefits and costs. Rather, it will be detailed and localized B/C estimates that take into account actual preferences of the communities concerned, including for equity, non-market valuations, and aversion to risk. This is because decisions about adaptation will not be taken in aggregate for whole economies (as might often be the case for mitigation), but sector by sector and locality by locality.

Arguably the most pressing need for understanding in the policy community relates to the effect of policy settings on adaptation. Existing policies can support adaptation, or be counter-productive and hinder adaptive responses. This implies that many aspects of the existing policy framework in any country will need to be examined for their likely effect on climate change adaptation. New policies will be needed in some areas, to support types of adaptive behavior that would otherwise not come about, and some existing ones will need to be scrapped. Much work will need to be done to understand where these needs are, and how they can best be met. This will include quantitative work about benefits and costs, but rarely at a highly aggregated level.

Bibliography

Ackerman, F., S.J. DeCanio, R.B. Howarth, and K. Sheeran, 2009: Limitations of integrated assessment models of climate change, *Climatic Change* **95**, 297–315

Baer, P., 2009: Equity in climate – economy scenarios: the importance of subnational income distribution, *Environmental Research Letters* **4**, 1–11

Bosello, F., C. Carraro, and E. De Cian, 2009: *Market- and Policy-Driven Adaptation*, chapter 6 in this volume

Ciscar, J.C., A. Soria, A. Iglesias, L. Garrote, S. Pye, L. Horrocks, P. Watkis, R. Nicholls, R. Roson, F. Bosello, L. Feyen, R. Dankers, A. Moreno, B. Amelung, J.M. Labeaga, X. Labandeira, O.B. Christensen, C. Goodess, and D. van Regemorter, 2009: Effects of climate change in Europe: results from the PESETA study, *IOP Conference Series Earth and Environmental Sciences* **6** (322002)

Garnaut, R., 2008: *The Garnaut Climate Change Review*, Cambridge University Press, Cambridge

Nordhaus, W.G., 2007: *The Challenge of Global Warming: Economic Models and Environmental Policy*, Yale University, New Haven, CT

Nordhaus, W.D. and J.G. Boyer, 2000: *Warming the World: The Economics of the Greenhouse Effect*, MIT Press, Cambridge, MA

Parry, M., N. Arnell, P. Berry, D. Dodman, S. Fankhauser, C. Hope, S. Kovats, R. Nicholls, D. Sattherwaite, R. Tiffin, and T. Wheeler, 2009: *Assessing the Costs of Adaptation to Climate Change: A Critique of the UNFCCC Estimates*, International Institute for Environment and Development and Grantham Institute for Climate Change, London

Quiggin, J., 2008: Stern and his critics on discounting and climate change: an editorial essay, *Climatic Change* **89**(3–4), 195–205

Richardson, C. *et al.*, 2009: *Climate Change: Global Risks, Challenges and Decisions*, Synthesis Report from the Copenhagen Climate Change Congress, University of Copenhagen/International Alliance of Research Universities

Schellnhuber, J., 2009: Terra quasi-incognita: beyond the 2°C line, Presentation at conference *4 Degrees and Beyond*, Oxford, September 28–30

Spash, C., 2007: The economics of climate change impacts à la Stern: novel and nuanced or rhetorically restricted?, *Ecological Economics* **63**(4), 706–13

Stern, N., 2007: *The Economics of Climate Change: The Stern Review*, Cambridge University Press, Cambridge

Tol, R.S.J., 2005: The marginal damage costs of carbon dioxide emissions: an assessment of the uncertainties, *Energy Policy* **33**, 2064–74
    2009: Carbon dioxide mitigation, chapter 2 in this volume

Weitzman, M.L., 2009: On modeling and interpreting the economics of catastrophic climate change, *The Review of Economics and Statistics*, **91**(1), 1–19

Technology-Led Climate Policy

ISABEL GALIANA AND CHRISTOPHER GREEN*

Introduction

Evidence mounts that humankind is changing the Earth's energy balance. The change in energy balance is attributable to the build-up in the atmosphere of greenhouse gases (GHGs) that partially trap outgoing long-wave radiation – that is, radiation given off by the Earth as a result of absorbing solar (short-wave) radiation. There is still some debate as to how much of the change in energy balance has shown up to date in the form of changes in climate-related variables such as global average temperature and precipitation–evaporation patterns. But there is overwhelming evidence that some GHG-induced change has occurred, as distinct from changes attributable to natural phenomena (solar or volcanic) or factors affecting long-term variability in the earth's climate (Solomon *et al*. 2007). We also know that at least some (perhaps half) of the imbalance is temporarily hidden – stored in the oceans (Hansen and Nazarenko 2005). Almost certainly as the twenty-first century progresses the climatological evidence of human-induced change will mount – and so will the impacts on the environment and vulnerable aspects of the economy and society.

There are ongoing attempts to frame a climate policy to succeed the Kyoto Protocol. Unless there is an epiphany in climate policy thinking, the emphasis will be on how *much* to do in the next period, rather than how *to* do it. Predictably, the word "targets" will be heard early and often, and used at least an order of magnitude more times than

the word "technology." Commitments to "ends" (emissions reductions) will dominate discussion. Little or no consideration will be given to whether the "means" of cutting emissions are sufficient to achieve the emission-reduction "ends." The idea of committing to "means" (actions) rather than "ends" will be far from the policy makers' thoughts, even though such a commitment is likely to be both more credible (Scheling 1992, 2005) and effective than commitments to "ends" (results). There will be much talk about the need for a price on carbon, and what it can allegedly do, with little consideration of the important things a carbon price cannot do.

This chapter attempts to fill a void. It attempts to make a serious case for a technology-led climate policy. The logic is that if *global* emissions are to be cut 50–80% by 2050 and 2100, respectively, doing so will require Herculean efforts to: (i) increase energy efficiency/reduce energy intensity, and (ii) develop the means of producing vast quantities of carbon emission-free energy in the next 50–100 years.

An example gives some idea of the magnitude of the challenge. Suppose by 2100 we wish to reduce *global* emissions by 75% from current levels. Suppose further, that over the course of the twenty-first century, the "trend" rate of global GDP growth in the absence of climate policy were 2.3%. (We ignore for the moment the effect on GDP of damages produced by climate change.) To achieve the emission reduction target and not lose more than *11%* of the cumulative output that would otherwise flow from a 2.3% per annum growth in global economic activity, would require that by 2100: (i) global energy intensity is reduced by two-thirds from the level in 2000, and (ii) carbon emission-free energy in 2100 is *two and a half times greater* than the level of *total* energy consumed globally

* We would like to thank Valentina Bosetti, Gregory Nemet, Vernon Smith, and an anonymous reviewer for their very helpful comments on an earlier draft of the chapter. We also wish to thank Soham Baksi, Francisco Galiana, John Kurien, and Roger Pielke, Jr. for useful conversations.

in 2000. (In 2000, global energy consumption was ~420EJ/year, 85% of which was supplied by fossil fuels. Of the carbon-free energy produced, 95% was nuclear and hydroelectric.)

Here is another example. In order to reduce *global* emissions by 50% from current levels by 2050 (an oft-discussed target) and 80% by 2100, the *average annual* rate of de-carbonization of *global* output (i.e. the *rate of decline* in the carbon intensity of output (CIO), or GDP) must be raised from its "historic" (the last thirty years) rate of 1.3% to over 4.0%. *Not only must the rate of decarbonization triple, but most of the increase will have to come from a de-carbonization of energy which "historically" has declined, in global terms, at a 0.3% rate*. Most of the long-term decline in the de-carbonization of output is associated with a decline in energy intensity (1.0%), attributable chiefly to improvements in energy efficiency and, to a much lesser extent, *global* shifts in the composition of output. (We shall make considerable use of the rate of decline in the carbon intensity of output (RCIO) later in the chapter.)

These calculations are not a mistake! But for many we suspect they may come as a surprise. They may seem at variance with the conclusions reached by IPCC Working Group III that the barriers to stabilizing climate are socioeconomic and political, but not technological (Metz *et al.* 2001, 2007). The calculations may also appear at variance with the estimate of *The Stern Report* that the cost of stabilizing climate is around 1% of GDP (Stern 2007, 2008).

At the same time, the calculations should not provide solace to those who wish to ignore the climate change threat – which is real. Nor do the calculations suggest that in benefit-cost (B/C) terms the *long-term* rise in atmospheric carbon concentration and global average temperature need only be reduced moderately (e.g. Nordhaus 2008). As controversial as are *The Stern Report* (Stern 2007) estimates that climate change damages range from 5 to 20% of global world product (GWP), a climate policy that would only reduce the rise in global average temperature from, say, 4.0°C to 3.5°C a century from now ought to convince no one that such a policy reduces substantially the possibility of large potential damages to the global environ-

ment and economy. *Nevertheless, the calculations do imply that a new route to emission reductions must be found.*

The calculations suggest, then, that if we are going to do something significant in terms of "stabilizing climate" we will have to rethink *how* to proceed. In particular, we need to recognize that the key variables in climate stabilization involve energy technology changes. One set of changes is in the form of very large energy efficiency improvements that could make possible a two-thirds reduction in *global* energy intensity in the face of a development process in populous, developing countries that is, and for the foreseeable future will be, energy-intensive (Green *et al.* 2007; Pielke *et al.* 2008). The other is in the form of technological breakthroughs that would make possible a vast expansion in carbon emission-free energy.

It is the technology imperative that drives us to propose a climate policy in which research and development (R&D) are front and center, at least in the initial stages. Given the lags in capturing the total productivity increase of new technologies (diffusion and learning new techniques), it becomes all the more important to act quickly in developing them. But lest there be any misunderstanding, *this chapter is about mitigation, but mitigation in which technology development policies that make deep emission reductions possible are in the forefront.*

In sum, the chapter proposes a technology-led approach to mitigating GHG emissions. Climate change will impose increasing costs, but there are no quick or easy solutions such as those the US Environmental Protection Agency (EPA) imposed on emitters of sulfur dioxide (SO_2) or nitrogen oxides (NOx). The technologies to achieve SO_2 and NOx reductions were ready and scalable, something that is not currently true of CO_2. Instead for CO_2 mitigation, the accent is placed on *energy technology research, development, and testing*. For this reason we think the role of carbon pricing should initially be limited to a low (as global as possible to avoid leakages and to ensure broad commitment) carbon tax or fee that is used to finance energy R&D. Over time the tax should be allowed to rise slowly in order to send a "forward price signal" that would induce deployment

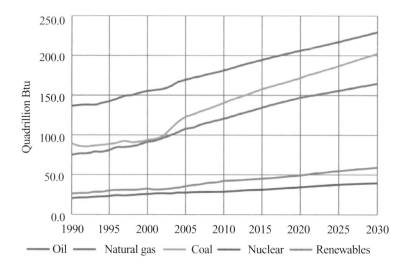

Figure 7.1 *World energy use, by fuel type*

Sources: EIA and world energy projections plus (2008).

of *effective, scalable, cost-competitive technologies* as they reach "the shelf" (i.e. become ready to deploy).

Confronting the proximate cause of climate change via *attempts to directly control emissions* is defective for several reasons:

1 The amount of carbon emission-free energy required to "stabilize" climate is huge – at least 15 to 20 times more than current levels, almost all of which is supplied by nuclear and hydro-electric.

2 Alternative energy sources are currently neither ready nor (just as important), as yet, scalable, and in most cases still require basic R&D.

3 Relying on carbon pricing to cut global emissions substantially is *neither* likely to be politically *acceptable nor* economically *time-consistent.* Carbon pricing alone, or as the main policy tool, is not an effective means of inducing long-term commitments to undertake and pursue endemically uncertain (of success) basic R&D. But as we shall see, carbon pricing has two important *ancillary roles* to play.

4 In the modern world, *energy is a necessity.* In the twentieth century, energy consumption increased sixteen-fold. Under the best of circumstances (improved energy efficiency, conservation, and the elimination of wasteful use) global energy consumption will double by 2050 and triple by 2100. Any attempt to reduce carbon emissions by artificially reducing the availability of energy will not be accepted, at least not for long. For energy use to substantially increase while carbon emissions are substantially reduced requires that there must be a suite of good non-carbon-emitting energy substitutes. Except for nuclear electric, current candidates are, in technological terms, still severely limited (MacKay 2009).

5 Currently, 85% of global energy requirements are met by fossil fuels. This is so for both technological and economic reasons. Fossil fuels consumption is likely to increase for the next few decades (see figure 7.1) and these fuels will continue to be important deep into the current century. By 2050, global energy demand will at least *double*. To reduce the share of fossil fuels by 50% by 2050 will be a daunting technological task. And even if it could be achieved, it would still leave carbon emissions unchanged from current levels, unless carbon capture and storage (CCS) can be quickly ramped up, itself a daunting task.

6 On the face of it, attempts to directly control *global* carbon emissions will not work, and

certainly not in the absence of ready-to-deploy, scalable, and transferable carbon emission-free energy technologies. The technology requirements cannot be wished, priced, assumed, or targeted away. A technology-led climate policy is a means of breaking the knot.

7 To be clear, a technology-led policy is an alternative approach to mitigation. To make possible substantial, continuing emission reductions, it is necessary, we think, to focus on basic and applied research, development, and testing of alternative energy technologies, and infrastructure to make them both viable and less expensive. A technology-led policy is not a recipe for subsidies to energy production, such as those given to the owners of wind farms and solar energy arrays. In general, these subsidies are often wasteful and do not solve key technological problems.

8 In short, if efforts to *de-carbonize the global economy* are to be effective, they need rethinking. The blinders that have distorted climate policy to date need to be replaced by a hard-headed appreciation of the nature and magnitude of the technological task ahead.

The chapter proceeds as follows. In the first section, we present measures of the size of the technology challenge posed by climate stabilization. Current technological readiness and what might be achieved with current technologies is considered in the second section. In the next section, we examine the implications of a failure to tackle the technology challenge directly. The following section sets out the character of the technology-led proposal and the ancillary, but important, role of carbon pricing. The political economy of reliance on carbon pricing, especially as it relates to energy-intensive industries, is discussed in the fifth section. The sixth section addresses the institutional factors that increase the likelihood that a technology-led policy will be "incentive-compatible." Some specifics of the technology-led approach are set out in the seventh section. In the penultimate section, we turn to a benefit-cost analysis (BCA), using three different methods to assess the relative benefits and costs associated with a technology-led approach to climate policy. Some concluding thoughts are presented in the final section.

The Magnitude of the Technology Challenge

By any measure, the magnitude of the challenge posed by stabilizing the atmospheric concentration of GHGs in the atmosphere at an acceptable (non-dangerous) level (hereafter "stabilizing climate") is huge. One measure is the cumulative emissions that need to be reduced by energy-efficiency improvements and shifts to less energy-using activities, and the introduction of carbon emission-free ("carbon neutral") technologies. Pielke *et al*. (2008) estimate these for the scenarios used by the Intergovernmental Panel on Climate Change (IPCC). These cumulative emissions estimates and demands on carbon-neutral technologies are much greater than would be inferred from the emissions scenarios employed by the IPCC or *The Stern Report*. The IPCC uses emissions scenarios that already build in 57–91% of the emission reductions attributable to technological change as baselines for measuring the size of the challenge. This is shown in figure 7.2, where the lightest grey portions of the bar represent the emissions reductions that are built into the emission scenarios. The issue here requires further explanation.

In assessing what it will take to stabilize atmospheric GHG concentrations (in cost and technology terms), models usually employ no-climate policy emission scenarios as references or baselines. However, using emission scenarios as baselines for assessing climate stabilization creates a huge understatement of the technological change needed (and, by extension, economic cost incurred) to stabilize climate (Pielke *et al*. 2008). The problem is that built into most emission scenarios are very large, primarily technologically driven, emission reductions that are assumed to occur automatically.

By building into their emissions scenarios very large technology-generated emission reductions, analysts (with important exceptions including Battelle Memorial Institute 2001; Edmonds and Smith 2006; Fisher *et al*. 2007: 220); Wigley *et al*. 2007; Pielke *et al*. 2008) are assuming that the technology challenge is measured by the medium grey portions of the bars in figure 7.2 – that is, by the difference between the emissions scenario baseline and the stabilization path. The result is to

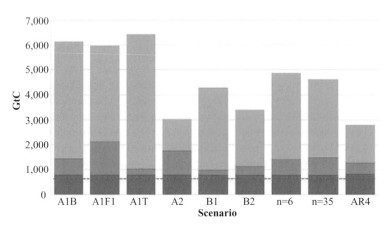

Figure 7.2 *Cumulative emissions and technology in IPCC scenarios*

substantially understate the magnitude of the energy technology challenge.

To get around the problem posed by using an emission scenario baseline for assessing the magnitude of the technology challenge, one can use a "frozen technology" baseline (Edmonds and Smith 2006; Pielke *et al.* 2008). For a slightly different usage of the "frozen technology" concept, see Greene *et al.* (2009). A "frozen technology" baseline is an estimate of future emissions as if they were produced using today's energy technology – hence the technology is "frozen." (Frozen technology baselines were used in constructing figure 7.2.)

While no one expects technology to be/remain "frozen," a hypothetical "frozen" technology baseline allows complete transparency in assumptions about future technologies, innovation, and the processes that will lead to such innovation, crucial issues that are obscured by emission scenario baselines. Assessing the technology challenge from a "frozen" technology baseline also avoids the potential for "double-counting" technologies, once in the emission scenario and again in the movement from the emission scenario to the stabilization path.

The IPCC B2 scenario can serve as an example to illustrate the magnitude of the challenge to stabilizing climate even in scenarios that are relatively modest in terms of cumulative emissions (see figure 7.2 and table 7.1). In B2, the GDP growth rate 2010–2100 of 2.0% (market exchange rates, MER) – 1.77% (PPP) – is quite modest, yet carbon dioxide (CO_2) emissions rise from the current level of about 8 GtC to almost 14 GtC in 2100. This occurs even though built into the B2 scenario is (i) substantial average annual rates of energy intensity decline, and (ii) a large increase (2010–2100) in carbon-free – or carbon-neutral – energy consisting of a thirteen-fold growth in nuclear power, a six-fold increase in biomass (much more if one only considers "new" biomass, see notes to table 7.1), and a twenty-fold increase in other renewables (including hydro).

The energy technology change built into the B2 emission scenario will require many technological improvements and some technological breakthroughs. For example, breakthroughs would be needed in: (a) the production of biomass fuels to assure they are low-carbon-emitting on a life-cycle basis; (b) storage for intermittent solar and wind energy which must make up a large portion of the growth in "other renewables"; and (c) generation IV and newer generations of closed-cycle nuclear electric reactors (using reprocessed nuclear fuel) in order to make possible a huge increase in nuclear electricity, given limits to U-235 and waste storage capacities. These examples make clear why it is important to consider *technology built into an emissions scenario as well as that required to move from an emissions scenario baseline to a stabilization path* (Pielke *et al.* 2008).

Another way to measure the stabilization challenge is to directly estimate the amount of carbon-neutral energy (or power) that will be needed by 2050 or 2100 to get on to a stabilization path. This

Table 7.1 The IPCC B2 scenario (1990–2100)

	1990–2100	2010–2100	2010–2050	2050–2100
GDP growth rate %MER (%) (PPP)	2.2 (2.0)[a]	2.0 (1.77)	1.61 (2.22)	1.53 (1.42)
Rate of decline in E/GDP (%) MER (PPP)	0.97 (0.77)	0.84 (0.62)	1.12 (0.73)	0.64 (0.53)
Rate of decline of C/GWP (%) MER (PPP)	1.44 (1.24)	1.40 (1.16)	1.76 (1.36)	1.11 (1.01)
Rate of increase of CO_2 emissions	0.76	0.61	0.85	0.41
Cumulative CO_2 emissions (GtC)	1157	998	395	602
Cumulative emissions (GtCO$_2$)	4245	3661	1451	2210
Built into the B2 scenario are (EJ/year)				
– Nuclear {7}[b]	135	131	50	81
– Biomass {46}	269	269	59	210
– Other renewable (incl. hydro) {8}	204	190	85	105
Total	608	590	194	396

Notes:
[a] The numbers in parentheses (()) mean that GWP is measured in purchasing power parity (PPP) terms.
[b] The numbers in brackets ({}) are EJ/year supplied by the energy source in 1990 (IPCC 2000: tables, B2 "Message" emission scenario). Note that almost all of the 46 EJ/year of biomass is "old" biomass, including wood for domestic fuel, charcoal, and burning of dung, mostly by poor communities without access to electricity or other commercial energy. Old biomass is replaced by carbon-neutral "new" ("plantation") biomass for generating electricity or producing biofuels.
Source: IPCC (2000).

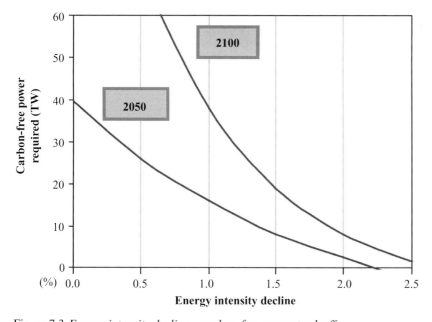

Figure 7.3 *Energy-intensity decline – carbon-free power tradeoff*

is the approach undertaken by Hoffert *et al.* (1998). For a global average GDP growth rate of 2.4% (1990–2100) estimates of the carbon-free power required by 2100 generally fall in the range of 25–40 terawatts (TWs), the amount depending on the global average annual rate of energy-intensity decline (see figure 7.3). A TW is 10^{12} watts – a TW over the course of a year is 8760 TW hours – or 8.76 trillion kilowatt hours. (One TW equals 31.56 exajoules (EJ) of energy per year.) Currently, the world's consumption of energy measured in power terms is 16.5 TWs. Of this amount less than 2.5 TWs are carbon-neutral, almost all of it derived from nuclear and hydroelectric power.

Even with only modest growth in the demand for energy (based on the assumption of huge improvements in energy efficiency) the world will consume upwards of 30 TWs in 2050. (At a 1.5% growth rate energy consumption in 2050 would be about 31 TWs. A more likely 2.0% growth rate would raise energy consumption in 2050 to almost 39 TWs.) To get on to a stabilization path, at least half of the energy used in 2050 will have to be carbon emission-free; by 2100 almost all of it would have to be carbon emission-free. Assuming 31 TWs of power will be needed in 2050 implies a six-fold rise in carbon-neutral energy to 15 TWs by 2050. Is that feasible? And what is needed to make it feasible? By 2100 upwards of 30 TWs of carbon-free power will be required. The Hoffert *et al.* (1998) analysis, and the accompanying figure 7.3, are taken up again in the next section.

Technological Readiness

Technological readiness implies deployable (on-the-shelf) technologies that are as scalable and as cost-competitive as possible. Assessments of "technological readiness" require comparing the magnitude of the technology challenge with the capabilities of current carbon emission-free (or "carbon-neutral") energy technologies. On this basis we are nowhere near ready to reduce global emissions substantially by the mid-century, much less achieve climate stabilization by the end of the century. Let us look at several energy technologies/sources and their potential contributions by 2050. (An excellent complement to the technology-readiness assessment below is Barrett 2009).

Hydroelectricity

Sites for hydroelectric power are limited. A doubling of the present capacity is probably the best we can do. Doubling capacity would add about 340 GWe and eliminate the need for construction

of about 700 500 MWe coal-fired plants with total emissions of 2.4 billion tonnes of carbon dioxide (or about 0.65 GtC). The addition of time-of-day pricing, thereby raising the capacity factor of hydro from around 50% to 75%, might eliminate another 0.65 GtC.

Nuclear Electric

Nuclear energy has been and will likely be an important contributor to non-carbon-emitting electric power generation. But in its current technological form there are resource and storage limits to its scalability (MIT 2003). There are currently 439 nuclear reactors in the world producing an estimated 390 GWe. Many existing plants are approaching the end of their useful life. Assuming that all the existing reactors are replaced when they wear out, it would require adding fifteen reactors every year from 2010 to 2050 to raise nuclear-generating capacity to 1 TW. The *additional* 600 GWe of nuclear capacity would replace coal-fired electric capacity emitting 1.1 GtC a year (MIT 2003).

Carbon Capture and Storage (CCS)

Currently a lot of weight (hope) is being placed on the CCS option. There is little choice given the huge amount of coal-fired electric capacity now churning out a substantial fraction of global emissions (MIT 2007). Moreover, the slow ramp-up of nuclear capacity and huge hurdles to large-scale, baseload energy from solar and wind (see below) suggest continued heavy reliance on coal to meet the rapid growth in electricity demand in many parts of the world, especially the developing world and parts of the developed world.

But ramping up CCS will be slow (Edmonds *et al.* 2007) for several reasons: (a) CCS has not yet been applied to a coal-fired electricity-generating plant; (b) the only examples of operational CCS involve relatively small-scale operations, the best known being the Sleipner field project that stores about 1 Mt of CO_2 (or 270,000 tones of carbon) each year from Norway's North Sea natural gas operations; (c) it would take the equivalent of 3,500

Sleipner fields (Pacala and Socolow 2004) to store 1 GtC each year, and that means a large amount of geological investigation to assure the existence, safety, and security of the required geological sites; (d) pipelines would have to be built from the source of CO_2 to the designated geological sequestration sites; (e) capture technologies have not yet been perfected, and those that are operational at a test-site level would not only increase plant capital costs but would exact an energy penalty of 20%–40%, depending on technology. If by 2050, CCS can be ramped up to 1 GtC ($\sim$3.7 GtCO$_2$), the *net* reduction in energy-related emissions (emissions from electricity generation net of the added energy needed to capture emissions) would be about 0.7 GtC.

Biomass

Although biomass has been counted on as major carbon-neutral source, recent experience has greatly lowered expectations – at least from "first-generation" biofuels such as corn ethanol and soybean-based biodiesel. There are several reasons why biomass, at least in its current forms, is unlikely to produce in the future the large amounts of (*net*) carbon-neutral energy expected just a few years ago. These reasons include: (a) the effect on food stocks and prices caused by devoting large amounts of cropable land to energy crops (Pimentel *et al.* 2009; Wise *et al.* 2009); (b) the enormous amounts of water that large-scale biomass production will require (Bernedes 2002; Gerbens-Leenes *et al.* 2009); (c) evidence that on a life-cycle basis the net energy from biofuels output is not much greater – and in some cases may be less – than the energy inputs into producing the biofuels (Pimentel and Patzek 2005; Farrell *et al.* 2006); (d) indications that converting land from pasture to energy crops may release carbon from the soils in amounts that substantially outweigh any prospective reductions in emissions that the conversion from fossil to biofuels is expected to produce (Fargione *et al.* 2008; Searchinger *et al.* 2008).

The realization that "first-generation" biofuels may do little or nothing to reduce emissions (or energy use) has led in two directions: (i) to focus on biomass as a *solid* energy source for generation of electricity rather than as a *liquid* biofuel for use in vehicles; (ii) R&D into the possibility of "second-generation" biofuels from cellulosic by-products of primary feedstocks, to switchgrass and to algae. Each has potentially important limitations. In the case of solid biomass, finding sufficient forest that can be dedicated to electricity production may be limited to a few places in the world. In the case of "second-generation" biofuels, these will require technological breakthroughs and even then their scalability is in doubt. We will be hard put to produce enough *net* energy from biomass by 2050 to reduce emissions by an estimated 0.3 GtC.

Solar and Wind

Currently, these two potentially substantial sources of energy supply only a tiny fraction (less than 1%) of the world's energy. The reasons go far beyond their higher costs (Love 2003; Love *et al.* 2003; Denholm and Margolis 2007a, 2007b). Beyond reducing production costs, three big hurdles (in ascending order of difficulty and importance need to be overcome). (1) Direct current lines need to be constructed to carry solar and wind energy from the areas of highest insolation and wind speeds to the populous areas where most consumers are located – often 1,000 km or more distant. (2) More flexible, "smarter" grids will be needed to cope with the variability inherent in wind and solar power. (3) Because of their intermittency and variability, even with "smart grids," solar and wind power are unlikely to be able to supply much more than 10–15% of grid-based electricity (net of energy used in "spinning reserve" back-up) without the development of utility-scale storage. To overcome scalability barriers, scientific and technological breakthroughs will be needed (Lewis 2007a). Assuming that in the next couple of decades sufficient investment is put into the electric grid infrastructure and into researching and developing large-scale storage for solar and wind-powered electricity generation, it is possible/conceivable that by 2050 these two renewable sources could together supply 500–700 GWe, and displace up to 1.5GtC.

Geothermal

Geothermal power is an excellent source of power in those few areas (such as Iceland) where hot springs are abundant. Not surprisingly, it is currently a very limited source of power. With technological changes it is possible to increase the availability of geothermal for generation of electricity. Moreover, if new buildings are fitted with proper piping at the time of construction, geothermal could eventually become a widespread means of space-conditioning – moderating, to a degree, the growth of demand for electricity. By 2050, it may be conceivable that geothermal could displace 0.5 to 1 GtC.

Ocean Wave Energy

There is growing interest in harvesting electric power from ocean waves. The amount of energy in the oceans' waves is large, but it is very dilute and only a fraction is economically viable – assuming that the many technological problems can be overcome. One estimate of the viable resource in the USA is equal to about 6% of *current* electricity demand. Because wave energy is concentrated at low frequencies, efficient conversion and transmission to a grid is difficult (Scruggs and Jacob 2009). The marine environment creates other problems, including seawater corrosion, marine organism fouling, and large loads imposed by big storms on wave energy converters. Ocean wave energy might displace 0.1 GtC.

Taken altogether, current energy technologies, if hugely scaled up, might get us halfway toward a stabilization path by 2050 – but only a fraction of the way toward achieving stabilization by 2100. One way to see why this is so is to refer back to table 7.1 and note there the large amount of carbon-free energy built into the B2 emission scenario. That the B2 scenario is not atypical is evident from figure 7.2. Unfortunately, perceptions differ. One reason is that many analysts assume that rates of energy efficiency improvement and energy intensity decline much more than can be sustained globally over an extended period of time. As Hoffert *et al.* (1998) demonstrate, there is a tradeoff between the amount

of carbon-free energy required to stabilize climate and the rate of energy intensity decline. Figure 7.3, based on Hoffert *et al.* (1998), indicates the relationship. (Note that the tradeoff in figure 7.3 is based on an assumed global rate of GDP growth of a little over 2.4% (1990–2100) and an atmospheric CO_2 stabilization target of 550 ppm.)

As figure 7.3 indicates, the amount of carbon-free energy required to achieve stabilization is very sensitive to the global rate of energy intensity decline. The amount required for a 1.0% rate of energy intensity decline is twice that required of a 1.5% rate of decline, which in turn is approximately twice the level required if the rate of energy intensity decline is 2.0%. Hoffert *et al.* (1998) thought that the *global* economy might achieve a 1.0% rate of energy intensity decline for the 110-year period 1990–2100, a rate that reflects past trends. But many scenarios utilize no-policy rates of energy intensity decline substantially in excess of 1.0%. Here are some facts:

1 In general, the IPCC's Special Report on Emission Scenarios (SRES) build in high rates of energy intensity decline. Of the forty scenarios (from four basic families, A1, A2, B1, and B2) thirty-two had 110-year (1990–2100) built-in energy intensity declines greater than the 1.0 %/year rate used in the BAU IS92a scenario. It is likely that, on balance, the energy intensity declines in many of the SRES scenarios are highly unrealistic. If so, they have contributed to a major understatement by the IPCC of the magnitude of the energy technology – and, by extension, climate stabilization – challenge.

2 Baksi and Green (2007) have devised a method, using mathematically exact formulas, for computing aggregate energy intensity decline from changes over time in the efficiency of different energy-using sectors and their relative contributions to GDP and energy use. They found that, *even after applying stabilization policies*, it would be difficult to substantially exceed a 1.0 %/year *global, average*, rate of energy intensity decline over 1990–2100 – or about 1.1% on a 100-year (2000–2100) basis. Yet 80% of the *pre-policy* SRES scenarios build in 110-year global average annual rates of energy intensity decline that exceed 1.0 %/year (and 75% exceed 1.1%).

3 The Baksi–Green calculations of an approximately 1.0% rate of decline in energy intensity (1990–2100) assume global average energy efficiency increases in industry, commerce, and transportation of 200% (300% for cars and light trucks), 300% in residential uses, but less than 100% in the efficiency with which electricity is generated (Lightfoot and Green 2001). The calculations also assume that over the course of the twenty-first century there are very large reductions in the GDP and energy shares of energy-intensive industries, a rise in the energy share for electricity generation, and a substantial rise in the GDP share of the commercial sector, reflecting the increasing importance of services.

4 The formulas generated by Baksi and Green (2007) can be used to demonstrate that only about 20% (bounds of 10 and 30%) of the *global* energy intensity decline can be contributed by sectoral shifts from higher to lower energy-intensive uses. The rest must come from energy efficiency improvement, which means widespread adoption of the best available technology plus technological change. While at the individual country level sectoral shifts can contribute considerably more than 20% of energy intensity decline, at the global level there is a lot of canceling out as energy-intensive industries move from one part of the world to another.

5 Baksi and Green (2007) also demonstrate that achieving very high, century-long, rates of energy intensity decline (ones that would substantially reduce the amount of carbon-free energy needed for stabilization) require improvements in energy efficiency that are almost surely physically impossible. For example, Baksi and Green (2007) show, in their table 4, that a 2.0% rate of decline (the B1 marker scenario has a 2.13% average annual rate of decline, 1990–2100), requires sectoral energy efficiency improvements ranging from 450 to 1100%.

The IPCC (2001) technology-readiness claims were contested by Hoffert *et al.* (2002). One reason for the clash is that the methodology developed by Hoffert *et al.* (1998) avoids the trap of "built-in" emission reductions endemic to the IPCC emission scenario baselines. In figure 7.3, the calculation of

carbon-neutral energy requirements is based on the rate of growth of global GDP, *given the explicitly accounted-for average annual rate of decline in energy intensity*. In this way, the baseline in figure 7.3 (the 2100 curve) is the equivalent of a "frozen" technology baseline.

The second reason revolves around the *scalability* of current carbon-neutral technologies. The scalability issue, emphasized by Hoffert *et al.* (2002), recognizes that while some technologies are not yet scalable because they are still at the R&D stage, others although apparently "on-the-shelf" are nevertheless not yet scalable. In some cases, scalability is limited because of the lack of an "enabling technology." An example of an "enabling" technology is grid integration and storage for intermittent and variable solar and wind power. These potentially large, but dilute energy sources are not only land-intensive (Lightfoot and Green 2002), but of limited use without storage. Electric utilities generally will not be able to meet any more than about 10% of non-peak electricity demand from directly supplied, intermittent, or variable sources. While pumped hydro, hydrogen, and compressed air energy storage can provide some storage potential, we are still very far from a good, reliable, and scalable means of storage for electricity generation and supply.

Similarly, CCS faces scalability issues on the storage side. While studies suggest that there is potentially plenty of storage capacity for CO_2 emissions captured and geologically sequestered in the foreseeable future (Herzog 2001; IPCC 2005), as a practical matter each geological storage site needs to be checked for leakage potential. This will require a potentially time-consuming effort by a large number of geologists. Detailed examinations cannot be ignored: CO_2 leakage would not only limit the effectiveness of CCS, but create a public hazard because CO_2 in concentrated form is an asphixiant that disperses slowly if a leak occurs, especially if the wind is not blowing. It is true that there are a number of small-scale examples of CCS, but there is nothing even remotely approaching the scale required for CCS to contribute significantly to reducing future net CO_2 emissions. Finally, "conventional," once-through, nuclear fission is not only limited by Uranium 235 supplies (MIT 2003), but faces political and technological limitations with respect to storage of the large amounts of radioactive waste that would be generated even if nuclear simply maintained its current 17% share of global electricity generation. A real breakthrough may come if the integral fast reactor (IFR) project, terminated for apparent political reasons in 1994, is reactivated and proves as potent an energy source as some nuclear scientists and writers think (Blees 2008).

Storage is not the only "enabling" technology that is required to make a number of carbon-neutral energy technologies viable. Other examples include *retrofit* technologies for the large and rising number of coal-fired plants, especially those in China, India and the USA or, as an alternative, CO_2 capture from the air (Lackner 2003; Pielke 2009c). While nuclear electric generation is an obvious low carbon-emitting alternative to coal, large-scale expansion will greatly increase the incentive to reprocess nuclear "waste." However, doing so will require some means of "spiking" the resulting plutonium to make it too hot to handle by terrorists, and a means of preventing nuclear proliferation. While the latter clearly involves political ingenuity, it also involves science and engineering developments, as is indicated by the apparent technological as well as political hurdles ahead for the US-promoted Global Nuclear Energy Partnership (GNEP) (Tollefson 2008). (The key reprocessing facet of GNEP has recently been cancelled.)

Once the scalability problem is understood, it is easier to see why there is still a large technology gap between usable carbon-neutral energy with current technologies and the amount required for climate stabilization. Green *et al.* (2007), build on Hoffert *et al.* (1998), in an attempt to measure the "advanced energy technology gap" (AETG), the gap between the carbon-neutral energy required for stabilization and the carbon-neutral energy that could be supplied from "conventional" carbon-neutral sources. "Conventional" carbon-neutral energy technologies include: hydroelectricity (subject to site limitations); once-through nuclear fission (subject to Uranium 235 supplies as well as security, political, and waste storage limitations); solar and wind without storage; some biomass, geothermal, tidal and wave (ocean) energies. The

authors found that "conventional" carbon-neutral energy sources might, at a stretch, supply 10–13 TW by 2100. Liberally assuming 13 TW from these "conventional" sources, we still need 15–25 TW of power from advanced technologies (the AETG) to reach the 28–38 TW of carbon emission-free energy required by 2100 to stabilize at 550 ppm, assuming a 2.4% rate of growth of GDP (1990–2100). These findings support the Hoffert *et al.* (1998, 2002) claim that major breakthroughs in new as well as existing energy technologies and sources will be required for stabilization at 550 ppm, and even more so for stabilization at 450 ppm.

Implications of Failing to Address the Technology Challenge

If, as seems likely, the SRES emissions scenarios have made CO_2 stabilization appear much easier than it will be (Green and Lightfoot 2002; Pielke *et al.* 2008), then there are important implications for climate policy. First and foremost, achieving large reductions in global CO_2 emissions requires a veritable energy technology revolution. A first implication is the need for a technology-based climate policy.

A second implication involves the relationship between a carbon-price policy and a technology policy. Instead of the carbon-price policy carrying the main load of emission reduction, carbon prices should be viewed as playing two supportive roles: (a) as a means of raising revenues to finance the publicly financed component of the energy technology race without which stabilization is unachievable; and (b) as a way of sending a *forward price signal* that will be increasingly powerful as the carbon price slowly rises and as new technologies appear "on-the-shelf" (a form of what Yohe *et al.* 2008, term "when" flexible mitigation). These considerations suggest a carbon tax that starts low and rises very gradually over time.

In thinking about climate policy, an important distinction should be made between technologies that are "on-the-shelf" and therefore are deployable now (if it were economically advantageous to do so), and those that either (a) require further development before deployment is possible; or (b) are still at the basic R&D stage; or (c) have not yet been thought of (Sandén and Azar 2005). Carbon prices are likely to be effective in inducing deployment of technologies that are "on-the-shelf", but may well be ineffective inducements to invest, long-term, in technologies that still require basic R&D. The success of basic R&D is typically *uncertain*. This will have a major impact on the market's evaluation of it. Even if R&D proves an initially uncertain technology to be viable, it may take decades before it is ready for deployment.

Many climate policy modelers give an important role to induced technological change (ITC). The basic idea is that a strong carbon price will induce the private sector to make investments in energy R&D and technological changes that allow firms to reduce their carbon emissions. The payoffs from these investments is the carbon emission permits that do not need to be purchased (or if allocated can be sold to other firms) and/or the carbon taxes avoided.

In our work we make an important distinction where ITC is concerned. Carbon prices are given a central role in the adoption of on-the-shelf, ready-to-deploy technologies. But we are much more skeptical about the role of ITC where basic R&D and the testing of untried technologies are concerned – especially where the time frames are many years or even decades rather than a few years and success is highly uncertain. An excellent discussion of the reasons why the ITC mechanism is likely to be weak is in Nemet (2009a), the Perspective paper related to our chapter.

Our distinction between the role of market-based policies where technologies are "on-the shelf" and those requiring basic R&D (see Sandén and Azar 2005) is mirrored in a recent paper by Blanford *et al*. (2009). Blanford puts the issue nicely:

> Market-based abatement policies are effective mechanisms for bringing about the diffusion of existing technologies and can even spur incremental improvements through learning and induced applied R&D. Thus abatement policy is a mechanism for getting technologies "off-the-shelf." However, because of long time frames and limited appropriability in basic research, a second mechanism is required to put new abatement technologies

"on-the-shelf." The implementation of a technology strategy for a long-term environmental problem such as climate change is a challenging policy task.

The distinction that Blanford *et al*. (2009) is making can be framed as the difference between "demand-side" and "supply-side" influences on energy technologies. The demand side is found to be strong where "on-the-shelf" technologies are concerned, but for longer-term breakthrough technologies a supply-side, technology-based policy approach is required. The demand-side vs. supply-side distinction is the basis for a very useful paper by Nemet (2009b).

Nemet's (2009b) paper is one of the very few we have found that moves beyond theory and indirect empirical evidence to an actual case study. Nemet (2009b) examines the role of "demand-pull" and "technology-push" impacts on investments in the development of wind turbine technology. Nemet finds little evidence of a demand-pull influence. Most of the technology development appears to have been a response to government programs in the early and mid-1970s to pursue energy independence and reduced reliance on foreign oil, and to have preceded increased demand for wind power. Citing earlier work by Dosi (1988) and Kemp (1997), Nemet concludes that "These results fit with earlier work suggesting incremental innovation is more likely to respond to demand-pull than to technology-push and that non-incremental innovation is more responsive to technology-push" (Nemet 2009b: 707).

Finally, a study by Hoffmann (2007) found little impact in the first round (2005–7) of the EU emission trading scheme (ETS) on large-scale, long-term investments by the German electricity industry. Like Blanford and Nemet, Hoffmann found that the German electricity industry "does make low carbon investments with limited risks" such as retrofits or "investments with an inherent option character (R&D)" (Hoffmann 2007: 472). Perhaps the weakly applied first phase of the EU ETS does not allow us to pass judgment on the long-term R&D and infrastructure effects of a much tighter set of emission caps than those that evolved in the 2005–7 period. Still, Hoffmann's findings resonate with the view that carbon-pricing is unlikely to provide a strong inducement to the private sector to undertake long-term, inherently risky and uncertain investments in the development of breakthrough technologies.

An interesting question is whether price-induced technological change has led to any major technological breakthrough. An answer in the affirmative is not supplied by any of the climate–energy–economy literature we have seen. Yet, Held *et al.* (2009) state that "the inclusion of endogenous technological change led to results showing remarkably low mitigation costs for ambitious climate protection targets." Why is this so, given the apparent lack of empirical evidence linking technology breakthroughs to either targets and/or carbon-pricing?

An answer may reside in an ITC modeling comparison study carried out by Edenhofer *et al.* (2006). Eight of the ten models explored by Edenhofer *et al.* include "learning-by-doing" (LBD). LBD can be a powerful influence in reducing costs as the scale of production increases. This is evident from a series of case studies (none referred to by Edenhofer *et al.* 2006) including: airframe production (Wright, 1936); "Liberty ships" (Searle 1945; Lucas 1993); and semiconductors/microchips (Scherer 1996). However, these studies apply chiefly to manufacturing operations. It is a huge (and probably unjustified) leap to applying LBD to many of the activities most critical to the appearance of new energy technologies: research, development, testing, and demonstration.

It may also be significant that, in the Edenhofer *et al.* (2006) model comparison, six of the ten models include a backstop technology. Including a backstop technology effectively solves the energy technology problem by assumption. What the "backstop" technology assumption does is to ensure that raising the carbon price sufficiently will bring forth an unlimited supply of carbon emission-free energy. When the LBD and backstop assumptions are taken together, it is not surprising that ITC appears powerful even though no evidence for such an influence has been induced. It is all by assumption!

There is an additional problem. As Montgomery and Smith (2007) have demonstrated, private funding of long-term R&D encounters a "dynamic" (time) inconsistency. Generally, current governments cannot tie the hands of future governments to cover the potentially large (as well as uncertain) upfront R&D investment costs for technologies that may or may not prove successful and deployable decades hence. The Montgomery and Smith (2007) and Sandén and Azar (2005) papers therefore imply that "induced technical change" may be less important than one might gather from the IPCC Working Group III, chapter 11 (Barker *et al.* 2007). Further, to these considerations we may add a "political" time inconsistency between a four to five-year election cycle and the decades-long time scale for the development of deployable and scalable carbon-neutral energy technologies. The nature of the R&D required, and the time inconsistencies inherent in a long-term investment problem, suggest that the price system has limitations as a tool of climate policy.

Current climate policies appear to be influenced by a perception that the technologies required for stabilization are already "on-the-shelf," or almost so. In 2001, in its Summary for Policy Makers (SPM) the IPCC Working Group III argued that "most model results indicate that known technological options could achieve a broad range of atmospheric CO_2 stabilization levels, such as 550 ppmv, 450 ppmv, or below over the next 100 years, but implementation would require associated socio-economic and institutional changes" (IPCC 2001: 8). The IPCC defined "known technological options" as "technologies that exist in operation or pilot plant stage today" (2001: 8n.). In 2007, with only slightly more caution, the IPCC Fourth Assessment Report states in the SPM of its Synthesis Report (SYR) that "There is *high agreement* and *much evidence* that all stabilization levels assessed can be achieved by deployment of a portfolio of technologies that are currently available or expected to be commercialized in coming decades" (IPCC 2007b: 20, emphasis in the original).

In contrast to these general assessments, there are numerous reports and studies that detail what needs to be done to *current* carbon-neutral technologies to make them ready to be deployed and/or scalable. (Some of the findings are summarized in the preceding section, see also Barrett 2009.) The inconsistency between careful analyses of technological readiness and the claims of the IPCC Working Group III is traceable to a number of factors. One is the crucial issue of how scalable they are. A pilot plant operation may not be a good indicator of scalability. Another factor is that some technologies that are deemed ready, such as nuclear electric and post-combustion CCS, face *long ramp-up* times – and cost is a nagging concern, as well. In still others, such as wind and solar, scalability awaits "enabling" technologies such as grid integration and storage. In short, there is a large gap between current readiness and deployability on the scale required for substantial reductions in global emissions. That "gap" has important implications for attempts to quickly push down global carbon emissions in the absence of the ready-to-deploy, scalable technologies (what we term "brute force" mitigation).

A "thought experiment" can help to illustrate. Suppose the emission reduction target is an 80% reduction in global emission from current levels by 2100. To reach the 2100 target requires a 1.8% average annual rate of decline in carbon emissions. Now suppose the expected "trend" rate of growth in GWP from 2010 to 2050 is 2.2%. To avoid a reduction in the growth rate of GWP would require a 4.0% average annual RCIO. (The calculation of 4.0% for RCIO is based on $C = GDP \times C/GDP$, a reduced form of the Kaya Identity where the terms are converted to rate-of-change terms and RCIO is the rate of change of the C/GDP term.)

If a policy of reducing emissions by "brute force" is adopted, irrespective of technical feasibility, even an increase in the average annual RCIO to 3.6% from its "historic" rate of 1.3% (a very unlikely event in the absence of a technology-led policy) implies a reduction in the growth rate of GWP from the 2.2% "trend" rate to 1.8% for the period 2010–2100. Such a reduction would cost (an *undiscounted*) $86 trillion in 2100 alone and (an *undiscounted*) $2280 trillion *cumulative* over the ninety-year interval. (It is assumed that GWP in 2010 is $41 trillion, measured in MER terms.) And even these huge reductions in GWP

would not do the trick (meet the emission target) if we cannot push the rate of *decline* in C/GWP up to 3.6% (which is almost triple the "historic" rate).

This "thought experiment" casts serious doubt on the credibility of estimates of the cost of stabilizing climate. Estimates in the 1–3% of global GDP range – or lower (IPCC 2007a; Stern 2007) are not credible *unless* there is a *prior focus* on reducing the technology gap. The low-cost estimates reflect a variety of self-serving assumptions. Some models employ an emission scenario baseline that builds in large, automatic improvements in energy technology. Other models include a *carbon-free backstop technology* (often generic) that assumes that once the carbon price reaches a specified level there is an unlimited supply of carbon emission-free energy forthcoming. Still others have very high implicit rates of energy intensity decline, ones that would almost surely be physically impossible to achieve. Finally, some models make very optimistic assumptions (generally inconsistent with the evidence) about the availability and readiness of carbon-neutral technologies and/or the responsiveness of successful innovation of new energy technologies to carbon prices.

None of these modeling conveniences or assumptions contributes to a *reliable* approach to estimating the cost of mitigation. Perhaps the most deceptive are models that build in a backstop carbon-free energy technology, because this effectively assumes away what *is* the problem. Unless a specific effort is made to research and develop, test, and make ready-for deployment scalable carbon emission-free technologies, the cost of mitigation is likely to be as much as an order of magnitude, or more, higher than has been reported.

The route to an effective climate policy would appear, then, to run through technological change. For all intents and purposes that means a technology policy that can assure the long-term R&D of scalable carbon-neutral technologies. This in turn requires committed effort and financing plus a commitment to deploy the technologies when effective, scalable, and competitive ones reach "the shelf." How this might be accomplished is taken up in the next section.

Characteristics of a Technology-Led Climate Policy

The magnitude of the challenge, the lack of readiness of the required energy technologies, and the limitations of mitigation policies that are "brute force" in character suggest that current approaches to stabilizing climate will not work. Specifically, what is needed is a *realignment* in the *time-related* mix of mitigation, adaptation, and technology (R&D). Until now the emphasis has been on up-front mitigation, with adaptation and R&D adapting to needs as they arise. But such a policy mistakes the real character and magnitude of the climate problem. Climate change is a technology problem, and the size of the problem is huge.

Although there is increasing pressure for big emission reductions soon (in part a response to concern about "tipping points"), our assessment suggests that, globally, large emissions reductions are not attainable without the development of new technologies and infrastructure. And, realistically, this will take time. Thus the pressure is likely to increase for a "quick fix" such as that which some form of geoengineering (GE) *might* supply (Crutzen 2006; Wigley 2006). Whatever the possible merits of GE (and we believe adoption of any such policy beyond research and possibly a local experiment is premature), it is time to think of an alternative approach to how climate policy approaches GHG mitigation.

We suggest that climate policy dispense with date-specific, national emission reduction commitments. Instead, climate policy should aim at inducing technologically capable countries (which include many countries not covered by current emission reduction mandates) to undertake energy R&D and infrastructure *commitments*. The aim of these commitments is to develop scalable, deployable energy technologies that are capable of displacing fossil fuels at prices that are not significantly greater (and conceivably could eventually be somewhat below) that of fossil fuels – the prices of which are likely to rise in the meantime. A second set of commitments would be to deploy effective, scalable, reasonably competitive technologies as they reach the shelf and are ready to be deployed. Such a policy requires frank acceptance that the rate at which global emissions decline will depend on the uncertain (*ex ante*) rate of success in developing carbon-neutral technologies.

An important missing ingredient in the realignment of commitments just described is that it lacks a mechanism to *fund the R&D* and then *induce deployment of new, scalable energy technologies when they are ready* – that is "reach the shelf." Even if the new technologies are relatively competitive (a version of Nordhaus 2008's "low-cost backstop"), inertia and transaction costs could substantially delay deployment. There are a number of possibilities, including technology regulations and standards. Although in some cases technology standards may be appropriate, particularly for appliances, building codes, and to some extent for vehicles, too, we believe that a *modified version of carbon-pricing* has an important role to play.

As climate policy has evolved, putting a price on carbon has become an increasingly widely accepted means of mitigating CO_2 emissions. For most economists, putting a price on carbon has become the *sine qua non* of an effective climate policy (e.g. Stern 2007, 2008; Nordhaus 2008; Metcalf 2009). Carbon-pricing can be undertaken *directly* by placing a tax (or charge or fee) on the carbon content of energy fuels or *indirectly* by the issuance or auction of a limited number of carbon emission permits ("cap and trade"). The economic logic behind carbon-pricing rests on the incentives a carbon price creates to reduce consumption of energy, or at least carbon emitting energy, and on its putative stimulus to the development and deployment of carbon-neutral energy technologies. (An "optimal" carbon price is tied to the SCC, see Tol 2009.)

Although the *logic* of carbon-pricing appears impeccable, it is only a means to an end – an end which in the case of climate stabilization is achievable only with the appearance of new energy technologies that first must be researched and developed. We have already explained why we believe a price on carbon is too weak an instrument to induce the requisite R&D. A carbon price can induce deployment of "on-the-shelf" technologies and perhaps their prior commercialization, but the market alone is an ineffective means of stimulating, financing, and sticking with R&D of technologies whose success is uncertain and which, in any event, may take many years, even decades, to reach "the shelf."

We therefore recommend a modified approach to carbon-pricing. Instead of relying on carbon-pricing as a first-line approach to reducing emissions, we suggest that in the first instance carbon-pricing be used to finance the R&D of effective, scalable energy technologies and the infrastructure required to deliver them. For this purpose, a low tax on each tonne of CO_2 is all that is needed to raise tens of billions of dollars globally. A $5.00 per ton of CO_2 tax would raise $30 billion *a year* in the USA, about the same in China, almost as much in the EU, and lesser but significant amounts in Russia, India, and in other countries. Annually, as much as $150 billion could be raised in this way worldwide. A $5.00/t$CO_2$ tax or "fee" is a relatively unobtrusive method of raising funds. Its use in energy R&D could also be publicly popular in an energy-hungry world. The sums raised over time would easily finance a vigorous international energy technology race and leave monies available for up-dating national grids and other energy infrastructure.

In short carbon-pricing although it would not be the centerpiece of climate policy would nevertheless play two important *ancillary* roles. First, a low carbon fee or tax would be an ancillary (although important) appendage to the main up-front objective: developing the technological and infrastructure means to reduce emissions substantially in the future. Then while technological progress is being made, the carbon fee or charge should slowly, gradually, and by agreement, *automatically* rise. This could be achieved by a commitment to increase the fee slowly so that it doubles every ten years or fifteen years. Here then is the second ancillary role of carbon-pricing. As the carbon fee slowly rises it would take on the character of a "forward price signal," generating incentives to deploy new technologies as effective, scalable, and increasingly cost-competitive ones "reach the shelf." At this point a "virtuous circle" may develop.

There are great advantages in the modified form of carbon-pricing just described. By starting as a means of financing energy technology development and infrastructure, a carbon tax could (and

should) be very low. It therefore has a chance of being adopted by all or almost all *leading* emitters and many smaller ones, too. This would permit a widespread "harmonization" of carbon prices, something that would be impossible if carbon prices start high and/or rapidly rise from a modest level. Even if initially the rate adopted by developing countries were lower (say $3.00/ton of CO_2) than the $5.00 rate adopted by developed countries, the carbon price would be approximately "harmonized" across most of the world.

Harmonizing carbon prices is important if substantial "carbon leakage" is to be avoided. Carbon leakage occurs as energy-intensive activities shift to countries that only loosely regulate or price carbon, if they do so at all. Although good estimates of carbon leakage have been hard to come by (most estimates suggest about 20% for energy-intensive industries (Metz *et al*. 2007; Aldy and Pizer 2009), the leakage problem is bound to increase as emission reduction mandates/carbon prices begin to bite in some parts of the world but not in others. Carbon leakage undermines global emission-reducing efforts, and is an inevitable consequence of "brute force" mitigation. But as important as the "harmonization" and "carbon leakage" issues are, they pale by comparison with another issue that has not yet been given sufficient attention.

The heightened concern over climate change is occurring at a time when momentous events are taking place in the developing world. One of the coincidental facts of recent history is that at the same time as humankind is grasping the idea that it is changing the earth's climate, dramatic advances in economic growth and development are taking place in the most populous parts of the world. Beginning in the last decades of the twentieth century, much of what was once the poorest part of the world (Asia), accounting for almost half the world's population, has begun to free itself from the bonds that had contributed to endemic poverty. The nature of growth in East and South Asia has been for a time concealed by the fact that as markets were freed up countries with very large populations oriented themselves toward labor-intensive activities and away from the production of capital-, and typically energy-intensive production. This was abundantly apparent in China, where economic reform began in the late 1970s. For example, China experienced an annual average rate of decline in energy intensity of about 3.5% during the first two decades of reform (1978–98).

For most of its first two decades, reform in China was characterized by the production of labor-intensive goods (clothing, furniture, household electronics, and the like), much of it carried out by town and village enterprises located outside China's cities where 80% of the population lived (Naughton 2007; Brandt and Rawski 2008). But as an increasing portion of the population became richer, moving into a new middle class, many began to relocate to cities. Given the high population to land ratios, if life in shanties and squalid slum areas was to be avoided then residential living as well as commercial activity would have to take place chiefly in high-rise buildings.

But middle-class urbanization requires a shift in production toward the materials used in high-rise buildings and supporting infrastructure, including urban and interurban transportation systems and equipment, utilities, and fresh and waste water-related facilities. The materials used in these projects – steel, cement, flat glass, and aluminum – are highly energy-intensive, among the most energy-intensive products in the world – ten times or more energy-intensive than most other manufactured products (Lightfoot and Green 2001). (These industries, not coincidentally, are the same ones that EU countries have largely exempted from their carbon taxes and that Germany has asked to be exempted from permit auctioning under a post-2012 ETS design.)

To gain some measure of what is happening consider that, in 2006, China's share of world production of cement was 48%, flat glass 49%, steel 35%, and aluminum 28% (Rosen and Houser 2007). Not surprisingly, their production is accompanied by rising emissions, and may account for much of the tripling in the annual rate of change in *global* emissions from 1.1 %/year in the 1990s to 3.1 %/year in 2001–6. Thus the development success story, particularly that coming out of Asia, is associated with a huge shift in the location and relative importance of energy-intensive industries, ones which rely heavily on power generated from combusting coal. China provides a "model."

The energy intensity of urbanization is not limited to the materials used in buildings. Not only must buildings be space-conditioned in a climate as harsh as China's, but streets need to be broad, crisscrossed with overpasses and underpasses. These, too, require large amounts of cement and structural steel. And so do the railways and subways that are required to transport people around the city, to say nothing of the materials required for roads and airports, viaducts for transporting water and sewage, and for the construction of huge power- and water-diversion projects. In short, over the next few decades, as China brings most of its 1.3 billion population into the middle class, huge amounts of energy-intensive materials will be needed (and the energy to produce them) on a scale never before seen in human history. And what is now happening in China will happen to a substantial extent in other populous South East and South Asian countries, from Vietnam to India.

There are very important lessons here. The rapid development of a large region with half of the world's population is a huge success story for the countries themselves. It is also a success story for the world as a whole, which helped make it possible via an open trade system. It is unfortunate that a by-product of this success is rapidly growing CO_2 emissions. Nevertheless, the countries that are making the historic transition from poverty to increased well-being, opportunity, and fulfillment will not give up what they are in the process of earning, whether or not they are asked to do so. (This in fact was recognized in section 2 of the UNFCCC and again by the Kyoto Protocol, in its distinction between the thirty-eight developed and transitioning countries that would take on emission reduction mandates and the much larger number of developing countries that would have no such obligations.)

As a result, we have only begun to see the surge in global energy use that the transformational development process now implies. And with that development process and energy surge will come growing GHG emissions that will only cease with a transformation of the world's energy systems. Not only will that transformation be a long, slow process, but the required energy technologies, for the most part, are not yet ready or *scalable*. And when

Table 7.2 Carbon tax equivalents

	CO$_2$ (tones)	at \$5/t CO$_2$ (\$)	2009 Price (\$) per unit
Tonne of coal	2.86	14.30	16–110
Barrel of oil	0.37	1.85	45–70
Gallon of gasoline	0.0088	0.044	2.00–2.50
1,000 ft^3 of NG	0.055	0.22	10–11
1,000 m^3 of NG	2.025	8.10	~400

Note: NG = Natural gas.

they are ready and scalable, it will likely require a huge technology transfer to the developing world before there is a substantial payoff in global CO$_2$ emissions reductions.

To summarize, there are four main elements of any technology-led climate policy proposal:

(1) First and foremost would be long-term *commitments* by technologically capable countries to undertake individually or in groups research, development, and testing of carbon-neutral technologies, with ultimate emphasis on their scalability (more on this in the next section).
(2) The *financing* of those commitments could best be achieved by adopting a "fee" or tax on carbon emissions. Such a "fee" or tax would start very low – say at \$5.00/tCO$_2$. (Table 7.2 indicates what a \$5.00/tCO$_2$ fee/tax implies in terms of carbon fuel prices.)
(3) Developing countries would be expected to levy the fee as well – even those who do not undertake energy R&D. In the case of the latter, the revenues would be used to help purchase *competitive technologies* when they become deployable on a wide scale.
(4) A second set of commitments would be to *double the "fee" or tax* – say, every ten years or so. Building in a slowly rising carbon price *both* increases financing for R&D and energy-related infrastructure and, more important, provides a price signal to induce commercialization of new technologies and their deployment when they are ready and scalable.

We should, however, step back and acknowledge that there is nothing to guarantee that this technology-led policy will succeed. Monies can be spent on R&D, but we cannot ensure discovery. The search may fail in the sense that R&D will not produce an adequate carbon-intensity-reducing return. If this is the case, modifying the policy by accompanying it by stronger mitigation controls, or in the worst case aborting the policy, may become necessary. But at this juncture there is no reason to believe a technology-led policy will fail, while there is plenty of evidence that alternative mitigation approaches would either be hugely costly – and still not ensure success – or have no chance of stabilizing climate at an "acceptable level. (Policies that would assure a rise in global average temperature of at least 3°C, and probably a good deal higher, would not appear to be acceptable – at least at this juncture.)

What, then, would the technology-led proposal achieve?

(a) It will set in motion the sort of technology program/race without which (and perhaps with which) it will not be possible to reduce global emissions substantially.
(b) It provides a steady and reliable means of R&D financing, at low cost.
(c) It holds out the best hope of bringing down the cost of new energy sources and technologies (as characterized in figure 7.4). After all, much of the value of energy R&D is in reducing the cost of the new energy technologies that we will need to displace the current carbon-emitting ones (Edmonds *et al.* 2004).
(d) It would embed a "forward price signal" via commitments to raise the carbon fee/tax at a slow/gradual rate, doubling every ten years or so.
(e) It holds out the best hope of attracting developing countries to the fold. There are no emission reduction commitments. The fee (or tax) per ton of carbon is low and only very gradually rises. Those countries such as China, India, Brazil, and Korea, which are certainly technologically capable would (with enthusiasm, we expect) make contributions to energy technology development. Other developing countries would slowly accrue funds that would allow them to defray at least part of the cost of transferring carbon-neutral technologies as

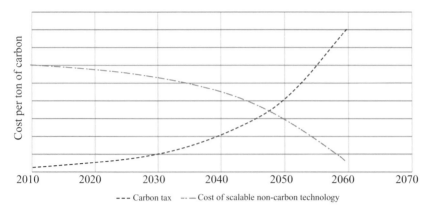

Figure 7.4 *Rising carbon tax and falling cost of carbon-free technology*

they become available and ready to deploy. (Blanford *et al.* 2009 demonstrate the importance of developing countries making some commitment.)

Together, these components add up to a means of reducing the cost of carbon emission-free energy over time, while slowly raising the price of carbon emitting energy. In graphical terms the declining cost per unit of carbon-neutral technology approaches the gradually rising price of carbon emitting energy (figure 7.4). The role of a carbon fee/tax facilitates the decline in cost in the desired technology while slowly raising the price of the carbon emitting technologies which we wish displaced by carbon-neutral ones.

Carbon Pricing and Technology: A "Chicken and Egg" Problem?

The technology-led approach to climate policy sketched out in the preceding sections is prone to misunderstanding. Indeed it is evident that some misunderstanding exists. One reaction is that it is concerned with R&D but not mitigation. Another is that what little there is of carbon-pricing is very "weak." Still another is that our proposal differs in extent but not kind from orthodox proposals that make carbon-pricing a central feature of climate policy.

In this section we clarify some of the issues that the technology-led approach raises, and then

delve deeper into the reasons why we think that this approach has a good chance of succeeding where other more orthodox approaches do not. The issues treated in this section revolve around the relationship between carbon-pricing, on the one hand, and technology, on the other. We begin by comparing our approach with the more orthodox "economic" approaches.

A Comparison of Approaches

Our proposed approach to mitigation has technology leading carbon-pricing. As explained previously, carbon-pricing plays two roles: *first*, the role of *funding* R&D and technology change; and, *second, committing to* the slow but steady *increase* in the *carbon price*, thereby *signaling adoption* of on-the-shelf, ready-to-deploy, and scalable energy technologies. Thus, in our approach, technology and mitigation are linked via carbon pricing. But they are not linked via today's carbon price, but rather via *tomorrow's* carbon price. In our approach, carbon-pricing *initially* plays a largely passive and ancillary, albeit important, role. But as time passes, carbon-pricing plays a more active role by sending a "forward price signal" as a result of commitments to its slow but steady rise.

Thus it is *not* correct to interpret our approach as emphasizing R&D but not mitigation. The two are inextricably intertwined. Also it is not correct to say that we all but ignore carbon-pricing. Indeed, carbon-pricing plays two important roles – one financial, one signaling. Moreover, a carbon price that starts at $5.00/tCO$_2$ in 2010, and doubles every ten years, reaches $80/tCO$_2$ in 2050.

Still, our approach is the reverse of the typical carbon-pricing story. In that story carbon-pricing leads – or in some cases operates in parallel with – technological change. In the typical story, carbon-pricing operates in two different ways. First, it creates an incentive to adopt ready-to-deploy technologies that are more expensive to use than carbon emitting ones in the absence of the carbon price. Second, carbon-pricing is said to *induce technological change* by creating incentives to research, develop, commercialize, and ultimately deploy new non-carbon emitting, or "carbon-neutral" technologies. Before explaining why we are reversing the standard story, it is interesting to compare our approach with the more orthodox approaches taken by Nordhaus (2008) and Bosetti *et al.* (2009).

Nordhaus (2008) sets out and analyzes a suite of policies to curb GHG emissions, evaluating each in rigorous benefit-cost (B/C) terms. Nordhaus' "optimal" policy is a pure carbon-pricing policy which maximizes net present value (NPV). Although presumably new technologies are adopted, the "optimal" policy makes no reference to energy technology and/or R&D policies. What is most interesting from our standpoint is that among the suite of policies the one with the highest benefit-cost ratio (BCR) (by far) is the one that Nordhaus terms a "low-cost backstop". Indeed, Nordhaus (2008: 88) states that:

> Although it might not be currently feasible, the high value of the low-cost backstop technology suggests that intensive research on such energy sources is justified.

It is interesting that the rising carbon price so fundamental to Nordhaus' analysis only reaches $24 per tonne of CO$_2$ (90 per tonne of carbon) in 2050 in his "optimal" policy. In the Nordhaus model, even in the policy that would limit temperature increase to 2°C, the carbon price only reaches $80/tCO$_2$ in 2050.

Bosetti *et al.* (2009) model the role of R&D and energy technologies in climate change miti-

gation. Their work is important not least because it gives a central role to "major technological breakthroughs." These breakthroughs make possible backstop technologies which they define as "a compact representation of a portfolio of advanced technologies" (Bosetti *et al.* 2009: 9). This is a very useful statement! However, they then insist that to "achieve major technological breakthroughs, a strong price signal is still needed to spur the necessary investments" (Bosetti *et al.* 2009: 6). While we agree that a price signal is a useful means of deploying on-the-shelf technologies, we disagree this would be the case for technological breakthroughs. (See also chapter 6 in this volume.)

The "Chicken and Egg" Problem

There are three reasons why we have strong reservations about the necessity for a strong price signal to bring about "major technological breakthroughs." First, most "major technological breakthroughs" require basic R&D for which government funding will be needed for the usual public good/appropriability reasons. Second, there is little or no evidence that carbon-pricing will induce major technological breakthroughs. It is perhaps telling that a major paper on scenario development by authors deeply involved with the report of the IPCC Working Group III (IPCC 2007a), conclude that "from all the variables... involved in the climate change debate, technology emerges as a particularly important area worth further study." The authors go on to say that technology "represents a more 'maleable' variable for *directed* [our emphasis] policy interventions" (Riahi *et al.* 2007: 930–1).

A third reason for skepticism is of a somewhat different nature: it is that a high and/or rapidly rising price will be unacceptable to many of the most important emitters. Here we pursue our third claim. To understand the Achilles' heel of high and/or rapidly rising carbon prices look at table 7.2. A $5.00/tCO$_2$ carbon tax/price implies a $13.80 charge on a ton of coal. For every $10 added to the price of carbon, add $28.60 to a ton of coal. Current coal prices range from as low as $10.00 per ton up to a little over $100 per ton. The higher part of the range is for coking coal. Most coal for electricity-generating purposes sells for less than $60.00/ton at the mine mouth. Thus a carbon price of $20/tCO$_2$ implies an *approximate doubling of carbon prices*. A $100/tCO$_2$ carbon price would represent an increase in coal prices of between 400 and 500%.

In Bosetti *et al.* (2009) the carbon price path without backstops reaches about $425/tCO$_2$ in 2050, and even with backstops almost $200/tCO$_2$. Not only is it uncertain that "major technological breakthroughs" will emerge (Bosetti *et al.* 2009; Bosetti and Tavoni 2009), but the resultant carbon price with breakthroughs/backstops nevertheless implies a "tax" on coal of over $550 per ton. So not only are high and rapidly rising carbon prices required according to the Bosetti *et al.* (2009) modeling exercise, but even after being more than halved by backstops the carbon price will still be very high.

Among the major emitters (some more important than the others) are the USA, China, India, Russia, Australia, and Poland. What each of these countries has in common is a heavy dependence on coal. Why would any of these countries accept rapidly rising carbon prices without some assurance that these will soon be alleviated by technological breakthroughs? This important question has been obscured by the emphasis on emission-reduction commitments rather than technologies.

Thus the Bosetti *et al.* (2009) paper reflects a fundamental problem. The authors comprehend the technology imperative, but their insistence on a primary, up-front role for carbon-pricing is not only debatable, but appears to give rise to a "chicken and egg" problem. The carbon pricing "chicken" may be ineffective without the technology "egg."

The problem goes well beyond that of coal used in electricity generation. It infects several broad industry groups that are very energy-intensive. Five of these sectors – ferrous metals, non-ferrous metals, pulp and paper, petrochemicals, and non-metallic minerals (cement, glass) – have energy intensities that are, on average, an order of magnitude higher than the average of other manufacturing industries. For many countries these industries are considered important. There is now some experience with the application of carbon-pricing to these industries, and the results are instructive.

Some Cases in Point

Carbon-Pricing in Europe

The story begins more than a decade ago after several European nations (including Sweden, Norway, the Netherlands, Germany, Finland, and Denmark) enacted carbon taxes (Metcalf 2009). Within a short time after the introduction, it became necessary to substantially reduce (by up to 90%), or eliminate entirely, the carbon tax on energy-intensive industries such as steel, cement, aluminum, and flat glass. The concern was that the competitiveness of these industries would be harmed by high carbon taxes. Unless the tax were substantially reduced or eliminated, firms in these and other energy-intensive industries might move operations to and/or make any new investments in countries that did not have a carbon tax – or a much lower one. Not only would there be a loss of production activity and jobs at home, but there would be a "leakage" of carbon emissions abroad, muting the reduction in global emissions. The claim by the IPCC Working Group III (IPCC 2007a) that "carbon leakage" would be minor, affecting no more than 20% of emissions reduction is not reflected in the response of parliaments.

EU emission permits

The story continues with the EU ETS. In the first three years of the ETS, 2005–7, most EU country allocated emissions permits so as to minimize the impacts of emission caps on vulnerable industries. The result was that the emission caps were hardly binding. As the second phase (2008–12) of the ETS program began, the global recession took the pressure off the emission caps and has allowed emission permit prices to remain low. Nevertheless, concern about the impact on energy-intensive industry continues. For example, Germany has moved to ameliorate the impact of rising electricity prices on such electricity-intensive industries as aluminum, copper, and zinc by subsidizing their electricity bills (*Financial Times*, May 26, 2009), and at the EU summit in December 2008, Germany successfully argued that energy-intensive industries should not be required to buy emission permits between 2013 and 2020.

The Waxman–Markey bill

Similar pressures have come to bear on the US Congress' Waxman–Markey (WM) Bill. The original bill had called for a 20% reduction in US emissions from their 2005 level with 100% auctioning of emission permits. However, to attain sufficient support in a Committee in which Democrats outnumbered the Republicans almost two to one, many concessions had to be made to representatives from coal states and districts. When WM emerged from Committee it did so with: (i) a reduced emission reduction target (17%); (ii) a provision to allocate rather than auction 57% of total permits to the electric power and energy-intensive industries; and (iii) a provision that would allow domestic *uncapped* sectors and international sources to sell up to 2 billion tons of "offsets" per annum to capped firms and industries in return for undertaking their own emission reductions. (The "offsets" are controversial. They are virtually impossible to monitor; are a recipe for fraud; add carbon price uncertainty; will provide a field day for lobbyists and "rent-seekers"; generate public distrust; erode political capital; and blemish the integrity that the Obama Administration has worked hard to restore in Washington.)

By the time WM passed the US House of Representatives, by a 219–212 vote, on June 26, 2009, 85% of the permits had been allocated. Moreover, some of the limited revenues from the 15% of permits auctioned would be used to subsidize consumers who found their electricity and gas bills rising too rapidly. Little was left to underwrite basic R&D. As a result, the Breakthrough Institute (2009) estimated that the cap in WM is not binding – it is a cap in name only. Estimates made by the Congressional Budget Office and the EPA, that the cost per household would be low, are another indication that actual emission reduction will come nowhere near the WM caps. Thus, in its present form, WM appears unlikely to do much to curb US emissions, while diverting funding and attention away from crucial R&D.

A carbon-pricing plan for Canada

Another glimpse into the potential ramifications of attempting to achieve emission reduction targets via "brute force" methods comes from Canada. The Conservative government has proposed the adoption of emission cuts from current levels of 20% and 65% by 2020 and 2050, respectively. Although no means to achieve these targets has been adopted, Canada's National Roundtable on the Environment and the Economy (NRTEE) has worked on a plan to meet them. The NRTEE circulated a report with a carbon-pricing proposal, the goal being to achieve the 20% and 65% cuts in carbon emissions using a "cap-and-trade" approach to carbon-pricing (NRTEE 2009).

In producing its report, the NRTEE commissioned modeling analyses of their proposal to gauge its implications. The findings are instructive – although perhaps not in the way the NRTEE report contemplated. For our purposes, there are two striking facts about the NRTEE plan. First, to achieve the 2020 target of a 20% reduction in emissions from current levels, carbon prices (in Canadian dollars) rise from $15/tCO_2$ in 2010 to $115/tCO_2$ in 2020. Secondly, the rapidly rising carbon price has a substantial impact on some energy-intensive industries. How great depends on the carbon-pricing policies of other countries (see table 7.3).

Bataille *et al.* (2009) investigate the implications of the NRTEE's carbon-pricing proposal for Canada's competitiveness. Their findings are indicated in table 7.3. Bataille *et al.* (2009) consider three cases: Canada acts alone; only OECD countries follow Canada's example; all countries "cooperate – that is, follow Canada's example. Since a $115/tCO_2$ carbon price implies a $329 tax per ton of coal (coal prices are typically well under $100/tonne), we can assume that neither the USA nor China will "cooperate," and the same is likely true of many other countries that use coal for energy purposes (e.g. Poland and Germany). Thus, if Canada adopted the NRTEE proposal, effectively Canada would be acting alone, even though Canada's goals (targets) were not dissimilar to those nominally adopted by many OECD countries.

Table 7.3 indicates that several energy-intensive industries would experience substantial declines in physical output from "business-as-usual" (BAU) levels. This is clearly the case for "industrial minerals," such as cement, limestone and the silicates

Table 7.3 Projected changes in physical output of energy-intensive sectors (2020)

Industry	Canada acts alone	OECD cooperates (% change)	Globe cooperates
Chemical products (tonne)	−10	−3	−2
Industrial minerals (tonne)	−50	−27	−14
Iron and steel (tonne)	0	0	0
Metal smelting (tonne)	−3	−2	−1
Mineral mining (tonne)	−1	−1	0
Paper manufacturing (tonne)	−9	−4	−2
Other manufacturing (2005 $ GDP)	−4	−1	−1
Petroleum refining (m³)	−28	−28	−27
Petroleum crude extraction (barrels per day)	0 to −8	0 to −6	0 to −5
Natural gas extraction (m³)	0 to −16	0 to −15	0 to −14

Note: For petroleum and NG extraction, the first estimate is with economic rents, the second is with no economic rents.
Source: Authors' calculations from CIMS; Bataille *et al.* (2009).

used in glass production, and for petroleum refining. If Canada acts alone (and the watering down of WM suggests the USA would not begin to contemplate anything like a $100/t price for CO_2), paper manufacturing and chemicals will be hit hard, too. Somewhat surprising is the apparent lack of impact on Canada's steel industry and the very small impact on the metal-smelting sector. These may reflect Canada's relative abundance of hydro power and lack of dependence on coal for electricity generation.

Few countries, in fact, are likely to follow the steeply rising carbon-pricing policy that Canadian modelers indicate is required to achieve the NRTEE plan. Not only is there a huge rise in carbon prices in the first ten years, the plan calls for carbon prices that continue their steep rise for another decade, reaching $300/t CO_2 (or about $860/tonne of coal) a little after 2030. While Canada's large hydro electric resources limit its dependence on coal for electricity generation, and provide some insulation from the NRTEE schedule of carbon prices, few other countries are so fortunate. For the others, carbon prices in the hundreds of dollars are not thinkable until there are reliable, plentiful, and cost-competitive non-carbon emitting sources of energy available and deployable, in which case high carbon prices are unnecessary. Here is another reason why a carbon-pricing policy should follow rather than lead a technology-led policy.

Targets and Non-Credible Commitments

Still another way to understand the logic behind the technology-led approach is as a means of freeing ourselves from the straitjacket of targets and non-credible commitments to them. Like it or not, climate policy is still ruled by the setting of emission reduction targets and pressures on countries to commit to them. Here are some examples:

(a) The original WM, drawn up by the US House of Representatives Energy and Environment Committee, called for a 20% cut in US emissions from 2005 levels by 2020, and 80% by 2050. When the bill emerged from Committee, the targets had been slightly modified to 17% by 2020 and 83% by 2050.

(b) The UN target of a 20–25% cut in developed country emissions below 1990 level by 2020. (Developing countries have called for the cut to be 40% by 2020 before they would sign up to emission reduction responsibilities.)

(c) The G8 target of reducing *global* emissions 50% by 2050 from 2007 levels. This was approved at the meeting of the G8 in 2007, and a similar target was agreed to at the G8 meeting in 2009.

(d) A 2100 target calling for a global emission cut from current levels of 80%.

We can provide some metrics by which we can evaluate the feasibility of these targets:

(a) The *original* version of WM would have auctioned virtually all of the emission permits. To reach its targets from domestic reductions, the bill would have required a 4.5% rate of de-carbonization (RCIO, C/GDP) if the US economic growth from 2010 to 2020 were 2.3%; if the growth rate were a more robust 3.3%, the required rate of de-carbonization (rate of reduction in the carbon intensity of output, CIO) would be 5.5%. Assuming a \$14 trillion US economy in 2010 and a 2.3% rate of growth 2010–20, achieving the WM emission reduction target would mean a cumulative loss of US GDP of \$8.9 trillion, even if the RCIO of GDP averaged a phenomenal 3.2% from 2010 to 2020. (From 1980 to 2006, the US RCIO was about 2.2%, more than 50% higher than the 1.3% average for the world as a whole.)

(b) The UN 2020 targets have been the object of much recent discussion and cajoling. The United Nations wants these targets to provide the basis of the commitments made at the 2009 Copenhagen Summit to finalize a successor to the Kyoto Protocol. Each developed country has been formulating plans. Japan has said it would commit to a cut of 15% below 2005 levels – or 6% below 1990. The Japanese plan has been roundly (and unfairly) criticized (Pielke 2009a) as much too little – although it may be feasible. The UK Climate Change Act 2008 calls for cuts of 34% below 1990 by 2022 and 80% below 1990 by 2050. The UK plan would require 4.0% + RCIO between 2007 and 2022, and a 5.5% rate between 2007 and 2050 (Pielke 2009b) – rates which are almost certainly infeasible.

(c) The G8 target of reducing global emissions 50% by 2050 from current levels would require a 4.2% rate of de-carbonization of the global economy if the rate of growth of GWP from 2010 to 2050 were 2.5% per annum. By 2050, the *world as a whole* would have to have a CIO of GDP similar to that currently enjoyed by Switzerland.

(d) For the USA to cut its emissions 80% by 2050 would require an average annual rate of de-carbonization of 6.0%, if US GDP grows at a rate of only 2.0% over 2010–50. The required

average annual RCIO of GDP would have to be 6.5% if the GDP growth rate averaged 2.5%.

(e) If global emissions are to be cut by 80% by 2100, then the average annual rate of de-carbonization must be 4.0%, assuming that the global economy grows at a 2.2% rate from 2010 to 2100. If the global growth rate were 2.7%, the average annual rate of global de-carbonization in the remainder of the twenty-first century would have to be 4.5%.

These rates are unattainable with current technologies. It would take a technological revolution to make possible the longer-term targets. Faced with these conditions, we think the rational strategy would dispense with emission reduction targets and reverse the time-related roles of energy R&D and carbon-pricing. The main aim of current climate policy would focus on (i) raising the RCIO, and (ii) bringing down the cost and raising the reliability, effectiveness, and scalability of the means of achieving large future reductions in emissions. An initially low carbon price would first finance and, second, as the carbon price slowly rises, send a "forward price signal." This is the logic behind a technology-led climate policy.

An "Incentive-Compatible" Technology Race

It is much easier to talk about a technology policy than it is to carry out an effective one. It is much easier to spend on R&D than ensure that the monies are well spent. This is especially so where the market cannot be counted on to exercise the sort of discipline that should avoid the most egregious waste. The problem is that neither the private sector nor the public sector alone possesses the appropriate incentives to create the sort of energy technology revolution that is required to stabilize climate. Both private and public investment in R&D will be needed. Although we focus on the latter, we assume there will be growing private interest (as there already is) in investments in carbon-free energy R&D, especially at the applied development and commercialization stages. Here, however, we focus on publicly funded R&D which will be especially important at the basic stages. Before proceeding further, it is useful to look briefly at the R&D literature.

What do we know about inducements to R&D and innovation? Much of the relevant economic literature on R&D, patents, and innovation is found in the field of industrial organization. A dominant theme has been whether monopoly power or competition is "better" for innovation. Schumpeter (1942) postulated that market power provided both the finance (profits) and the incentive (future market power and profits) for undertaking risky and uncertain investments in R&D and other innovative activities. But Arrow (1962) demonstrated that, in principle, a firm in a competitive (price-taking) industry has more to gain from a process of (cost-reducing) invention/innovation than does a monopoly firm. Of course, market power does not imply that a firm has a monopoly, and competition can be between a small number of rivals, not just among a large number of price-taking firms who ignore rival behaviour. Dasgupta and Stiglitz (1980) demonstrated that firms in a market with a small number of rivals (an "oligopoly") had the greater need, and therefore incentive, to innovate because the very survival of each might depend on at least some innovative success. The Dasgupta–Stiglitz paper is useful in thinking about the organization of a technology race.

There is a growing literature on technology change, particularly as it relates to climate. Popp *et al.* (2009) review a large literature on energy, environment, and technological change. Clarke *et al.* (2006) address the sources of technological change, finding empirical support for the importance of spillovers and LBD. The IPCC (Metz *et al.* 2007) reviews a modeling literature in which the impact of carbon prices on (induced) energy technology change is central. Jaffe *et al.* (2005) address the market failures that surround technology and environmental policy. They conclude that these provide a strong rationale for a "portfolio of public policies" to deal with both technology development and emissions reduction. Baker and Adu-Bonnah (2008) investigate how climate change *uncertainty* affects the *optimal* amount of investment in risky R&D programs (it increases it!).

Nordhaus (2002) uses his DICE model to assess the likely impact of induced innovation, and finds it to be relatively small. While Nordhaus has not placed much emphasis on R&D policies in the context of climate change, it is perhaps significant that in Nordhaus (2008), he finds that low-cost "backstop" technology(ies), if it (they) could be developed (as opposed to assumed as in many economy–environment models), dominates all other policy options. Although he appears doubtful any "low-cost backstops" will appear, Nordhaus (2008: 88) says that "intensive research on energy sources is justified." Finally, the literature on "mechanism design" and "implementation theory" (Maskin 2008) may yield some practical applications to "incentive-compatible" R&D programs.

The industrial organization literature also considers the role of patents and the implications of patent races. Patents explicitly raise issues of the effectiveness of: (i) incentives to innovate; (ii) the ability to appropriate the payoffs from successful innovation; and (iii) the ability to "invent around" patents, or "reverse engineer." Because patents may not provide incentive enough to invest in R&D where uncertainty is high and appropriability is low, some have thought about alternative incentives. The problem is further complicated where public funding is required.

It is well known that where governments replace markets in the generation of R&D, a series of problems can arise. These include a tendency to "pick winners" – an exercise that often fails. There is also the possibility of "lock-in" to what turns out to be an inferior technology because of the government's wish to get a quick return (before the next election) for the taxpayer/voter money it has laid out (Arthur 1989). There is the ever-present problem of bureaucratic (turf) "infighting," and decisions tainted by the exercise of lobbyist influence. Many of these problems could be reduced, if not eliminated, by some form of competition.

A modicum of competition can be injected into a government-funded process in a number of ways. One approach is that of prizes (Wright 1983; Montgomery and Smith 2007). Here one might conceive of a "tournament" in which a prize(s) is given to "winners" in a contest to innovate. Still another initiative might take the form of awarding research contracts on the basis of creativity and perceived chances of success – but the choice of awardee could come perilously close to picking winners. In either case "incentive compatibility" can be increased if governments commission a set of independent (of government, and hopefully other political influence) experts to pick areas for an R&D competition and then to act as judges of the results (a version of the Gates Foundation model). This approach would minimize the problem of picking winners and "lock-in" (there could be a decision that no one won, and that another round of competition is in order), and the process should be able to avoid both bureaucratic turf battles and lobbyist influence.

Another approach applies the model of an energy technology race and team work to the international arena. Here the idea is one of competing international consortia or teams acting as participants in an energy technology "race" (Green 1994). One can conceive of several consortia, each made up of three or four countries capable of contributing science, engineering, and other technology-related talent to one or more projects. Individual countries could be members of more than one consortium.

A technology "race" between competing consortia could capture public interest and imagination in its own right. It would also place a premium on creativity in developing effective, competitive, deployable carbon-neutral technologies rather than requiring the sacrifices that would inevitably accompany "brute force" mitigation. In this way, energy technology commitments could rally the current younger and future generations in a way that "brute force" mitigation would not. However that may be, there would have to be intellectual property protection for new inventions, and also some agreement for cross-licensing potential users at reasonable rates.

An energy technology race that includes leading developing countries such as China, India, Brazil, and South Korea could also obviate at least part of the technology transfer (TT) problem (see also chapter 8 in this volume). If these countries are part of international consortia with developed countries, they could share in successes. If they succeed on their own, they have something with which to "trade" with developed countries.

Whatever the means of introducing competition into the energy innovation process, it is crucial that the R&D be well funded, and consistent. Nothing could stunt the development of new technologies more than underfunding, or funding that is uncertain, stop and go. Although the funding of any specific venture should be held to account, with funding terminated after failure is inevitable, there must be adequate funds to continue a wide variety of R&D initiatives and to start new ones as older ones are terminated. The question is how to assure sufficient and consistent funding that is free (or largely free) of political interference or influence.

There are at least three possibilities. Only the last meets the tests of *sufficiency, consistency*, and relative *freedom from political influence*. The first is funding out of general funds, which could fail the test on all three grounds. Such funding is inherently subject to political discretion, could be diverted to other uses when tax revenues fall, and may never be sufficient in amount.

The second approach is funding with a carbon tax, but if the tax revenue is not isolated from the general budget it is prey to political interference or diversion to other uses. This approach would fail the *consistency* criterion.

The third approach would use a "dedicated" carbon tax, the revenues from which would be placed in a "trust fund" managed by "trustees" independent of Congress and the Administration in power. The "model" might be the US Interstate Highway Trust Fund created during the Eisenhower Administration to build and maintain America's Interstate highways. It is funded by an 18 ¢ per gallon federal gasoline tax. Because that tax is viewed as providing clear benefits to a large part of the electorate and thus to most taxpayers, it has not generated the hostility that many other taxes have generated. A low-carbon tax, dedicated to improving and strengthening the energy system, the funds for which are isolated in a "clean energy" trust fund, could be expected to be similarly welcome –or at least not too unwelcome to pass political muster.

Who or what will manage or direct a "clean energy" "trust fund" is important. It is not just Congressional influence that is of concern, but political influence in general. The point of holding the carbon tax revenues in a trust fund is to increase the incentive compatibility of the R&D technology program. Thus it would make sense to draw the

Board of Directors of the fund from the private sector as well as the public sector. The Directors would have to be given full oversight of how the funds are being used – hopefully drawing on engineering and scientific expertise in making choices and allocating funds. There would also have to be provision made for intellectual property protection and the allocation of patent rights. There would also have to be agreement on a reasonable rate at which other countries are cross-licensed for new technologies that are developed.

A further embellishment would be to allow private entities to invest (put equity into) publicly funded energy R&D projects. Doing so would add to the total R&D funds. As long as the Board of Directors of the "clean energy" trust fund make the choices of R&D projects, the injection of private equity should not affect the direction of innovation other than through the indirect influence of where they place their funds. One might anticipate substantial additional funding from fossil fuel producers as they attempt to diversify their portfolios – and their risks.

The Technology-Led Proposal

The preceding five sections have attempted to make a systematic case for a technology-led climate policy. In this section we set out proposed means of carrying through a technology-led strategy for the next ten years (and beyond). The two strategies that (i) focus on "enabling" technologies, and (ii) focus on "breakthrough" technologies, are not mutually exclusive, and in some cases new technologies fit both categories. But their emphasis is different. While both strategies should be pursued at least to some extent, our analysis tries to answer the question: Which should be emphasized over the next decade?

The two strategies are strongly supported by the current state and long-term capabilities of current carbon-neutral energy technologies. Some technologies are close to being ready but cannot be scaled up without the development or supply of an "enabling" technology. Examples include: (a) grid integration and upgrade; smarter grids; DC lines for long-distance transmission; and most important

of all utility-scale storage for intermittent solar and wind energy (the last would constitute a "breakthrough" technology as well); (b) more energy efficient retrofits for fossil fuel-fired electricity generating plants to allow for CO_2 capture from existing plants; (c) identification of safe and secure geological storage sites for captured CO_2; (d) methods to "spike" plutonium produced as part of a nuclear electric closed-cycle process (which would greatly economize on Uranium 235 and substantially reduce nuclear waste).

Examples of "breakthrough" technologies include: (a) a class of widely usable "breeder" reactors; (b) nuclear fusion; (c) deep geothermal; (d) a worldwide "superconducting" grid; (e) air capture of CO_2; and (f) many of the steps required to make a "hydrogen economy" feasible. Most of these may be decades away from being operative. The list is not inclusive of all possibilities.

The two strategies are interdependent, particularly over longer time horizons than a decade. Many current carbon-free technologies require "enabling" technologies to become scalable. But these will not be enough. Breakthrough technologies will be needed – and research on these needs to start early. Thus in the CBA, we do not, and can not, distinguish between the two.

Before proceeding to an evaluation of the technology-led proposal, there are a number of issues to consider; each provides a glimpse of how we view the technology problem from an economic standpoint.

(a) The technology cost function

$$C = F + vq, \qquad (1)$$

where
C = total costs
F = fixed costs, most of which are up-front, sunk costs in R&D
v = long-run average variable production costs
  In average cost (AC) terms, we have:

$$C/q = F/q + v, \qquad (2)$$

which implies that AC declines as the quantity of output (q) (or length of production run) increases.

If, in addition, there are LBD effects in the production of the output, we have:

$$C = F + vq^\alpha,$$

where

$$\alpha < 1 \qquad (2')$$

In our view, the economics of researching, developing, and testing new, uncertain-of-success, energy technologies is wrapped up in the fixed-cost factor, F. We agree with the analysis of Montgomery and Smith (2007) that there is a "dynamic" *time inconsistency* that makes it highly doubtful whether the market will supply the funds for the large "up-front" sunk cost component of F, particularly when "up-front" may mean decades away rather than years. In other words, if the time from basic R&D is long, the private sector is very unlikely to make risky investments which are uncertain of success and whose payoffs are in the distant future. Moreover, the inability of current governments to tie the hands of future governments to do anything more than cover the costs of production of a successful innovation makes the *time inconsistency* complete. The publicly funded R&D "socializes" the risks inherent in the "F" term.

(b) Competition

But that is not the end of the economic problem: the way in which the R&D funds are used should maximize the opportunities for success and minimize outright "waste." (The failure of a particular scientific initiative to bear fruit is *not* "waste.") We have briefly described some means of getting the most out of the funds designated for energy R&D. We believe that the injection of competition in technological pursuit (as opposed to competition for funds) is important, and would contribute to what we term "incentive compatibility."

(c) The "technology return" to R&D investment

Here we introduce a construct that we believe is very useful in undertaking the CBA and in comprehending our estimates. To do so, we develop what we believe is a new (and valuable) concept: the "carbon intensity-reducing return to R&D investment" (or CIR³D). The "carbon intensity" referred

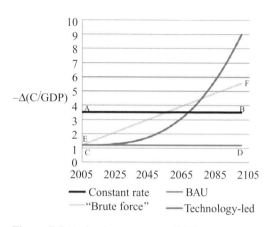

Figure 7.5 *Technology return to R&D*

to here is the ratio of carbon emissions to output (GDP – or, globally, GWP). The idea underlying CIR³D is that effective investments in energy R&D lead to reductions in the CIO ($CIO = C/GWP$). They do so via reductions in: (i) the energy intensity of output (EI) ($EI = E/GDP$) and/or (ii) the carbon content of energy ($CCE = C/E$). The rate at which R&D investments are translated into carbon-reducing technologies (accounting for the decline in CIO) is crucial to estimating the benefits of a technology-led approach.

We define the following:

(a) **"Baseline" or "historic" rate of CIO reduction**: this is the rate of reduction in CIO (C/GDP) which would occur "naturally" – that is, without the help of R&D investments in "enabling" and or "breakthrough" energy technologies (line CD in figure 7.5).

(b) **The *constant* required average annual rate of decline in CIO**: that is the rate required to achieve climate stabilization, *given* a trend rate of GWP growth and an emission level consistent with a "targeted" atmospheric concentration of CO_2 (or more generally GHGs) (line AB in figure 7.5).

(c) **The required "straight-line" rate of decline in CIO**: this is the *constant rate of increase* in the decline in CIO that achieves the targeted atmospheric concentration level (line EF in figure 7.5. On average, the rates indicated by EF are equal to the constant rate indicated by line AB).

(d) **The "trend" rate of growth in GWP (g)**: this is needed in order to derive (b) and (c).

(e) **The carbon intensity-reducing R&D curve**, the slope of which at any point is the CIR^3D (this is the upward-sloping curve in figure 7.5).

To give some feel for what we doing, it is useful first to set out the well-known Kaya Identity (3) and a further simplified form of the Identity (3′):

$$C = P \left(\frac{GWP}{P} \right) \left(\frac{E}{GWP} \right) \left(\frac{C}{E} \right), \qquad (3)$$

where

$C = CO_2$ emissions
$P = $ population
$GWP = $ gross world product
$E = $ energy consumption

Canceling the P and E terms, we have

$$C = GWP \left(\frac{C}{GWP} \right). \qquad (3′)$$

Taking the time derivative of the natural logs of (3′),

$$\%\Delta C = \%\Delta GWP + \%\Delta \left(\frac{C}{GWP} \right), \qquad (4)$$

where $\%\Delta$ is the average annual rate of change.

Let us introduce some numbers. The "historic" RCIO is approximately 1.3%. If the growth rate of GWP is 3.0%, CO_2 emissions would grow at an average annual rate of 1.7% – which, in fact, is the rate of growth of CO_2 experienced from 1980 to 2006.

Now let us look forward. Suppose the long-term (2010–2100) "trend" rate of growth in GWP (g) is 2.2%. Suppose, further, that in order to avoid going above a specified atmospheric concentration of CO_2, emissions must be reduced from about 8 GtC in 2010 to 2.5 GtC in 2100. The rate of reduction in C implied by reducing emissions from 8 to 2.5 GtC in 90 years (2010–2100) is 1.3%.

Using (4): we can derive the *required average annual rate of decline* in C/GWP:

$$-1.3\% = 2.2\% + \%\Delta(C/GWP),$$

implying that the last term (CIR^3D) is −3.5%.

In other words, the rate of decline in C/GWP must average −3.5%, which gives us a value for (b) above.

To derive the "*baseline*" rate of decline in CIO, the rate that would "naturally" hold in the absence of "enabling" or "breakthrough" technologies, we can draw on: (i) historical evidence; (ii) predictions based on what is known about the scope for EI decline (Baksi and Green 2007), and the current capabilities of carbon-free energies (Hoffert *et al.* 2002; Caldeira *et al.* 2003; Green *et al.* 2007; Lewis 2007b); (iii) what is assumed in emission scenarios. (A problem with the scenario approach is that most of the IPCC emission scenarios are highly unrealistic about what might occur naturally. However, one of the scenarios, B2, is close enough to be used in our CBA.)

Using historical evidence would give a baseline decline in RCIO of 1.3% (Hoffert *et al.* 1998). Evidence based on analyses of EI and CCI (ii) yield a rate of 1.2% (Baksi and Green 2007; and Green *et al.* 2007), and the IPCC B2 scenario (1.4%). Pulling together the required rate of decline in CIO (line *AB* in figure 7.5) and the baseline rate of decline in CIO (line *CD* in figure 7.5) allows us to derive the "straight-line" rate of the CIO decline line, *EF*. The curve reflects the impact of a technology-led policy on the CIO.

Some examples may illustrate how figure 7.5 can help us to understand the implications of various emission reduction proposals:

(a) The G8 target of reducing global emissions by 50% by 2050 from current levels would require a 4.2% rate of de-carbonization if the global economy is to grow at an annual average rate of GWP of 2.5% per annum (2010–50).

(b) If global emissions are to be cut by 80% by 2100, then the average annual rate of de-carbonization must be 4.0%, assuming the global economy is to grow at a 2.2% rate from 2010 to 2100.

The examples highlight a simple fact. The widely discussed targets for emission reductions require a huge increase in the rate of decline in the RCIO. To come even close to achieving a long-term global average RCIO greater than 2.0, to say nothing of

3.0, 4.0, or higher rates, will require a thoroughgoing transformation in the way in which individual economies and the world produce and transform energy. The required rates of de-carbonization imply a transformation of energy systems so large and rapid that their achievement requires developing a fleet of scalable carbon-free energy technologies.

(d) "Crowding out"

Before moving on, we should acknowledge an issue that has been raised by William Nordhaus (Nordhaus 2002) and his former student David Popp (Popp 2004). Nordhaus and Popp have argued that directing R&D spending to the climate change problem is likely to misallocate scarce scientific and engineering resources. There are many competing uses for scarce scientific talent, implying that an extra dollar spent on energy R&D is likely to yield benefits much smaller than spending that R&D dollar on an alternative project. In effect, investment in R&D would "crowd out" other even more worthwhile R&D.

We think, however, that this argument is flawed in three ways. First, energy R&D spending, in the USA and globally, has significantly declined over the last quarter-century (Nemet and Kammen 2007; Barrett 2009). Second, climate change and the energy R&D spending that will be needed has a much longer time dimension than other scientific R&D. Thus the diversion of scientific resources away from other important uses is likely to be smaller than is implied by the Nordhaus analysis. Third, and most important, whatever concerns a large and growing global population may generate it has meant an increase in brain power (Johnson 2000). An increasing amount of that brain power and scientific talent is in East and South Asia.

The Nordhaus–Popp analyses appear to overlook the increasing supply elasticity of scientific and engineering talent when viewed in global terms. Moreover, their analyses overlook the attraction of new (or additional) talent toward a real and meaningful technology race. Such attraction may provide an outlet for the "best and the brightest" of our mathematical and physics talent, many of whom were hired (until 2007) to use their mathematical capabilities to craft financial "weapons of mass destruction" on behalf of those who helped bring on a world recession.

Benefit-Cost Analysis

We now come to what is by far the most difficult part of the project: calculating benefit-cost ratios (BCRs) for the proposed "solutions." There are a number of reasons why carrying out a CBA analysis is difficult and debatable, especially where technology "solutions" are concerned. One obvious problem is that, because time is of the essence in climate change-related CBA analysis, it makes a difference when new, scalable energy technologies will become available, and ready for deployment. It is really impossible to predict success, much less a date of success, in advance, although with a large number of technological initiatives underway one might find that the law of large numbers comes in handy. Another problem we face is comparing specific technology "solutions" such as an emphasis on "enabling" technologies, or alternatively on "breakthrough" technologies. We therefore assume in our CBAs that both types of R&D will be undertaken, while recognizing that success in developing enabling technologies may give an earlier return to R&D investments.

A second issue is that as a matter of current decision making the "default" climate policy is no longer one of BAU or "no-policy." Rather, the "default" position at Copenhagen 2009 has been a policy along the lines of the emission reduction targets and commitments adopted by most developed countries at Kyoto. This has implications for the role of "climate damages" in benefit-cost (B/C) evaluations. In the standard CBA, the benefits are climate damages avoided. However, when comparing the technology-led proposal with another approach to mitigation, the main benefits are the abatement costs avoided by adopting the technology-led approach.

Given competing ways in which the benefits of a technology-led policy might be interpreted and evaluated, we have chosen three different ways of generating BCRs. These are:

A. **The standard CBA approach using an emission scenario baseline.** To carry out an evaluation of climate damages and those avoided by a technology-led policy, we used William Nordhaus' DICE model.

B. **An analysis using *cumulative emissions* to indicate how close the technology-led approach comes to limiting the global rise in temperature to 2°C using cumulative emissions out to 2100.** Here we make use of the findings of Allen *et al.* (2009) and Meinshausen *et al.* (2009).

C. A third approach is estimates based on **comparisons of a technology-led approach with a "brute force" mitigation approach** to achieve widely discussed emission reduction targets.

Before turning to the three B/C evaluations, we should say a word about how we approached the first of these – (A), the standard CBA. As we do not have our own climate economy model, we have used the well-known DICE model developed by William Nordhaus. There are two reasons for using the DICE model. One is that Nordhaus makes the model widely available (it is easily downloaded from his website). The second is that it is relatively easy (after a few weeks' work) to learn to use by persons not already familiar with climate modeling.

However, the initial damage function parameter values[1] appear likely to underestimate climate damages (see Nordhaus 2008). For example, the estimate of the PV of climate damages from a policy of delaying climate change action 250 years is only $22.5 trillion. At the same time the PV of climate change damages if warming could be limited to only 1.5°C is $9.95 trillion (Nordhaus 2008, table 5.1). What this means is that even if a technology-led policy of spending $100 billion a year for the next ninety years were able to limit warming to 1.5°C (something no one, including us, believes is achievable by any policy) the BCR would only be 3.9.

Benefit-Cost Calculations

In this section we undertake three different ways in which to derive benefit-cost ratios.

Standard CBA

The goal here is to assess the cost-effectiveness of a technology-led policy with respect to a BAU baseline scenario. To this end, we use the well-known integrated assessment model (IAM) DICE-2007 designed by William Nordhaus[2] and apply a standard benefit-cost ratio:

$$BCR = \frac{\text{NPV of climate damages not incurred}^{3}}{\text{NPV R\&D expenditures}}$$

Climate damages not incurred (the numerator in *BCR*) is the difference between climate damages suffered under a baseline scenario and those not prevented under a technology-led policy. Throughout this section we assume R&D expenditures of $100 billion/year for the period 2010–2100, an amount we believe sufficient to produce major reductions in CIO. The procedure is as follows.

(1) Estimating the technological return to R&D
First, the emission path is estimated under our proposed technology-led policy by accelerating the historic RCIO. A functional form for the relationship between R&D expenditures and CIO is estimated as:

$$\mu = 1 - \frac{\alpha (RD)^2}{\beta + (RD)^2}$$

where μ is the reduction in CIO due to R&D, α and β are parameters that may be adjusted depending on beliefs about R&D success; and, lastly, (RD) is cumulative expenditure on R&D.

This particular functional form allows for an initially slow return to R&D, followed by a period of breakthroughs and lastly decreasing returns. Although DICE is a reasonably flexible model that allows modification of its numerous parameters, only those directly related to carbon intensity have been modified in order to maintain the integrity of the BCRs and assure verifiability. The corresponding parameters in DICE used to simulate the technology-led policy are: CO_2-eq

---

[1] Damage function parameters are not modified to obtain the BCRs in this chapter.
[2] http://nordhaus.econ.yale.edu/DICE2007.htm.
[3] R&D expenditures refer to all stages of innovation and deployment.

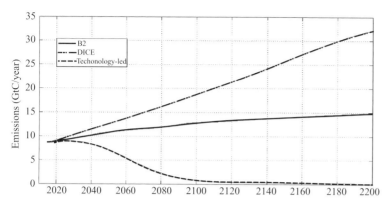

Figure 7.6 *Emissions, by baseline/policy*

Note: B2 extrapolated beyond 2100 by the authors.

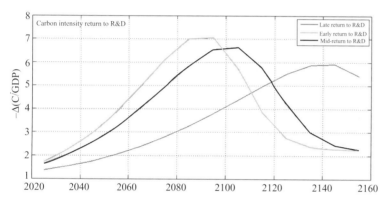

Figure 7.7 *Carbon intensity return to R&D: alternative profiles*

emissions GNP ratio 2005 (SIG0); initial growth of sigma per decade (GSIGMA); decline rate of de-carbonization per decade (DSIG); and quadratic term in de-carbonization (DSIG2).[4]

Three alternative rates of technological return to R&D are considered, an early, mid, and late return (figure 7.6). Figure 7.7 depicts the carbon intensity return to R&D of the three technology-led scenarios while figure 7.8 depicts the resulting declining CIO. We consider a successful technology-led policy to be one which achieves cumulative emissions consistent with a 50% probability of remaining below a 2°C increase in temperature (Meinshausen *et al.* 2009) such as in the scenarios of early and mid- R&D returns. The primary difference between the early and late returns cases is that the former reaches its peak acceleration of

carbon intensity decline at around 2035 while the latter reaches peak acceleration of carbon intensity decline only around 2060. In the case of the least successful R&D program, where R&D returns on carbon intensity reductions are delayed, only a 32% reduction of *cumulative* emissions is achieved from the B2 scenario for the 2010–2100 period vs. a 57% reduction in the successful case. Furthermore, the successful case presented here is by no means meant to be considered a best-case scenario, but rather a highly plausible one.

(2) Simulation of baselines and technology-led policies   Two alternative baselines are considered in order to assess the avoided damages of a

---

[4] Please contact the authors for specific parameter choices.

Table 7.4 Early return to R&D

2010–2110	Discount rate (%)	Damages avoided from baseline (NPV)[a]		R&D costs (NPV)	BCR by baseline (R&D/damages avoided)	
		B2	DICE[b]	Tech.-led	B2	DICE[c]
2010–2110	1.4	26.57	38.57	5.1	5.21	7.56
	3	7.75	11.27	3.1	2.5	3.64
	4	3.77	5.49	2.43	1.55	2.26
2010–2200	1.4	191.38	294.80	5.10	37.52	57.80
	3	24.00	36.15	3.1	7.74	11.66
	4	7.96	11.86	2.43	3.28	4.88

Notes:
[a] NPVs are expressed in $ trillion PPP.
[b] We would like to emphasize that we do not use the full power of the DICE model in that we are only using the climate damages calculated by DICE to obtain the BCRs.
[c] DICE BCRs are included to complement B2 BCRs given that DICE damages are used.

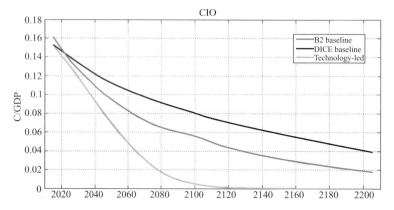

Figure 7.8 *CIO, by scenario*

technology-led policy: the IPCC B2 scenario, discussed in detail on p. 296 and the DICE "no-policy for 250 years" baseline (see figure 7.6).

The latter is included principally as a sensitivity analysis and to highlight the importance of baseline choice. Moreover, including the DICE baseline ensures that at least one of the baselines used is well calibrated to the model. Furthermore, as the B2 baseline is only available to 2100, B2 emissions are extrapolated out to 2200 (extended scenario), with a higher rate of decline in CIO than the 2010–2100 trend would imply so as not to favorably bias our BCRs. Although the B2 scenario is characterized by assumptions on GDP growth rates, technological change, sectorial change, etc., only the CO$_2$ intensity parameters are modified to simulate the B2 emissions path and thus obtain climate dam-

ages. Each of the three technology-led scenarios, the B2 extended scenario, and the DICE baseline are then simulated altering the already mentioned parameters in DICE. Finally, climate damages are read from the output.

(3) Benefit-cost ratios The BCRs of NPVs are calculated from the base year 2010 for the period 2010–2110, as well as for the period 2010–2200. Discount rates of 3% and 4% are used and, as a comparison, Stern's 1.4% discount rate. We extend to 2200 because, in the case of a technology-led policy, the major costs are incurred up-front, while the benefits are experienced for an extended period of time. The resulting BCRs are shown in table 7.4 (early return

Table 7.5 Mid-return to R&D

	Discount rate (%)	Damages avoided from baseline (NPV)		R&D costs (NPV)	BCR by baseline (R&D/damages avoided)	
		B2	DICE	Tech.-led	B2	DICE[c]
2010–2110	1.4	23.35	35.29	5.1	4.58	6.92
	3	6.74	10.25	3.1	**2.17**	**3.31**
	4	3.25	4.98	2.43	1.34	2.05
2010–2200	1.4	178.31	281.72	5.10	34.96	55.24
	3	21.78	33.93	3.1	**7.03**	**10.95**
	4	7.10	11.00	2.43	2.92	4.53

Table 7.6 Late return to R&D

	Discount rate (%)	Damages avoided from baseline (NPV)		R&D costs (NPV)	BCR by baseline (R&D/damages avoided)	
		B2	DICE	Tech.-led	B2	DICE[c]
2010–2110	1.4	12.28	24.22	5.1	2.41	4.75
	3	3.40	6.91	3.1	**1.10**	**2.23**
	4	1.59	3.31	2.43	0.65	1.36
2010–2200	1.4	128.83	232.25	5.10	25.26	45.55
	3	14.49	26.64	3.1	**4.67**	**8.59**
	4	4.39	8.29	2.43	1.81	3.42

to R&D), table 7.5 (mid-return to R&D), and table 7.6 (late return to R&D).

BCRs (at 3%) for the early return and mid-return to R&D range from 2.17 to 3.64 for the first ninety years and 7.03 to 11.66 for the entire 190-year period. The DICE baseline produces higher BCRs given its lower built-in rate of carbon intensity decline, and perhaps, a more suitable damage function calibration within the model.

The BCRs may reveal something about the private sector's willingness to invest in break-through technologies. Long-term investments of $100 billion/year are unlikely to appear as induced technological change. In order to induce the required amount of technological change, the price of carbon would need to be set far above the social cost for the return to basic innovation to become potentially profitable in the private sector.

Cumulative emissions analysis

Building on the notion that the global community appears determined to tackle the climate change challenge and the current discussion emphasizes limiting temperature change to 2°C, we propose an alternative approach to the CBA. We put forward the idea that the climate change debate has evolved beyond the point of whether a policy should be implemented to one which asks which policy is the most economically efficient and environmentally effective. Moreover, Allen et al. (2009) establish a fundamental notion of climate change on which our cumulative emissions analysis relies: peak increases in global temperature are dependent only on cumulative emissions, and not the emission path.

We therefore propose a CBA based on cost-effectively limiting cumulative emissions to those consistent with limiting temperature increase to 2°C. Meinshausen et al. (2009) provide us with the basis for this alternative approach to a CBA. Meinshausen et al. (2009) surveyed a large number of climate models to determine the cumulative emissions over 2007–50 that would be good predictors of whether the global temperature increase can be limited to no more than 2°C over the course of the twenty-first century. They provide cumulative global CO_2 emission budgets to 2050

Table 7.7 Cumulative emissions comparisons

	2010–2100	2010–50
B2	997	395
Allen *et al.* (2009) and Meinshausen *et al.* (2009)	572	328
Early R&D	464	337
Mid-R&D	520	350
Late R&D	727	388

Table 7.8 Cumulative emissions: a policy comparison

	"Brute force"	Technology-led
Prob. $\Delta T < 2°C$	50	50
Cumulative CO_2	1,203	1,203
Cumulative GtC eq	327.79	327.79
Increase in rate of decline of CIO (%) [(C/GDP)]	0	3.50
2050 emissions (GtC/year)	7.45	5.31
Growth rate GDP (%)	1.3	2.5
NPV GDP (3%) ($)	1,182.77	1,526.21
NPV GDP (4%) ($)	983.30	1,246.63
NPV GDP (5%) ($)	830.37	1,035.16

and the related probabilities of a temperature increase that exceeds 2°C (see also Schmidt and Archer 2009). Allen *et al.* (2009) go one step further than Meinshausen *et al.* (2009) and specify cumulative emission over an indefinite time period and their associated peak warming. Particularly, in order to maintain warming to below 2°C, Allen *et al.* (2009) find allowable emissions to be 2050–2100 Gt CO_2 (or 572 GtC). Table 7.7 contrasts B2 scenario emissions, for the periods 2010–50 and 2010–2100, with the cumulative emission allowances from Allen *et al.* (2009), Meinshausen *et al.* (2009), as well as the three technology-led proposals from the previous section.

The early and mid-technology-led proposals remain, easily, within the cumulative emission budgets for the period 2010–2100. For the period 2010–50 the technology-led proposals fall only slightly short of the Meinshausen *et al.* (2009) 2°C indicator targets, but given the rapidly declining rate of emissions in the early and mid-technology-led profiles, as they apply to 2050–2100, they are largely in line with the limits established by Allen *et al.* (2009).

To produce a benefit-cost ratio, we compare the cost-effectiveness of a technology-led policy with a "brute force" policy in achieving a $\Delta T < 2°C$ with 50% probability. Based on our analysis of achievable rates of decline in CIO, the 50% case may be the only one that is realizable given the considerable increases in the rate of de-carbonization required for the other, higher-probability, cases. In line with the Meinshausen *et al.* (2009) period, our CBA for the cumulative emissions case is limited to the forty-year period, 2010–50. In table 7.8, we determine the required rate of acceleration of de-carbonization, maintaining a constant growth rate of GDP that would be consistent with the

probabilities of limiting cumulative emissions to or below desired levels. For a 50% probability of $\Delta T < 2°C$, the emission budget for the period 2010–50 is limited to 327 GtC. Consequently, to maintain a 2.5% growth of GDP we require an acceleration of de-carbonization of 3.5 percentage points.

The findings of Meinshausen *et al.* (2009) are summarized in rows two and three of table 7.8. These rows indicate the CO_2 and carbon (C) emission budgets and their respective probabilities of limiting $\Delta T < 2°C$. The fourth row indicates the required annual rise in the RCIO which would maintain global emissions over 2010–50 within their total budgets. (Note that in the "brute force" case the zero value of the annual rise in the de-carbonization rate is not to be confused with a "frozen technology" baseline but rather a frozen rate of improvement in technology.) We then used the Kaya Identity (set out in the preceding section of the chapter) for a given emission budget to determine the required reduction in GDP growth consistent with a BAU rate of de-carbonization growth.

We then used the NPV of global world product for 2007–50 (the last three rows of table 7.8) to calculate BCRs for "brute force" and technology-led policies.

In the comparison of technology-led and "brute-force" mitigation policies, we need make no claim on the value of the damages avoided. In both cases, damages avoided are *assumed* comparable as we consider identical cumulative emissions for the same period (2010–50).

(a) Technology-led policy

(i) The benefits are the climate damages avoided out to 2050

(ii) The costs are the discounted value of $100 billion a year of R&D for forty years (2010–50), $2.31 trillion at a 5% discount rate

(iii) BCR = damages avoided/2.31.

(b) "Brute force" policy

(i) The benefits are the climate damages avoided out to 2050

(ii) The costs of a "brute force" policy are the lower PV of GDP: $1,035 – $830 = $205 trillion (at a 5% discount rate)

(iii) BCR = discounted values of damages avoided/cumulative R&D expenditures = $205 trillion/2.31 trillion (we then divided this result by 4).

One could argue that the BCR for the "brute force" case is a *worst*-case scenario (reflected in no change in the rate of decline of *C/GDP* from its "historic" 1.3% rate). At the same time, the results for the technology-led approach assume that the R&D will be sufficiently successful in the next two or three decades to achieve by 2050 the Meinshausen *et al.* (2009) 50% Prob $\Delta T < 2C$. Let us assume, then, that the technology-led BCR has been overstated four-fold. Even then, the comparison of the technology-led policy with the "brute force" policy produces a BCR of 22.

Comparison of technology-led and "brute force" mitigation policies

We also carried out a CBA comparing a technology-led policy with a policy to achieve emission reduction targets without consideration of whether the technologies were available to meet the targets at a reasonable cost. We have termed this "brute force" mitigation, and have elaborated on its implications earlier in the chapter.

We think it is important that some comparison be made between a technology-led policy and a policy that effectively calls for mitigation by "brute force." Recall that a policy is "brute force" if it requires polities to meet emission caps when the required energy technologies have not been developed. In these cases, it is necessary to either reduce the emission reduction goals or risk very large reductions in GDP (or GWP).

In previous sections, we noted several examples of proposed targets. Some were global, some for the USA. Some applied to 2020, others to 2050 or 2100. Each involved average annual reductions in the CIO (GDP or GWP) of at least 4.0% if the targets are to be met, and met via emission reductions at home. In contrast, the "historic" RCIO is 1.3% and that of the B2 scenario 1.4%. Thus the gap between the *required* increase in RCIO and the historic or BAU RCIOs is huge. (Note that the RCIO baseline is *not* a "frozen technology" baseline, but one based on historic RCIO or B2 levels – each of which includes substantial productivity and technological change. See the description of the B2 scenario in table 7.1, and the discussion of the RCIO.

In our CBA comparison between "brute force" and technology-led policies we need to make some assumptions about the response of RCIO in the case of "brute force" policies. How much could a "brute force" policy raise RCIO as a result of pressure to stay within or meet demanding emission caps? It is inconceivable to us that carbon pricing *alone* could do more than make a small dent in the gap between the "historic" RCIO (globally averaged at ~1.3 %/year) and the required average annual RCIO which, as the examples in the chapter indicate, are around, or in excess, of 4.0%/year.

So, how much might a "brute force" policy increase the decline in RCIO above the "historic" or BAU level? A generous estimate would be 50% by 2050 and perhaps 100% (at the outside) by 2100. But if that is all a "brute force" policy could do, either the emission reduction targets would be missed by a very wide mark, or the GWP growth rate would be reduced to zero or negative levels, as (4) on p. 320 implies.

Thus, in our CBA comparisons of a technology-led with a "brute force" policy, we decide to set the "brute force" average annual RCIO at 100% of "historic" levels (2010–50) to 2.6%, and at 150% of "historic" levels (2010–2100) to 3.3%. We regard these *as wildly and unrealistically high* in the absence of a technology-based climate policy. But they serve our purpose by *understating*, by a considerable margin, the mitigation costs of a "brute force" policy. (They also serve the purpose of providing a rough indicator of a "feasible brute force" policy, one in which the emission reduction targets are substantially lower and the RCIOs are more realistic.)

Assumptions of technology-led vs "brute force" B/C calculations

In making our calculations of the benefits and costs of a technology-led policy that is evaluated against a "brute force" mitigation baseline, the following assumptions or calculations apply (further details and explanations relating to these assumptions and the BCRs to follow are found in the appendix, p. 333):

(i) The "trend" or BAU rate of growth of GWP is 2.3%, 2010–2050; and 2.0%, 2010–2100

(ii) The targets for a "brute force" policy: *global* CO_2 emissions from energy-related sources are to be cut 50% by 2050, from current levels of 8 GtC, and by 80% by 2100. These cuts require average annual rates of decline in CO_2 emissions of *1.7%* (2010–50) and *1.8%* (2010–50).

(iii) From (4) above, the *required* rate of decline in the RCIO is 4.0% for 2010–50 (2.3% + 1.7%); and 3.8% for 2010–2100 (2.0% + 1.8%), if substantial reductions in GWP are to be avoided.

(iv) However, if the maximum achievable RCIOs are 2.6% (for 2010–50) and 3.3% (for 2010–2100) under a "brute force" policy, then a rearrangement of (4) implies that to achieve the emission reduction targets will require limiting the GWP growth rate of GWP to 0.9% (2010–50) and 1.5% (2010–2100).

$$\%\Delta\, GWP = \%\Delta C + \%\Delta\, (C/GWP), \quad (4)$$

where RCIO $= \%\Delta\, (C/GDP) = 0.9 = -1.7\% + 2.6\%$ (2010–50).
    $1.5\% = -1.8\% + 3.3\%$ (2010–2100)

(v) Avoidable climate damages not avoided: there will be some climate damages under either a "brute force" or a technology-led policy. Let us conservatively assume that (i) the "brute force" policy achieves its emission reduction target, albeit at the expense of

(large) reductions in the GWP growth rate; and (ii) the technology-led policy, because it works at emission reduction more slowly (at least initially), results in somewhat higher temperature and higher climate damages than does a "brute force" policy that is able to achieve the emission reduction target. The difference in damages between the two we term "avoidable climate damages not avoided."

(vi) Avoidable climate damages that are not avoided under a technology-led policy are assumed to be 2% of GWP. This figure is consistent with 1.5°C additional (2.5°C total) warming (see (4)).

(vii) R&D expenditures are $100 billion a year over 2010–2100. The marginal cost of public funds used to finance the R&D is 25%. (That is, the cost of R&D spending is grossed up by 1.25.)

(viii) The deployment costs of carbon emission-free (or carbon-neutral) energy technologies are 1% of GWP with a technology-led policy.

(ix) Discount rates are: 4.0, 3.0, or 1.4%. We use both "high" and "low" discount rates to sidestep a debate that threatens to turn the "economics of climate change" into a debate over the appropriate discount rate (Nordhaus 2007a, 2007b; Weitzman 2007, 2009; Heal 2009; Stern 2008).

The BCR formula

BCR = Mitigation costs avoided − Avoidable climate damages not avoided divided by R&D expenditures (1 + marginal cost of public funds) + deployment costs

The BCR estimates

We have not estimated a range of BCRs for the technology-led vs. "brute force" comparison. The reason is that the assumptions made in the analysis were favorable to the performance of the "brute force" policy (the high RCIOs and the assumption that the policy could actually achieve its emission-reduction target), and/or unfavorable to the technology-led policy (it would result in substantial climate damages as a percentage of GWP

Table 7.9 Technology-led vs. "Brute force" BCRs

Discount rate (%)	2010–50	2010–2100
4.0	15.7	10.0
3.0	16.7	**12.4**
1.4	18.0	16.5

relative to "brute force policy" and it incurs substantial deployment costs). *Thus, the BCRs presented in table 7.9 should be considered as the lower end of the range.* Alternative assumptions would have produced higher BCRs.

Of some note is the fact that the assumptions appear to have produced somewhat lower BCRs for 2010–2100 than for the shorter period 2010–50. This may tell us that using "brute force" policies to attempt to achieve a 50% reduction in *global* emissions by 2050 would be especially damaging relative to a technology-led policy.

Concluding Comments

We have employed three different ways of evaluating our technology-led proposal. The main estimates are those with a BAU baseline, using the DICE model to estimate damages avoided. The BCRs are, with one exception, greater than 1, and are larger for the 2010–2200 period than the 2010–2100 period. The two other means of evaluating the technology-led proposal, one based on cumulative emissions with a 2°C warming benchmark, and the other a comparison of a technology-led policy with a "brute force" mitigation policy, produce high BCRs. Overall, a successful technology-led policy appears "robust" in benefit-cost terms.

Some Concluding Thoughts

In our concluding section, we pose and then answer five questions. Then we summarize the main message of the chapter.

The Questions

1 Are there any parallels to the technology challenge posed by climate stabilization?
2 Why hasn't a technology-led policy been adopted?
3 Why will there be continued resistance to a technology-led climate policy?
4 Why is it likely a technology-led policy will *eventually* be adopted?
5 What is the relation between a technology-led climate policy and some and other climate-related policies?
6 What are the implications of "tipping points" and "catastrophic" climate change?

Some Answers to the Questions

Are there any parallels to the technology challenge posed by climate stabilization?

In their paper in *Nature*, Hoffert *et al.* (1998). concluded that "researching, developing, and commercializing carbon-free primary power . . . could require efforts . . . pursued with the urgency of the Manhattan Project or Apollo space program" (Hoffert *et al.* 1998: 884). In 2002, Hoffert *et al.* wrote that "combating global warming by radical restructuring of the global energy system could be the technology challenge of the century" (Hoffert *et al.* 2002: 372). Not all observers have found the Manhattan/Apollo analogs useful. Our view is closer to that of Hoffert *et al.* (2002), in that we can find no anolog in terms of time frame, required infrastructure change, and the physics of energy to the technology challenge posed by climate stabilization. The magnitude is so large and encompassing that the challenge of researching, developing, testing, and deploying a whole new energy system on a worldwide scale has no parallel.

Why hasn't a technology-led policy been adopted?

The answers to this question occupied parts of the second and third sections of the chapter. There we took up (and rejected) claims that: (i) the required technologies are available; and (ii) carbon-pricing will provide sufficiently strong inducements to technological change to assure that the required technologies become available without inordinate delay. Earlier, we took up another concern: the effectiveness of government-funded R&D. Because this is a legitimate concern that needs to be addressed, we set out some ways in which energy technology R&D can be made "incentive-compatible."

There has been another roadblock to a technology-led policy. The obsession with (emission reduction) targets puts the emphasis on emission reductions rather than on the technological means of achieving them. Further, date-specific targets are incompatible with a technology-based approach because the success of new technologies, much less the time at which success occurs, cannot be predicted ahead of time. In our view, more than any other factor, emission reduction targets have straitjacketed climate policy – and any discussion of alternatives (Prins and Rayner 2007). Target obsession has led to an insufficient focus on R&D-initiated technology change. As a result, potentially large emission reductions that could eventually follow successful technology breakthroughs are given up in a vain attempt to achieve near-term emission reduction certainty.

Why will there be continued resistance to a technology-led climate policy?

One reason is human obstinacy. Belief in the efficacy of emission reduction targets has not yet been dulled by the failure of commitments to them. Evidently, the response to the failure of the Kyoto Protocol to make any real difference to the course of global, or OECD, emissions is to call for even more demanding reductions and timetables. An irony is that whereas technology is about "success breeding success," a target-based climate policy reflects "failure breeding failure."

A new, and perhaps more compelling, factor has entered the picture. There is growing scientific evidence that climate is changing more quickly, and the probable impacts of change may be larger, than was contemplated a decade ago. The new evidence has fueled the argument that substantial emission reductions must begin now – and cannot await a technological revolution. Some argue that the government must put "science" ahead of "politics," even in the face of parliamentary or Congressional resistance. (Of course, in democracies, "science before politics" may not be possible, especially when the science is uncertain, the politics involve large costs, and those who

make the decisions are unlikely to survive the next election.)

The science-based compulsion for immediate action is increasingly tied to predictions that warming beyond some threshold will lead to increased probability of reaching "*tipping points.*" These may occur when beyond some atmospheric carbon concentration threshold, climate change, its impacts, and the damages it produces are predicted to accelerate. The policy implication is that mitigation of GHGs must proceed sufficiently quickly that low-probability events with big (undesirable) consequences do not become high- (or higher-) probability events. But the argument, while convincing in principle, runs up against the fact that the required pace of emission reduction may exceed what is technologically possible, and only can be achieved by *substantially reducing* economic growth or even reducing overall economic activity. In these circumstances, many are likely to conclude that the cure is worse than the disease, at which point "politics" will trump "science" in most democracies.

Still, those who admit to political realities, and the large economic costs of "brute force" mitigation, may feel uneasy about any policy which appears to lack accountability for emission reductions. A technology-led policy does not allow us to say when emissions will be reduced – because "when" depends heavily on the development of new technologies. Many may view any vagueness about the path of technology-led emission reductions as unacceptable. They may believe that a technology-led approach blurs responsibility for ultimate reductions to the point where the policy lacks the necessary degree of "accountability." The accountability issue may in fact be the basis for the enduring popularity of targets. Yet, there is nothing less accountable than politically grandstanded, non-credible, emission-reduction commitments, the responsibility for which lies well beyond the next election cycle.

Why is it likely that a technology-led policy will *eventually* be adopted?

The short answer is that there is no other choice. But that answer does not really address the question. To begin, the chapter has attempted to lay bare the huge costs of a target-based, "brute force" approach to mitigating emissions. Still it is likely to take more time for this realization to sink in. As a result, it may take another (wasted) decade of target-based policy, now with the USA involved, before failure is admitted, and another several years before an alternative with much greater chances of succeeding is adopted.

The delay is sad and unnecessary. There is, however, one silver lining: the growing appreciation that the technology challenge to climate stabilization is huge. That recognition is spurring several countries, including China, Japan, Korea, and the USA, to put considerable resources into energy technology R&D.

Sooner or later the proponents of the current target-led and carbon-pricing-based climate policy options will have to concede that putting the "cart" (large cuts in emissions) before the "horse" (the technological means for making the cuts) is a doomed approach. In the process of awakening, it will dawn that there is little logic in trying to price carbon emissions out of the market instead of developing good, cost-effective carbon-neutral alternatives that can be *priced in*. At the point of discovery, the world may finally turn to a technology-led climate policy.

The relationship between technology-led climate policy and some and other climate-related policies

While we think that the case for a technology-led climate policy is very strong, and made all the stronger by comparison with a policy of "brute force" mitigation, a technology-led policy cannot stand alone. We view a technology-led policy as at least partially (or potentially) *complementary* to other policies. These include: (i) adaptation (e.g. Pielke 2007); (ii) "alternative mitigation" policies; and (iii) investments in researching, and possibly testing, proposals for *geoengineering*. Each has a role to play in a portfolio of climate policies.

Both climate and technology change are imbued with uncertainty. As a result, the timing and extent of climate change, and the pace of tech-

nology development, not only make some adaptation inevitable but greater adaptation may be required if, as is widely anticipated, emissions overshoot the targeted stabilization level (Parry *et al.* 2009).

Mitigation is not limited to CO_2. Some examples of "alternative mitigation" are forest carbon sequestration and black carbon and CH_4 mitigation. Afforestation would reduce the approximately 20% of CO_2 emissions that occur as a result of changes in land use. Methane is a powerful GHG that is in good part associated with animal husbandry, agriculture (especially rice cultivation), and landfill. Black carbon is particularly associated with inefficient use of diesel fuel. If cost-effective means of reducing these sources of carbon are available then "alternative mitigation" is clearly both desirable and a *complement* to a technology-led climate policy.

There may also be need for a *palliative* in the event of rapid climate change. The role of the "palliative" would be to limit climate change while the technological means are developed to substantially reduce emissions (Barrett 2009). One category of "palliatives" is encompassed by the term "GE." Proposals to "geoengineer" climate (by reducing solar insolation with stratospheric aerosols or other reflective particles) are now being contemplated and researched (see, for example, Crutzen 2006; Wigley 2006; Matthews and Caldeira 2007 and chapter 1 in this volume), although one hopes, given uncertainty about effects and effectiveness, they will never have to be used.

What about mitigation policies that call for a price-, rather than a quantity-based, mechanism to cut emissions while placing emphasis on *upfront energy technology R&D*? Such an approach, under a gradually rising carbon price, would not be altogether dissimilar to our technology-led policy, the difference being more a matter of emphasis than of kind. That said, what is clear is that a technology-led policy is *incompatible* with a policy of "brute-force" mitigation, typified by demanding, time-specific emission reduction targets. After all, the case for a technology-led policy of mitigation is in large part the case against "brute force" mitigation. But "brute force" mitigation aside, we need an arsenal of policies, ones that are *complementary* (or similar in kind) to a technology-led policy.

Thus the *certainty* of some climate impacts will require investments in adaptation, the need for a portfolio of mitigation possibilities suggests the wisdom of "alternative mitigation," and the *possibility* of rapid climate change calls for researching, and possibly testing, proposed means of GE climate. Each should be considered, therefore, part of an *arsenal* of climate policies *with a technology-led policy at its center*.

What are the implications of "tipping points" and "catastrophic" climate change?

We return once more to the argument that the *science* of climate change indicates that the world cannot wait for the fruits of a technology-led policy to appear. To wait for technology, the argument goes, invites "disaster." A slower pace of emission reduction than the science "demands," so the argument goes, could mean huge damages, the possible transgressing of "tipping points," and the "catastrophic" changes that doing so might entail.

The first point we would note is that so far as "tipping points" and catastrophic climate change are concerned we are still in the domain of "low probability of high consequence" events. The second is that while the argument has some merit, especially in so far as discounting and the interpretation of cost-benefit analysis (CBA) is concerned (Weitzman 2007, 2009), it has less merit when the technology challenge to climate stabilization is considered.

A major problem is the rhetoric of the climate change debate. In the run-up to Copenhagen 2009, we heard much about "catastrophic" climate change and "saving the planet." We can do little about such "emotive" language, but would note the following. The debate is *not* about climate change and its scientific basis. These are firmly established. The debate is *not* about whether to act – virtually all are agreed we must act. The debate is about *how* to act: about what action is *appropriate*. What is appropriate is as much, or more, a matter of technology, behavior, economics, and politics as it is climate science. The technology-led proposal

made in this chapter is different from currently favored approaches precisely because it is driven neither by climate science nor by the axioms of economic theory. Rather, given what climate science is telling us, it is driven by our understanding of technology, behaviorally and institutionally based economics, and political and development realities.

The proposal is also driven by a sense that there are few more important things to human *survival* than *energy*. We obviously need air to breathe and water, but after these, virtually all of the requirements of life (including clean air and clean water) will depend in a highly populated world on abundant energy (Smalley 2005). And survival from real "catastrophes," such as an asteroid hitting the earth, the eruption of a "super-volcano," or a nuclear conflict, clearly will require all the *concentrated* energy at our disposal – and that currently means mainly fossil fuels and nuclear energy.

"Catastrophes" usually happen suddenly and unexpectedly. Response must be rapid with little or no time for "adaptation." In this respect, it is not clear under what circumstances climate change would be "catastrophic." But if we really do face "tipping points" and "catastrophic" climate change, we will still need all the energy – human and especially non-human – that we can marshall. And that brings us back to technology. Whatever we do to meet the climate change challenge we will need to maintain and improve the quantity and quality of the world's energy resources and technologies.

Summing Up

We will be brief. Our technology-led proposal is easy to describe. More complicated is why it is the best approach. Much of the chapter has been needed to demonstrate why the major alternative, and current favorite, will not work. Here we distill the main points:

(i) Human-induced climate change is a problem that, left unattended, will become more serious as the century progresses.
(ii) To substantially reduce global GHG emissions will require a technological revolution.

(iii) Our chapter demonstrates that:
  - The *magnitude* of the technology challenge is huge.
  - The required technologies are not ready – and many still require *basic* R&D.
  - A policy of *"brute force" mitigation* to meet arbitrary and time-specific emission reduction targets will not work. One cannot cap CO_2 emissions unless there are good, non-carbon-emitting energy and/or energy technology *substitutes*.
  - A policy that *sets aside targets* and puts the up-front emphasis on energy R&D, infrastructure, and deployment of ready technologies, is intuitively sensible and workable.
  - *Carbon pricing* has an important *ancillary* role to play – first as a means of long-term financing of energy R&D, technology testing, and energy infrastructure development and renewal, and second as a means of sending a "forward price signal" as an (initially) low carbon price (say $5.00/tCO_2$) slowly and steadily rises over time (doubling, say, every ten years).

(iv) Using a BAU baseline, the BCRs range from 1.1 to 11.66 with an outlier at 0.65 in the case of a low return to R&D and non-inclusion of damages avoided from 2100 to 2200. In benefit-cost terms, a technology-led policy dominates a policy of "brute force" mitigation, with BCRs ranging from 10–18 regardless of the assumed level of climate damages.

(v) A technology-led climate policy could generate an energy technology race that would challenge the creativity of the younger generations while minimizing sacrifice in lost economic activity or a weakened energy system. In contrast, "brute force" mitigation would require large sacrifices with no assurance of a stronger and more resilient energy system.

(vi) Although we have neither discussed nor placed a value on spillovers from energy technology R&D into non-energy uses, it is likely that an energy technology race could generate many external benefits and possibly prove to be as important as the contribution to reducing GHG emissions.

Appendix: Calculation of BCRs with a "Brute Force" Mitigation Baseline

Here we set out how we calculated the BCRs for the comparisons of technology-led and "brute force" mitigation policies.

Benefits of a Technology-Led Policy

Typically the "benefits" of a policy to abate GHG emissions are the damages that such action avoids (see Tol 2009 for an excellent assessment of the economic costs of climate change). But in our comparison of the two policies, the "benefits" of a technology-led policy are largely the abatement costs avoided by *"brute force"* mitigation.

Nevertheless, in calculating the "benefits" of a technology-led policy we need to take into consideration the possibility that the slower pace of emission reduction may lead to higher avoidable damages than under a "brute force" policy that achieved its targets, albeit at great economic cost. The avoidable damages that are not avoided by a technology-led policy should be netted out of the benefits arising from abatement costs avoided.

Costs of a Technology-Led Policy

The costs associated with a technology-led approach include the following:

  I. R&D expenditures
  II. Demonstration projects and testing
  III. Deployment costs
  IV. The value of other R&D "crowded out" by carbon emission-free energy R&D
  V. The marginal cost of public funds.

The first two components are largely unrelated to the growth in GWP, while the second two are likely to be a small percentage of GWP.

Benefit-Cost Ratios

The "benefits" (numerator) and costs (denominator) when evaluating a technology-led policy against a "brute force" baseline can be expressed as follows:

BCR = Mitigation costs avoided − Avoidable climate damages not avoided/R&D expenditures

(1 + marginal cost of public funds)

+ deployment costs

Calculation of Mitigation Costs Avoided

To calculate mitigation costs avoided, we: (i) rearrange (4) and (ii) use the concept of the *required* average annual rate of decline in C/GWP – the one that allows the emission reduction target to be met, given (that is, without reducing) the "trend" rate of growth in GWP:

$$\%\Delta GWP = -\%\Delta C - (-\%\Delta C/GWP), \quad (A.1)$$

where $\%\Delta$ is the average annual rate of change.

To the extent that the actual rate of decline in C/GWP falls short of the required rate of decline, the rate of growth of GWP must adjust downward, *assuming that brute force policy single-mindedly keeps to the emission reduction target*. The *mitigation costs avoided* term in the B/C formula are calculated as the discounted sum by which cumulative GWP is reduced as a result of the lower (than "trend") rate of its growth.

Calculation of Avoidable Climate Damages not Avoided

For our estimate of avoidable damages of a technology-led climate policy when compared with a "brute-force" mitigation policy, we use a simplified approach rather than taking estimates directly from the DICE model. The estimates for avoidable damages that we use in these calculations are, if anything, higher than those in DICE, and thus will lower our estimated BCRs.

Cline (1992) and Nordhaus (1992, 1994) adopted a simple but powerful representation of the aggregate damages from climate change. The function takes the following form:

$$D(t) = d_0[W(t)/S°]^\alpha, \quad (A.2)$$

where $D(t)$ is damages as a percentage of GDP; d_0 is damages as a percentage of GDP attributable to a doubling of the atmospheric concentration of CO_2eq (CO_2e); $W(t)$ is warming as a result of the increased atmospheric concentration of CO_2e. The magnitude of W depends on "climate sensitivity," S, the response of global average temperature to a *doubling* of atmospheric CO_2. The parameter value of S is highly uncertain. It is estimated to fall in the range of 1.5°C to 4.5°C+. $S°$ is the anticipated or expected average climate sensitivity, typically 2.5 or 3.0°C.

Although (A.2) provides an estimate of damages, it is not a measure of damages avoided by mitigation of GHG emissions. Since CO_2 (the major component in CO_2e) is already 40% above its preindustrial level, some damages from the resultant equilibrium warming (climate change) are now unavoidable. This problem has been dealt with in the past by arbitrarily assuming that some percentage of the damages, say 20% (Cline 1992), is *unavoidable*. However, this approach to bridging the gap between damages and estimates of the climate changes that are avoidable seems to us increasingly ad hoc. We therefore suggest a modified form of the damage function.

Unavoidable damages from climate change are assumed to be those associated with a build-up of atmospheric CO_2 from preindustrial levels to 400 ppm. All increases in CO_2 beyond 400 ppm are assumed to be avoidable, although this may be a stretch given that we already are at 386 ppm. Further it is assumed that the rise in CO_2 from preindustrial levels of ~275 ppm to 400 ppm will raise the *equilibrium* global average temperature by 1°C, with long-term damages equal to 1.0% of GWP. For climate change damages associated with increases in atmospheric CO_2e concentration *in excess of* 400 ppm, we suggest the following function and parameter values:

$$D'(t) = d'[W']^\alpha \quad (A.3)$$

where D' = damages as a percentage of GWP; d' = 1.0; W' = warming over and above the 1°C associated with CO_2e in excess of 400 ppm; and α = 1.5. Thus damages from additional warming of 1.5°C (over and above the initial 1°C associated with an atmospheric concentration of 400 ppm) would be

1.8% of GWP. For additional warming of 3°C and 5°C, additional damages would be 5.2% and 11.2% of GWP, respectively. These damage estimates do not include possible adjustments that might be made to take into consideration income inequalities and loss aversion (see Stern 2007, 2008).

Bibliography

Aldy, J. and W. Pizer, 2009: The competitiveness impacts of climate change mitigation policies, Pew Center on Global Climate Change

Allen, M.R. *et al.*, 2009: Warming caused by cumulative carbon emissions towards the trillionth tonne, *Nature* **458**(7242), 1163–6

Arrow, K.J., 1962: Economic welfare and the allocation of resources for invention, in *The Allocation of Economic Resources*, Stanford University, 609–26

Arthur, W.B., 1989: Competing technologies, increasing returns, and lock-in by historical events, *Economic Journal* **99**(394), 116–31

Baker, E. and K. Adu-Bonnah, 2008: Investment in risky R&D programs in the face of climate uncertainty, *Energy Economics* **30**(2), 465–86

Baker, E. *et al.*, 2008: Technical change and the marginal cost of abatement, *Energy Economics* **30**(6), 2799–2816

Baksi, S. and C. Green, 2007: Calculating economy-wide energy intensity decline rate: the role of sectoral output and energy shares, *Energy Policy* **35**(12), 6457–66

Barker, T., I. Bashmakov, A. Althari, M. Amann, L. Chifuentes, J. Drexhage, M. Duan, O. Edenhofer, B. Flannery, M. Grubb, M. Hoogwijk., F.I. Ibitoye, C.J. Jepma, W.A. Pizer, and K. Yamaji, 2007: Mitigation from a cross-sectional perspective, in *Climate Change 2007: Mitigation – Contribution of Working Group III to the Fourth Assessment Report of the Intergovernmental Panel on Climate Change*, B. Metz, O.R. Davidson, P.R. Bosch, R. Dave, and L.A. Meyer (eds.), Cambridge University Press, Cambridge

Barrett, S., 2009: The coming global climate technology revolution, *The Journal of Economic Perspectives* **23**, 53–75

Bataille, C., B. Dachis, and N. Rivers, 2009: *Pricing Greenhouse Gas Emissions: The Impact on Canada's Competitiveness*, C.D. Howe Institute, Toronto, Canada, February

Battelle Memorial Institute, 2001: *Global energy Technology Strategy: Addressing Climate Change*, Joint Global Change Research Institute, College Park, MD

Bernedes, G., 2002: Bioenergy and water – the implications of large-scale bioenergy production for water use and supply, *Global Environmental Change* **12**(4), 253–71

Blanford, G.J., 2009: R&D investment strategy for climate change, *Energy Economics* **31**(Supplement 1), S27–S36

Blanford, G.J., R.G. Richels, and T.F. Rutherford, 2009: Feasible climate targets: the roles of economic growth, coalition development and expectations, *Energy Economics* **31**, S82–S93

Blees, T., 2008: Prescription for the Planet, www.booksurge.com

Bosetti, V., C. Carraro, R. Duval, A. Sgobbi, and M. Tavoni, 2009: The role of R&D and technology diffusion in climate change mitigation: new perspectives using the WITCH model, OECD Economics Department Working Papers **664**, OECD, Paris

Bosetti, V. and M. Tavoni, 2009: Uncertain R&D, backstop technology and GHGs stabilization, *Energy Economics* **31**(Supplement 1), S18–S26

Brandt, L. and T. Rawski, 2008: *China's Great Economic Transformation*, Cambridge University Press, Cambridge

Breakthrough Institute, 2009: http://breakthrough. org

Caldeira, K. *et al.*, 2003: Climate sensitivity uncertainty and the need for energy without CO_2 emission, *Science* **299**(5615), 2052–4

Clarke, L., J. Weynant, and A. Birky, 2006: On the sources of technological change: assessing the evidence, *Energy Economics* **28**(5–6), 579–95

Cline, W.R., 1992: *The Economics of Climate Change*, Institute for International Economics, Washington, DC

  2004: Climate Change, in chapter of *Global Crises, Global Solutions*, (ed.) Bjorn Lomborg, Cambridge University Press, Cambridge

Crutzen, P., 2006: Albedo enhancement by stratospheric sulfur injections: a contribution to resolve a policy dilemma?, *Climatic Change* **77**(3), 211–20

Dasgupta, P. and J. Stightz, 1980: Uncertainty, industrial-structure, and the speed of R&D, *Bell Journal of Economics* **11**(1), 1–28

Denholm, P. and R.M. Margolis, 2007a: Evaluating the limits of solar photovoltaics (PV) in electric power systems utilizing energy storage and other enabling technologies, *Energy Policy* **35**(9), 4424–33

2007b: Evaluating the limits of solar photovoltaics (PV) in traditional electric power systems, *Energy Policy* **35**(5), 2852–61

Dosi, G., 1988: Sources, procedures, and microeconomic effects of innovation, *Journal of Economic Literature* **26**(3), 1120–71

Edenhofer, O., K. Lessmann, C. Kemfert, M. Grubb, and J. Kohler, 2006: Induced technological change: exploring its implications for the economics of atmospheric stabilization, *The Energy Journal*, Special Issue, 57–107

Edmonds, J.A., C. Green, and J. Clark, 2004: The value of energy technology in addressing climate change, in M. Thorning and A. Illaronov (eds.), *Climate Change Policy and Economic Growth: A Way Forward to Ensure Both*, International Council for Capital Investment and the Institute for Economic Analysis, Brussels

Edmonds, J.A. and S.J. Smith, 2006: The technology of two degrees, in *Avoiding Dangerous Climate Change* (eds.), H.J. Schnellnhuber *et al.*, Cambridge University Press, Cambridge, 385–92

Edmonds, J.A., M.A. Wise, J.J. Dooley, S.H. Kim, S.J. Smith, P.J. Runci, L.E. Clarke, E.L. Malone, and G.M. Stokes, 2007: *Global Energy Technology Strategy: Addressing Climate Change*, Global Energy Technology Strategy Program, Battelle Memorial Institute, College Park, MD

Farrell, A.E. *et al.* 2006: Ethanol can contribute to energy and environmental goals, *Science* **311**, 506–8

Fargione, J. *et al.*, 2008: Land clearing and the biofuel carbon debt, *Science* **319**(5867), 1235–8

Fisher, B.S., N. Nakicenovic, K. Alfsen, J. Corfee Morlot, F. de la Chesnaye, J.-Ch. Hourcade, K. Jiang, E. La Rovere, A. Matysek, A. Rana, K. Riahi, R. Richels, S. Rose, D. van Vuuren, and R. Warren, 2007: Issues related to mitigation in the long term context, in *Climate Change 2007: Mitigation* – Contribution of Working Group III to the Fourth Assessment Report of the Intergovernmental Panel on Climate Change, B. Metz, O.R. Davidson, P.R. Bosch, R. Dave, and L.A. Meyer (eds.), Cambridge University Press, Cambridge

Gerbens-Leenes, W. *et al.*, 2009: The water footprint of bioenergy, *Proceedings of the National Academy of Sciences* **106**(25), 10219–23

Green, C., 1994: The greenhouse effect and environmentally induced technological change, in Y. Shionoya and M. Perlman, *Innovations in Technology, Industries, and Institutions: Studies in Schumpeterian Perspectives*, University of Michigan Press, Ann Arbor, MI

Green, C., S. Baksi, and M. Dilmaghani, 2007: Challenges to a climate stabilizing energy future, *Energy Policy* **35**, 616–26

Green, C. and H.D. Lightfoot, 2002: Making climate stabilization easier than it will be: the report of IPCC WG III, *C^2GCR Quarterly*, 2002–1

Greene, D. *et al.*, 2009: Advanced technologies to achieve US energy goals for climate change and oil independence, Oak Ridge National Laboratory, TN

Hansen, J. *et al.*, 2005: Earth's energy imbalance: confirmation and implications, *Science* **308**(5727), 1431–5

Heal, G., 2009: Climate Economics: a meta-review and some suggestions for future research, *Review of Environmental Economic Policy* **3**(1), 4–21

Held, H. *et al.*, 2009: Efficient climate policies under technology and climate uncertainty, *Energy Economics* **31**(Supplement 1), S50–S61

Herzog, H., 2001: What future for carbon capture and sequestration?, *Environmental Science and Technology* **35**(7), 148–53

Hoffert, M.I. *et al.*, 1998: Energy implications of future stabilization of atmospheric CO_2 content, *Nature* **395**(6705), 881–4

2002: Advanced technology paths to global climate stability: energy for a greenhouse planet, *Science* **298**(5595), 981–7

Hoffmann, V.H., 2007: EU ETS and investment decisions: the case of the German electricity industry, *European Management Journal* **25**(6), 464–74

Intergovernmental Panel on Climate Change (IPCC), 2000: *Emissions Scenarios*, N. Nakicenovic and R. Swart (eds.), Cambridge University Press, Cambridge

2001: *Contribution of Working Group III, Climate Change 2001: Mitigation*, Cambridge University Press, Cambridge

2005: *Special Report, Carbon Dioxide Capture and Storage*, Cambridge University Press, Cambridge

2007a: *Working Group III, Climate Change 2007: Mitigation of Climate Change*, Cambridge University Press, Cambridge

2007b: *Fourth Assessment Report, Climate Change 2007b: Synthesis Report*, Summary for Policy Makers, Cambridge University Press, Cambridge

Jaffe, A.B. *et al.*, 2005: A tale of two market failures: technology and environmental policy, *Ecological Economics* **54**(2–3), 164–74

Johnson, D.G., 2000: Population, food, and knowledge, *American Economic Review* **90**(1), 1–14

Kemp, R., 1997: *Environmental Policy and Technical Change: A Comparison of the Technological Impact of Policy Instruments*, Edward Elgar, Cheltenham, UK

Lackner, K.S., 2003: A guide to CO_2 sequestration, *Science* **300**(5626), 1677–8

Lewis, N.S., 2007a: Toward cost-effective solar energy use, *Science* **315**(5813), 798–801

2007b: Powering the planet, *Engineering and Science* **2**, 13–23

Lightfoot, H.D. and C. Green, 2001: Energy intensity decline implications for stabilization of atmospheric CO_2, Centre for Climate and Global Change Research Report, **2001–7**, McGill University, October

2002: *An Assessment of IPCC Working Group III Findings of the Potential Contribution of Renewable Energies to Atmospheric Carbon Dioxide Stabilization*, Centre for Climate and Global Change Research, Report 2002–5, McGill University, November

Love, M., 2003: *Land Area and Storage Requirements for Wind and Solar Generation to Meet US Hourly Electric Demand*, University of Victoria, BC, Canada, MSc Thesis

Love, M., L. Pitt, T. Niet, and G. McLean, 2003: Utility scale renewable energy systems: spatial and storage requirements, Institute for Integrated Energy Systems, University of Victoria, BC, Canada

Lucas, R.E., 1993: Making a miracle, *Econometrica* **61**, 251–72

MacKay, D.J.C., 2009: *Sustainable Energy – Without the Hot Air*, UIT, Cambridge

Maskin, E.S., 2008: Mechanism design: how to implement social goals, *American Economic Review* **98**(3), 567–76

Massachusetts Institute of Technology (MIT), 2003: *The Future of Nuclear Power*, MIT Press, Cambridge, MA

2007: *The Future of Coal*, MIT Press, Cambridge, MA

Matthews, H.D. and K. Caldeira, 2007: Transient climate-carbon simulations of planetary geoengineering, *Proceedings of the National Academy of Sciences* **104**(24), 9949–54

Meinshausen, M. *et al.*, 2009: Greenhouse-gas emission targets for limiting global warming to 2degC, *Nature* **458**(7242), 1158–62

Metcalf, G.E., 2009: Market-based policy options to control US greenhouse gas emissions, *The Journal of Economic Perspectives* **23**, 5–27

Metz, B., P. Bosch, R. Dave, O. Davidson, and L. Meyer, 2007: *Climate Change 2007: Mitigation – Contribution of Working Group III to the Fourth Assessment Report of the Intergovernmental Panel On Climate Change*, Cambridge University Press, Cambridge

Metz, B., O. Davidson, R. Swart, and J. Pan, 2001: *Intergovernmental Panel on Climate Change, Third Assessment Report, Working Group III, Climate Change 2001: Mitigation*, Cambridge University Press, Cambridge

Montgomery, D. and A. Smith, 2007: Price, quantity and technology strategies for climate change Policy, in M.E. Schlesinger, H.S. Kheshgi, J. Smith, F.C. de la Chesnaye, J.M. Reilly, T. Wilson, and C. Kolstad (eds.), *Human-Induced Climate Change: An Interdisciplinary Assessment*, Cambridge University Press, Cambridge

National Round Table on Environment and Economy (NRTEE), 2009: *Achieving 2050: A Carbon Pricing Policy for Canada*, Ottawa

Naughton, B., 2007: *The Chinese Economy: Transitions and Growth*, MIT Press, Cambridge, MA

Nemet, G., 2009a: Perspective paper 7.2 in this chapter

2009b: Demand-pull, technology-push, and government-led incentives for non-incremental technical change, *Research Policy* **38**(5), 700–9

Nemet, G.F. and D.M. Kammen, 2007: US energy research and development: declining investment, increasing need, and the feasibility of expansion, *Energy Policy* **35**(1), 746–55

Nordhaus, W.D., 1992: An optimal transition path for controlling greenhouse gases, *Science* **258**(5086), 1315–9

1994: *Managing the Global Commons: The Economics of Climate Change*, MIT Press, Cambridge, MA

2002: Modeling induced innovation in climate change policy, in A. Grubler, N. Nakicenovic, and W.D. Nordhaus (eds.), *Modeling Induced Innovation in Climate Change Policy*, Resources for the Future Press, Washington, DC

2007a: Critical Assumptions in the Stern Review on Climate Change, *Science* **317**(5835), 201–2

2007b: A review of the Stern Review on the Economics of Climate Change, *Journal of Economic Literature* **45**(3), 686–702

2008: *A Question of Balance: Weighing the Options on Global Warming Policies*, Yale University Press, New Haven, CT

Pacala, S. and R. Socolow, 2004: Stabilization wedges: solving the climate problem for the next 50 years with current technologies, *Science* **305**(5686), 968–72

Parry, M. *et al.*, 2009: Overshoot, adapt and recover, *Nature* **458**(7242), 1102–3

Pielke, R.A., Jr., 2007: Future economic damages from tropical cyclones: sensitivities to societal and climate changes, *Philosophical Transactions of the Royal Society A – Mathematical, Physical, and Engineering Sciences* **2086**, 1–13

2009a: Mamizu climate policy: an evaluation of Japanese carbon emissions reduction targets, *Environmental Research Letters* **4**(4) –

2009b: The British Climate Change Act: a critical evaluation and proposed alternative approach, *Environmental Research Letters* **4**, 1748–57

2009c: An idealized assessment of the economics of air capture of carbon dioxide in mitigation policy, *Environmental Science & Policy* **12**(3), 216–25

Pielke, R.A., Jr., T.M.L. Wrigley and C. Green, 2008: Dangerous assumptions, *Nature* **452**(7187), 531–2

Pimentel, D. and T. Patzek, 2005: Ethanol production using corn, switchgrass, and wood: biodiesel production using soybean and sunflower, *Natural Resources Research* **14**(1), 65–76

Pimentel, D. *et al.*, 2009: Food versus biofuels: environmental and economic costs, *Human Ecology* **37**, 1–12

Popp, D., 2004: ENTICE: endogenous technological change in the DICE model of global warming, *Journal of Environmental Economics and Management* **48**(1), 742–68

Popp, D., R.G. Newell, and A.B. Jaffe, 2009: Energy, the environment, and technological change, NBER Working Paper **14832**

Prins, G. and S. Rayner, 2007: Time to ditch Kyoto, *Nature* **449**(7165), 973–5

Riahi, K., A. Grubler, and N. Nakicenovic, 2007: Scenarios of long-term socio-economic and environmental development under climate stabilization, *Technological Forecasting & Social Change* **74**, 887–935

Rosen, D. and T. Houser, 2007: China energy: a guide for the perplexed, Peterson Institute for International Economics, Washington, DC

Sandén, B.A. and C. Azar, 2005: Near-term technology policies for long-term climate targets – economy wide versus technology specific approaches, *Energy Policy* **33**(12), 1557–76

Schelling, T.C., 1992: Some economics of global warming, *The American Economic Review* **82**(1), 1–14

2005: What makes greenhouse sense?, *Indiana Law Review* **38**, 581–93

Scherer, F.M., 1996: Learning-by-doing and international trade in semiconductors, *Behavioral Norms, Technological Progress, and Economic Dynamics: Studies in Schumpeterian Economics* (eds.), E. Helmstadter and M. Perlman, University of Michigan Press, Ann Arbor, MI, 247–60

Schmidt, G. and D. Archer, 2009: Climate change: too much of a bad thing, *Nature* **458**(7242), 1117–18

Schumpeter, J.A., 1942: *Capitalism, Socialism and Democracy*, Harper & Bros, New York, 2nd edn., 1946

Scruggs, J. and P. Jacob, 2009: Harvesting ocean wave energy, *Science* **323**(5918), 1176–8

Searchinger, T. *et al.*, 2008: Use of US croplands for biofuels increases greenhouse gases through emissions from land-use change, *Science* **319**(5867), 1238–40

Searle, A.D., 1945: Productivity changes in selected wartime shipbuilding programs, *Monthly Labor Review* **61**, 1132–47

Smalley, R., 2005: Future global energy prosperity: the terawatt challenge, *Material Research Society (MRS) Bulletin* **30**, 412–17

Solomon, S. *et al.*, 2007: *Climate Change 2007: The Physical Science Basis – Contribution of Working Group I to the Fourth Assessment Report of the Intergovernmental panel on Climate Change*, Cambridge University Press

Stern, N., 2007: *The Economics of Climate Change: The Stern Review*, HM Treasury, London

2008: The Economics of Climate Change, *The American Economic Review* **98**(2), 1–37

Tol, R.S.J., 2009: The economic effects of climate change, *Journal of Economic Perspectives* **23**(2), 29–51

Tollefson, J., 2008: Nuclear fuel: keeping it civil, *Nature* **451**, 380–1

Weitzman, M.L., 2007: A review of the Stern Review on the Economics of Climate Change, *Journal of Economic Literature* **45**(3), 703–24

2009: On modeling and interpreting the economics of catastrophic climate change, *Review of Economics and Statistics* **91**(1), 1–19

Wigley, T.M.L., 2006: A combined mitigation/geoengineering approach to climate stabilization, *Science* **314**(5798), 452–4

Wigley, T.M.L. *et al.*, 1996: Economic and environmental choices in the stabilization of atmospheric CO_2 concentrations, *Nature* **379**(6562), 240–3

2007: Climate, energy and CO_2 stabilization, unpublished manuscript, March

Wise, M. *et al.*, 2009: Implications of limiting CO_2 concentrations for land use and energy, *Science* **324**(5931), 1183–6

Wright, T.P., 1936: Factors affecting the cost of airplanes, *Journal of Aeronautical Science* **2**, 122–8

Wright, B.D., 1983: The economics of invention incentives: patents, prizes, and research contracts, *The American Economic Review* **73**(4), 691–707

Yohe, G.W, R.J.S. Tol, R.G. Richels, and G.J. Blanford, 2009: Climate change, chapter 5 in B. Lomborg (ed.), *Global Crises, Global Solutions*, 2nd edn., Cambridge University Press, Cambridge

Technology-Led Climate Policy

Alternative Perspective

VALENTINA BOSETTI[*]

Introduction

This Perspective paper has a two-fold objective. The first is more general – that of commenting and shedding new light on the issue of R&D in energy technologies as a solution to climate change. The second is more specific – that of discussing the costs and benefits associated with R&D programs in a specific technology, carbon capture and storage[1] (CCS).

R&D in Energy Technologies and Climate Change

Much has been said on how to reduce current anthropogenic emissions with the aid of a portfolio of existing technologies. However, the stabilization of temperature to a safe level requires that over time net emissions fall to very low levels, if not to zero. There is only one way that this can be achieved in a manner that is acceptable to the majority of the world's citizens: through some kind of technological revolution. Extensive research and development (R&D) investments will be required to bring

about such a breakthrough. This will be specifically important for countries interested in maintaining both a leading position in climate negotiations and a first-mover advantage in earning the rents on innovation. Indeed, technological breakthroughs (and maybe, more importantly, the large-scale commercialization of these new technologies) will play an essential role in the competitiveness issue that has lately gained great relevance in the policy debate. On top of this, technological transfers to developing countries could be the key to solve the logjam affecting international negotiations. Innovation and technology treaties have been analyzed in the context of climate coalition formation, suggesting that they could improve the robustness of international agreements to control climate change (Barrett 2003; Burniaux *et al.* 2009; Hoel and de Zeeuw 2009).

While it is commonly agreed that we need extensive R&D efforts to reduce emissions in an efficient manner, less consensus characterizes the debate on whether relying on R&D policies alone might be sufficient to achieve the required reduction in emissions.

Many have argued that R&D policies alone will not be sufficient to achieve stringent targets and/or to minimize mitigation costs, because such an approach would provide no direct incentives for the adoption of new technologies and, by focusing on the long term, would miss near-term opportunities for cost-effective emissions reductions (Philibert 2003; Sandén and Azar 2005; Fisher and Newell 2007; Bosetti *et al.* 2009a).

Nonetheless, the argument that innovation and technology policies might be sufficient to solve the climate change problem has a strong appeal for policy makers (see, for example, the position of George W. Bush's second Administration on the role of technical change).[2] Some climate-related

[*] This chapter was written while the author was visiting fellow at the Princeton Environmental Institute (PEI), in the framework of cooperation between the EuroMediterranean Center on Climate Change (CMCC) and PEI. The hospitality and excellent working conditions offered there are gratefully acknowledged. The author also gratefully acknowledges useful comments from Massimo Tavoni and Shoibal Chakravarty. All usual disclaimers apply.

[1] When describing CO_2 in geological formations and oceans, the term "CO_2 storage" is used. It is now commonly accepted that the term CO_2 sequestration refers only to the terrestrial storage of CO_2.

[2] See for example the "Administration Actions to Advance Technologies for Addressing Global Climate Change," www.climatetechnology.gov/vision2005/, August 2005.

scientific and technology agreements have emerged, including the Carbon Sequestration Leadership Forum, the Asia Pacific Partnership (APP) on Clean Development and Climate, and the International Partnership for a Hydrogen Economy. Proposals of international technology agreements that would encompass domestic and international policies to foster R&D and knowledge-sharing, have been put forward (Newell 2008).

Recent empirical and numerical studies (see the next section) show that R&D investments, though essential to improve the efficiency of a climate policy, are typically induced by some carbon price signal. Conversely, stand-alone R&D policies will not be enough to produce the required halt in emissions. Revenues from the carbon policy (whether a tax or cap-and-trade system with fully or partly allocated permits) can be used to finance additional R&D investments, though the largest part of the investments will respond to the higher price of carbon.

Chapter 7 by Galiana and Green, on Technology-Led Climate Policy, rightly emphasizes this crucial role of R&D policies,[3] but at times underestimates the equally crucial role of carbon-pricing. Nonetheless, the chapter admits that R&D policy should be complementary to a carbon price policy. By arguing for a low carbon tax, doubling every ten years, to finance energy R&D spending, the chapter authors are supporting a mild environmental target (something in line with the stabilization of atmospheric CO_2 at 550 ppm levels[4]), a specific climate policy instrument, a carbon tax, and a specific recycling scheme.

On the one hand, depending on the assumptions concerning the discount rate, the magnitude of the damage, and the climate sensitivity parameter, one could agree on the optimality of a mild climate policy (although the underestimation of damages in earlier modeling exercises showed up in recent studies – as, for example, the revised estimates of sectoral impacts for the USA in Hanemann 2008). On the other hand, the whole discussion arguing against a cap-and-trade system is potentially misleading and, in the light of the EU trading system (ETS) and the Waxman–Markey (WM) proposal, basically unrealistic.

Although the basic message of chapter 7 is at times contradictory, it overall matches that of this Perspective paper on three basic issues:

1 R&D will be an essential part of any climate policy, independently of how stringent the optimal climate policy is believed to be
2 R&D policy alone will not do the trick, unless the goal is simply to diversify energy provision rather than significantly reduce emissions
3 When added to a carbon policy, a R&D policy (as, for example, an international fund for breakthrough technologies R&D) might lead to substantial efficiency gains and help to contain climate policy costs.

The next section will discuss these three points in detail.

The Large, but Limited, Power of Innovation Policies

The empirical analysis of the process of innovation is chiefly based on patent counts, employed to measure the output of innovation but also on the transfer of inventions across borders. One extensive study (Dechezleprêtre *et al.* 2008) shows how the Kyoto Protocol actually induced innovation. In particular, the increased innovation in carbon-free technologies that has taken place in Annex I countries that have ratified the Kyoto Protocol was not mirrored in Australia and in the USA. The link between environmental policy and induced innovation has been found in a large number of studies. The literature review on empirical studies in Vollebergh (2007) points to the clear impact, found across many studies, of environmental policy on invention, innovation, and diffusion of technologies.

---

[3] "Relying on carbon-pricing to cut global emissions substantially is neither likely to be politically acceptable nor economically time-consistent. Carbon pricing alone, or as the main policy tool, is not an effective means of inducing long-term commitments to undertake and pursue endemically uncertain (of success) basic R&D. But as we shall see, carbon-pricing has two important ancillary roles to play" (Galiana and Green 2009: 294).

[4] Throughout the Perspective paper, when discussing stabilization scenarios, I will be referring to ppm CO_2-only numbers.

Hence, the empirical evidence points towards the need for a climate policy to induce (and not only to finance) the required innovation. However, Dechezleprêtre *et al.* 2008) also find that there is no evidence that the Kyoto Protocol has increased the transfers or international spillovers of knowledge. Hence, there is room for improving the design of a climate policy by including some mechanisms to promote spillovers of knowledge (although this might be tricky as free-knowledge spillovers lower the rents on innovation and thus might discourage innovators). That technology transfers are a crucial point in negotiations is no big news, as manifested by the institution of an Expert Group on Technology Transfer within the UNFCCC framework.[5]

Many analysts have concluded that the current scale of energy R&D is inadequate for the climate challenge and propose more or less arbitrary increases to the level of effort. Both the USA and the European Commission envision large expansions of government energy R&D funding.[6] Nemet and Kammen (2007), claim that a five- to ten-fold increase in American energy R&D spending is both warranted and feasible. Using a rule of thumb, Stern (2007) recommends doubling all government energy R&D budgets.

Similarly, by using an Integrated Assessment model (IAM) with a fairly detailed description of endogenous technical change in the energy sector, Bosetti *et al.* 2009b, find that energy R&D is crucial if we aim to create a significant dent in carbon emissions. Investments in public energy R&D would need to return to at least the peak of the 1980s as a relative share of GDP. Expenditures should thus increase from today's 0.02% to 0.08% of world GDP, or equivalently from $8 billion to $40 billion. These extra investments should take place in the next twenty years, given the long lags that separate research from market adoption. In chapter 7, the authors look at different types of energy R&D, and find that public energy R&D should be targeted at innovative technologies that can contribute to the de-carbonization of energy indispensable for significant emissions cuts. The non-electric sector (transport above all) in particular needs breakthrough technologies that are not available today. The power sector needs innovation as well, but to a smaller extent. Only if the use of existing carbon-free technologies such as nuclear power, renewables, or CCS is limited by sociopolitical constraints is the development of alternative technologies necessary to prevent policy costs from increasing by up to 40%. Nonetheless, R&D may also contribute to improving the efficiency and safety of existing technologies.

In order to understand the potential benefit of R&D in breakthrough technologies one can estimate the additional cost of a climate policy assuming that no R&D program aimed at bringing down the cost of breakthrough technologies in both the electric and non-electric sectors is undertaken. As a result, the costs of these new technologies would remain as high as they are today in the coming years. Breakthrough technologies would become competitive twenty years later, without an R&D program, thus diffusion and learning-by-doing (LBD) mechanisms would be delayed as well. Table 7.1.1 reports figures relative to the increase in policy costs for two different policies – a mild climate target (550 CO_2 ppm) and a more stringent one (450 CO_2 ppm). In both cases, and

Table 7.1.1 Increase in climate policy costs without an R&D program aimed at breakthrough in low-carbon technologies, for two climate policy targets

	ppm	Discount rate	
		3% ($)	6% ($)
Increase in climate policy costs associated with the lack of a breakthrough R&D program. (Discounted trillion 2005 USD)	550	24	3
	450	63	9.5

[5] See for example the *Advance Report on Recommendations on Future Financing Options for Enhancing the Development, Deployment, Diffusion and Transfer of Technologies Under the Convention. Note by the Chair of the Expert Group on Technology Transfer. Subsidiary Body for Implementation*, Thirtieth session, Bonn, June 1–10, 2009.

[6] "National Commission on Energy Policy. 2004. Ending the Energy Stalemate: A Bipartisan Strategy to Meet America's Energy Challenges," National Commission on Energy Policy, Washington, DC, and "European Commission. 2009. Communication from the Commission to the Council, the European Parliament, the European Economic and Social Committee and the Committee of the Regions: Towards a Comprehensive Climate Change Agreement in Copenhagen," Section 3.3, European Commission, Brussels, 10.

independently of the discount rate, there is a sizeable increase in policy costs due to the lack of the induced breakthrough.

One should not forget that technological change is an uncertain phenomenon. In its most thriving form, ground-breaking innovation is so unpredictable that any attempt to model the uncertain processes that govern it is close to impossible. Despite the complexities, research dealing with long-term processes, such as climate change, largely benefits from incorporating the uncertainty of technological advance. Adu-Bonnah and Baker (2008), Blanford (2009), and Bosetti and Tavoni (2009), among others, model R&D as an uncertain phenomenon. Two of the main findings of this literature are: (i) that the optimal level of energy R&D investments should be higher in order to cope with climate change, if we acknowledge the uncertainty characterizing the innovation process; (ii) that a portfolio of technologies should be considered in order to hedge the risks of R&D program failures.

Additional evidence corroborating the call for R&D policies comes from the analysis of the international uneven distribution of R&D efforts and the recognition that social returns on R&D are higher than private ones. National and international R&D funds aiming to foster technology diffusion and to overcome the various innovation market failures, such as the underinvestment in R&D in the private sector, could be extremely beneficial. As investigated in Bosetti *et al.* (2009a), an R&D policy complementing a carbon policy could lead to visible efficiency gains, reducing policy costs by up to 10–15%.

But, however essential, R&D programs will not be sufficient. As underlined in Bosetti *et al.* (2009a), under fairly optimistic assumptions about the funding available for, and the returns to, R&D innovation policies alone cannot stabilize global concentration and temperature; a strong carbon price signal is indispensable. A very robust finding across a wide range of simulations is that the largest achievable reduction in cumulative emissions with respect to the baseline case is in the order of 13–16%. To put this in perspective, the reduction required to be consistent with a mild stabilization target (550 ppm CO_2) would be in the order of halving cumulative emissions.

Cost-Benefit Assessment of R&D in Carbon Storage as a Solution to Climate Change

We now shift the focus on a specific category of R&D, that is dedicated to the improvement in CO_2 CCS technologies. Among the many technologies available in the climate mitigation portfolio, CCS is considered central because it allows the continued use of fossil fuels while reducing the CO_2 emissions produced. CCS may therefore play an important role, especially in countries that heavily rely on coal for the generation of electricity, such as China and India. Low carbon electricity could also have an additional value if the de-carbonization of the transport sector follows an electrification path. CCS can also play a significant role in the event that a very stringent climate policy, such as that in line with a 2°C stabilization target, is enacted. Bioenergy coupled with CCS is the only way to obtain the negative emissions that might become unavoidable in the very long run.

On the other hand, unlike other technologies which present benefits unrelated to climate change (such as increasing energy security, decreasing local pollution, or producing electricity at lower cost), CCS is not meaningful outside the context of a climate policy, as it otherwise represents a decrease in plant efficiency and an increase in capital and operating expenses. In addition, CCS technologies present a whole set of non-technical difficulties, related to the long-term security of geological storage and social acceptance.

CO_2 is already being captured in the oil, gas, and chemical industries. Indeed several plants capture CO_2 from power station flue gases for use in the food industry.[7] However, only a fraction of the CO_2 in the flue gas stream is captured – to reduce emissions from a typical power plant by 75%, the equipment would need to be ten times larger. If capture is used to minimize CO_2 emissions from a power plant it would add at least 1.5 US ¢/kWh to the cost of electricity generation. In addition, the generating

---

[7] For more references on the technical description of CO_2 CCS and detailed information on current R&D programs, the reader is referred to the IEA Greenhouse Gas R&D Programme site, www.co2captureandstorage.info/.

efficiency would be reduced by 10–15 percentage points based on current technology. The widespread application of this technology is expected to result in developments leading to a considerable improvement in its performance. The cost of avoiding CO_2 emissions is 40–60 US$/ton of CO_2[8] (depending on the type of plant and where the CO_2 is stored), which is comparable to other means of achieving large reductions in emissions.

Having captured the CO_2 it would need to be stored securely for hundreds or even thousands of years, in order to prevent it from reaching the atmosphere. Major reservoirs, suitable for storage, have been identified under the earth's surface and in the oceans. Work to develop many of these options is still in progress.

As underlined in the IEA report on *CO$_2$ Capture and Storage* (IEA 2008), the next ten years will be critical for CCS development. By 2020, the implementation of at least twenty full-scale CCS projects in a variety of power and industrial sector settings, including coal-fired power plant retrofits, will considerably reduce the uncertainties related to the cost and reliability of CCS technologies. Given that the financial resources required to support these demonstration projects cannot be obtained from the market alone, one of the most crucial factors for the development of CCS technologies is the need for government finance to support the decisive early demonstration projects. Also, some additional effort by governments in designing adequate legal and regulatory frameworks is needed, as storage of CO_2 raises issues such as liability for CO_2 leakage and property rights. A similarly important endeavor will be needed to carry out a campaign to inform and raise public opinion awareness, as large-scale CCS might encounter strong public resistance. We refer the reader to IEA (2008) for a detailed description of R&D actually undertaken in OECD and fast-growing countries. Research projects currently in place range from the analysis of public acceptance,

to the availability of sites and the risks associated with CO_2 storage, to the optimal structure of the transport network.

Keeping in mind that the demonstration aspect is the top priority in preparing the avenue for large-scale deployment of CCS technologies, research investments, though secondary in this early stage, might play an important role later on. One important future breakthrough would, for example, concern the increase of the capture rate at a reasonable cost and with acceptable losses in plant efficiency.

Baker *et al*. (2009), focus on understanding how current investment in R&D has the potential to lower CCS costs forty–fifty years in the future. They perform an expert elicitation to identify areas where there is potential for significant progress or even breakthroughs and then to assess the probability of success and failure of R&D programs in these areas. Crucial areas of investigation are: pre-combustion carbon capture, alternative combustion, and post-combustion removal. They find that both post-combustion- and chemical looping-(alternative combustion) targeted R&D programs are characterized by serious disagreement over the probability of success. They also stress that the rationale of a large R&D investment in CCS technologies strongly depends on the likelihood of implementing CCS technologies at a large scale. Indeed, "if the likelihood of implementing CCS is not high, then it reduces the attractiveness of a broad R&D investment in this technology (and increases the importance of pursuing other lines of research)." The National Academy of Sciences (NAS) study on *Prospective Evaluation of Applied Energy Research and Development*[9] made a first attempt to assess this likelihood, but they recognize that it is a very complicated question as it involves technical issues about the viability and long-term security of geological storage plus a range of non-technical issues and social preferences.

Given the large sources of uncertainties we have discussed so far, concerning both the actual implementation of large CCS technologies and the probability of success of R&D programs, some heroic assumptions have to be made in order to evaluate the benefits and costs of R&D in CCS technologies as a solution to climate change. The basic idea of the exercise I have presented here is the following.

---

[8] It should be noted that the actual figure is uncertain and some sources talk about 100 US$/ton of CO_2.

[9] National Research Council, *Prospective Evaluation of Applied Energy Research and Development at DOE (Phase Two)*, The National Academies Press, Washington DC, 2007, www.nap.edu/catalog/11806.html.

Table 7.1.2 Technological parameters for traditional coal and IGCC–CCS power plants

	Investment costs (World average USD$_{2005}$/KW)	O&M World average USD$_{2005}$/KW	Fuel efficiency (%)	Load factor (%)	Plant lifetime (years)	Depreciation (%)
Coal	1,530	47	45	85	40	5.6
IGCC–CCS	3,170	47	40	85	40	5.6

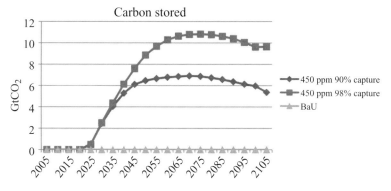

Figure 7.1.1 *Carbon stored*

Assuming that R&D investments can contribute to improving the capture rate of CCS technologies, it is possible to provide a rough estimate of the benefits associated with such an improvement in terms of decreased policy costs, and compare these with the potential costs of the R&D program.[10]

I use the WITCH model, an IAM first described in Bosetti *et al.* (2006), as it explicitly represents the optimal portfolio in energy technologies in the face of different climate policies. In WITCH, CCS can be applied to an integrated coal gasification combined cycle power plant (IGCC–CCS). IGCC–CCS competes with traditional coal, so that it replaces it for a sufficiently high carbon price signal. CCS transport and storage supply cost curves are region-specific and they have been calibrated following Hendriks *et al.* (2004). Costs increase exponentially with the capacity accumulated with this technology. The CO_2 capture rate is set at 90% and no after-storage leakage is considered. Other technological parameters such as efficiency, load factor, investment, and operation and maintenance (O&M) costs are described in table 7.1.2.

As CCS is not competitive in a baseline scenario, I will focus the investigation on two policy

scenarios where the objective is to stabilize CO_2 concentration at 450 ppm and 550 ppm levels by the end of the century, respectively. For each of the two policy scenarios I consider two cases: the basic case, where the capture rate is 90%, and a second case where, as a result of an R&D program, the capture rate is 98%, without any increase in electricity costs or efficiency loss. The effect of different capture rates on total stored carbon is significant, as shown in figure 7.1.1, for the 450 ppm policy case. During the second half of the century the climate target implies an increasing carbon price. The vented carbon that is not captured represents a cost for IGCC–CCS plants; hence, being able to reduce such a pricy by-product could decisively increase the potential of CCS technologies.

I compute the decrease in policy costs that would be associated with such a technological leap and use

[10] I have already discussed that what would be essential in the short run would be to concentrate the efforts more on the demonstration phase of innovation; however, computing the benefits and costs of direct investments in CCS would lose sight of the primary objective of this Perspective paper, so I concentrate more on the research part and on a longer-term vision of the problem.

Table 7.1.3 CBA of R&D in a CCS technologies program

	ppm	Discount Rate 3% ($)	6% ($)
Benefit as avoided policy costs (discounted USD trillion)	550	0.48	0.09
	450	0.92	0.20
Cost of R&D program (discounted USD trillion)		0.03	0.02

that as a measure of the benefit of a dedicated R&D program. Table 7.1.3 reports benefits, as decreased policy costs, for two discount rates and for two policies. By considering the two policy scenarios we mimic two cases, one where damages from climate change are higher (the 450 ppm stabilization case) and a second where climate change damages are lower (the 550 ppm stabilization case).

In order to provide an estimate of the R&D program costs, I assume that the expenditure on the R&D program on CCS is 10% of the overall energy R&D bill (which is endogenously calculated by the model) and that its duration spans 2010–45. Table 7.1.4 summarizes the BCRs.

The basic message that can be derived from this very preliminary analysis is that if we place some value on the reduction of the climate change threat then investing in an R&D program in CCS technologies passes the B/C tests.

Many simplifications are required to perform this analysis, hence results should be approached with due caution. In particular, it should be kept in mind that cost estimates are very rough as we assumed the probability of failure of the R&D program as equal to zero. However, the gap between benefits and costs is wide. To improve on this analysis, one should bear in mind the following caveats:

- Estimates do not take into account the additional benefits that result from these measures, such as the growth in markets, job creation, etc. On the

other hand, the extensive use of coal has many external costs – for example, those associated with mining – that we have not accounted for here.

- Institutional, legal, and social barriers can become a major issue in the large-scale deployment of CCS technologies. As we have seen, independently of the technological dimension, a large deployment of CCS might not take place.
- The analysis performed is deterministic. Baker *et al.* (2009) extensively discuss the uncertainties surrounding the effectiveness of such R&D programs. In order to diversify such risk, the portfolio of CCS R&D investments should cover different promising technologies, at least in the early stages.
- Deployment and demonstration projects are key to bringing about some reduction in costs; these are not considered in the present analysis.
- International spillovers of knowledge might speed up the breakthrough in capture technologies, thus lowering the actual costs of the R&D program, but they are not considered in the present analysis.

Conclusions

In July 2009, the G8 countries reiterated their commitment to take rapid and effective global action to combat climate change. The representatives of the largest developed economies recognized the need to set a 2°C limit to the increase in global average temperature above preindustrial levels. They also agreed that they would aim to reduce developed countries' emissions by 80% by 2050, and proposed a global objective of minus −50% by 2050.

Meeting these targets is going to require a monumental change in the energy system and in the whole economy, a change that only a series of technology revolutions can make possible. The question

Table 7.1.4 BCRs for R&D in a CCS technologies program

Discount rate	Low (3%)		High (6%)	
Climate change damage	Low (550 ppm)	High (450 ppm)	Low (550 ppm)	High (450 ppm)
BCR	16	30.7	4.5	10

then rests on whether technology-push or market-pull instruments will do the trick. In this Perspective paper I claim that both instruments will be required and that a hybrid policy will probably prove to be the most effective in both economic and environmental terms.

Induced and directly financed R&D investments should be diversified (over many technologies – such as solar, CCS, nuclear, etc. – and alternatives for each broad technology category as well – such as photovoltaic, solar thermal, etc.) – as only a portfolio of investments can hedge against the risks associated with the success of R&D programs. Innovation is highly uncertain and its dynamic poorly understood, and extensive efforts should thus be made to improve our understanding of how to measure and foster innovation.

Transport is the sector where carbon-free alternative technologies are the least competitive, therefore a large part of the R&D portfolio should be dedicated to existing promising technologies in order to cut the costs and start commercializing some of them.

CCS technologies could play a relevant role in the power sector. If electrification of the transport sector becomes one of the major responses to the quest for the de-carbonization of transport, then CCS could play an even larger role. Finally, if CCS technologies are coupled with biomass to produce both fuels and electricity, then CCS could have a crucial role in providing negative emissions as well. Assuming that all non-technical barriers to the large-scale diffusion of CCS technologies can be overcome, then investing in R&D in CCS technologies (as one of the options in a larger portfolio) would pass the B/C test.

The demonstration phase is now the top priority in preparing the avenue for large-scale deployment of CCS technologies; research investments to improve the capture rate and capture costs of CO_2, though secondary at this early stage, might play an important role later on.

Finally, one should keep in mind that stringent stabilization targets will require large shifts of investments in the energy sector and in the economy as a whole, figures which are an order of magnitude larger than R&D investments – a small, although important, portion of the overall picture.

Bibliography

Adu-Bonnah, K. and E. Baker, 2008: Investment in risky R&D programs in the face of climate uncertainty, *Energy Economics* **30**, 465–86

Baker, E., H. Chon, and J. Keisler, 2009: Carbon capture and storage: combining economic analysis with expert elicitations to inform climate policy, *Climatic Change*, Special Issue "The Economics of Climate Change: Targets and Technologies" **96**(3), 379–408

Barrett, S., 2003: *Environment and Statecraft*, Oxford University Press, Oxford

Blanford, G.J., 2009: R&D investment strategy for climate change, *Energy Economics* **31**(1), S27–S36

Bosetti V., C. Carraro, M. Galeotti, E. Massetti, and M. Tavoni, 2006: WITCH: a world induced technical change hybrid model, *The Energy Journal, Special Issue. Hybrid Modeling of Energy-Environment Policies: Reconciling Bottom-up and Top-down*, 13–38

Bosetti, V., C. Carraro, R. Duval, A. Sgobbi, and M. Tavoni, 2009a: The role of R&D and technology diffusion in climate change mitigation: new perspectives using the WITCH model, OECD Economics Department Working Paper, **664**, OECD, Paris

2009b: Optimal energy investment and R&D strategies to stabilise greenhouse gas atmospheric concentrations, *Resource and Energy Economics* **31**(2), 123–37

Bosetti, V. and M. Tavoni, 2009: Uncertain R&D, backstop technology and GHGs stabilization, *Energy Economics* **31**, 18–26

Burniaux, J.M., J. Chateau, R. Dellink, R. Duval, and S. Jamet, 2009: The economics of climate change mitigation: how to build the necessary global action in a cost-effective manner, OECD Economics Department Working Paper **701**, OECD, Paris

Dechezleprêtre *et al.*, 2008: Invention and transfer of climate change mitigation technologies on a global scale: a study drawing on patent data, www.cerna.ensmp.fr/index.php?option=com_content&task=view&id=192&Itemid=288, December

Fisher, C. and R.G. Newell, 2004: Environmental and technology policies for climate change and renewable energy, RFF Discussion Paper **04–05**, Resources for the Future, Washington, DC

Galiana, I. and C. Green, 2009: *Technology-Led Climate Policy*, chapter 7 in this volume

Hanemann, W.M., 2008: What is the cost of climate change?, CUDARE Working Paper **1027**, University of California, Berkeley, CA

Hendriks, C., W. Graus, and F. van Bergen, 2004: Global carbon dioxide storage potential and costs, **EEP-02001**, Ecofys, Utrecht

Hoel, M. and A. de Zeeuw, 2009: Can a focus on breakthrough technologies improve the performance of international environmental agreements?, NBER Working Paper **15043**, NBER, Cambridge, MA

International Energy Agency (IEA), 2008: *CO$_2$ Capture and Storage – A Key Carbon Abatement Option*, IEA, Paris

National Academy of Sciences (NAS), 2007: *Prospective Evaluation of Applied Energy Research and Development*, NAS, Washington, DC

Nemet, G.F. and D.M. Kammen, 2007: US energy research and development: declining investment, increasing need, and the feasibility of expansion, *Energy Policy* **35**(1), 746–55

Newell, R.G., 2008: International climate technology strategies, discussion Paper **08–12**, Harvard Project on International Climate Agreements, Cambridge, MA

Philibert, C., 2003: Technology innovation, development and diffusion, OECD and IEA Information Paper, **4**, International Energy Agency, Paris

Sandén, B. and C. Azar, 2005: Near-term technology policies for long-term climate targets – economy-wide versus technology specific approaches, *Energy Policy* **33**, 1557–76

Stern, N., 2007: *The Economics of Climate Change: The Stern Review*, Cambridge Press, Cambridge

Vollebergh, H.R.J., 2006: Differential impact of environmental policy instruments on technological change: a review of the empirical literature, Tinbergen Institute Discussion Papers **07–042/3**, Tinbergen Institute, University of Amsterdam

7.2 Technology-Led Climate Policy

Alternative Perspective

GREGORY NEMET

Introduction

This Perspective paper reviews cost-benefit (C/B) calculations on the effectiveness of energy research and development (R&D) as a means to mitigate climate change. It is generally supportive of claims that benefit-cost ratios (BCRs) are well above 1 and that these values are robust to the full range of assumptions about input values – particularly de-carbonization rates, discount rates, and the productivity of R&D. Special emphasis is placed on critically examining the argument that induced technological change (ITC) will enable adequate de-carbonization of the world economy. Weak ITC is probably the most important driver of the high BCRs found for energy R&D as a solution to climate change.

Chapter 7 by Galiana and Green clearly establishes: (1) the inexorable growth in demand for energy services over the current century, (2) the magnitude of the technological revolution required to address climate change, and (3) the inability, for various reasons, of *on-the-shelf* technologies to adequately fulfill the required technological change. This Perspective paper generally agrees with their conclusion that comparing a Technology-led policy to "brute force" mitigation produces BCRs well above 1. However, this Perspective paper makes several points that are central to their calculations, and to consideration of climate R&D in general, and require further elaboration:

1 A *carbon price signal* is insufficient to induce the technology development investments required to limit global temperature increase, for two reasons:
   – first, voters have a low willingness-to-pay (WTP) to avoid climatic damages and
   – second, knowledge spillovers make the private returns to R&D investments low.

2 The technology-led policy will shift the bulk of *technological decision making* from the private sector to the public sector; several challenges need to be resolved to achieve the BCRs described, including: reliance on fewer decision makers, institutional capacity, unstable social priorities, and risk aversion.

3 Full acknowledgment of the inherent stochasticity of the returns to R&D investments makes the "brute force" mitigation policy alternative best described as a *highly risky* choice, rather than dismissible as a futile one.

4 Collective action problems associated with international cooperation on R&D would produce unproductive *duplication of effort*, analogous to a patent race.

5 Mediation of *crowding out effects* could improve BCRs.

6 Alternative policies that involve low abatement costs and high climate-related damages are more likely in many large-emitter countries than is "brute force" mitigation. Consideration of *modest mitigation, modest technology investment* policies would clarify the crucial role of technology investment.

7 Incorporation of *health-related co-benefits* associated with changes to the energy system would produce substantially higher benefit values for all options that involve mitigation.

This Perspective paper describes these issues and comments on their implications for the BCRs estimated in the chapter.

Price Signals are Insufficient for Inducing Technology Investment

A central argument behind mitigation oriented approaches to climate change is that policy-driven

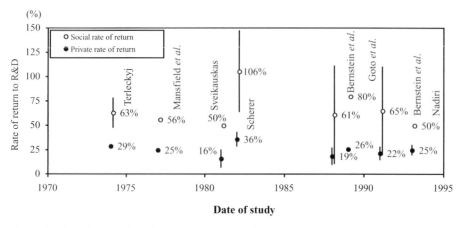

Figure 7.2.1 *Private and social rates of return to R&D*

changes in prices will stimulate development and adoption of improved low-carbon technologies. There are two distinct reasons why a greenhouse gas (GHG)-based price signal is insufficient to induce the required investments in technology development: (1) knowledge spillovers associated with technology development investments are high and (2) the public apparently has a low WTP for climate change mitigation. The scope of the changes to the energy system is indeed vast but in itself does not justify the technology-led policy; it is these two reasons that make carbon prices inadequate. There is a wide body of literature on ITC making the case that changes in input prices, and expectations about future markets, direct investments in innovation. Weaknesses in that argument provide the strongest justification for the technology-led policy described in the chapter.

Knowledge Spillovers Make Payoffs Too Low

Knowledge spillovers arise because firms underinvest relative to the socially optimal level of R&D (Nelson 1959; Arrow 1962; Teece 1986). Firms are unable to capture the full value of their investments in R&D because a portion of the outcomes of R&D efforts "spills over" to other parties as freely available knowledge – e.g. other firms can reverse engineer new products (Griliches 1992). Jones and Williams (1998) found that the social rate of return to R&D is four times larger than the private rate of return. Okubo *et al.* (2006), in an effort to estimate the macroeconomic asset value of R&D expenditure, surveyed previous work comparing the social and private rates of return to R&D. In figure 7.2.1, I display the data in the surveyed studies – (Terleckyj 1974; Mansfield *et al.* 1977; Sveikauskas 1981; Scherer 1982; Bernstein and Nadiri 1988, 1991; Goto and Suzuki 1989; Nadiri 1993) – to show that the public rate of return consistently exceeds the private rate of return and to show the dispersion in estimates. The average private return to R&D across these studies is 25% whereas the public return is 66%. While spillovers, *per se*, are beneficial since they expand access to the outcomes of R&D efforts, inappropriability prevents firms from receiving the full incentive to innovate and thus discourages them from investing as much in R&D as they otherwise would.

The inability of firms to appropriate the returns to their investments in innovation is an even more severe problem for early-stage technologies, such as would be necessary to catalyze the energy technology revolution called for in the chapter. Much of the technical progress in early stages easily becomes shared knowledge, is difficult for inventors to patent or easy for others to patent around, and is less amenable to becoming embodied in physical devices and manufacturing equipment and processes. Because knowledge spills over, *price signals alone*, even in combination with strong intellectual property protection, fail to provide sufficient incentives for private sectors to invest in

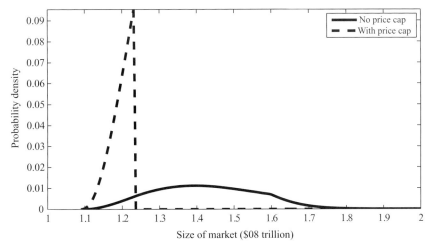

Figure 7.2.2 *Size of a market for a low-carbon energy technology with and without carbon price caps*

developing the technologies needed to transform the energy sector.

Willingness-to-Pay for Avoided Climatic Damages is Too Low

A second deficiency of the argument that mitigation policy will induce technological change is the low likelihood of carbon prices high enough to provide incentives for the required investments. Uncertainty in expectations about future policies increases the risk in investing in low-carbon energy technologies. If expectations about the level – or existence – of mitigation policies several years in the future are uncertain, then firms will discount the payoffs resulting from these future policies and underinvest in innovation.

Firms rationally discount proposed policies and resulting prices. A dominant characteristic of public policy related to energy over the past four decades has been volatility; targets are set and are changed; the electoral cycle reshuffles supportive political coalitions. Even for policies that involve "long-term" targets, out-clauses, and options for non-compliance undermine longer-term incentives in the name of "flexibility" and "cost containment." For example, "safety valves" included in an array of proposed GHG reduction policies impose price caps on carbon prices, thereby limiting the payoffs

to investments in innovation. Figure 7.2.2 shows an example of the reduction in payoffs for a hypothetical technology investment imposed by proposed legislation in the US Congress (Nemet 2010). It shows a probability density function (PDF) showing the size of the market for a zero-carbon technology (trillion current dollars) assuming a distribution of possible future carbon prices. The solid line shows the PDF of market size when no price cap is in place and the dashed line shows the PDF of market size with a price cap in place at $29/tCO_2.

The low expected likelihood of high carbon prices is often attributed to "political infeasibility." While this assessment is probably accurate, it is perhaps more helpful to consider the source of this infeasibility. There are two likely candidates, both of which exist for the same reason: the public has a low WTP to avoid climate damages.

One source of "infeasibility" is that the public, while supportive of climate policy in general, is simply not willing to pay more than a small premium on their energy consumption; in the USA something in the range of a 10–15% increase appears tolerable. While there is a dearth of work in estimating this parameter, WTP is almost certainly far less than the costs that would be imposed under the "brute force" mitigation strategy discussed in the chapter. It is also well below the marginal climatic damages of future emissions, estimated as

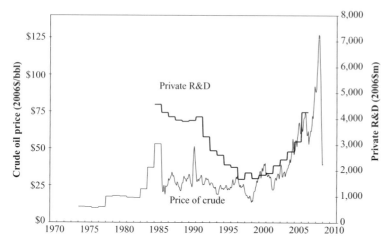

Figure 7.2.3 *US private sector energy R&D and crude oil prices*

somewhere in the range of $10–30/tCO₂ (Tol 2009). In a contingent valuation study of willingness of US residents to pay for the Kyoto Protocol, Berrens *et al.* (2004) estimate that US households valued the benefits at just under $191/household/year. With average household CO_2 emissions of approximately 50tCO₂/household/year, WTP appears to lie in the mid-single digits of $/tCO₂. $5/ton of CO_2 is far below the price level needed to catalyze the technology investments required to achieve climate stabilization; it is, however, well aligned with the technology-led policy the chapter authors recommend.

A survey-based contingent valuation study in the USA found WTP for energy R&D of $137/household, which amounts to about $16 billion/year nationally (Li *et al.* 2009). This amount is about one-sixth of the amount the chapter's authors recommend for worldwide R&D in the technology-led strategy. If national R&D contributions in the proposal are prorated based on current national GDP, US WTP for energy R&D of $16 billion/year is not far from the required proportional contribution of $24 billion/year. *The technology-led policy fits much more closely to the preferences of voters than does "brute force" mitigation.*

A second source of political infeasibility is that the incidence of carbon-pricing is likely to be concentrated among a small group of industries with such large economies of scale that they will be especially vulnerable to carbon-pricing. As a result, these firms are able to wield an influence in negotiations over legislation that is disproportionate even to their large size. The best way to accommodate the concerns of these influential firms is simply to compensate them for the cost of making the transition from a world of free emissions to one of costly emissions (Bovenberg *et al.* 2008). However, the WTP described above severely limits the feasibility of shifting abatement costs from carbon-intensive firms to consumers.

Finally, consider the data that show US private energy R&D's responsiveness to price signals (figure 7.2.3). Crucially, these are price signals *not* associated with political decisions to raise prices to capture externalities; rather, they result from transactions in the global oil market. While there appears to be some price elasticity of energy R&D investment, the level of investment ($2–4 billion/year) is strikingly low compared with the technology-led policy's proposed government funding of $100 billion/year. For example, the doubling of private sector energy R&D from 2000 to 2006 was associated with a tripling of world oil prices. The CO_2 price required to raise the cost of oil from $25/bbl to $75/bbl would be roughly $115/tCO₂, which is about twenty times the WTP described above. Inducing private sector R&D sufficient to fund the development of low-cost low-carbon technologies appears an unlikely prospect.

In short, price signals are insufficient because knowledge spillovers make the payoffs to investors too small and because low WTP by the public will keep prices low. This combination makes the technology-led approach essential. The small set of contingent valuation surveys that exists suggest that the political feasibility of the technology-led approach is far greater than the "brute force" mitigation approach.

Shifting Technological Decision Making to the Public Sector

The discussion in the previous section supports the chapter's claim that direct government support of technology investment is needed. It must be clear, however, that there are important implications in shifting technological decision making from the private to the public sector. These implications need to be addressed in program design. If they are not, the BCRs are likely to overstate the advantages of the technology-led strategy.

Centralized Decision Making

An important advantage of a price-induced technology strategy – such as "brute force" mitigation – is that decision making about technology development is dispersed among a large set of actors. The thousands of important decisions related to the funding, continuation, and abandonment of technology development would occur among actors that presumably should be able to incorporate vast amounts of information obtained from diverse sources. In contrast, direct involvement by governments in supporting technology development necessarily shifts a substantial portion of decision making to the government itself. A much smaller group of individuals will be involved in the vetting of technology decisions. They will be challenged with assimilating large amounts of costly information of varying reliability about the ultimate prospects for promising technologies.

Examples of difficult choices that will be increasingly made by the public sector instead of the private sector include:

- Assessing technical viability and market acceptability at early stages
- Determining when to switch from exploring alternatives to focusing resources on individual technologies and initiating demonstration and deployment
- Diversifying technology investments – especially in the early-to-middle stages of the development process
- Canceling unpromising development programs before they become expensive
- Assessing the critical scale for the research program, to avoid overdiversification by funding too many programs at low levels
- Enabling intertechnology knowledge flows, by supporting collaboration, incorporating new knowledge from outside existing R&D programs, and dispersing knowledge to other programs.

All of these decisions will need to be made amid interest group pressure and inevitable competing social priorities.

The challenge is to preserve some aspects of private sector decision making within the public sector. Changes to the intellectual property system, such as adjusting patent length and breadth, provide one avenue to incorporate private sector knowledge. Prizes that allow flexibility in deciding means to achieve government-prescribed technological ends are another. Establishing industry consortia, such as Sematech, as well as R&D subsidies for private sector research are yet another. Since the BCRs are highly sensitive to assumptions about the outcomes of the proposed technology program, a program design that enables governments to manage difficult technical decisions is essential to achieving the high B/C values estimated in the chapter.

Expertise and Institutional Capacity

A consequence of the shift in the loci of decision making is that governments themselves will require substantial increases in their capacities to make decisions about nascent technologies. Governments will need to get smarter. They can draw on private sector knowledge but ultimately, if energy-related R&D is to increase by a factor of ten, the intellectual capacity within governments will have to increase as well. The notion that "governments should not pick winners" is typically used to denigrate the government's ability to participate in technology decisions. And examples of poor picking abound (Cohen and Noll 1991). But if one accepts the arguments for a technology-led policy made in the chapter and in the previous section of this Perspective paper, then the suggestion provides little normative guidance for policy makers. As a result, governments need to improve their ability to "pick" and the option to abdicate responsibility for doing so will not be viable.

Vulnerability to "Pork," Linked Issues, and Shifting Social Priorities

Large government R&D programs, especially in the energy sector, are notoriously vulnerable to political vagaries that are unrelated to the objective of the programs themselves. The authors make reference to an important concept – that strict emissions limits are unlikely to stimulate low-carbon investment because governments cannot credibly commit future administrations to strict adherence to costly climate policies. But why would this time-consistency problem not also exist for an R&D program that involves hundred of billions of dollars to be invested over forty years with the same international collective action problems as in mitigation? A large R&D program will need to address these issues – especially in the context of the history of volatility in energy R&D spending, lack of successful experience in international technology development cooperation, international knowledge spillover problems, and government budget reviews that typically treat technical failures in R&D programs as evidence of poor resource allocation. Such a program would almost certainly come under budgetary pressures in the face of inevitable competing social priorities over four decades.

Performance Management and Risk Aversion in Governments

Governments are often assumed to have longer time frames and more concern for social welfare than private firms, which in part leads to them employ social discount rates that are typically less than half the private discount rates used by firms. Yet governments increasingly adopt performance management techniques that reward measurable outcomes over discrete time periods. As a result, governments may actually find it *more* difficult to tolerate the inevitable technical dead-ends that will result from such a large R&D endeavor than would the private sector. Tolerance of many small failures in the effort to produce a few large successes is a hallmark of innovation and has been perhaps most successfully employed by the venture capital industry. Governments will have to change in order to persevere with large investments in technology development in the face of inevitable failures. There will be failures; the technology-led policy depends on the ability of governments not only to tolerate them, but also to learn from them.

The Case for a Small Price Signal

Another reason for implementing a low CO_2 price – rather than none at all – is the need to create an initial market for these technologies, feedback from the market, and selection mechanisms for which of the outputs from the R&D program are most promising. It is clear that a positive feedback exists between R&D and deployment. Knowledge is gained through the experiences of producers and users through learning-by-doing (LBD) and learning-by-using; this feedback informs the direction of the R&D program. Pursuing an R&D strategy and a modest market creation strategy simultaneously allows connection of technical opportunities (from R&D) and market opportunities (from demand). This feature allows some of the decision making to be done by the private sector, especially for later-stage technologies.

An important conclusion underlying this proposal is that the sum of the problems arising from shifting a substantial portion of technological decision making from the private to the public sector amounts to less of a concern than does dependence on ITC in response to carbon price signals. Still the BCRs presented in the chapter are sensitive to the outcomes of the proposed R&D programs, which in turn depend on governments resolving the challenges described above. The extent to which one considers the BCRs reliable depends in part on the ability to mediate the problems associated with the shift in decision making from the private to the public sector.

Returns to R&D are Stochastic

As in nearly every study that compares government R&D spending to other policy options, the attractiveness of R&D ultimately hinges on the expected returns to R&D investment.

The BCRs presented in the chapter depend on the investment of $100 billion/year successfully delivering low-cost low-carbon technologies. An investment of $100 billion/year in R&D over forty years amounts to $3 trillion in present value terms. That investment allows the deployment of low-carbon energy technologies sufficient to achieve the 2100 target with no impact on GWP and only a 1% of GWP extra cost for deployment on top of a carbon price of $80/tCO$_2$ by the mid-century.

The level of R&D investment seems reasonable given previous work on this issue. The deterministic relationship between R&D and deployment costs as well as emissions is, however, concerning; even at $100 billion/year the program might not succeed in producing adequate technologies. The case for R&D would be much easier to make if the probability of success were 100% – but even at such large amounts, it is surely not. The authors' sensitivity analysis allays some of the concern about assumptions, but not entirely. First, the results show that the claim for BCRs that are much greater than 1 is not entirely robust to the three assumptions on the timing of R&D returns and discount rates. Second, there really is not much empirical or theoretical evidence for the assumed acceleration of de-carbonization due to the R&D investment. The authors have little choice in developing BCRs for the R&D option: still, reliability of the results is an issue. Third, it is not clear that the sensitivity analysis, which consists of three assumptions of the timing of R&D returns, adequately spans the full range of possible outcomes of the R&D program.

Three assumptions are most important in evaluating the reliability of the BCRs presented in the chapter:

- *Rate of de-carbonization*: one of the most important assertions in this cost-benefit analysis (CBA) is that global twenty-first century de-carbonization needs to be −4.0%/year and that, even in a most favorable case under "brute force," will not exceed −3.3%/year. While the authors make a strong case for the former, the reliability of the second assumption is much more difficult to ascertain. This estimate is crucial since the BCR results are dominated by the GWP loss that directly results from the gap between these two figures. The BCRs depend on the extent to which this −3.3% de-carbonization limit is a lower bound on how much the world economy can de-carbonize under climate policy.
- *Mitigation costs avoided*: it comes across clearly that this value dominates the BCRs reported since it is an order of magnitude larger than the other three. Also, as mentioned above, this value depends directly on the expected rate of de-carbonization under "brute force" mitigation. As suggested above, the size of this value also seems sensitive to the assumption that the de-carbonization shortfall gets expressed as a GWP loss rather than an excess of emissions
- *Climate damages*: since the authors' BCRs depend on the value of climate damages (S) and the timing of them, it is not obvious that BCRs remain well above 1 at all levels of climate sensitivity.

The inherent stochastic aspect of R&D investments implies that one should at least acknowledge the presumably low, but non-negligible, probability that the technology-led strategy may fail to deliver the necessary technologies – not just that they are delayed. Conversely, there must be some probability that the "brute force" mitigation strategy succeeds in achieving sufficient de-carbonization. In short, the chapter's policy conclusions would be more convincing if it discussed "brute force" mitigation as a highly risky strategy rather than dismissing it as a futile one.

International Cooperation on Technology Development

The collective action problems that appear to paralyze global cooperation on emissions reductions also exist in the technology-led policy. The best case made for the policy in this regard is that the investments at risk of free-ridership are smaller. Still, international cooperation on technology development has very little precedent. A likely result is that investment strategy will be competitive rather than cooperative. Competitive R&D development will increase the BCR to the extent that national-level decision making is superior to coordinated decision making, and will decrease it to the extent that it leads to technology races and duplication of effort.

Mediation of Crowding Out Effects

The chapter rightly acknowledges the issue of crowding out effects. While some previous analyses see this as a central problem for any R&D program (Goolsbee 1998) others find mixed results when surveying empirical work (David *et al.* 2000). The chapter authors point out that crowding out is not a serious issue at present – certainly not when less than $12 billion/year is spent on energy-related R&D worldwide. But at a proposed $100 billion/year, this program would constitute about 12% of current global R&D across all sectors. At that level there would likely be some economic cost to this redeployment of scientific and engineering talent away from other productive ends. Any crowding out above 0% would have a negative effect on GWP and would decrease the BCRs.

The authors point to the *supply* of scientific and engineering talent as a reason to expect low crowding out effects. Rapid economic development in East and South Asia provides one avenue for mediation of crowding out. This reason, however, assumes that opportunities for technical advance in non-energy fields grow more slowly than does education. A more purposive means with which to remedy crowding out is to increase the supply directly – by devoting a portion of the technology-led strategy to education, or perhaps by enlarging the program. This plan would raise the cost of the program but would reduce the adverse GWP impact described above.

Modest Mitigation, Modest Technology Investment

The Expert Panel should consider the BCRs in the light of alternative policies, not discussed in the chapter, that involve *lower* abatement costs and *higher* climate-related damages. The authors assert strongly, throughout the chapter, that "brute force" mitigation is the most likely policy direction at present, and thus deserves to be the basis for B/C comparisons. The CBA shows the technology-led strategy to be superior to "brute force" mitigation, a result that is robust to a large range of assumptions.

"Brute force" is a seriously considered option in a few countries, mainly in Europe. But important emitter countries such as China, the USA, Canada, Australia, and perhaps even Japan are far more likely to proceed along a path of what the authors at one point call "feasible brute force." We might call this path, *modest mitigation, modest technology investment* – small near-term emissions reductions combined with limited technology development investment. An example might be the legislation passed in the US House of Representatives (HR 2454, The American Clean Energy and Security Act) that includes soft emissions reductions targets and a technology-funding component that amounts to approximately 1% of the level proposed in the chapter's technology-led strategy. HR 2454 is a modest policy: modest mitigation, modest technology investment.

If the marginal cost of abatement is greater than marginal climate damage costs then why shouldn't governments just exceed the emissions limits? Is part of the reason that the BCRs are so high in the technology-led vs. the "brute force" mitigation scenario due to the assumption that governments are strictly unwilling to exceed their emissions targets? In BCR terms modest mitigation, modest technology investment – as compared with "brute force" mitigation – would have: lower mitigation costs, higher climate damages, lower R&D costs, and lower deployment costs.

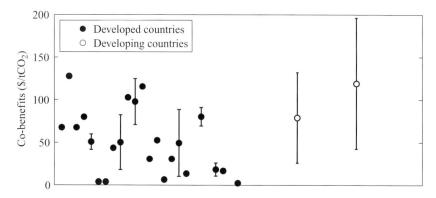

Figure 7.2.4 *Estimates of the value of air-quality co-benefits*

Given the appealing B/C characteristics of the technology-led strategy, the most relevant concern is not that governments will impoverish their constituencies by making draconian emissions reductions. Rather, it is that governments will choose to pursue a combination of modest abatement and inadequate technology investment. It is important that the analyses of the energy R&D solution area are used not only to reject "brute force" mitigation but to inform choices between a technology-led strategy and a policy strategy that is modest in both near-term abatement and in funding for technology development.

Air-Quality Co-Benefits

One should also consider the non-climate benefits associated with transformation of energy production and use. The deep uncertainty associated with the damages from climate change has shaped climate policy making so that it involves selecting emissions *targets* rather than valuing the benefits of emissions *abatement*. One consequence of this emphasis on cost minimization, rather than CBA, is that it discourages full consideration of the ancillary benefits that accrue to human health through air-quality improvement – even though these *co-benefits* are substantial, more local, nearer-term, and less uncertain. These co-benefits, however, are not easily compared to those of climate change mitigation. Differences in the characteristics of associated risks, valuation issues, epistemic communities,

and institutional arrangements reinforce the barriers to consideration of the two benefits simultaneously. As a result, the air-quality co-benefits of climate change abatement, while generally acknowledged, are treated as a windfall or serendipitous result of climate change activities. The contentiousness of climate change policy – heightened by the combination of distant and diffuse benefits with concentrated and immediate costs – implies that policy makers are unlikely to value air-quality co-benefits fully unless they can be compared on an equivalent basis. While they have been asserted as providing a hedge against uncertainty in the benefits of climate change abatement, air-quality co-benefits may actually be dependent on better valuation of climate damages in order to positively affect decisions regarding policy stringency and international cooperation.

The magnitudes of the air-quality co-benefits of mitigation are non-trivial and relatively certain. Positive co-benefits have been estimated across a large set of studies and across varied geographies, time periods, and sectors. One paper has surveyed thirty-seven studies that provided forty-eight estimates of the economic value of air-quality benefits of climate change mitigation (Nemet *et al.* 2010). In figure 7.2.4 studies of developed countries are shown on the left and those of developing countries on the right. Within each category, data are reported from left to right by date of study (1991–2006). The values for developed countries are in black and those for developing countries in white. For the twenty-two values in the twenty-four

Table 7.2.1 Implications for BCRs

		Benefits	Costs	BCR
1	ITC is weak	+		+
2	Public decisions	−		−
3	Stochastic R&D	−		−
4	R&D free-riding	−	−	−
5	Avoid crowding out	−	+	
6	Compare to modest policy	+	+	−
7	Air quality	+		+

developed country studies the range was $2–128/tCO_2$, the median was $31/tCO_2$, and the mean $44/tCO_2$. For the seven values in the thirteen developing country studies the range was $27–196/tCO_2$, the median was $43/tCO_2$, and the mean was $81/tCO_2$. Inclusion of the value of co-benefits would increase benefits in both the technology-led strategy and in the "brute force" mitigation scenarios.

Summary

Table 7.2.1 summarizes the effect of these comments on the BCRs in chapter 7. The formulation used for the BCRs is as follows:

$$BCR = \frac{Abatement_{avoided} - Damages_{not\ avoided}}{R\&D + Deployment}$$

(7.2.1)

The first point – that one should not expect adequate technology improvement as a result of ITC – is directed not at the authors but at those who are optimistic about the innovation and decarbonization that would result from a "brute force" mitigation policy. Acknowledging the weakness of ITC improves BCRs under a technology-led policy by reducing abatement costs, and thus increasing abatement avoided. Second, the shift of decision making to the public sector has the potential to have an adverse impact on the BCRs by making the program not as effective as intended. Retaining the full value of the BCRs in the technology-led policy depends on mediating the challenges associated with concentrating technology decisions in the public sector. Third, acknowledging the inherently stochastic aspects of R&D allows for the possibility that the technology program may be less successful than anticipated. Given some appropriate soci-

etal risk aversion to this outcome, treating R&D returns in an expected value framework may overstate the BCRs. Sensitivity analysis in the chapter allays some of these concerns. Fourth, the possibility of international collective action problems on R&D investment is likely to create higher abatement costs, as the R&D investments levels in the proposal are not reached. Because the returns to R&D are positive, the reduction in benefits exceeds that of costs and the BCRs fall. Fifth, avoiding crowding out effects – for, example by investing in education – will increase the BCRs by reducing the social opportunity cost of R&D spending. Sixth, comparison of the technology-led strategy with a modest suite of policies that target little mitigation but also little technology investment, would likely reduce the BCRs; costs would be higher due to the lack of R&D and deployment in the modest policy; abatement costs avoided would be lower given a modest climate policy; damages not avoided would be negative. As the authors show in a sensitivity analysis, the net effect is lower BCRs, albeit a quite minor change. Finally, the valuation of air-quality co-benefits will increase the BCRs by providing additional health benefits that should be included in the numerator of (7.2.1).

This Perspective paper supports chapter 7's primary claim that the technology-led strategy is highly preferable to "brute force" mitigation. This alternative perspective has attempted to clarify the arguments in favor of technology investment, particularly the reasons why relying on price signals to induce technological change appear unlikely and risky. Just as in mitigation oriented policies, the implementation details of a technology investment strategy are crucial to the ultimate outcomes. Realizing the benefits that drive the high BCRs depends on these details and will require important changes to how governments interact with the technology innovation system.

Bibliography

Arrow, K., 1962: *Welfare and the Allocation of Resources for Invention. The Rate and Direction of Inventive Activity: Economic and Social Factors*, Princeton University Press, Princeton, NJ

Bernstein, J.I. and M.I. Nadiri, 1988: Interindustry R&D spillovers, rates of return, and production in high-tech industries, *American Economic Review* **78**(2), 429–34

1991: *Product Demand, Cost of Production, Spillovers, and the Social Rate of Return to R&D*. National Bureau of Economic Research, Cambridge, MA

Berrens, R.P., *et al.*, 2004: Information and effort in contingent valuation surveys: application to global climate change using national internet samples, *Journal of Environmental Economics and Management* **47**(2), 331–63

Bovenberg, A.L. *et al.*, 2008: Costs of alternative environmental policy instruments in the presence of industry compensation requirements, *Journal of Public Economics* **92**(5–6), 1236–53

Cohen, L.R. and R.G. Noll, 1991: *The Technology Pork Barrel*, Brookings Institution, Washington, DC

David, P.A. *et al.*, 2000: Is public R&D a complement or substitute for private R&D? A review of the econometric evidence, *Research Policy* **29**(4–5), 497–529

Goolsbee, A., 1998: Does government R&D policy mainly benefit scientists and engineers?, *American Economic Review* **88**(2), 298–302

Goto, A. and K. Suzuki, 1989: R&D capital, rate of return on R&D investment and spillover of R&D in Japanese manufacturing-industries, *Review of Economics and Statistics* **71**(4), 555–64

Griliches, Z., 1992: The search for R&D spillovers, *Scandinavian Journal of Economics* **94**(S), 29–42

Jones, C.I. and J.C. Williams, 1998: Measuring the social return to R&D, *The Quarterly Journal of Economics* **113**(4), 1119–35

Li, H. *et al.*, 2009: Public support for reducing US reliance on fossil fuels: investigating household willingness-to-pay for energy research and development, *Ecological Economics* **68**(3), 731–42

Mansfield, E. *et al.*, 1977: Social and private rates of return from industrial innovations, *Quarterly Journal of Economics* **91**(2), 221–40

Nadiri, M.I., 1993: *Innovations and Technological Spillovers*, National Bureau of Economic Research, Cambridge, MA

Nelson, R.R., 1959: The simple economics of basic scientific-research, *Journal of Political Economy* **67**(3), 297–306

Nemet, G., 2010: Cost containment in climate policy and incentives for technology development, *Climate Change*, http://dx.doi.org/10.1007/s10584-009-9779-8, in press

Nemet, G., T. Holloway, and P. Meier 2010: Implications of incorporating air-quality co-benefits of climate change policymaking, *Environmental Research Letters* **5**(1), 014007

Okubo, S. *et al.*, 2006: R&D satellite account: preliminary estimates, Bureau of Economic Analysis/National Science Foundation, Washington, DC

Scherer, F.M., 1982: Inter-industry technology flows in the United States, *Research Policy* **11**(4), 227–45

Sveikauskas, L., 1981: Technological inputs and multifactor productivity growth, *Review of Economics and Statistics* **63**(2), 275–82

Teece, D.J., 1986: Profiting from technological innovation – implications for integration, collaboration, licensing and public policy, *Research Policy* **15**(6), 285–305

Terleckyj, N., 1974: *Effects of R&D on the Productivity Growth of Industries: An Explanatory Study*, National Planning Association, Washington, DC

Tol, R.S.J., 2009: The economic effects of climate change, *Journal of Economic Perspectives* **23**(2), 29–51

Technology Transfer

ZILI YANG

Introduction

Climate change is an ongoing challenge faced in the twenty-first century and beyond. Economic activities since the industrial revolution, mainly fossil fuel combustions and agriculture, have emitted huge amounts of greenhouse gases (GHGs) into the atmosphere. The anthropogenic GHG emission is the main source for measureable atmospheric temperature increases over past decades (IPCC 2007). Economists predict that global GHG emissions will keep increasing in the future, which will lead to further temperature rises. The climate that human beings have been used to for centuries will change drastically (IPCC 2007).

The detrimental impacts of climate change have long-lasting, sometimes irreversible, consequences. To alleviate these impacts, international cooperation on GHG emission reduction is urgently called for. The United Nations Framework Convention on Climate Change (UNFCCC), established in 1992, has been the grand institutional setting for potential international cooperation. Technology transfer (TT) as the means for international cooperation and a concrete approach to GHG mitigation have been at the center of policy debates and at the negotiation table.

The international community has recognized the vital importance of TT in coping with climate change. Without TT, "it may be difficult to achieve emission reduction at a significant scale" (IPCC 2007). TT should be a key component of any effective GHG mitigation strategies, therefore comprehensive studies of TT issues are crucial to GHG mitigation policy designs and implementations.

In this chapter, I survey the scope of issues surrounding TT in the context of climate change and conduct some rudimental cost-benefit analysis

(CBA) on a few options. The remaining parts of the chapter are organized as follows: The second section contains general discussions and surveys on TT; the next section is the CBA of TT issues under different assumptions and policy backgrounds; the final section contains some concluding thoughts.

Describing Technology Transfer

TT is an encompassing theme in policy discussions of climate change. In the text of UNFCCC, "transfer of technologies" is identified as the means for mitigating GHG emissions and adapting the impacts of climate change (UNFCCC, Articles 4, 9, and 11). In the subsequent fourteen sessions of the Conference of Parties (COP), decisions made on "development and technology transfer" appeared twelve times (all but COP 6 and COP 9). The clean development mechanism (CDM), an important channel for potential transfers of GHG mitigation technologies, is in the treaty contents of the Kyoto Protocol (Article 14). Since the inception of the IPCC in 1988, TT has been a perpetual theme on its agenda. All four assessment reports of the IPCC (IPCC 1992, 1996, 2001, 2007) contain detailed analysis of TT issues. In 2000, the IPCC published a special report, entitled *Methodological and Technological Issues in Technology Transfer* (IPCC 2000a). This volume, of over 400 pages, with over 200 contributors, is the most comprehensive study on TT in the context of climate change.

Transfer of environmentally sound technologies (EST) from developed to developing countries plays a key role in mitigation and adaptation in climate change; technology diffusions among developed countries also enhance the effectiveness of

GHG mitigation efforts. COP documents and IPCC reports demonstrate the vital importance of TT in dealing with climate change.

The Definition of Technology Transfer

The concept of TT can be very broad. Here we quote the definition from IPCC (2000a): "[TT is] a broad set of processes covering the flows of know-how, experience and equipment for mitigating and adapting to climate change amongst different stakeholders such as governments, private sector entities, financial institutions, NGOs and research/education institutions." Nevertheless, TT may convey varied connotations among scholars or decision makers in different contexts. The above definition is a balanced one.

The description of the TT concept contains several components. In TT processes there are providers/donors and recipients: providers/donors are generally from developed countries, recipients are in developing countries. A TT process takes place across borders. The entities (stakeholders) in a TT process can be governments, NGOs, international agencies, or private sectors. In this chapter, we use developed countries (North) and developing countries (South) as "proxies" for entities in TT processes.

TT processes involve primary and dual flows. The primary flow is tangible technologies or intangible "know-how" from developed to developing countries; the dual flow is the money that finances the TT. While the sources and destinations of the primary flows are transparent (from North to South), the directions of the dual flows can be complicated. If developed countries fund the TT process, money flows from North to South; if the TT process is a part of an international trade transaction, money flows from South to North.

The institutional setting and market structure of TT processes are diverse. Both governments and international organizations sponsor and channel TT. Some exemplary TT projects have governmental backing on both sides. For example, many EST projects have been launched under the auspices of the OECD/IEA (Philibert 2004); sizeable TT projects are under way under the framework of CDM (de Connick et al. 2007). In these circumstances, North is properly called "donor." Nevertheless, many TT activities are involved in commercial trades or are parts of foreign direct investment (FDI) (Less and McMillan 2005). In such a setting, technology is sold to developing by developed countries. Thus North is a "provider" (of technology) not a "donor."

The implementation of TT includes a litany of possible projects and measures in many sectors in developing countries. The tangible TTs take place in energy supply, transportation, agriculture, and many other industries; the intangible TTs are spreads of knowledge on more effective energy usage, protecting the global environment, etc. The intangible TTs permeate from North to South through education and exchange of ideas.

Transfer Issues in the Literature

Over the past decade, studies of TT issues in the context of climate change have been extensive. Hundreds, if not thousands, of scholarly papers, reports, and documents have been devoted to the subject. The literature on transfers can be categorized in four strands:

(i) *The publications* by the IPCC. Discussions of TT in Assessment Reports (IPCC 1996; 2001; 2007) and a Special Report (IPCC 2000a) represent a collective understanding of TT in climate change by international communities. They also offer policy guidelines for implementing TT projects, IPCC (2000a), in particular, is a rich source of TT literature. The bibliographies in its chapters include hundreds of articles and documents on all relevant aspects.

(ii) *Independent studies of TT issues in the context of climate change.* Many peer-reviewed articles as well as reports assess the TT issues outlined in UNFCCC and the Kyoto Protocol. For example, Ellis et al. (2004) reviewed the progress and outlook of CDM; Brewer (2008) examined the institutional and legal aspects of TT issues; Saggi (2002) surveyed the relationship between trade, FDI, and TT; Martinot et al. (1997) engaged in country studies of TT in climate change. The literature on broad issues related to technologies is a huge reservoir: a comprehensive survey would require many volumes.

(iii) *TT issues in international environmental agreement (IEA) studies.* Climate change stimulates the studies of IEA by game-theorists and environmental economists. Transfers in IEA studies, an abstract monetary transfer that is broader than TT as defined in UPCC (2000a), are widely adopted to ensure the formation of IEA. In numerical simulations of IEA models, the transfer amounts are quantified; in this line of the literature, the timing and intensity of transfers are not under consideration. The amounts and directions of transfers are sometimes questionable from policy perspective.[1]

(iv) *Transfer issues in integrated assessment modeling of climate change.*[2] In most integrated assessment models (IAMs), various financial transfer mechanisms are introduced to calculate "efficient" GHG mitigation policies. Economic theories state that a global GHG mitigation policy is "efficient" when the marginal costs (MC) of GHG mitigation are equalized across regions. Such MC equalization requires financial transfers. The interpretation of the material flow counterpart of such transfers is TT. The transfer amounts and directions in IAMs are much more reasonable than those in (iii). Nevertheless, the speed of TT, or the "absorptive capacity" of recipients, is not considered in these models. Such restrictions always exist in real economies (Borensztein *et al.* 1998).

Technology Transfer in Practice

The history of TT is as long as that of international trade. TTs targeted at coping with climate change have grown in the past decade, and many projects between developed and developing countries are under negotiation. IPCC (2000a) includes

thirty case studies of TT in GHG mitigation and adaptation of climate change. The diversity of these projects shows the promising potential of TT in the future international cooperation. Nevertheless, the scope and magnitude of TT projects fall far short of the demanding tasks of global GHG mitigation.

Cost-Benefit Analysis of Technology Transfer under Different Assumptions and Policy Backgrounds

Backgrounds and Assumptions

TT is an important and all-inclusive option for GHG mitigation and adaptation of climate change. All perceivable international cooperation on climate change is necessarily implemented through TT, directly or indirectly. Cost-effective GHG mitigation policies require that mitigation costs are equal at the margin for all regions. When developed countries help developing countries in their GHG mitigation efforts with money, TTs are behind such financial transfers. Regardless of institutional setting, such as CDM, joint implementation (JI), or FDI in the private sector, TTs are the material counterparts of all financial transfers from North to South.

Due to the "all-inclusive" characteristics of TT issues, TT as a "solution" option for climate change always encompasses other "solutions." If (traditional or alternative) GHG mitigation and adaption measures take place domestically, the TT does not occur; if GHG mitigation in developing countries is supported by technologies from developed countries, TT is in play. In the latter case, TT offers incremental benefits to solving climate change. In figurative terms, we try to quantify the cost and benefit of the second "T" in "TT" while treating the first "T" as a precondition. However, it is difficult to credit a share of potential gains of trans-boundary mitigation activities to TT or to mitigation itself. A conventional CBA approach is not valid here.

It is widely recognized that the scope and costs of TTs are very difficult to quantify; IPCC (2000a) concluded that "little is known about how much climate-relevant hardware is successfully

[1] The literature in this field is abundant. Because it is not connected to the issues in this chapter, we do not survey it here.
[2] For a comprehensive descriptions of major IAMs, see EMF 22, http://emf.stanford.edu.

'transferred' annually." Cost estimates on individual TT projects are hard to aggregate at a regional or global level. Intangible TTs, such as capacity-building and education, are not quantifiable monetarily, especially their potential benefits in the long run. In addition, the future of technological progress often turns out to be unpredictable. Therefore, a CBA of TT as a "solution" for climate change cannot be based on a plethora of project evaluations. In other words, a direct engineering approach is not feasible for such assessment: we must adopt an indirect economic approach.

The tentative analysis provided in this chapter is established on the "dual" side of the material flows of TT: we follow the related financial transfers. In the literature, financial flows are accepted as "proxies" for TT, with qualifications (IPCC 2000a). In IPCC (2000a), "financial resource flows" are used to track historical trends and patterns of TT in climate change. In fact, any financial flows in the context of climate change necessarily have material flow counterparts. Such material flows are TTs, as defined in the previous section. However, there are caveats to this approach. To make the analysis more credible and to avoid misunderstanding, the assumptions for the analysis framework need to be elaborated:

(i) *Intangible TTs are not included in the analysis.* From a CBA point of view, the costs of spreading "know-how" are very low but the "intangible" benefits are huge. For the reasons stated previously, such benefits are difficult to quantify. Furthermore, the impacts of intangible TTs do not flow directly or immediately into GHG mitigation measures. Having said that, the potential contribution of intangible TTs to GHG mitigation and adaptation to climate change can be significant in the long run, as can the impacts of technology spillovers on other dimensions of societies (Keller 2004).

(ii) *Financial transfers are efficient.* This assumption implies that a unit of financial transfer is backed by TT at a competitive market price. In addition, TT is applied to the most cost-effective sectors for GHG mitigation or adaptation of climate change in developing countries. Thus, the financial transfers represent the

efficient allocations of mitigation technologies worldwide. Admittedly, such a "low-hanging-fruit" principle may not be the case in real life. For example, the EU and China are negotiating on transferring advanced carbon sequestration technologies, despite a large portion of Chinese energy suppliers continuing to use out-of-date inefficient technologies.

(iii) *A broad interpretation of financial transfers follows category (iv), not category (iii), in the literature reviews* (see p. 361). In IEA studies, TTs are used as tools to facilitate the formation of a coalition. Institutional reality and practicality of transfers are not considered in these types of models. In the numerical simulations of coalition models, the transfer values (at billions or even trillions of dollars) can flow into any region on an annual basis. On the other hand, IAMs are more attentive to data calibration and are policy oriented. Forecasting scenarios and policy solutions in IAMs are based on the best knowledge of the modelers and consensus among their peers. In these IAMs, transfer channels are set up in such a way that the sole purpose of their presence is to ensure cost-effective GHG mitigation globally. The magnitude and directions of transfers are much more realistic in IAMs. Consequently, transfers in IAMs are the best reflection of TT. Nevertheless, "absorptive capacity" is not considered in most models.

(iv) *Optimal TTs are policy-dependent or policy-driven.* Different GHG mitigation policy scenarios require different transfer regimes. Particularly when regions fulfill their international GHG mitigation obligations, such as those set by the Kyoto Protocol, they may offer transfers (developed countries) or receive transfers (developing countries) to collectively achieve their mitigation targets. Transfer amounts and directions are determined by policy targets. Assessing GHG mitigation policies, TT is a part of a larger picture, and seldom the whole picture. For example, many pilot projects under the CDM framework of the Kyoto Protocol are parts of donors' and recipients' cooperation on GHG mitigation. One cannot say that CDM projects represent the entirety of donors' or recipients' GHG mitigation policy.

Methodologies

As mentioned above, the CBA here is targeted at "transfer," not "technology." There are at least three measurements of the benefit-cost ratios (BCRs) of TT. The first is defined as follows: the costs are measured at total mitigation cost under the TT scheme; the benefits are measured as total mitigation cost reduction without the transfers. Here "B" avoids high mitigation costs and "C" actually incurs mitigation costs.

Optimal TT always has net gains, otherwise it does not happen. Based on these observations, the second measurement of the BCRs of TT is as follows: the cost is measured as a total financial transfer amount, T; the benefit is the net gains in mitigation cost reductions from the TT scheme. The BCR is defined as: $(B_1 - C_1)/T$. Here B_1 and C_1 are benefit and cost calculated in the first measurement. In this chapter, we present both measurements.

The third measurement is broader, but probably more vague. We use reduced climate damage (compared with the business-as-usual (BAU) scenario) in the policy scenario as the benefit of TT; the transfer amounts incurred in the policy are the cost. Here the BCR is defined as $\Delta D/T$.

Because TT is associated with GHG mitigation activities and particular policies, separating the cost and benefit of the TT from mitigation itself can be tricky. TT should not take credit for the total benefit of the whole mitigation effort. The contribution of TT may be large, or may be small. For example, in two mitigation policy scenarios involving TT, the inferior one with a lower overall BCR might have a higher BCR from TT. In a simple arithmetic expression: when $B_{t,1} > B_{t,2}$, it does not imply that $B_1/C_1 > B_2/C_2$ (here, $B_{t,1}$ and $B_{t,2}$ are benefits from TT in policy 1 and 2; B_1, C_1 and B_2, C_2 are total benefits and costs of policy 1 and 2, respectively).

We use the following hypothetical example to illustrate the first two B/C measurements. Suppose achieving a certain mitigation target incurs $10 million of cost globally without TT. A TT scheme with $1 million transfers, combined with domestic mitigation efforts, reduces the global mitigation cost to $8 million. The benefit (avoided high costs) $B = \$10$ million; the cost (actually incurred) $C = \$8$ million; the BCR is: $B/C = 10/8 = 1.25$ in the first measurement. The net gains from TT here are $(10 - 8) = \$2$ million. The second measurement of the BCR is $(B_1 - C_1)/T = (10 - 8)/1 = 2$.

There are different approaches to assessing a project or policy in a CBA (Layard 1994). It is difficult to claim that one BCR measurement is always superior to another, at least in the case here. The three BCR measurements cover different aspects of TT. Their complementary nature renders them all useful.

Using these three measurements of the BCR, this chapter conducts a CBA of TT in the RICE model developed by Nordhaus and Yang (1996). Estimation methodologies are heavily reliant on Yang (1999) and Yang and Nordhaus (2006). The RICE model is a multi-region extension of the aggregate DICE model (Nordhaus 1994, 2008). A regional breakdown of RICE is essential for modeling TT issues, because transfers flow across borders.

To set up the model for dealing with TT issues, financial transfers are introduced in the RICE model in such a way that the transfer costs of donors (developed countries) are deducted from their GDP; the transfers go into the GHG mitigation functions of the recipients (developing countries). The purpose of the transfer is to mitigate GHG emissions more cheaply in recipient countries. The donors benefit from the transfer through reduced climate change impacts. Such a model structure rules out the effects of pure welfare transfers where monies go directly into developing countries' treasury and nothing happen to GHG mitigation.[3] The modeling methodology is relevant for connecting the financial transfer to TT. Finally, the transfer amounts are endogenous. The model solution reflects the optimal transfers under a given policy scenario.

For this chapter, two policy scenarios (solution categories) are proposed for a CBA of TT. The first is the Kyoto Protocol-like scenario that lasts for

---

[3] In most IAMs, the distributional (wealth) effects of transfers are not separated from GHG mitigation cost-reduction effects. Therefore, financial transfers in these models probably overestimate optimal TT volumes.

the entire modeling horizon. In this case, developing countries are not obligated to reduce their baseline GHG emissions. They will mitigate GHG emissions if developed countries pay them to do so, through CDM, JI, or FDI. From the technology aspect, all mitigation efforts in South use the technologies provided by North. North takes credit for the outcome. The final outcome of such a TT scheme is the equalization of the marginal costs of GHG mitigation across all regions. Using the "fruit" metaphor, all "fruits," low-hanging or high-hanging, in the Southern orchard, are picked with North technologies and financed with Northern money. The harvesting activities in the Southern orchard will go on until all untouched fruits hang at the same height as the remaining fruits in Northern orchard.

In the second scenario, developing countries shoulder certain GHG mitigation obligations that are compatible with their own incentives (considering the climate change impacts on them).[4] After developing countries have fulfilled their mitigation obligations, the developed countries will help them with further GHG mitigation, through TT, to achieve a globally cost-effective GHG mitigation outcome. A scenario of international cooperation on GHG mitigation such as this has been a target sought by some developed countries in the post-Kyoto negotiations. GHG mitigation commitment by the major developing countries, such as China, India, and Brazil, has been the focal agenda in the Copenhagen COP 15. In this scenario, developing countries with their indigenous technologies will exploit the "low-hanging-fruit" mitigation opportunities. For example, it is not necessary to use advanced technology from Europe to replace all old coal-burning technologies in China. Equipment with mature technologies and manufactured in China can improve fuel efficiency sufficiently.

In the calculation of TTs in the above two scenarios, how much the global community wants to spend on TT is based on the optimal solution of the model under the given policy scenario, not prescribed. We cannot phrase the TT issue by treating the amount of TT as exogenous, such as "what is the cost and benefit of spending $1 billion on TT in a year?" Our calculations of BCRs are *ex post* or side calculations after policy-driven TTs, along

with other control and state variables, have been solved endogenously.

Many other TT scenarios can be proposed. The above two are probably located at the polar ends of the potential role of TT. Due to space limitations the technical aspects of modeling are not fully explained here; readers can find the detailed modeling methodologies in RICE related to the scenarios here in this perspective paper and Nordhaus (1994, 2008).

Calculation Results of Cost-Benefit Analysis

The time frame of the calculation follows the guideline set for chapters in the Copenhagen Consensus project. The costs and benefits are expressed as the present values (PVs) of the flows of costs and benefits for 100 years (2005–2105) at a given discount rate ($r = 3\%$ and $r = 5\%$). In addition, the current values (CVs) of these flows are also calculated. The TTs, costs, and benefits associated with them are flows over time. Policy scenarios may affect the timing and volume of TT. The CV is therefore a useful piece of information.

We summarize the simulation scenarios as follows:

(i) *Scenario 1*: The Kyoto Protocol-like case at $r = 3\%$ and 5%. In this case, the global GHG mitigation outcome is stringent. Much of the initial GHG mitigation burdens fall on developed countries.

(ii) *Scenario 2*: A full-cooperation case based on the willingness-to-pay (WTP) principle at $r = 3\%$ and 5%. In this case, the global GHG mitigation outcome is less stringent, compared with scenario 1. All regions are obligated to GHG mitigation based on their mitigation costs and climate damage situations.

The numerical calculations are based on a six-region version RICE model used in Yang (2008). In this version, three regions (USA, EU, and other high-income countries (OHI)) are donors/providers

---

[4] More specifically, the policy scheme is close to the Lindahl equilibrium outcome in Yang (2008). Each region's initial mitigation obligation is based on their respective WTP principle.

Table 8.1 PV of total global benefits and costs of TT in 100 years (the first measurement) (billion 2000 US$)

Scenario 1								
Benefit and costs	$r = 3\%$ Benefit	Costs	CV Benefit	Costs	$r = 5\%$ Benefit	Costs	CV Benefit	Costs
	2,523	805	13,688	4,355	347	112	4,000	1272
BCRs	3.134		3.143		3.098		3.144	
Scenario 2								
Benefit and costs	$r = 3\%$ Benefit	Costs	CV Benefit	Costs	$r = 5\%$ Benefit	Costs	CV Benefit	Costs
	339	262	2,160	1,688	42	32	637	498
BCRs	1.294		1.280		1.312		1.279	

Table 8.2 PV of total global benefits and costs of TT in 100 years (the second measurement) (billion 2000 US$)

Scenario 1								
Benefit and costs	$r = 3\%$ Benefit	Costs	CV Benefit	Costs	$r = 5\%$ Benefit	Costs	CV Benefit	Costs
	1,718	470	9,333	2,551	236	66	2,728	754
BCRs	3.655		3.659		3.576		3.618	
Scenario 2								
Benefit and costs	$r = 3\%$ Benefit	Costs	CV Benefit	Costs	$r = 5\%$ Benefit	Costs	CV Benefit	Costs
	77	70	472	445	10	8.5	139	134
BCRs	1.10		1.061		1.17		1.037	

Table 8.3 PV of total global benefits and costs of TT in 100 years (the third measurement) (billion 2000 US$)

Scenario 1				
Benefit and costs	$r = 3\%$ Benefit	Costs	$r = 5\%$ Benefit	Costs
	1,221	470	190	66
BCRs	2.60		2.88	
Scenario 2				
Benefit and costs	$r = 3\%$ Benefit	Costs	$r = 5\%$ Benefit	Costs
	746	70	112	8.5
BCRs	10.66		13.18	

in TT schemes; the remaining three regions (China (CHI), Former Soviet Union and Eastern European countries (EEC), and the rest of the world (ROW)) are recipients. The model's baseline GHG emission prediction is in the mid-range of the IPCC's emission scenarios (IPCC 2000b). The optimal solutions in Yang (2008) are moderate, compared with other IAMs.

The results of a CBA of TT are presented in tables 8.1–8.3. The calculations procedure is outlined as follows: first, the optimal solutions without transfers in different scenarios are obtained. This step is as if each region is picking the "low-hanging fruits" in GHG mitigation opportunities within its borders, according to their obligations specified by the policy. In the solutions, North always reaches higher "fruits." This implies that North incurs higher mitigation costs than South. Second, the necessary (minimum) amounts of transfers that enable the equalization of marginal mitigation costs

are obtained through a set of side calculations. The outcome reflects the fact that North explores "low-hanging-fruits" opportunities in South through TT, "returns" some "high-hanging fruits" in North, and the total numbers of harvested "fruits" remain the same globally before and after TT. Third, addition calculations are conducted to obtain B/C values according to the definitions discussed on p. 364.

Other relevant results are presented in figures 8.1–8.3. Figure 8.1 contains the optimal TT flows over time in different scenarios; figure 8.2 depicts B/C flows, as defined in the first measurement in Scenario 1; figure 8.3 is the same flows as in figure 8.2 for Scenario 2. Figures 8.2 and 8.3 capture the shift of mitigation flows caused by TT.

These tables and graphs outline the general overviews of TT as reflected by the RICE model. In Scenario 1, both the magnitude of and potential gains from TT are huge. Valued with both B/C measurements, the BCRs are greater than 3 in

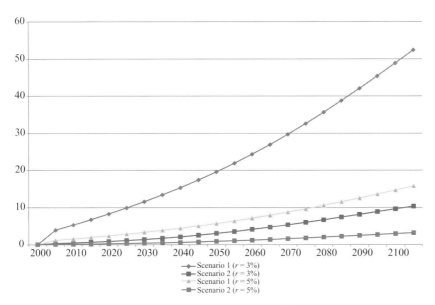

Figure 8.1 *Optimal transfer amounts (billion 2000 US$)*

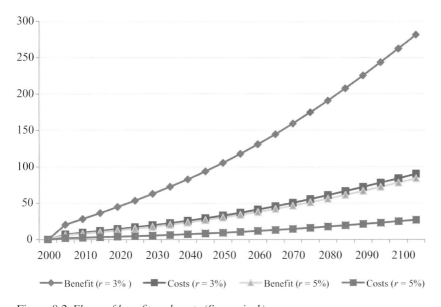

Figure 8.2 *Flows of benefit and costs (Scenario 1)*

Scenario 1. In this case, the developed countries have to rely on TT to reduce the costs of their GHG mitigation burdens. TT plays the most important role in reducing global GHG mitigation costs. Most GHG mitigation activities in developing countries are financed by developed countries and use imported technologies. Given a burden-sharing rule like this, one would wonder why developed countries would agree to such an arrangement in the first place. TT reduces the mitigation costs of developed countries; TT attracts developing countries joining in the global cooperation in GHG mitigation. However, the initial policy setting is not the most desirable one for some regions.

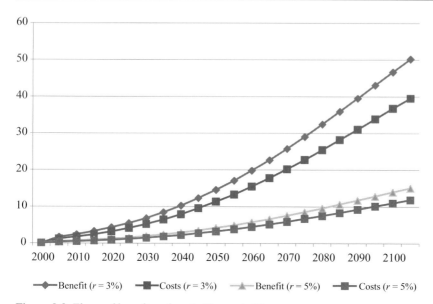

Figure 8.3 *Flows of benefit and costs (Scenario 2)*

Scenario 2 represents the case in which all "low-hanging fruits" of GHG mitigation options in the developing countries are exploited domestically with indigenous technologies. Such voluntary actions are based on common concerns about climate change by all regions. Maximum participation by developing countries has been pushed very hard by some developed countries, such as the USA, in post-Kyoto negotiations. On top of domestic efforts with indigenous technologies in developing countries, advanced TTs take place to equalize the MC of mitigation costs across all regions after all cheap options have been exhausted domestically. In Scenario 2, both the magnitude of and the gains from TT are much lower than they are in Scenario 1. The BCRs under the two measurements are slightly higher than 1. Such a result shows that the marginal gains from "picking high-hanging fruits" are small. The result in this scenario does not imply that technology has little to do with GHG mitigation. It indicates a scenario in which transfer amounts could be moderate. Domestic GHG mitigations need technologies.

The three measurements are basically consistent with one another. They all show that implementing TT creates a win–win outcome for both donors/providers and recipients when compared

with no-TT results. Here we should also indicate that climate damages are predicted to be more severe beyond the 100-year time span for this chapter. Mitigation efforts in this century are, in part, aimed at reducing climate damage beyond 100 years. B/C measurement truncated in time may underestimate the true benefits, compared with a longer time-horizon calculation. This caution is also applicable for the third measurement.

The two scenarios reflect two extreme situations involving TT. The calculated transfer amounts are the minimal/optimal transfers that equalize the marginal mitigation costs globally. Net of values of intangible TT, the estimated total B/C of TT and transfer amounts in Scenario 1 should be at the upper bound of the potential scope of TT in the next century; those in Scenario 2 should be at the very bottom of the potential scope of TT in the next century. The actual outcomes of TT are probably somewhere in the middle of the two scenarios. If the values of intangible TT could be included in the estimation, the net benefits of TT in the long run would be much higher.

Finally, all TTs are motivated by specific mitigation policies or international negotiation outcomes. We cannot draw any conclusions on the policies or infer BCRs of those policies based on a CBA of

TT alone. In this chapter, we do not claim that the policies behind Scenario 1 are superior to those in Scenario 2 because the gains from TT are larger. As we have emphasized repeatedly, the evaluation of TT has to be in connection with other parts of mitigation and/or adaptation processes. For any GHG mitigation policy, incorporating TT can reduce the aggregate costs further. Therefore, TT should be ubiquitous in optimal GHG mitigation policies.

Conclusions

TT, in conjunction with other "solution categories," is an effective option for GHG mitigation and adaptation of climate change. Despite this, it is never a stand-alone solution. TT is a part of all meaningful GHG mitigation policies from a global perspective. Technology progress is the key for the challenges human beings will face in the future, and climate change is one of such challenges. Since climate change is a global phenomenon, international cooperation that involves all nations is necessary for cost-effective GHG mitigation. TT is a combination of technology and international cooperation and is therefore an inseparable component of any climate change policy.

In this chapter, we conduct the CBA of TT in the context of climate change. A CBA of TT at a global level and in the long run is very difficult. Unlike the evaluation of individual CDM undertakings, where costs and benefits (to a lesser degree) are measurable, the aggregate effects of TT are not simple additions of individual projects. We hope that the indirect methods used here can shed some light on the evaluation of the effectiveness of TT in dealing with climate change.

Bibliography

Borensztein, E., J. De Gregorio, and J.-W. Lee, 1998: How does foreign direct investment affect economic growth?, *Journal of International Economics* **45**, 115–35

Brewer, T., 2008: Climate change technology transfer: a new Paradigm and policy agenda, *Climate Policy* **8**(5), 516–26

deConnick, H., F. Haake, and N. Van Der Linden, 2007: Technology transfer in the clean development mechanism, *Climate Policy* **7**, 444–56

Ellis, J. *et al.*, 2004: CDM: taking stock and looking forward, *Energy Policy* **32**(1), 15–28

IPCC, 1992: *Climate Change 1992: The IPCC Supplementary Report*, J. T. Houghton, B. A. Callanders, and S. K. Varney (eds.), Cambridge University Press, Cambridge

1996: *Climate Change 1995: The Second Assessment Report*, Cambridge University Press, Cambridge

2000a: *Methodological and Technological Issues in Technology Transfer*, Cambridge University Press, Cambridge

2000b: *Emission Scenarios*, Cambridge University Press, Cambridge

2001: *Climate Change 2000: The Third Assessment Report*, Cambridge University Press, Cambridge

2007: *Climate Change 2007: The Fourth Assessment Report*, Cambridge University Press, Cambridge

Keller, W., 2004: International technology diffusion, *Journal of Economic literature* **42**(3), 752–82

Layard, R., 1994: *Cost-Benefit Analysis*, Cambridge University Press, Cambridge

Less, C., and S. McMillan, 2005: Achieving the successful transfer of environmentally sound technologies: trade-related aspects, OECD Trade and Environment Working Paper, **2005–02**, OCCD, Paris

Martinot, E., J. Sinton, and B. Haddad, 1997: International technology transfer for climate change and the cases of Russia and China, *Annual Reviews of Energy and Environment* **22**, 357–401

Nordhaus, W. D. 1994: *Managing the Global Commons*, MIT Press, Cambridge, MA

2008: *A Question of Balance: Weighing the options on Global Warming Policies*, Yale University Press, New Haven, CT

Nordhaus, W. D. and Z. Yang, 1996: A regional dynamic general-equilibrium model of alternative climate-change strategies, *American Economic Review* **86**(4), 741–65

Philibert, C., 2004: *International Energy Collaboration and Climate Change Mitigation*, OECD/IEA, Paris

Saggi, K., 2004: Trade, foreign direct investment, and international technology transfer: a survey, *The World Bank Research Observer* **17**(2), 191–235

Yang, Z., 1999: Should North make unilateral technology transfers to South? – North–South cooperation and conflicts in responses to global climate change, *Resource and Energy Economics* **21**(1), 67–87

2008: Strategic bargaining and cooperation in greenhouse gas mitigations: an integrated assessment modeling approach, MIT Press, Cambridge, MA

Yang, Z. and W. Nordhaus, 2006: Magnitude and direction of technological transfers for mitigating GHG emissions, *Energy Economics* **28**, 730–41

Technology Transfer
Alternative Perspective

DAVID POPP

Introduction

Reducing carbon emissions without dramatic reductions in output and consumption requires the use of new technologies. These may be as simple as improvements in energy efficiency, or involve advanced technologies for generating electricity from solar power, or capturing and storing carbon emissions from coal combustion. Recent efforts to reduce emissions in developed countries have stimulated the development of many such technologies, as illustrated in figure 8.1.1. This figure shows dramatic increases in inventive activity for renewable energy technologies, measured by applications for renewable energy patents submitted to the European Patent Office (EPO), corresponding to both national policies and international efforts to combat climate change that followed the signing of the Kyoto Protocol in December 1997 (Johnstone *et al.* 2010). Similarly, the increased energy prices that accompany a carbon tax or emissions trading scheme (ETS) have led to innovation in both energy efficiency and alternative energy sources (Popp 2002).

As is the case with most research and development (R&D), this increased innovation has occurred primarily in the developed world (*Dechezleprêtre et al.* 2008a).[1] At the same time, carbon emissions from developing countries have become a greater concern. For instance, in 1990, China and India accounted for 13% of world carbon dioxide (CO_2) emissions. By 2004, that figure had risen to 22%, and it is projected to rise to 31% by 2030. Overall, the US Energy Information Administration (EIA) projects that CO_2 emissions from non-OECD countries will exceed emissions from OECD countries by 57% in 2030 (EIA 2007).

Due to the growth in emissions from developing countries, designing a policy that encourages the transfer of clean technologies to them has been a major discussion point in climate negotiations. Currently, the Kyoto agreement includes the clean development mechanism (CDM), which allows polluters in industrialized countries with emission constraints to receive credit for financing projects that reduce emissions in developing countries, which do not face emission constraints under the Kyoto Protocol. Because carbon emissions are a global public good, CDM can help developed countries reach emission targets at a lower total cost, by allowing developed country firms to substitute cheaper emissions reductions in developing countries for more expensive reductions in the home country. For developing countries, technology transfer (TT) and the diffusion of clean technologies may be an additional benefit from CDM.[2]

TT provides several potential benefits. By providing access to technologies not readily available in developing countries, it can take advantage of unused, low-cost emission reduction opportunities in developing countries. Taking advantage of these opportunities results in a lower total cost of emissions reductions, by allowing substitution from high-marginal-cost activities in developed countries to low-marginal-cost opportunities in developing countries. It is these cost-saving benefits that Yang captures in chapter 8.

Perhaps more important, however, are the potential dynamic gains that come from TT. By increasing the technology base of the recipient

---

[1] In 2006, global R&D expenditures were about $960 billion, with 85% of this R&D occurring in the OECD, and 50% in the USA and Japan alone (OECD 2008).

[2] Lecocq and Ambrosi (2007) provide a description of the CDM. Popp (2008) discusses its potential for TT.

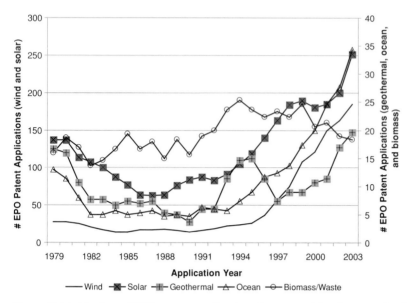

Figure 8.1.1 *Number of EPO patent applications for renewables, by type of technology*

Note: Number of applications for patents pertaining to various renewable energy technologies, sorted by year of application.

country, transfers of climate-friendly technology potentially lower the marginal abatement cost (MAC) curve of the recipient country, making future emissions possible at lower costs. When considering environmental policy, countries weigh the benefits of a cleaner environment against the costs of complying with the regulation. Technological advances lower the cost of compliance, making regulation more likely. For instance, Lovely and Popp (2008) show that access to better pollution-control technologies results in countries adopting environmental regulation at lower levels of *per capita* income over time. Exemplifying this, the 2006 *Report on the State of the Environment in China (State Environmental Protection Administration 2007)* declared scientific innovation the key to "historic transformation of environmental protection" and "leap-frog development." By lowering future carbon mitigation costs, TT can

provide important dynamic benefits by increasing the willingness of developing countries to commit to binding carbon emission reductions.

Challenges to Modeling Technology Transfer

Modeling the costs and benefits of international TT has many challenges. First, TT does not occur in a vacuum. Because carbon emissions are not priced in free markets, there is little incentive to reduce then in the absence of climate policies that reduce emissions, either through restrictions on emission levels or tax policies that place a price on carbon emissions.[3] This holds true for TT as well. With the exception of some energy efficiency technologies, clean technologies typically do not flow across borders unless environmental policies in the recipient country provide incentives to adopt clean technology. Given the need for continued development, developing countries are unlikely to enact policies requiring binding carbon emissions reductions at this time. Instead, incentives for these technology flows occur as a result of developed country commitments. For example, most transfers of climate-friendly technology to developing countries

[3] Note that some actions that reduce emissions, such as improving energy efficiency, may occur without policy, as they also provide private benefits. For example, firms investing in improved energy efficiency lower their energy costs. However, even these investments will be less than optimal without climate policy, as firms will not incorporate the external benefits of reduced carbon emissions in their decision making.

currently occur through the CDM, which allows developed country actors to meet emissions reduction limits by sponsoring projects in developing countries. This poses a challenge for estimating the costs and benefits of TT, as these transfers do not occur independent of other climate policies. To address this, chapter 8 looks at the incremental gains from TT, by considering the cost savings that result compared to a base case with comparable emissions reductions, but no TT. Nonetheless, when comparing the cost-benefit B/C estimates of the TT option to other policies, it is important to keep in mind that TT *by itself*, is not sufficient.[4]

Second, TT comes in many forms. As chapter 8 notes, TT can be direct or indirect. Direct transfers include those modeled in the chapter, in which developed countries finance carbon mitigation projects in developing countries. The mitigation technology is available for use in the recipient country only because of the financing provided by the developed country. Direct transfers could also come via international trade, particularly in the case where a technological advance is embodied in the product being traded. For the modeler, data on direct transfers are readily obtainable, and thus are straightforward to include in policy assessments.

In contrast, indirect TT involves disembodied knowledge. Examples include demonstration projects, training local staff, and local firms hiring staff from multinational firms operating in a developing country. Disembodied TT provides the well-known spillovers often cited in the productivity literature. Spillovers occur when the provider of a technology is not fully compensated for the gains realized by the recipient. To consider the importance of these spillovers, note that the use of advanced equipment provided to a recipient country (embodied TT) may allow the recipient country to reduce carbon emissions. However, such transfers do not necessarily give the recipient country the ability to replicate the technology on their own. In contrast, disembodied TT enables the recipient to develop skills that can be used in later projects initiated by the recipient country, providing a spillover benefit. Because spillovers come from a wide range of activities, they are more difficult to track.

This distinction is important because it affects the future potential of carbon emission reductions in developing countries. One criticism often raised by critics of TT schemes such as the CDM is the problem of "low-hanging-fruit."[5] The low-hanging fruit critique follows from the economic principle of diminishing returns. To the extent that TT to a developing country includes only direct transfer, low-cost abatement options will be used up, making future emission reductions more costly. Proponents of the "low-hanging-fruit" theory worry that if developed countries receive credit now for performing the cheapest emissions reductions options in developing countries, these options will be unavailable for later use by developing countries. As such, these countries will be worse off when later attempting to reduce emissions on their own, and will be less willing to agree to binding emissions reductions at a later date.[6] In essence, such projects move a country to a higher point on their MAC curve, as shown by MAC_0 in figure 8.1.2.

However, TT can counteract the impact of diminishing returns. While it is true that the costs of additional emissions reductions *at a given time* will increase as more projects are completed, the arrival of new technologies provides new opportunities for emissions reductions, so that the future costs of reducing emissions can be lower. In particular, disembodied TT shifts the MAC curve in, making future emission reductions less costly. This shift will partially (MAC_1 in figure 8.1.2) or completely (MAC_2 in figure 8.1.2) offset the low-hanging-fruit

[4] While not related specifically to TT, Popp (2006) finds similar results when studying the viability of R&D subsidies as a climate policy tool. Compared to a combined policy using both optimal carbon taxes and R&D subsidies, a policy using only the optimal R&D subsidy attains just 11% of the welfare gains of the combined policy. In contrast, a policy using only the carbon tax achieves 95% of the welfare gains of the combined policy.

[5] See, for example, references in n. 1 of Narain and van't Veld (2008).

[6] Note that developing countries can be compensated for future cost increases, so that CDM projects become mutually beneficial. Indeed, since such projects require the voluntary agreement of all parties, one would expect such compensation to take place (Rose *et al.* 1999; Narain and van't Veld 2008). However, even if compensation is received, so that the recipient country is not made worse off, the developing country recipient may still delay undertaking their own emissions reductions and participating in future treaties if the easiest options for lowering emissions have already been exhausted.

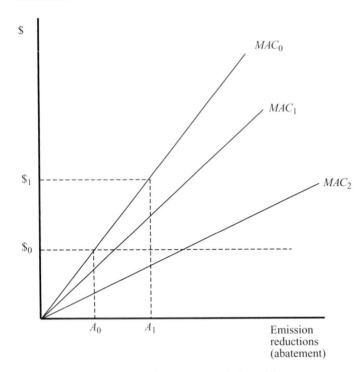

Figure 8.1.2 *Low-hanging fruit and knowledge spillovers*

Note: The MAC curve MAC_0 represents the costs associated with current technologies in developing countries. Initial abatement levels are A_0, with marginal costs $\$_0$. Financial transfers increase abatement to A_1, raising the MAC to $\$_1$. As a result, future abatement efforts by developing countries will cost more – the "low-hanging-fruit" effect. This cost increase can be offset if the transfers include spillovers that lower the MAC. Here, MAC_1 represents a shift that partially offsets the "low-hanging-fruit" effect, while MAC_2 represents a shift where new technologies completely offset the "low-hanging-fruit" effect, so that further abatement is possible at an MAC less than $\$_0$.

problem. By lowering future MACs, such TT also increases the possibility that developing countries will agree to future emission constraints.

Given these dynamic concerns, the potential benefits from knowledge spillovers are quite high, and are likely to exceed the benefits of direct TT. Nonetheless, the chapter's decision to ignore these spillover benefits, focusing instead on the direct transfer benefits, is defensible, given the third challenge of modeling TT – a lack of empirical evidence on the magnitude of spillovers across countries. Measuring direct flows is straightforward. International trade data, for instance, provide evidence of flows of technologies across countries. However, estimating the spillover benefits is more challenging. In the broader literature on technological change, economists consistently find that knowledge spillovers within countries result in a wedge

between private and social rates' return to R&D. Examples of such studies include Mansfield (1977, 1996); Pakes (1985); Jaffe (1986); Hall (1996); and Jones and Williams (1998). Typical results include marginal social rates of return between 30 and 50%, suggesting social rates of return about four times higher than private rates of return. However, few studies provide empirical evidence on the extent to which these gains flow to developing countries.

While estimates of the magnitude of spillovers to developing countries are hard to find, several studies provide evidence of the existence of spillovers to developing countries. The focus of these studies is oriented more towards microeconomics, making it difficult to directly incorporate the results into a macroeconomic climate model such as RICE. However, they provide some insight as to the potential for spillover benefits from TT to developing countries. The method of TT (e.g. via international trade or foreign direct investment (FDI) matters, with spillovers less likely to occur when FDI is the method of transfer, as firms choose FDI when they want to keep knowledge internal (Saggi 2000; Keller 2004). The absorptive capacity of the recipient country is also important. "Absorptive capacity" describes a country's ability to do research to understand, implement, and adapt technologies arriving in the country. It depends on the technological literacy and skills of the workforce, and is influenced by education, the strength of governing institutions, and financial markets (World Bank 2008). Countries with greater absorptive capacity are more likely to receive spillovers from TT.

Technology Transfer in Yang's Model: Implications and Evidence

Chapter 8 uses two scenarios to evaluate TT. In the first, only the developed countries face binding emission constraints. TT takes the form of financial transfers used to finance carbon abatement activities in developing countries. As these countries do no other abatement, the marginal costs of these sponsored abatement projects are low. In the equilibrium, MACs are equated across regions, resulting in minimized abatement costs. While labeled as "technology transfer," one could generate the same result by modeling a global tradable permit scheme in which developing countries were given sufficient permits to cover all current emissions.

In the second scenario, developing countries voluntarily undertake some emission reductions. Countries consider the damages that they face from climate change, and abate until their MAC equals *their own* marginal damage. Thus, abatement in developing countries is insufficient, as it ignores the value of damages to other countries, but is no longer zero. Given the chapter's setup, in which TT takes the form of financial transfers, the net benefits of TT are lower in this scenario, as some low-hanging fruit are picked by the developing country before transfers occur.

These two contrasting results illustrate the importance of omitting the dynamic effects of TT. By modeling TT as a simple movement along the MAC curve, the largest possible gains occur when developing countries take no unilateral action. However, allowing TT to shift the MAC curve could change the results. In this case, lowering MACs enables developing countries to take more unilateral action. This would be expected to close the gap in net benefits between these two scenarios. Whether the ordering of policies would change is unknown, and depends upon the magnitude of the shift of the MAC curve.

While acknowledging the importance of knowledge spillovers, chapter 8 omits these benefits from its analysis because of a lack of empirical evidence on their magnitude. This same lack of information makes it difficult to know whether the MAC shift described above would be sufficient to change the ordering of the two policy simulations. Nonetheless, there are two studies that provide some guidance as to the likely importance of spillovers.

The importance of absorptive capacity for spillovers is captured by Bosetti *et al.* (2008). This paper uses the WITCH model, which is based on the RICE model used in chapter 8. The WITCH model includes more technological detail than the RICE model, allowing for endogenous technological change that can potentially improve energy efficiency or the production of energy from various sources, including renewable energy. To model TT, WITCH includes a global stock of knowledge for each of the technology options described above. The ability to use this knowledge varies, depending on a country's absorptive capacity. The model treats technology differently in developed and developing countries. Innovation from developed countries sets the technological frontier, which is readily available to all high-income countries. Developing countries do not contribute to the technological frontier. Instead, each has its own knowledge stock which consists of knowledge absorbed from the frontier. The ability of a developing country to use knowledge from the frontier depends on its absorptive capacity. Such absorptive capacity varies, and is higher for countries whose own knowledge stocks are closer to the frontier.

In the WITCH model, TT takes a different form than in chapter 8's simulation. Bosetti *et al.* (2008) first simulate the effects of a global permit trading policy stabilizing CO_2 emissions at 450 ppm.

With the exception of having a more stringent standard than the chapter, this is otherwise identical to its TT case, as the trading results in MACs being equal across regions. TT is then considered as a special case, in which the permit market is augmented by a policy in which the revenue from permit sales is used to build absorptive capacity in developing countries. Such aid is analogous to the spillovers discussed earlier, as it results in a shift of the MAC curves of developing countries. The amount of aid made available varies over time, from $2 billion (in 1995 US dollars) in 2007, to $105 billion in 2062. Thus, the amount transferred is greater than in chapter 8's simulation, in which the optimal transfer peaks at just over $50 billion (in 2000 US dollars). The resulting increase in technology in developing countries reduces their climate stabilization costs by 2.3%.[7] To further ascertain the benefit of the spillovers themselves, the authors simulate the effect of lump-sum income transfers equal to the amount of TT funding. There, stabilization costs fall by just 1.55%, suggesting that TT resulting in spillovers is nearly 50% more effective at reducing abatement costs.[8]

While the WITCH model provides some evidence of the importance of spillovers, it is a simulation model, and must make assumptions about key TT parameters, given the lack of good empirical estimates on these spillovers. One study that provides some evidence is Dechezleprêtre et al. (2008b). These authors consider whether CDM projects have a TT component. They look at 644 CDM projects registered by the Executive Board of the UNFCCC, asking how many projects transfer "hardware," such as equipment or machinery, as opposed to "software," which they consider to be knowledge, skills, or know-how. That is, how often do CDM projects transfer knowledge and skills that not only allow a developed country investor to meet emission reduction credits, but also enable the recipient developing country to make continual improvements to their own emission levels? The results provide some insight as to the likelihood that the types of transfers modeled in chapter 8's simulation may result in knowledge spillovers.

Dechezleprêtre et al. (2008b) find that 279 projects, or 43%, involve TT. However, these projects are among the most significant CDM projects, as they account for 84% of the expected emissions reductions from registered projects. Of these projects, 57 transfer equipment, 101 transfer knowledge, and 121 transfer both equipment and knowledge. The percentage of projects involving TT varies depending on the type of technology used. For instance, all projects reducing trifluoromethane (HFC-23) involve transfer, but this is solely a transfer of equipment. Most projects reducing nitrous oxide (N_2O) and recovering methane (CH_4) also involve equipment transfer, as do renewable energy projects such as wind and solar. In contrast, energy efficiency measures are less likely to include TT. TT also varies by recipient country. Just 12% of the projects studied in India include TT, compared to 40% in Brazil, and 59% in China.

While the results of these two studies are not directly comparable to chapter 8's simulation, they do enable a "back-of-the-envelope" calculation of the additional gains that might arise in the chapter 8 model were spillovers considered. First, Bosetti et al. (2008) find that, compared to lump-sum income transfers, the spillovers that result from improving absorptive capacity reduce climate mitigation costs by 50%. From Dechezleprêtre et al. (2008b), approximately two-thirds of all emission reductions from CDM transfers involve knowledge transfer.[9] If we assume that the spillover results from Bosetti et al. (2008) apply to only two-thirds of the transfers in the chapter 8 model, the cost savings from TT would increase by one-third.

---

[7] Unfortunately, Bosetti et al. 2008 do not provide magnitudes of the cost savings, so that direct comparisons to the chapter's results are not possible.

[8] Note that a direct comparison to the chapter's result is not possible from the information in Bosetti et al. (2008). The chapter's simulation shows the benefits from equating MACs across regions. This, however, is Bosetti et al. (2008)'s base case. The 2.3% cost reduction in Bosetti et al. (2008) represents the gains from moving to a base case where MACs are equal across regions, to a world where MACs are equal *and* TT improves the absorptive capacity of developing countries.

[9] The calculation is as follows: projects involving TT account for 84% of the emission reductions from CDM projects. Of the 279 projects with technology transfer, 222 (80%) involve a transfer of knowledge. Multiplying these two percentages yields 67%.

Conclusions

Evaluating TT as a policy option has several complications. Most importantly, TT by itself is not a policy option. TT will not be effective unless countries face binding emissions constraints compeling them to reduce emissions, rather than using newly acquired technologies to increase output. This is consistent with other findings in the climate policy simulation literature showing technological advances augmenting the effects of policy, but in a secondary role. Instead, direct factor substitution in response to policy incentives is more important (Nordhaus 2002).

Moreover, while encouraging emissions reductions in developing countries is important, most near-term reductions will come from developed countries. The limited role for developing country reductions helps reconcile two results in the preceding section. While we find that accounting for spillovers would significantly lower chapter 8's cost estimates, it is also the case that Bosetti *et al.* (2008) find that simulating spillovers reduces mitigation costs by just over 2%. This is consistent with the small role that developing countries play. While spillovers may greatly decrease the cost of emissions reductions in these countries, emissions reductions in developing countries are still just a small share of global emissions reductions.

As a result, while TT should not be considered as an isolated option, it is likely to play an important role as part of a broader policy package. Carbon emissions from developing countries are growing at the same time that developed countries begin to reduce their own emissions. Given their need for continued economic growth, developing countries are unlikely to agree to constrain emissions without compensation from developed countries. TT provides one such form of compensation.

The effectiveness of this TT depends on the nature of the transfer. As modeled in chapter 8's simulation, TT provides significant short-term gains, as MACs are equalized across regions. However, as the financial transfers in chapter 8's model do not shift the MAC curve of developing countries, future emissions reductions from developing countries will cost more than they would have done

without the transfers. By raising the future abatement costs of developing countries, this makes their future participation in a climate treaty less likely. In contrast, TT that shifts the MAC of developing countries may offset the "low-hanging-fruit" problem described here, making future participation more likely. Because chapter 8 focuses on tangible gains that can be readily modeled, it acknowledges that such gains are not considered in its model. While this is reasonable given the need to produce a concrete number, it also suggests limitations to focusing only on measurable impacts of TT.

Bibliography

Bosetti, V., C. Carraro, E. Massetti, and M. Tavoni, 2008: International energy R&D spillovers and the economics of greenhouse gas atmospheric stabilization, *Energy Economics* **30**(6), 2912–29

Dechezleprêtre, A., M. Glachant, I. Hascic, N. Johnstone, and Y. Ménière, 2008a: Invention and transfer of climate change mitigation technologies on a global scale: a study drawing on patent data, Working Paper, CERNA, Mines ParisTech

Dechezleprêtre, A., M. Glachant, and Y. Ménière, 2008b: The clean development mechanism and the international diffusion of technologies: an empirical study, *Energy Policy* **36**, 1273–83

Energy Information Administration (EIA), 2007: *International Energy Outlook 2007*, US Department of Energy, Washington, DC

Hall, B.H., 1996: The private and social returns to research and development, in B.L.R. Smith and C.E. Barfield (eds.), *Technology, R&D and the Economy*, Brookings Institution and American Enterprise Institute, Washington, DC, 140–62

Jaffe, A. B., 1986: Technological opportunity and spillovers of R&D: evidence from firms' patents, profits, and market value, *American Economic Review* **76**(5), 984–1001

Johnstone, N., I. Hascic, and D. Popp. 2010: Renewable energy policies and technological innovation: evidence based on patent counts, *Environmental and Resource Economics* **45**(1), 133–5

Jones, C.I. and J.C. Williams, 1998: Measuring the social rate of return to R&D, *Quarterly Journal of Economics* **113**(4), 119–35

Keller, W., 2004: International technology diffusion, *Journal of Economic Literature* **42**, 752–82

Lecocq, F. and P. Ambrosi, 2007: The clean development mechanism: history, status, and prospects, *Review of Environmental Economics and Policy* **1**(1), 134–51

Lovely, M. and D. Popp, 2008: Trade, technology and the environment: why do poorer countries regulate sooner, NBER Working Paper **14286**

Mansfield, Edwin, 1977: social and private rates of return from industrial innovations, *Quarterly Journal of Economics* **91**, 221–40

  1996: *Estimating Social and Private Returns from Innovations Based on the Advanced Technology Program: Problems and Opportunities*, **NIST GCR 99–780**, National Institute of Standards and Technology, Gaithersburg, MD

Narain, U. and K. van't Veld, 2008: The clean development mechanism's low-hanging fruit problem: when might it arise and how might it be solved?, *Environmental and Resource Economics* **40**, 445–65

Nordhaus, W., 2002: Modeling induced innovation in climate change policy, in A. Grubler, N. Nakicenovic, and W. Nordhaus (eds.), *Technological Change and the Environment*, Resources for the Future Press, Washington, DC

Organisation of Economic Co-operation and Development (OECD), 2008: *Main Science and Technology Indicators*, vol. **2008**/1, OECD, Paris

Pakes, A., 1985: On patents, R&D, and the stock market rate of return, *Journal of Political Economy* **93**(2), 390–409

Popp, David, 2002: Induced innovation and energy prices, *American Economic Review* **92**(1), 160–80

  2006: R&D subsidies and climate policy: is there a "free lunch"?, *Climatic Change* **77**(3–4), 311–41

  2008: International technology transfer for climate policy, Center for Policy Research Policy Brief 39, Syracuse University

Rose, A., E. Bulte, and H. Folmer, 1999: Long-run implications for developing countries of joint implementation of greenhouse gas mitigation, *Environmental and Resource Economics* **14**, 19–31

Saggi, K., 2000: Trade, foreign direct investment, and international technology transfer, World Bank Policy Research Working Paper **2349**

State Environmental Protection Administration, 2007: Report on the state of the Environment in China, Information office of the state council of the People's Republic of China, Beijing

World Bank, 2008: *Global Economic Prospects: Technology Diffusion in the Developing World*, World Bank, Washington, DC

PART II

Ranking the Opportunities

Expert Panel Ranking

NANCY L. STOKEY, VERNON L. SMITH, THOMAS C. SCHELLING, FINN E. KYDLAND, AND JAGDISH N. BHAGWATI

The Goal of the Project

The goal of the Copenhagen Consensus on Climate was to evaluate and rank feasible ways to reduce the adverse consequences from global warming.

Individual proposals that would achieve this were examined under the eight solution headings of: Climate Engineering, Carbon Cuts, Forestry, Black Carbon Cuts, Methane Cuts, Adaptation, Energy Technology, and Technology Transfers (TTs).

Ranking the Proposals

A Panel of economic experts, comprising five of the world's most distinguished economists, was invited to consider these proposals and identify the proposals where investments would be most effective. The members were: Jagdish N. Bhagwati of Columbia University, Finn E. Kydland of the University of California, Santa Barbara (Nobel Laureate), Thomas C. Schelling of the University of Maryland (Nobel Laureate), Vernon Smith of Chapman University (Nobel Laureate), and Nancy L. Stokey of the University of Chicago.

The Panel was asked to answer the question:

If the global community wants to spend up to, say, $250 billion per year over the next 10 years to diminish the adverse effects of climate changes, and to do the most good for the world, which solutions would yield the greatest net benefits?

The sum of up to $250 billion per year was chosen by the Copenhagen Consensus Center because it is in the order of magnitude of spending that world leaders could commit to in the Copenhagen COP 15 negotiations, and is consistent with the relevant economic literature on the expected costs of dealing with global warming.

The basis for the Expert Panel's discussions and ranking were the eight chapters and thirteen perspective papers: new research commissioned from acknowledged authorities in each policy area.

The chapters review the existing frontier academic literature and present the economic costs and benefits of one or more relevant policy responses to global warming, as well as outlining the strengths and weaknesses in the applied methodology.

To ensure complete information on each category of solutions, all chapters are complemented by at least one Perspective paper, providing a critique of the assumptions and calculations used in the chapter.

During a roundtable meeting at Georgetown University in Washington, DC, the Expert Panel appraised the research in great depth, and engaged with the chapter and Perspective paper authors.

Based on this work, the Panel ranked the proposals, in descending order of desirability.

The Expert Panel Ranking

Rating	Solution	From chapter (no.)
Very Good	1 Marine cloud whitening research	Climate Engineering, Bickel and Lane (1)
	2 Energy R&D	Technology, Galiana and Green (7)
	3 Stratospheric aerosol insertion research	Climate Engineering, Bickel and Lane (1)
	4 Carbon storage research	Technology, Valentina Bosetti (7.1)

(cont.)

Rating	Solution	From chapter (no.)
Good	5 Planning for adaptation	Adaptation, Francesco Bosello, Carlo Carraro, and Enrica De Cian (6)
	6 Research into Air Capture	Climate Engineering, Roger Pielke, Jr. (1.1)
Fair	7 Technology transfers	Technology Transfer, Zili Yang (8)
	8 Expand and protect forests	Forestry, Brent Sohngen (3)
	9 Stoves in developing nations	Black Carbon, Robert E. Baron, W. David Montgomery, and Sugandha D. Tuladhar (4)
Poor	10 Methane reduction portfolio	Methane, Claudia Kemfert and Wolf-Peter Schill (5)
	11 Diesel vehicle emissions	Black Carbon, Robert E. Baron, W. David Montgomery, and Sugandha D. Tuladhar (4)
	12 $20 OECD carbon tax	Carbon, Gary Yohe and Richard Tol (research from Copenhagen Consensus 2008, not in this volume)[a]
Very Poor	13 $0.50 global CO_2 tax	Carbon, Richard S.J. Tol (2)
	14 $3 global CO_2 tax	Carbon, Richard S.J. Tol (2)
	15 $68 Global CO_2 tax	Carbon, Richard S.J. Tol (2)

Note:
[a] See chapter 5 in B. Lomborg (ed.), *Global Crises, Global Solutions*, 2nd edn., Cambridge University Press, Cambridge, 2009.

In ordering the proposals, the Panel was guided predominantly by consideration of economic costs and benefits. The Panel acknowledged the difficulties that cost-benefit analysis (CBA) must overcome, both in principle and as a practical matter, but agreed that the cost-benefit (C/B) approach was an indispensable organizing method.

In setting priorities, the Panel took account of the strengths and weaknesses of the specific C/B appraisals under review.

For some proposals, the Panel found that information was too sparse to allow a judgment to be made. These proposals, some of which may prove after further study to be valuable, were therefore excluded from the ranking.

Each expert assigned his or her own ranking to the proposals, and the Panel's ranking was calculated by taking the median of the individual rankings. The Panel jointly endorses this median ordering as representing their agreed view.

If one calculates the total cost of the "Very Good" and "Good" solutions, the expenditure proposed by the Copenhagen Consensus runs to around $110 billion a year from 2010 to 2020.

Notes on Solution Categories

Climate Engineering

The Expert Panel highly recommends research into climate engineering (CE) strategies. Of the strate-

gies that the Expert Panel considered, solar radiation management (SRM) methods – especially marine cloud whitening – appear to show the greatest promise. The Expert Panel notes that, compared with other solution categories, geoengineering (GE) reduces the risk of "pork barrel politics" and lowers transaction costs. In the case of a low-probability, high-impact situation, CE could play a crucial role because of its speed. The Expert Panel notes that a short-term focus on research into CE would be beneficial in establishing the limitations and risks of this technology, and the identification of these should happen sooner rather than later. They find that research into air capture (AC) would be useful as it appears to have potential as a backstop technology.

Technology

The Expert Panel believes that increased research into energy technology is vital to ensure a move away from reliance on fossil fuels. There is a significant energy technology challenge to stabilizing climate, demonstrated by the lack of readiness of current carbon-emission free energy technologies. The Expert Panel finds that there is a compelling case for greater research into technologies including (among others) storage for energy, batteries, nuclear energy and nuclear reprocessing technology, fusion, second-generation biofuels, wave energy, geothermal energy, and technology that increases the conversion rate of fossil fuels. They also find that research into carbon capture and sequestration (CCS, carbon storage) is very important because this technology has considerable potential as a "bridging technology" to a zero-carbon future.

Adaptation

Whatever other policy options are selected, adaptation will be needed because it is unlikely that all of the impacts of climate change will be avoided. Adaptation is thus unavoidable and may serve multiple purposes, including helping developing countries in terms of development, and non-climate-related disaster readiness. The Expert Panel finds that it is very important to ensure that planning occurs for future adaptation, focusing particularly on anticipatory (or preparatory) measures. In the long term, a combination of proactive and reactive adaptation is an effective means of reducing the damage from climate change. Because of the distribution of expected climate change effects, most adaptation expenditure will need to be beneficial to developing nations.

Technology Transfer

Technology transfer (TT) is a promising approach for dealing with climate change, because international cooperation on both greenhouse gas (GHG) mitigation and adaptation must involve transfers of technologies and dissemination of knowledge. While developed countries are beginning to constrain growth in carbon emissions, emissions from developing countries are growing, showing a requirement to ensure that knowledge on mitigation and adaptation strategies and implementation is shared.

Forestry

Ecosystems store approximately 1 trillion tons of CO_2 in the biomass of living trees and plants. Methods to increase this carbon efficiently in order to reduce the future damages of climate change include afforestation (planting old agricultural land with trees), reduced deforestation, and forest management. The Expert Panel agrees with the chapter 3 and Perspective paper 3.1 findings that these solutions would have benefits in terms of both reducing global warming and in terms of increasing biodiversity. The forestry solution was not given a higher ranking because it would be a relatively costly way of cutting carbon, and there are regulatory challenges relating to implementation and leakage to be overcome.

Black Carbon Mitigation

The Expert Panel heard that mitigating black carbon emissions would be beneficial for health improvements in developing nations as well as in climate change outcomes. However, there is a broad difference of scientific opinion regarding the role of black carbon in global warming, and the research into this field is relatively young. The non-climate, health benefits vastly outweigh the climate benefits, making it more of a health policy proposal. When looking at the proposal to reduce household black carbon emissions in the developing world, the Expert Panel found it difficult to locate large-scale, successful examples of programs, and the evidence suggests that there are both acceptance and transition issues. The costs of implementing vehicular technology solutions is high relative to the benefits. For these reasons, the Expert Panel gives solutions considered under the topic of black carbon mitigation a lower ranking.

Methane Mitigation

Methane (CH_4) is a major anthropogenic GHG, second only to carbon dioxide (CO_2) in its impact on climate change, but is challenging to regulate and control. It has many non-point sources that can be small, geographically dispersed, and not related to energy sectors. The most important single sector emitting CH_4 is livestock production, and the technical measures available to reduce emissions from livestock are limited. The Expert Panel observes that the short-term nature of CH_4 means that its mitigation is less relevant than other proposals to longer-run climate damage. The best options to regulate CH_4 in livestock and agriculture will face almost insurmountable obstacles in practice. For these reasons, the solution considered under the category of CH_4 mitigation was given a lower ranking. The Expert Panel notes that commercial-scale extraction of CH_4 clathrate would pose a serious issue, as it could lead to large leakage.

Carbon Mitigation

The Expert Panel finds that, while a well-designed, gradual policy of carbon cuts could substantially reduce emissions at a low cost, poorly designed or overly ambitious policies could be orders of magnitude more expensive. Very stringent targets may be costly or even infeasible. The Expert Panel finds that high levels of carbon tax, in the short term, will be a poor response to climate change. They note that the geographical spread of global warming damage – and its greater damage to developing nations – means that estimates of GDP loss should be treated with some caution, and that the low probability of high impact results from global warming should be taken into account when evaluating carbon mitigation.

In addition to the three global tax options in chapter 2 by Richard S.J. Tol, the Expert Panel finds it relevant to scrutinize the impact of a tax on developed nations alone. Therefore, they have considered a scenario from Yohe *et al.*'s chapter 5 in Lomborg (2009),[1] proposing a CO_2 tax of $20 on OECD nations. The Expert Panel has looked at carbon taxes, which are likely to be more efficient than a cap-and-trade scheme. The Panel notes that many politicians are opting for the latter, and that the use of such an emissions trading scheme (ETS) is likely to further diminish the returns of the solutions considered here. They also conclude that the costs and benefits of regulatory interventions (such as energy efficiency standards) to mitigate carbon deserve future examination.

Individual Rankings

NANCY L. STOKEY

Notes on Personal Ranking

Global warming has two groups of consequences. The first consists of the effects of slow but steady

increases in global temperatures. For example, warming affects agricultural yields, alters the crops that are planted, and affects heating and cooling costs. In addition, warming produces a (gradual and modest) rise in sea level, affecting coastal areas, and reduces the winter snow pack in certain mountain regions, affecting water supplies.

The second group consists of (possible) catastrophic events. The most important are the collapse of the West Antarctic ice sheet, which would trigger a large rise in sea level, and the release of large quantities of CH_4 from thawing permafrost in the Arctic, which would dramatically accelerate warming. The likelihood of these events is unknown, but they are possibilities that cannot be ignored.

How can we cope effectively with both groups of consequences? A portfolio of measures is needed: (1) development of technologies to avoid potential catastrophe, (2) policies to encourage research and development (R&D) of technologies that will be needed to eventually replace fossil fuels with alternative energy sources, and (3) policies to cope with current warming and to begin reducing GHG emissions.

Development of Technologies to Avoid Potential Catastrophe

Avoiding catastrophe is a high priority, and the two SRM technologies proposed here are tailor-made for this purpose. Both work by reducing the amount of solar energy warming in the Earth, in effect offsetting the additional warming caused by GHGs. SRM could be used rapidly and on a large scale if the threat of either catastrophe becomes too great. In addition, SRM could be used to mitigate the adverse effects of slow warming, in effect buying time to develop cost-effective alternatives to fossil fuels. Since the two mechanisms proposed here use currently available technologies, they could be developed quickly and at low cost.

The proposals are to invest in further development, including field trials, of marine cloud whitening and stratospheric aerosol insertion. While the term SRM seems frightening at first, these two technologies are much less threatening upon closer inspection.

[1] Lomborg, B., 2009: *Global Crises, Global Solutions*, 2nd edn., Cambridge University Press, Cambridge.

Marine cloud whitening involves deploying a fleet of unmanned "drone" vessels to sail around in circles, in a few carefully chosen regions of the ocean, stirring up seawater. The added seawater would whiten the clouds in these regions, increasing their reflectivity and thus reducing the amount of sunlight the earth Receives. A small reduction in solar absorption would offset the current excess of CO_2, and greater reduction could be achieved if faster cooling were desired. Moreover, the process is quickly reversible: the whitening agent, which is seawater, would precipitate out in a few weeks.

Stratospheric aerosol insertion involves mimicking the effect of a large volcanic eruption. Sulfur dioxide (SO_2) (or an alternative agent) is injected into the stratosphere, where it scatters sunlight back into space. The quantity of sulfur required would be small compared with the current emissions from power plants, so any additional pollution from this source would be minor. Aerosols have a rather short life (about a year), so this process is also reversible. Stratospheric aerosols would be substantially more expensive to deliver than marine clouds, but they could be targeted to either polar region, where the potential threats lie.

R&D for the Long Run

The second priority is developing technologies that will be needed in the medium and long run.

The world will continue to rely on fossil fuels for electricity generation for several decades, at least. A safe and cost-effective technology for capturing CO_2 at power plants and storing it underground (CO_2 storage), would allow the large electricity supplies from coal-fired power plants to continue flowing, without contributing to warming. The technologies here are very promising, so further development of CO_2 storage is a high priority.

Air capture involves removing CO_2 from ambient air, where the concentrations are much lower. Current technologies for AC are far from cost-effective, and they seem less promising than those for CO_2 capture at power plants. Nevertheless, AC would be useful if it could be done cheaply, so a moderate R&D investment seems warranted.

Many new and improved technologies will be needed for a greener planet in the long run. Among

these are various energy sources that will be alternatives to fossil fuels: cheaper solar panels, less expensive and more efficient wind turbines, fast-breeder reactors, geothermal energy, better biofuels, and so on. In addition, improvements in several ancillary technologies are also needed, including better batteries and other storage devices to accommodate the uneven nature of solar and wind power, and a smarter, more efficient energy grid system. Policies to promote development of these technologies (energy R&D) and public investment in the necessary infrastructure will be critical for long-term success, so they also have high priority.

Policies to Deal with Current Warming

The large stock of CO_2 already in the atmosphere means that warming will continue in the short run. Indeed, additional warming would occur even if global CO_2 emissions were (magically) to fall to zero immediately. Thus, policies are also needed to adapt to current warming and to begin reducing the level of GHG emissions.

First, consider adaptation. In the private sector, individuals and private firms will adapt by planting different crops, improving insulation to reduce air-conditioning costs, and so on. In developed economies, adaptation will occur without public intervention. In developing economies, assistance will be needed. The developing world will be more adversely affected by warming, because of geography (proximity to the Equator) and economic factors (larger shares of employment and output in agriculture), and will have less capability to deal with the problems that arise.

Public measures will also be needed. Specific projects in this category include investments in coastal protection (seawalls for the EU, better flood defenses in South Florida), adapting vulnerable infrastructure (harbors, bridges), and making provisions to deal with the higher caseloads of diarrheal disease, malnutrition, and malaria that can be expected.

Next, consider measures to reduce GHG emissions. Cost-effectiveness is the critical issue in deciding how much to do on this front, how quickly to do it, and by what means.

In the short run, a moderate carbon tax (or CO_2 cap) would be useful, for three reasons. First, it would provide better information about the potential for inexpensive reduction in CO_2 emissions using currently available technologies. In the absence of such a tax, estimates about that potential include a lot of guesswork. Second, it would give a strong signal to the private sector that investments in green technologies will be rewarded. If political opposition makes even a low tax infeasible, it may discourage investments that will bear fruit only if a much higher tax is implemented. Finally, some experimentation may be needed to reach a viable system for reducing emissions. While cap and trade has worked well for other pollutants, it may be less suitable for CO_2. If so, it would be useful to find out sooner rather than later.

A moderate carbon tax will accomplish these three tasks better than a very low one, but a high carbon tax is unwarranted at the present time. Drastically reducing CO_2 emissions in the short run is like digging the Panama Canal with a garden trowel: it is feasible, but very expensive. A better strategy is to begin by developing a steam shovel. A very high carbon tax would create the desired strong incentives, but the cost (deadweight loss) from such a tax would be intolerable. The incentives for R&D in energy technologies should be provided through other channels.

Other methods for reducing or offsetting GHG emissions in the short run include better forest management, reduction of CH_4 emissions, and reduction of black carbon. None of these has been tried, so their potential usefulness is hard to assess.

Consider forestry. A substantial fraction of global CO_2 emissions (17%) comes from burning tropical rain forests. Slowing the pace of deforestation would be useful, but it is not clear what mechanism will do it efficiently. The proposal here is to have a world body pay annual "rent" to keep intact tropical forests that would otherwise be cleared. But how does that body decide which forests should be slated for clearing? If the "rent" has to be paid on all tropical forests, the costs are much higher and the usefulness is correspondingly diminished.

CH_4 is a powerful but short-lived GHG. Although a few of the specific measures proposed

here could be implemented easily, most of the options seem difficult to enforce and seem to have only modest impacts on climate change.

Similar arguments apply to black carbon. Black carbon is even shorter-lived than CH_4, and none of the three major sources is an obvious candidate for effective intervention. Open burning produces aerosols as well as black carbon, and the former may offset (or more than offset) the latter. Persuading hundreds of millions of Chinese and Indian households to replace their cook stoves does not look like an easy task, and in any case it seems to have little to do with climate change. And retrofitting diesel vehicles is expensive compared with alternative methods of reducing GHGs.

Summary

Most of the public debate has centered on carbon taxes and cap-and-trade systems. A moderate tax (or cap) would be useful for the reasons described above, but policies to stimulate R&D in a variety of areas are even more crucial. Current technologies are simply not enough to address the problem of climate change. In particular, development and field testing of the two SRM technologies should be a high priority.

Personal Ranking: Nancy L. Stokey

Ranking	Solution	Solution category
1	Marine cloud whitening research	Climate Engineering
2	Stratospheric aerosol insertion research	Climate Engineering
3	Energy R&D	Technology
4	CO_2 storage research	Technology
5	Planning adaptation	Adaptation
6	AC research	Climate Engineering
7	TT	Technology Transfer
8	$20 OECD carbon tax	Cut Carbon
9	$3 Global CO_2 tax	Cut Carbon
10	$0.50 Global CO_2 tax	Cut Carbon
11	Expand and protect forests	Forestry

Ranking	Solution	Solution category
12	Methane reduction portfolio	Cut Methane
13	Diesel vehicle emissions	Cut Black Carbon
14	Stoves in developing nations	Cut Black Carbon
15	$68 global CO_2 tax	Cut Carbon

VERNON L. SMITH

Notes on Personal Ranking

Carbon (Dioxide) Emission Mitigation

This is the solution that I and the Panel rated lowest. I begin with this option because reducing carbon emissions is widely perceived by politicians, journalists, and many scientists (although skeptics abound) as necessary to reduce global climate change, and worth the cost. On the latter my view, given the state of current knowledge and clearly demonstrated in chapter 2 and Perspective papers 2.1 and 2.2, is that the cost in sacrificed human betterment and poverty reduction would be prohibitive in achieving reduced near-term effective atmospheric carbon inventories (new emissions have an uncertain half-life estimate of forty or many more years). Moreover, the certainty with which politicians often approach the need for carbon cuts are papering over much more uncertainty in many areas of climate modeling, and this again necessarily affects the rankings and decisions made here. My purpose in noting this is not to demean the enormous recent advances in climate science, but to emphasize that our ignorance of global dynamics continues to be overwhelming.

Cut Black Carbon

I rated this fairly high essentially because of the recent scientific claims that these particulate emissions may account for much of lower atmospheric temperature increases and particularly the regional warming associated with loss of Arctic and glacial ice. This may turn out to be a promising breakthrough, or just one more dead end, but it is

worth aggressive investigation. Since the Asian stove sources are also a health hazard, black carbon merits cutting in any case; the principal problem has been to implement a change in stove use.

Ramanathan *et al.* (2007) consider recent issues in black carbon (soot) and related brown cloud science:

> Here we use three lightweight unmanned aerial vehicles that were vertically stacked . . . over the polluted Indian Ocean . . . [that] . . . deployed miniaturized instruments measuring aerosol concentrations, soot amount and solar fluxes . . . [making] . . . it possible to measure the atmospheric solar heating rates directly. We found that atmospheric brown clouds enhanced lower atmospheric solar heating by about 50 per cent . . . brown clouds contribute as much as the recent increase in anthropogenic greenhouse gases to regional lower atmospheric warming trends. We propose that the combined warming trend of 0.25 K per decade may be sufficient to account for the observed retreat of the Himalayan glaciers.[2]

And again:

> We conclude that decreasing concentrations of sulphate aerosols and increasing concentrations of black carbon have substantially contributed to rapid Arctic warming during the past three decades.[3]

Planning Adaptation

I rated this solution very high. Regardless of the causes of climate change, the trend in global warming, sea level rise, and loss of glacial and ocean ice

---

[2] V. Ramanathan *et al.* 2007: Warming trends in Asia amplified by brown cloud solar absorption, *Nature* **448**, 575–8. See also V. Ramanathan and G. Carmichael, 2008: Global and regional climate changes due to black carbon, *Nature Geoscience* **1**, 221–7; and J.R. McConnel *et al.* 2007: 20th-century industrial black carbon emissions altered Arctic climate fording, *Science* **317**, 1381–4.

[3] Dr. Shindell and G. Faluvegi, 2009: Climate response to regional radiative forcing during the 20th century, *Nature Geoscience* **2**, 294–300.

for the last 20,000 years is likely to continue. If carbon is a principal new cause, its accumulated effects are thought to be already built in and irreversible short of an unanticipated natural reverse "tipping." Thus:

> Climate warming is expected to result in [a] rising sea level. Should this occur, coastal cities, ports, and wetlands would be threatened with more frequent flooding, increased beach erosion, and saltwater encroachment into coastal streams and aquifers. Global sea level has fluctuated widely in the recent geologic past. It . . . was 120m lower at the peak of the last ice age, around 20,000 years ago . . . A . . . clearly-defined accelerated phase of sea level rise occurred between 14,600 to 13,500 years before [the] present . . . termed a "melt water pulse" . . . when [the] sea level increased by some 16 to 24m.[4]

The failure to respond efficaciously to Hurricane Katrina shows clearly the need to ask whether, and in what way, adaptive planning can be implemented. We need also to ask if New Orleans or other cities located below sea level should be protected, rebuilt if lost, or simply moved with migration assistance.

Adaptation also makes sense because in intervals of tens of thousands of years the ice core temperature record going back 420,000 years shows that warm episodes have been rare and short-lived, on the order of a few thousand years, with carbon concentrations lagging temperature changes.

Climate Engineering

I rated research on cloud whitening highest, aerosol insertion lower. Cloud whitening is scalable, subject to relatively controlled experiments and so far as we know reversible. It appears therefore to chart an incremental low-cost learning path in which unintended consequences can be identified

on a small scale before using it more aggressively to counteract anticipated damages from warming. Aerosol insertion is less attractive on these measures since it is less incrementally controlled, but research seems justified because of the prospect that it could act more quickly than carbon mitigation. Even cloud whitening, however, is fraught with incredible uncertainties that are just elementary reflections of our broader scientific ignorance: "Despite decades of research, it has proved frustratingly difficult to establish climatically meaningful relationships among the aerosol, clouds and precipitation.[5]

These options have merit only because they offer promising new increases in our practical knowledge, not because they can be assured of rescuing us if that is necessary. Nevertheless, both of these technologies may have risks, whose origins are precisely the same as those governing the causes of global warming: we know precious little about systems as complex as that of the global climate, and we should proceed with caution to avoid unintended harm.

R&D

In line with the 2004 and 2008 Copenhagen Consensus meeting conclusions, I am persuaded that if target anthropogenic GHG reductions are necessary to reduce global climate change – a distinctly speculative proposition – then the "brute force" approach with existing technology is not feasible. If there is any effective means of reducing GHG emissions, it rests with R&D discoveries that will enormously increase energy savings (or, alternatively, finesse the whole issue through CE, as above). But we cannot assure discovery; we can only commit to trying, and chapter 2 and Perspective papers 2.1 and 2.2 cautiously recognize this potential outcome. Even in the absence of a carbon tax and public R&D it is easy to underestimate the extent to which rising relative energy prices for long periods will induce innovations that will increase energy efficiency, as is evident by simply looking back to 1830 when kerosene-from-coal – "coal oil" – was the response to the high price of whale oil.

---

[4] See www.giss.nasa.gov/research/briefs/gornitz_09/.
[5] B. Stevens and G. Feingold, 2009: Untangling aerosol effects on clouds and precipitation in a buffered system, *Nature* **461**, 607.

Personal Ranking : Vernon L. Smith

Ranking	Solution	Solution category
1	Marine cloud whitening research	Climate Engineering
2	Planning adaptation	Adaptation
3	Stratospheric aerosol insertion research	Climate Engineering
4	Energy R&D	Technology
5	Expand and protect forests	Forestry
6	TT	Technology Transfer
7	Stoves in developing nations	Cut Black Carbon
8	Methane reduction portfolio	Cut Methane
9	CO_2 storage research	Technology
10	AC research	Climate Engineering
11	$20 OECD carbon tax	Cut Carbon
12	Diesel vehicle emissions	Cut Black Carbon
13	$0.50 global CO_2 tax	Cut Carbon
14	$3 global CO_2 tax	Cut Carbon
15	$68 global CO_2 tax	Cut Carbon

THOMAS C. SCHELLING

Notes on Personal Ranking

My principal comment is that "carbon mitigation" got a bad review. Chapter 2 did not recognize the seriousness of climate change for the developing world, and proposed a few trivial solutions. It acknowledged that it gave no attention to "equity" in assessing damages. It did not recognize that in doing so it ignored most of the potential damages due to climate change. Its measure of the seriousness of climate change was the impact on global GDP. As a result, precisely because the poor countries have low GDP, it ignored the impact on them – their GDP is too small to matter!

There are a billion people with incomes less than $2 per day, many with less than $1. If they suffered a loss of half their income – i.e. if the poorest one-sixth of the world's population suffered disastrous losses – the impact on world GDP would be less than $365 billion per year, less than 1% of world GDP. In terms of global GDP, because they are poor they don't count!

Because the main damages, though possibly catastrophic in human terms, left the richer areas of the world not seriously damaged (according to chapter 2 that estimated damages due to climate change), TT to permit developing nations to participate in carbon mitigation without too much impact on their continuing development got too little emphasis. We did not consider, as chapter 2 did not really provide estimates, how much might constructively be spent in the first decade on either preparation for adaptation or investment in developing countries on things like carbon capture.

One of the difficulties with the "expenditure" or "budgetary" approach to these subjects is that proposals for taxation to provide incentives don't "cost" but promise to yield revenue, i.e. "negative" expenditure. We have never managed to overcome that difficulty.

Personal Ranking : Thomas C. Schelling

Ranking	Solution	Solution category
1	CO_2 storage research	Technology
2	Stratospheric aerosol insertion research	Climate Engineering
3	Marine cloud whitening research	Climate Engineering
4	AC research	Climate Engineering
5	Energy R&D	Technology
6	TT	Technology Transfer
7	Planning adaptation	Adaptation
8	Expand and protect forests	Forestry
9	Methane reduction portfolio	Cut Methane
10	$20 OECD carbon tax	Cut Carbon
11	Diesel vehicle emissions	Cut Black Carbon
12	Stoves in developing nations	Cut Black Carbon
13	$3 global CO_2 tax	Cut Carbon
14	$0.50 global CO_2 tax	Cut Carbon
15	$68 global CO_2 tax	Cut Carbon

FINN E. KYDLAND

Notes on Personal Ranking

Before writing my own personal comments on the rankings of the proposed Climate 2009 solutions, I had the pleasure of reading those by fellow Panel member Nancy L. Stokey (see p. 384). I agree with most of what she says. Rather than making a similar assessment of the solutions, which to a large extent would be a repeat of her analysis, I thought I would in part give a sense of the process for how we, or I at least, ended up with the final ranking.

We were provided with the chapters well in advance of our meeting in Washington, DC. Bjørn Lomborg requested that we all provide a preliminary ranking the week before that meeting. Then, at the meeting, we listened to presentations by the authors of the chapters, followed by the Perspective papers. Immediately after each such group of presentations, the Panel members discussed the solutions among themselves and had the option to call back the chapter and Perspective paper authors for further clarification. Each session ended with the Panel members adjusting their individual rankings of the solutions they had heard up until that point, including adding on the ranking tablet the new solutions just presented. After the last set of solutions, we had a few minutes to finetune our rankings.

I thought it might be interesting to talk about which solutions changed the most in ranking from my preliminary to my final version. Especially interesting is that of TT. In my preliminary ranking, I placed it first. This ranking may in part reflect my general economic belief, namely that there is a lot of technological knowledge available in general which can be transferred to less-developed nations, and that making use of it, perhaps with the addition of some R&D to adapt to local circumstances, is how those nations can hope to make progress in narrowing the income gap with the more well-to-do nations. So why wouldn't something like that work also in the context of technology to abate global warming?

Admittedly, in my notes after reading chapter 8 and Perspective paper 8.1, I had jotted down a few

questions. Indeed, because of my inclination to rank this solution highly, but at the same time having nagging questions, I brought to the Panel meeting these notes, so as to remind myself to get some of these points clarified.

> Theory states mitigation is efficient when marginal costs are equalized across regions. The opportunities for cost-effective carbon mitigation, considering existing technology, are found today in the developing world.

Two policy scenarios are considered. First is the Kyoto Protocol scenario, where developing countries are not obligated to reduce their emissions. They will do so if developed countries pay them. They will use the technologies from the North (developed countries). The second scenario involves developing countries agreeing to certain GHG emission reductions that are incentive-compatible for them. Once these are met, the North countries will help the South countries through TT for further emissions reduction. In the first scenario, full responsibility is on developed countries, while the second scenario relies on full cooperation.

BCRs were calculated in three different ways, I think the one that makes the most sense is the second, where the cost calculated is the cost of the transfers, and the benefit is the net gains in mitigating costs as a result of them. The latter are calculated from the baseline case where the same reductions must be met, but without transfers. The results are robust to different discount rates, so I will use the $r = 3\%$ case. The Kyoto scenario yields a BCR of 3.66 and the full-cooperation scenario yields a BCR of 1.10. As the second scenario represents the case with more moderate TT, these results imply that TT is important for cost-effective emissions abatement.

[Yang and Popp] admit that these are both extreme cases, and what will play out in reality is probably somewhere between these two scenarios. It seems that the BCA has been carried out on a global scale. It is recognized in the literature that developing countries will face the largest damages with respect to global warming, so I wonder if the incentives are there for developed countries to agree to the first scenario? Also, wouldn't institutional weaknesses in developing countries make these transfers less effective? What about monitoring costs to assure that transfers are spent on emission mitigation or technologies are indeed adopted? Perhaps this was considered in the analysis, but I would ask the authors about implementation costs and feasibility.

Popp emphasizes the dynamic gains that come from TT. Chapter 8 is likely *underestimating* the positive externalities gained from TT. Important to note is that merely targeting emissions abatement possible at lowest marginal costs in the developing world makes further abatement costlier for developing countries. What TT should be accomplishing is shifts in the MAC curve, rather than just movements along the curve, if we expect developing countries to continue with the policy in the future.

With the exception of some energy efficiency technologies, clean technologies typically do not flow across borders unless environmental policy is providing incentives. TT by itself is not sufficient; it must be coupled with binding emission reductions for developing countries.

While the argument for shifting the MAC curve rather than moving along the curve makes sense, Popp does not suggest what kind of technologies would fall into the previous category or the latter. I would ask for specific examples. Also, *what kind of environmental policy should be enacted in the developing world to spur the changes required, and be cost-effective?*

It seems unanimous that developing countries will need the aid of richer countries to reduce emissions and contain global warming. There are many political wrinkles to be ironed out, but generally I think this policy should receive priority on an international level, and I would put this in my top 5 recommendations.

So the main problem to me and, as it turned out, to the other Panel members as well, was the lack of concreteness. The whole thing seemed too much like a "black box." We wanted some examples of how it would work in practice. What are examples of technology improvements that shift the MAC curve, rather than simply moving along it? We even called the chapter authors back, but in the end I, at least, felt my questions by and large were still unan-

swered. As a consequence, I lowered this solution in my ranking.

In retrospect, I've come to think that my negative impression may have been overblown. I still have some faith in TT, even if not backed up by hard and fast examples. If I had done the ranking today, I would have upped that solution a couple of spots. This, of course, would not have substantially changed its position.

I'd like to comment also on the Panel's informal classification of the solutions into "Very Good", "Good", "Fair," and so on. Based on the Panel's rankings, TT is listed as "Fair." Research into AC is listed as "Good." They were tied in terms of median ranking. TT was barely lower in mean ranking, and the standard deviation of its rankings is much lower than that of AC. Though these classifications will be of assistance in understanding the Panel's work, it is clear that there is some blurring at the edges.

Research into AC is actually one of the three solutions among those considered whose rankings I *raised* by at least 2 from my preliminary to my final version, from sixth to fourth, primarily as a consequence of clarification accomplished during the presentations and the Panel discussion. The other two are Stratospheric aerosol insertion, which I moved from ninth to fifth, and Expand and protect forests, from tenth to seventh.

Finally, I need to comment on my ranking of the $0.50 global CO_2 tax, which is substantially higher than the overall Panel ranking. In my mind, this solution is related to Energy R&D, even though chapter 2 does not make that clear. I like the Energy R&D solution very much. I like this use of revenue from a carbon tax. I like the proposal of making the tax start out low, but then rise steadily year by year into the foreseeable future. This rising feature of the tax, if credible (a big question, of course), would provide an incentive for forward-looking economic actors to start making investments in carbon-reducing technologies more or less right away. The main problem with the presentation of the $0.50 global CO_2 tax solution, to me at least, was that the use of the revenue was not very clear. So I simply decided that, with benign assumptions about revenue use, that solution couldn't be that far behind the Energy R&D solution. Hence I ranked it only 3 spots behind. This is perhaps a little optimistic, especially with political credibility issues looming. Moreover, I have made a subjective assessment which admittedly is not backed up by hard-and-fast numbers. So if I had done the ranking today, just as I would have raised the TT ranking a couple of spots, I would have lowered my ranking of the carbon tax solution by a couple of spots.

While the relatively low tax is cost-effective even under conservative estimates, the same is not the case, according to chapter 2, with the more drastic carbon-tax solutions. I didn't learn anything during the Panel presentations and discussion to lead me to alter my preliminary low rankings for these solutions. For them, my final rankings ended up being almost in line with the Panel's ranking.

Personal Ranking: Finn E. Kydland

Ranking	Solution	Solution category
1	CO_2 storage research	Technology
2	Marine cloud whitening research	Climate Engineering
3	Energy R&D	Technology
4	AC research	Climate Engineering
5	Stratospheric aerosol insertion research	Climate Engineering
6	$0.50 global CO_2 tax	Cut Carbon
7	Expand and protect forests	Forestry
8	TT	Technology Transfers
9	Planning adaptation	Adaptation
10	Stoves in developing nations	Cut Black Carbon
11	Diesel vehicle emissions	Cut Black Carbon
12	Methane reduction portfolio	Cut Methane
13	$3 global CO_2 tax	Cut Carbon
14	$68 global CO_2 tax	Cut Carbon
15	$20 OECD carbon tax	Cut Carbon

JAGDISH N. BHAGWATI

Notes on Personal Ranking

My ranking was based on the evidence presented at the Copenhagen Consensus on Climate meeting in Washington, DC. Most of the chapters were skillfully executed and discussed, with many of the Panel's questions answered by video in an interactive fashion.

The proposals we looked at were a mix of mitigation and adaptation measures. I was aware that some NGOs feel that any attention to adaptation implies that mitigation to reduce global warming will be neglected. As far as we were concerned, that was simply wrong: we considered both. The rank ordering reflected, at least on my part, a sense of what the urgency may be in regard to our understanding of the unfolding time profile of the global warming problem as we confront it now.

In general, I felt that, given the urgency of the problem of global warming, a scientific judgment which I took as a "given" from outside the Panel (even though there are many scientific uncertainties at all levels of causality in this area), I felt that planning adaptation had to be given a high priority.

Similarly, mitigation in the form of stoves in developing countries like India which emit black carbon, had big externalities in terms of the health of the poor people who use these stoves, but whose carbon emissions are transient; I felt this should get high priority because we needed some forms of mitigation which had immediate, even if not lasting, effects on carbon accumulation.

I fully shared the view of the Panel that the major payoffs will come from research into mitigation technologies such as carbon storage, AC, marine cloud whitening, and stratospheric aerosol insertion, all of which received a high ranking from me. These will take some time, though – these are informed guesses, of course – the expected time for payoffs inevitably varies among them.

I put forestation and prevention of deforestation high because this was low-hanging fruit where I felt we were more likely to get agreement at the Copenhagen climate summit in December 2009 and the actual resources absorbed in promoting reforestation in particular are likely to be small. India is pushing for an international agreement and target-setting on this at Copenhagen, and this seems to me to be sensible.

TT, especially on mitigation, has a high payoff and we need to back it.

In the end, the important question is how we are going to put funds like $250 billion a year together. Gordon Brown has talked of $100 billion annually. Barack Obama has promised nothing. I have invoked the US domestic practice of the Superfund, requiring firms that have caused damage through hazardous discharges make compensatory payments, to suggest that the USA and other developed countries extend the practice to the carbon case and make "tort payments" into a Superfund at the international level. Once we have funds collected through such a rationale, the Copenhagen Consensus exercise can kick in.

Personal Ranking : Jagdish N. Bhagwati

Ranking	Solution	Solution category
1	Planning adaptation	Adaptation
2	Expand and protect forests	Forestry
3	Energy R&D	Technology
4	CO_2 storage research	Technology
5	TT	Technology Transfer
6	Stoves in developing nations	Cut Black Carbon
7	AC research	Climate Engineering
8	Marine cloud whitening research	Climate Engineering
9	Stratospheric aerosol insertion research	Climate Engineering
10	Diesel vehicle emissions	Cut Black Carbon
11	Methane reduction portfolio	Cut Methane
12	$3 global CO_2 tax	Cut Carbon
13	$68 global CO_2 tax	Cut Carbon
14	$0.50 global CO_2 tax	Cut Carbon
15	$20 OECD carbon tax	Cut Carbon

Conclusion

BJØRN LOMBORG

Reading the research in this volume – written by some of the top climate economists working in this field today – it is easier to understand why a single-minded focus on drastic carbon emission reductions has failed to work.

Of course, where it is possible to make relatively cheap reductions in carbon emissions through more efficient energy use it is a perfectly reasonable thing to do. However, Tol has starkly shown in chapter 2 that even a highly efficient global CO_2 tax aimed at fulfilling the ambitious goal of keeping temperature increases below $2°C$ would reduce annual world GDP by a staggering amount – around 12.9%, or $40 trillion, in 2100. The total cost would be about fifty times that of the avoided climate damage. And if politicians choose less-efficient, less-coordinated cap-and-trade policies, the costs could escalate a further 10–100 times.

Thus, the Expert Panel has found that drastic carbon cuts would be the poorest way to respond to global warming. There are important implications for policy makers here. Although carbon taxes and a "cap-and-trade" scheme should, in theory, have very similar outcomes, the latter produces a much higher opportunity for "pork-barrel politics" and waste. So cap-and-trade schemes – which many politicians are considering implementing today – would be even less effective than taxes, possibly 10–100 times worse.

At the same time, Galiana and Green in chapter 7 have demonstrated that the magnitude of the energy technology challenge to climate stabilization is huge – much larger than is widely appreciated.

Taken together, this is crucial knowledge. With such a serious challenge ahead of us, we do not have the money to waste, nor the time to spend, pursuing bad strategies.

The economic lessons are underpinned by real-world experience. In 1992, industrialized nations promised with a great fanfare in Rio de Janeiro to cut emissions to 1990 levels by 2000. Emissions in OECD nations overshot the target by 12%. In Kyoto, leaders committed to a cut of 5.2% in forty industrialized signatory countries below 1990 levels by 2010. While some of these countries may reach their targets thanks to the collapse of the Soviet Union, the economic decline in eastern European countries in the 1990s, and the 2009 recession, OECD nations will overshoot their emissions by 20%, and the failure for the world as a whole is even more spectacular, with global emissions increasing by an additional 40% on top of 1990 levels.

Undaunted by these failures, leaders gathered again in Copenhagen in 2009. This time, though, the political divisions and economic challenges proved too great, and no binding deal was struck.

This failure could be a blessing in disguise if it jolts politicians into considering other options, rather than attempting to implement the same ineffectual solutions – again and again. Of course, it is likely that the effective policy response should consist of a portfolio of effective options, of which many have been presented in this volume. However, Kyoto has shown the futility of betting everything on rapid cuts in carbon emissions to very specific targets and timetables.

In this context, I believe the Expert Panel's findings are particularly deserving of serious attention. There needs to be greater investment on research into climate engineering (CE) to explore its potential as a possible short-term response, and more research into non-carbon-based energy as a longer-term response to global warming.

As Bickel and Lane demonstrated in chapter 1, some proposed CE technologies – in particular, marine cloud-whitening technology – could be cheap, fast, and effective. Even if one approaches this technology with concerns – as many of us do – we should aim to identify its limitations and risks sooner rather than later.

It appears that CE could buy us some time, and it is time that we need if we are to make a sustainable and smooth shift away from reliance on fossil fuels. Non-fossil fuel energy sources will – based on today's availability – get us less than halfway towards a path of stable carbon emissions by 2050, and only a tiny fraction of the way towards stabilization by 2100.

As Galiana and Green argue, politicians need to invest significantly more in R&D – we would then have a much greater chance of getting this technology to the level where it needs to be. And, because it would be cheaper and easier than carbon cuts, there would be a much greater chance of reaching a genuine, broad-based – and thus successful – international agreement.

Carbon pricing could and should play an ancillary role – it could be used to finance R&D, and to send a price signal to promote the deployment of effective, affordable technology alternatives. Investing about $100 billion annually would mean that we could essentially resolve the climate change problem by the end of this century.

It is also clear, as Bosetti outlined in Perspective paper 7.1, that one of the central technologies we need to explore is CO_2 capture and storage (CCS). This allows the continued use of fossil fuels while reducing the CO_2 emissions produced and may therefore be hugely helpful, especially in countries, like China and India, that heavily rely on coal for the generation of electricity.

Ultimately this volume aims to provide the international political community with a better foundation for climate change response decisions. This means considering policies that are not at the top of the political agenda at the present time.

It is unfortunate that so many policy makers and campaigners have become fixated on cutting carbon in the near term as the chief response to global warming. It is heartening to read the research in this volume, and realize that there are meaningful, effective alternatives. The next step must be ensuring that sensible, smart responses to global warming receive more attention. I hope that this volume helps to serve that purpose.

If world leaders do not change track, they will be doing us – and future generations – a huge disservice. They will do much less good at much higher cost. If we care about the environment and about leaving this planet and its inhabitants with the best possible future, we actually have only one option: we all need to start seriously focusing, right now, on the most effective ways to fix global warming.

Index